Monographs in Electrochemistry

Surprisingly, a large number of important topics in electrochemistry are not covered by up-to-date monographs and series on the market, some topics are even not covered at all. The series “Monographs in Electrochemistry” fills this gap by publishing in-depth monographs written by experienced and distinguished electrochemists, covering both theory and applications. The focus is set on existing as well as emerging methods for researchers, engineers, and practitioners active in the many and often interdisciplinary fields, where electrochemistry plays a key role. These fields range – among others – from analytical and environmental sciences to sensors, materials sciences and biochemical research.

Antonio Doménech-Carbó ·
María Teresa Doménech-Carbó

Electrochemistry for Cultural Heritage

Springer

Antonio Doménech-Carbó
Departament de Química Analítica
Universitat de València
Burjassot, Spain

María Teresa Doménech-Carbó
Deparatament de Conservació i Restauració de Bens Culturals
Institut de Restauració del Patrimoni
Universitat Politècnica de València
València, Spain

ISSN 1865-1836 ISSN 1865-1844 (electronic)
Monographs in Electrochemistry
ISBN 978-3-031-31944-0 ISBN 978-3-031-31945-7 (eBook)
https://doi.org/10.1007/978-3-031-31945-7

This Springer imprint is published by the registered company Springer Nature Switzerland AG
The registered company address is: Gewerbestrasse 11, 6330 Cham, Switzerland

To Leire and Gonzalo.

Series Editors' Foreword

In 2009, the authors of this monograph have already published the monograph *Electrochemical Methods in Archaeometry, Conservation and Restoration.*[1] Now, after 14 years, the present monograph *Electrochemistry for Cultural Heritage* is a completely new treatise. It shows how electrochemistry and electroanalysis have matured as means to study objects of cultural heritage, not only by analyzing their chemical composition, but—in some cases—even to determine their age. The progress, which applications of electrochemistry and electroanalysis made during the last 2–3 decades, is almost entirely due to the immense research activities of the siblings María Teresa and Antonio Doménech-Carbó. María Teresa is Professor at *Universitat Politècnica de València, Facultad de Bellas Artes, Departamento de Conservación y Restauración de Bienes Culturales*, and Antonio is Professor at *Universitat de València, Departamento de Química Analítica*, Spain. Both have a Ph.D. in chemistry. Clearly, the cooperation between the two has long been an excellent basis for developing a research that needs expertise both in electrochemistry and in cultural heritage. Developing such a cross-sectional topic like *Electrochemistry for Cultural Heritage* requires experts in the two research fields, while simultaneously being able to understand the language of science of the partner. The great progress, which María Teresa and Antonio Doménech-Carbó have achieved in the interdisciplinary science 'electrochemistry/cultural heritage', is also based on fundamental new findings in the single field of electrochemistry and cultural heritage. Thus, they have made tremendous progress in developing solid-state electroanalysis, so that this may become a standard technique in archaeometry, conservation, and restauration. They are both *the* pioneers of electrochemical age determination of metals and other objects.

This monograph can serve two different and so far seldom interacting scientific communities, that of electrochemists and electroanalysts and that of scientists working in conservation, restoration, and analysis of objects of cultural heritage.

[1] Doménech-Carbó A, Doménech-Carbó MT, Costa V (2009) Electrochemical Methods in Archaeometry, Conservation and Restoration, In the series Monographs in Electrochemistry, F. Scholz editor, Springer, Berlin, ISBN 978-3-540-92867-6, http://www.springer.com/chemistry/electrochemistry/book/978-3-540-92867-6.

Because the authors have written the book with the clear goal to be understandable for scientists of rather different branches of science, they had to include some fundamentals of analytical chemistry, electrochemistry, archaeometry, conservation, and restauration of cultural heritage. The editors hope that this monograph is a true step toward a better mutual understanding and cooperation between scientists of theses realms.

Greifswald, Germany
Budapest, Hungary
January 2023

Fritz Scholz
László Peter

Preface

Electrochemical techniques were applied along the 20th century mainly in the field of metallic heritage focused on two aspects: the diagnosis of alterations due to corrosion and the application of stabilization and protection protocols. The intersection among archaeometry, conservation, and restoration fields with electrochemistry was significantly expanded at the end of the century by exploiting the capabilities of the voltammetry of immobilized particles, a solid-state technique developed by Fritz Scholz and his coworkers in the University of Greifswald which allows to analyze amounts of sample at the microgram-nanogram level. The latter is an essential demand for analytical techniques to be applied in the study and preservation of cultural heritage goods making the voltammetry of immobilized particles an interesting analytical tool for complementing the variety of existing techniques.

A first recompilation of the contributions of electrochemistry to the scientific study of works of art and archaeological objects was published by us in collaboration with Virginia Costa in 2009 (*Electrochemical Methods in Archaeometry, Conservation and Restoration*—Monographs in Electrochemistry Series, Scholz F, Ed—Springer, Berlin, Heidelberg). Since then, the implementation of electrochemistry into the cultural heritage domain has increased significantly, covering pigments and paintings, ceramic, glass and glazed materials, fibers, woods, and, of course, metals and alloys. Recent contributions include not only theoretical and experimental expansions of the voltammetry of immobilized particles, but also important developments in impedance techniques and excursions in other electrochemical techniques such as open circuit potential and scanning electrochemical microscopy. All these 'purely electrochemical' contributions are accompanied by the display of archaeometric information relative to provenance tracing, manufacturing techniques, and dating. The presentation of this scenario to a general audience required a reexamination and a reorganization of the contents so that an entirely new text rather than a second edition of the previously mentioned book results from the consensus of authors and editors.

The current book is aimed to present a general overview of all the contributions of electrochemistry to cultural heritage making it accessible to a non-electrochemist reader. Accordingly, a first chapter presents the context of available non-electrochemical techniques, and the three following chapters are devoted to the description of the essential aspects of electrochemistry, the voltammetric techniques, and their analytical applications. In these chapters, the mathematical treatment has been limited to a minimum accessible to graduate or undergraduate level. The next chapters present the specific contributions in the study of pigments and paintings, glass and glazes, pottery, organic materials and metallic heritage, here separating voltammetric, impedance and open circuit techniques, and, finally, dating techniques. The book also includes several tentative approximations to the electrochemical dating of ceramic and organic materials, including charcoal. These are matters under study, and the reader should be aware of the preliminary nature of these proposals.

We are indebted to the kind revision of the manuscript made by the editors of the series, Profs. Fritz Scholz and Lászlo Péter and the work of our colleagues and students in the University of València and the Polytechnical University of València, as well as other colleagues with whom we have been collaborating for the last twenty years. We hope that this book can contribute to facilitate the interest for the electrochemistry within people working on the study and protection of cultural heritage and open new approaches in this field.

Burjassot, Spain — Antonio Doménech-Carbó
Valencia, Spain — María Teresa Doménech-Carbó

Acknowledgments The preparation of the manuscript was aided by financial support from the Spanish R+D+I Project PID2020-113022GB-I00 supported by MCIN/AEI/10.13039/501100011033 and Project AICO/2021/095 which is supported with Generalitat Valenciana and Fondo Europeo de Desarrollo Regional (ERDF).

Contents

About the Authors

Antonio Doménech-Carbó (València, Spain, 1953) is a professor at the Department of Analytical Chemistry, Universitat de València (Ph.D. 1989). His research is focused on solid-state electrochemistry with particular emphasis on the study of porous materials and the development of electroanalytical methods for archeometry, conservation, and restoration. He is the author of over 250 articles including one IUPAC's technical report and several books; among them, *Electrochemical Methods in Archeometry, Conservation and Restoration* (Springer, 2009), *Electrochemistry of Immobilized Particles and Droplets* (2nd ed. Springer, 2014), and *Electrochemistry of Porous Materials* (Taylor & Francis, 1st ed. 2010, 2nd ed. 2021). Currently, he is a member of the editorial board of *ChemTexts* (Springer) and the topical editor of the *Journal of Solid State Electrochemistry* (Springer). He is a reviewer of the European Research Council and several national research agencies and a referee of over 180 indexed journals.

María Teresa Doménech-Carbó, B.Sc., D.Phil. in Chemistry (Universitat de València), since 1999 is a professor in Science of Conservation, Universitat Politècnica de València (UPV), since 2005 until 2016 was the director of the Institut Universitari de Restauració del Patrimoni (IRP) of the UPV, and the editor-in-chief of *Arché*, the Journal of IRP Research. In a research career spanning 25 years, she has published over 200 papers and books on chemical and physical methods of analysis of artworks including *Electrochemical Methods in Archeometry, Conservation and Restoration* (Springer, 2009). She has made over 150 presentations of her research work at international seminars and conferences. She has directed over 12 regional, national, and European R+D. She has supervised 20 research students successfully for the degrees of Ph.D. in chemistry and cultural heritage conservation.

About the Series Editors

Fritz Scholz is the emeritus professor at the University of Greifswald, Germany. Following studies of chemistry at Humboldt University, Berlin, he obtained a Dr.rer.nat. and a Dr.sc.nat. (habilitation) from that University. In 1987 and 1989, he worked with Alan Bond in Australia. His main interest is in electrochemistry and electroanalysis. He has published more than 300 scientific papers, and he is the editor and the co-author of the book *Electroanalytical Methods* (Springer, 2002, 2005, 2010, and Russian Edition: BINOM, 2006), the co-author of the book *Electrochemistry of Immobilized Particles and Droplets* (Springer 2005), the co-editor of the *Electrochemical Dictionary* (Springer, 2008; 2nd ed. 2012), and the co-editor of volumes 7a and 7b of the *Encyclopedia of Electrochemistry* (Wiley-VCH 2006). In 1997, he has founded the *Journal of Solid State Electrochemistry* (Springer) and served as the editor-in-chief until 2021. In 2014, he has founded the journal *ChemTexts—The Textbook Journal of Chemistry* (Springer). He is the editor of the series "Monographs in Electrochemistry" (Springer) in which modern topics of electrochemistry are presented. Scholz introduced the technique 'Voltammetry of Immobilized Microparticles' for studying the electrochemistry of solid compounds and materials, he introduced three-phase electrodes to determine the Gibbs energies of ion transfer between immiscible liquids, and he has extensively studied the interaction of free oxygen radicals with metal surfaces, as well as the interaction of liposomes with the surface of mercury electrodes in order to assess membrane properties. He is also interested in the history of science and the co-editor of the English translation of Wilhelm Ostwald's autobiography (Springer 2017) and the editor of the book *Electrochemistry in a Divided World: Innovations in Eastern Europe in the 20th Century* (Springer 2015).

László Péter is a scientific advisor at the Wigner Research Centre for Physics (Budapest, Hungary). He graduated as a teacher of physics and chemistry at the Eötvös University of Budapest in 1992 and obtained the Ph.D. degree at the same university in 1995. After spending 2 years in the USA as a postdoctoral fellow and 1 year in Japan as a visitor scientist, he returned to his home country and started working in the predecessor of the Wigner Research Centre. His main interest is electrochemistry; in particular, the formation of solid phases in electrochemical processes, the composition depth profile and physical properties of electrodeposited materials, and the electrochemical background of corrosion. Besides electrochemistry, he deals with various fields of experimental physical chemistry and research aspects of industrial problems. His publication list includes more than 100 research papers, two chapters, and one monograph. He is the founding secretary of the conference series called International Workshops on Electrodeposited Nanostructures (EDNANO). In 2013, he became the doctor of the Hungarian Academy of Sciences. From 2020, he is one of the topical editors of *Journal of Solid State Electrochemistry* (Springer).

Abbreviations

AAS	Atomic absorption spectroscopy
AES	Atomic emission spectroscopy
AFM	Atomic force microscopy
ALS	Airborne laser scanner
ATR	Attenuated total reflectance
CE	Capillary electrophoresis
CHNX	Elemental analysis of carbon, hydrogen, nitrogen, and heteroatom
CT	Computerized tomography
CXRF	Confocal micro-X-ray fluorescence
DNA	Desoxyribonucleic acid
DPMS	Direct pyrolysis mass spectrometry
DRIFT	Diffuse reflection Fourier transform infrared spectroscopy
DSC	Differential scanning calorimetry
DTA	Differential thermal analysis
DTMS	Direct temperature-resolved mass spectrometry
EA	Elemental analysis
EDTA	Ethylenediaminetetraacetic acid tetrasodium salt
EELS	Electron energy loss spectrometry
ELISA	Enzyme-linked immunosorbent assays
EPMA	Electron probe microanalysis
ESEM	Environmental scanning electron microscopy
ESPI	Electronic speckle pattern interferometry
ESR	Electron spin resonance
FAB	Fast atom bombardment
FESEM	Field emission electron microscopy
FIB	Focused ion beam

FIB-SEM-EDX	Focused ion beam–scanning electron microscopy–energy dispersive X-ray microanalysis
FID	Flame ionization detector
FORS	Fiber optics reflectance spectroscopy
FTIR	Fourier transform infrared spectroscopy
FUN	Substitution of fluorine, uranium, nitrogen
GC	Gas chromatography
GC-MS	Gas chromatography–mass spectrometry
GD-MS	Glow discharge ion source mass spectrometry
GPR	Ground-penetrating radar
GPS	Global positioning system
HI	Holographic interferometry
HPLC	High-performance liquid chromatography
HPLC-DAD	High-performance liquid chromatography–diode array device
HPLC-FD	High-performance liquid chromatography–fluorescence detector
HPLC-MS	High-performance liquid chromatography–mass spectrometer
HPLC-PAD	High-performance liquid chromatography–photodiode array
ICP	Inductively coupled plasma
ICP-SMS	Inductively coupled plasma sector field mass spectrometry
ICP-TOF-MS	Inductively coupled plasma time-of-flight mass spectrometry
IFM	Immunofluorescence microscopy
INS	Inertial navigation system
IRR	Infrared reflectography
ISE	Ion sensitive electrode
ITPFT	Isothermal plateau fission track
LIBS	Laser-induced breakdown spectroscopy
LIDAR	Light detection and ranging or laser imaging detection and ranging
MALDI	Matrix-assisted laser-induced desorption/ionization mass spectrometry
MC-ICP-MS	Multiple collectors inductively coupled plasma mass spectrometry
MLS	Mobile laser scanner
MRI	Magnetic resonance imaging
MS/MS	Tandem mass spectrometry
MS	Mass spectrometry
NAA	Neutron activation analysis

nanoLC-nanoESI-Q-qTOF-MS-MS	High-performance liquid chromatography–atmospheric pressure chemical ionization (tandem) mass spectrometry
NEXAFS	Near-edge X-ray absorption fine structure
NMR	Nuclear magnetic resonance
NRM	Natural remanent magnetization
OSL	Optically stimulated luminescence
OTC	Optical coherence tomography
Pb-IRM	Pb-isothermal remanent magnetization
PIGE	Particle-induced γ-ray emission
PIXE	Particle-induced X-ray emission
PTTL	Phototransferred thermoluminescence
Py-GC-MS	Pyrolysis–gas chromatography–mass spectrometry
Q-ICPMS	Quadrupole inductively coupled plasma mass spectrometry
RBS	Rutherford backscattering spectroscopy
SCLF	Single-crystal laser fusion
SE	Spectroscopic ellipsometry
SEC	Size exclusion chromatography
SEM	Scanning electron microscopy
SEM-EDX	Scanning electron microscopy–energy dispersive X-ray microanalysis
SEM-WDX	Scanning electron microscopy–wavelength dispersive X-ray microanalysis
SERS	Surface-enhanced Raman scattering
SIMS	Secondary ion mass spectrometry
SIMS-SS	Secondary ion mass spectrometry–surface saturation
SIPS	Sputter-induced optical spectrometry
SPME	Solid-phase microextraction
SRXRD	Synchrotron radiation X-ray diffraction
SRXRF	Synchrotron radiation X-ray fluorescence spectrometry
STEM-EDX	Scanning transmission electron microscopy–energy dispersive X-ray microanalysis
TEM	Transmission electron microscopy
TEM-EDX	Transmission electron microscopy–energy dispersive X-ray microanalysis
TGA	Thermogravimentric analysis
THz-TDS	Tetrahertz time-domain spectroscopic imaging
TIMS	Thermal ionization mass spectrometer

TL	Thermoluminiscence
TLC	Thin-layer chromatography
TLS	Terrestrial laser scanner
ToF	Time-of-flight
TRM	Thermoremanence
UV-Vis	UV-Vis spectrophotometry
VIMP	Voltammetry of immobilized microparticles
VOC	Volatile organic compound
XANES	X-ray absorption near-edge structure
XAS	X-ray absorption spectrometry
XFM	X-ray fluorescence microscopy
XPS	X-ray photoelectron spectroscopy
XRD	X-ray diffraction
XRF	X-ray fluorescence
XRR	Macroscopic X-ray radiograph
ZAF	Atomic number effect (Z), self-absorption effect (A), and fluorescence effect (F)
μ-FTIR	Micro-FTIR spectroscopy
μ-PIXE	Micro particle-induced X-ray emission
μ-SRXRD	Micro synchrotron-based X-ray diffraction
μ-SRXRF	Micro synchrotron-based X-ray fluorescence
μ-XRD	Micro-X-ray diffraction
μ-XRF	Micro-X-ray fluorescence

Chapter 1
Application of Instrumental Methods in the Analysis of Historical, Artistic, and Archaeological Objects

Although scientific methods have been applied in this field since the eighteenth century, it was in the twentieth century when most instrumental techniques were developed, and chemical analysis emerged as a valuable tool at the service of archaeology and the conservation of cultural heritage. Nowadays, a wide range of chemical and physico-chemical techniques provide varied information on the art and archaeological objects. Microscopy techniques and many non-invasive techniques that generate images of the object or a specific part of it provide morphological information. Information on the alteration processes that take place in the outer μm that form the object's surface is obtained from surface analysis techniques. X-ray diffraction is the most common method for crystallographic characterization. Colorimetry helps to characterize optical properties such as color. Elemental composition of inorganic materials is determined by neutron activation analysis or spectroscopic techniques such as X-ray spectrometry, and atomic spectrometry, whereas infrared and Raman spectroscopies and nuclear magnetic resonance detect the molecular structure of materials. Mass spectrometry alone or coupled with chromatographic or inductively coupled plasma devices is commonly used for obtaining the chemical composition of diverse materials (polymers, complex mixtures, etc.). Thermoanalytical methods provide information on the composition of the material through the effect that temperature changes induce in them. In parallel, a wide range of dating methods have been developed that have progressively increased their accuracy.

In this context, electroanalytical techniques have emerged as an interesting alternative to the other analytical methods. The main advantages of electroanalytical techniques are:

- The high sensitivity, allowing to reduce the amount of sample to a few μg or ng;
- The easiness of sampling and pretreatments, which may consist of an abrasive transfer from the object surface to the electrode;
- The portability of the equipment, enabling on-site analyses.

A. Doménech-Carbó and M. T. Doménech-Carbó, *Electrochemistry for Cultural Heritage*, Monographs in Electrochemistry,
https://doi.org/10.1007/978-3-031-31945-7_1

However, the most important contribution of electroanalytical techniques is the variety of applications and types of information that they provide:

- Identification of electroactive species in pigments used in canvas, panel, and wall painting;
- Characterization of pigments, corrosion processes and speciation studies of electroactive species in ceramics and glasses, identification of dyes in textiles, corrosion studies of metal objects, biodeterioration processes in paints;
- Discrimination studies of historical glasses, provenance studies of archaeological ceramics and coins, dating of gold, lead and copper objects, etc.

This book aims to describe the potential of electroanalytical techniques in cultural heritage and archaeometrical studies.

1.1 Archaeology and Conservation of Cultural Heritage

This section describes the origin and evolution of archaeometry and conservation of cultural heritage until now. The evolution of the concept of cultural goods or cultural heritage is also described as well as the singularity of this kind of objects that has determined a particular methodology of physico-chemical analysis.

1.1.1 Archaeology, Archaeometry, and Archaeological Science

Archaeology, a term derived from the Greek whose literal meaning is "the study of ancient history" [1], can be defined as the branch of social sciences and humanities disciplines devoted to the study of human activity in the past. For this purpose, archaeologists study and analyze all kind of artifacts, architectural remains, biofacts, and landscapes. The origins of archaeology are uncertain. Some authors [2–4] have considered the Babylonian kings as the first documented people, who dealt with tasks of archaeology. In particular, in the Neo-Babylonian Empire (556 BC) king Nabonidus carried out excavations before beginning the construction works of temples for studying the remains of old buildings to reproduce their style. The study of artifacts of past civilizations attracted the attention of Greek historians such as Herodotus (ca. 484–ca. 425 BCE). There are registers of studies of archaeological artifacts during the Chinese Empire since the ninth century. During the Renaissance, the interest evolved into collections of ancient artifacts among wealthy and noble persons. This antiquarian activity promoted the study of ancient civilizations by examining their remains so that it would become a more rigorous discipline of archaeology over time.

From the point of view of the conservation of archaeological materials, the term artifact has been defined in 1991 by the U.S.A. Department of Interior, the United States Code of Federal Regulations 36, Part 79, section 4, Curation of Federally-Owned and Administered Archaeological Collections as "*artifacts, objects, specimens, and other physical evidence that is excavated or removed in connection with efforts to locate, evaluate, document, study, preserve, or recover a prehistoric or historic resource*".

The term archaeometry is more recent and is linked to the journal also named Archaeometry and set up in 1958 by the Research Laboratory for Archaeology and the History of Art of the University of Oxford [5]. Although several definitions are found in the literature, "*measurements made on archaeological material*" [6] or "*application and interpretation of natural science data in archaeological and art historical studies*" [7], the editorial line followed by the journal opts for restricting the meaning of the term archaeometry to the physico-chemical analysis of materials. The Journal of Archaeological Science started in 1974, and it includes also biological, botanical, and zoological works. This journal also contributed to restricting the term archaeometry to the fields of analytical chemistry and physics.

1.1.2 Conservation of Cultural Heritage

The essential values that identify and symbolize a community or civilization in each historical period are tied to culture, religion, aesthetics, and technology, among others, and all that constitutes the cultural property. Safeguarding these values as a contribution to the universal acquis of humanity for future generations is a duty of the current generation. Therefore, the activity of conservation of cultural heritage stems from the need of preserving the heritage assets from damages due to the interaction with the environment and human activity. The growing social awareness toward the culture, in general, and cultural heritage, more concretely, has promoted the emergence of the activities of conservation and restoration from a traditional craft as a novel activity with a category of a social and scientific discipline that progressively has gained momentum. Therefore, it is pertinent to describe how this discipline is currently understood.

1.1.2.1 Conservation, Preventive Conservation, and Restoration

The ICOM-CC in a Resolution adopted at the 15th Triennial Conference, held in New Delhi in 2008 [8] has defined different activities devoted to the protection of cultural heritage:

> "**Conservation** - all measures and actions aimed at safeguarding tangible cultural heritage while ensuring its accessibility to present and future generations. Conservation embraces preventive conservation, remedial conservation, and restoration. All measures and actions should respect the significance and the physical properties of the cultural heritage item."

"**Preventive conservation** - all measures and actions aimed at avoiding and minimizing future deterioration or loss. They are carried out within the context or on the surroundings of an item, but more often a group of items, whatever their age and condition. These measures and actions are indirect – they do not interfere with the materials and structures of the items. They do not modify their appearance."

"**Remedial conservation** - all actions directly applied to an item or a group of items aimed at arresting current damaging processes or reinforcing their structure. These actions are only carried out when the items are in such a fragile condition or deteriorating at such a rate, that they could be lost in a relatively short time. These actions sometimes modify the appearance of the items."

"**Restoration** - all actions directly applied to a single and stable item aimed at facilitating its appreciation, understanding, and use. These actions are only carried out when the item has lost part of its significance or function through past alteration or deterioration. They are based on respect for the original material. Most often such actions modify the appearance of the item."

The term "conservation of cultural heritage" strictly refers to any indiscernible intervention on the object to ensure no or minimal alteration. In preventive conservation, the action is not applied to the object as it is aimed at adjusting the environmental conditions for maintaining the object in its current state. Remedial conservation and restoration are activities aimed at maintaining or returning to the object the loss or damaged perceptible features it had in its original or prior state. Conservation differs from restoration and remedial conservation as the item's appearance is not modified.

1.1.2.2 From Antiquities to the Modern Understanding of Cultural Heritage

The concept of cultural heritage or cultural property has evolved, adopting different meanings in different historical periods. In the past, the meaning of cultural heritage was linked to the characteristics of the objects on which conservation and restoration activities were carried out (Fig. 1.1). Since a craft activity aimed at repairing antiquities carried out in ancient times, as in the case of king Nabonidus [9], the concept of cultural heritage was progressively acquiring a broader meaning. In further historical periods, the aesthetic character of artworks [10] and the historical and historiographic character of archaeological remains determined which objects were subjected to conservation activities [11]. Nevertheless, in the twentieth century, this concept underwent a notable change, extending its meaning to a wider one of cultural heritage or property. Cultural goods represent the essential values of a community [12]. Thus, tangible objects are not only the subject of conservation activities but also the intangible ones that can bear witness of the community's values and culture [9, 11]. More recently, the cultural heritage concept has been revised, and its meaning extended to any object with a symbolic message representative of individuals or communities [13].

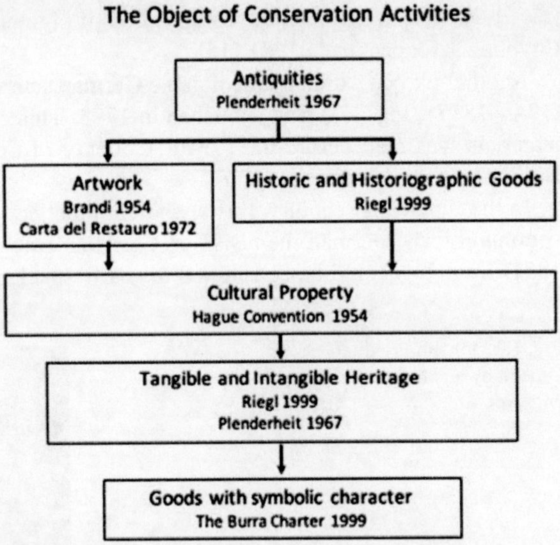

Fig. 1.1 Scheme showing the relationships between the different definitions proposed for the objects subjected to conservation activities [9–14]

1.1.3 A Brief History of the Scientific Analysis of Cultural Heritage

The origin of the scientific examination and analysis of cultural heritage is located in the eighteenth century with the dissemination of the new historiographic model of Johann Winckelmann (1717–1768) about the study of old civilizations by analyzing not only documents but also the archaeological remains [15]. Thus, a new period began in which archaeologists and historians started to use simple analytical methods for obtaining the composition of archaeological remains, which brought information about the materials and manufacturing techniques used by the artists and artisans of old civilizations. Similarly, most artists and antiquarians[1] used simple analytical tests such as the wipe (solubility) test to establish the best method for removing an old darkened varnish that disturbs the contemplation of a painting or cleaning an altered painting [15]. The first scientific studies date back to the eighteenth century.[2] Apart from a limited number of chemicals for performing the reactions, magnifying glasses were the unique instrumentation used for carrying out these assays. The earliest published account of a chemical analysis of historical materials of paintings has been attributed to the German scientist Johann Friedrich Gmelin (1748–1804)

[1] The role of antiquarianism flourished in Europe since the Renaissance, a period in which the classical antiquities were rediscovered and outstanding artists such as Michelangelo, Verrocchio, Donatello, Lorenzetto, or Cellini were involved in conservation and restoration of archaeological objects [16].

[2] Reported contributions of geology to archaeological studies go way back to the year 1720 when the English astronomer and scientist Edmond Halley (1676–1742) carried out a microscopy analysis of stones samples from the Stonehenge site ref. [17].

(Fig. 1.2), who determined the composition of pigments and binding medium of an Egyptian sarcophagus in 1781 [15].

Another early account by the notable German scientist Martin Heinrich Klaproth (1743–1817) (Fig. 1.3) was published in 1798. This chemical study of numismatic materials was read before the Royal Academy of Sciences and Belles-Lettres of Berlin three years before [18–20].

In the nineteenth century, this analytical practice was extended to the scientific community. The German chemists Karl Christian Traugott Friedemann Göbel (1794–1851) (Fig. 1.4) could be considered the earliest scientist who, in 1842, developed

Fig. 1.2 Johann Friedrich Gmelin by an anonymous engraver

Fig. 1.3 Martin Heinrich Klaproth by Ambroise Tardieu after original portrait by Eberhard-Siegfried Henne

Fig. 1.4 Friedemann Göbel (1841), oil painting by Aleksander Hripkov. Art Collection of the University of Tartu Library. Courtesy of Tartu University

extensive and systematic analytical research on archaeological materials. This scientist published a booklet in which he made an explicit reference and a detailed description of the use of analytical methods for characterizing archaeological metal objects. This booklet is considered a classic archaeological chemistry publication [16, 19, 20].

Significant contributions at this time were due to the English Sir Humphry Davy (1778–1829) and Michael Faraday (1791–1867), the French chemists Jean-Antoine Claude, Comte Chaptal de Chanteloup (1756–1832) and Jean d'Arcet (1724–1801), the German Friedrich August Walchner (1799–1865) and Otto Wilhelm Eduard Erdmann (1834–1905), the Swede Jöns Jacob Berzelius (1779–1848) and the Irish-born US chemist John William Mallet (1832–1912), among others, were also involved in analytical studies on archaeological objects and artworks [19]. The first analytical study in which an optical microscope was used has been reported by the German architect and art critic Gottfried Semper (1803–1879) in 1834 [15].

In the nineteenth century, a change of paradigm in the activities of conservation of cultural heritage took place with the introduction of the concept of scientific conservation by the Italian architects Camillo Boito (1836–1914) and Gustavo Giovannoni (1873–1947). According to Boito and Giovanonni, conservation is an activity founded on scientific theories, which is aimed at applying technical procedures and selecting materials in interventions of cultural heritage [21].

In this context, the German chemist Friedrich Rathgen (1862–1942), appointed first director of the State Museum of Berlin (*Chemisches Labor der Königlichen Museen zu Berlin*) in 1888, was the first scientist that applied physico-chemical principles in the conservation and restoration treatments. This work was published in a treatise entitled *Die Konservierung von Altertumsfunden* (The conservation of ancient discoveries) [22].

In the late nineteenth century and the first half of the twentieth century, the technological advances boosted the development of scientific studies of cultural heritage. The German physicist Walter König (1859–1936) performed the first X-ray radiograph on a painting in 1896 [23]. The Scottish chemist Arthur Pillans Laurie (1861–1949) introduced in 1914 the preparation of pictorial samples as cross-sections for their microscopic examination and the performance of spot tests for identifying pigments [24]. In 1924 the Bavarian State Painting Collections (*Bayerische Staatsgemäldesammlungen)* in Munich was the first museum to install an X-ray radiographic instrument. Other museums and research centers, such as the *Centre de Recherche et de Restauration des Musées de France* and the *Laboratory of the British Museum*, also started systematically using radiographic instruments to examine objects [24]. Harold Barker (1919–2015) was a scientist pioneer in developing the ^{14}C as a dating method for archaeological objects after developing this technique in 1947 [22]. Other instrumental techniques such as spectroscopy, chromatography, electroanalysis, mass spectrometry, X-ray diffraction, and microscopy were also progressively introduced in the heritage conservation field in the twentieth century [25]. In particular, the use of optical emission spectroscopy in archaeometry started in 1931 [26], and the application of IR spectroscopy for characterizing the composition of paintings' materials was pioneered by Robert L. Feller (1919–2018) in 1954 [27]. The use of electron probe microanalyzers (EPMA) and scanning electron microscopes coupled with energy dispersive X-ray microanalysis systems (SEM–EDX) started in the 1950s and 1960s [28].

In the second half of the twentieth century, technological advances in informatics led to the standardization of computerized platforms in analytical instrumentation. New advanced techniques such as atomic force microscopy, laser desorption systems, voltammetry of microparticles, and high-resolution liquid chromatography were developed and applied to the study of cultural heritage. As a result of the combination of two or more instruments, other new techniques were developed and successfully applied in this field such as tandem mass spectrometry, micro-Fourier transform infrared and micro-Raman spectroscopy, micro-X-ray diffraction or inductively coupled plasma-mass spectrometry. This period was also characterized by the miniaturization of analytical instrumentation that boosted the design of portable instruments that enable non-invasive and in situ analysis.

In the twenty-first century, the development of new instrumentation for surface analysis has enabled the determination of chemical composition (elemental and molecular) spatially resolved in two dimensions (2D) and three dimensions (3D) maps. Also, microsampling methods adaptable to a variety of instruments or miniaturized chemical and electrochemical microfluidic devices have improved the analysis of cultural heritage by reducing the sample size without losing confidence in the results. A new generation of spot tests that uses microfluidic or smartphone spectroscopy has been developed. Nanotechnologies have been essential for the development of immunoassays that can be used for the analysis of organic materials present in objects. On the other hand, the reduction in sample size and the miniaturization of analytical devices have given rise to green analytical chemistry in agreement with the current trends of sustainability and protection of the environment [29].

1.2 Role of the Analytical Methods in Archaeometrical and Cultural Heritage Research

Archaeometry and conservation of cultural heritage are multidisciplinary activities in which scientists of different fields work together and combine their findings efficiently to ensure the recovery of lost heritage and its sustainability for the future. The information collected about heritage assets differs depending on the field from which they are obtained, historical, architectural, mathematical, biological, geological, chemical, and aesthetical. The coordination of the multidisciplinary team is essential for avoiding misinterpretations of the results if they are not studied as a whole and are separately processed by each specialist.

Inside the multidisciplinary team, several scientific disciplines use physico-chemical analytical methods to obtain information on the materials composing the object under study, especially with regard to chemistry, physics, and engineering. Due to the subject of this book, the physico-chemical analytical methods used for heritage study are the focus of the discussion in this chapter.

It is worth mentioning that the modern concept of cultural heritage has extended this figure not only to tangible properties but also to intangible ones. At first, one could think that the analytical methods of the experimental sciences, such as chemistry, physics, geology, or biology, are restricted to studying tangible objects. However, a closer look at this subject reveals that intangible heritage can also require the support of analytical methods. For instance, in the early twentieth century, the famous Hungarian composers Béla Bartók and Zoltán Kodály collected rural songs on gramophone records. Nowadays, the old songs transmitted through oral tradition in primitive tribes are stored in modern CD records to safeguard their traditional values. Nevertheless, gramophone and CD records have material support that can be damaged over time, leading to the loss of their content. A similar issue is found with extinct ritual dances or traditional rites registered in old movies recorded on unstable celluloid films. Again, the intervention of scientists and conservators is required.

This section presents the different aspects that define the role of the analytical methodologies used mainly by chemists, physicists, and engineers (but also by archaeologists, architects, biologists, and geologists). Those methods are essential tools in archaeometrical studies and in the conservation of cultural property procedures for providing information on the material structure and composition of the objects.

1.3 Information Provided by the Analytical Research

It is worth remarking, first of all, the great complexity of the analytical studies in this field. That is due to four main factors:

- The variety of materials that conform to the objects: metals, ionic, molecular compounds, and covalent solids.

- The heterogeneity of the object, formed as a multilayered structure in which inorganic and organic compounds are distributed in each layer that can reach a thickness of a few nm or μm.
- Heritage objects are dynamic systems with active reactivity among the endogenous materials that compose the object and reactivity with exogenous compounds present in the surrounding environment.
- The restrictions in the sampling and use of instrumentation, even though it does not require sampling, can induce damages to the object that compromise its integrity or alter its figurative or historical symbolism.

The information provided by the analytical methods applied to the study of cultural heritage falls into three categories: examination and analysis of materials, study of the interaction between the object's materials and the environment, and study of the interaction between methods and products used in conservation and restoration treatments. As shown in Fig. 1.5, the first category is of interest in both fields of archaeometry and conservation of cultural heritage, whereas the other two studies are specifically useful for the conservation of cultural heritage [30].

The analytical studies on cultural heritage provide information on the state of conservation (materials, alterations, and interactions) of the object in the present moment (Fig. 1.6). These data enable knowing the state of conservation of the object for applying proper conservation, remediation, or restoration treatments. Nevertheless, the analyst must extrapolate the data back, to respond to the questions raised by the archaeologist on how the object was when it was made. Conversely, the extrapolation into the future of data on the present composition of the objects answers the questions of conservators concerning the object's fate in the short, medium, and long-term. This issue falls into the aims of preventive conservation.

Analytical methods to the service of the studies in cultural heritage have not only the object as the target of their experimental procedures. The study and characterization of the environmental factors that influence the state of conservation of

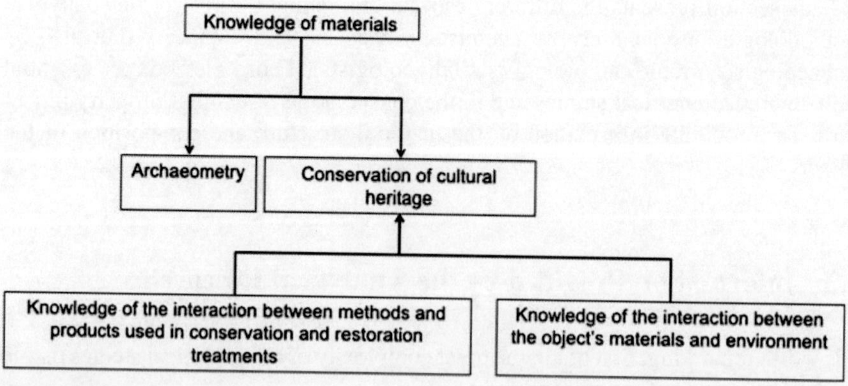

Fig. 1.5 Scheme of the information provided by analytical methods in cultural heritage and archaeometry

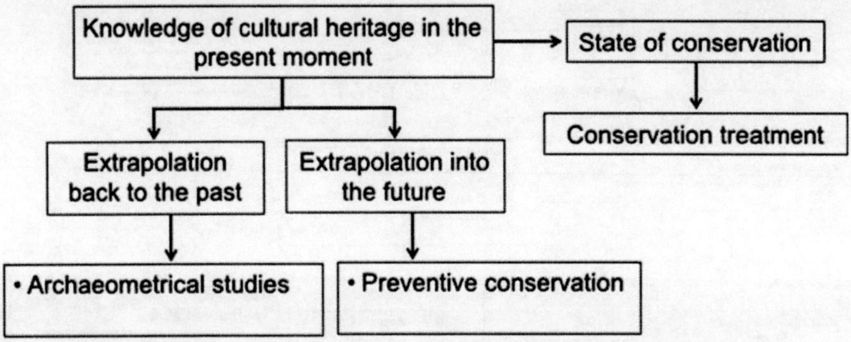

Fig. 1.6 Scheme of the objectives of the analytical methods in cultural heritage studies

the objects are also of great interest as they are mainly responsible for the aging and deterioration of the objects. The products and methods used in the conservation interventions on the objects are also a target of the analysis. This is because the materials composing the object can interact with these products yielding by-products and promoting secondary and undesired mechanisms of alteration that can produce additional damage to the object.

Figure 1.7 schematizes the relationship between the three different ways in which the analyst interprets the data, the analysis targets, the first physico-chemical models obtained, and the final applications achieved in the specific field in which the study is framed. Characterizing the original materials used by the artist or the craftsman for creating the object and the alteration products formed over time enables the analyst to identify the alteration mechanisms that have acted on the object. The data provide a complete picture of the state of conservation of the object and the basis on which the conservator can adequately intervene in the object. Additionally, the results obtained can provide valuable information for developing new materials and methods for the remediation and restoration of the objects. The data obtained on the materials identified in the artifacts can be interpreted and processed by the analyst to elucidate the artistic or manufacturing technique. Application of multivariate methods on compositional data of a series of objects together with a comparison with reference materials of known origin enables the analyst the discrimination of them by provenance from different geographic locations. The application of dating methods based on mathematical models of variation of specific physico-chemical properties of the materials enables the analyst to determine the chronology of the object. Identification of anachronic materials or comparison with other objects with well-known authorship enables the authentication of the object. The combination of compositional data of the object and physico-chemical parameters of the surrounding environment enables the analyst to develop theoretical predictive models for describing the alteration mechanisms to which the object will be subjected in the short, medium, and long-term. The conservator uses this information to estimate the damage risk at which the object will be subjected in the future and thus apply a reasonable control of conditions to minimize these alterations.

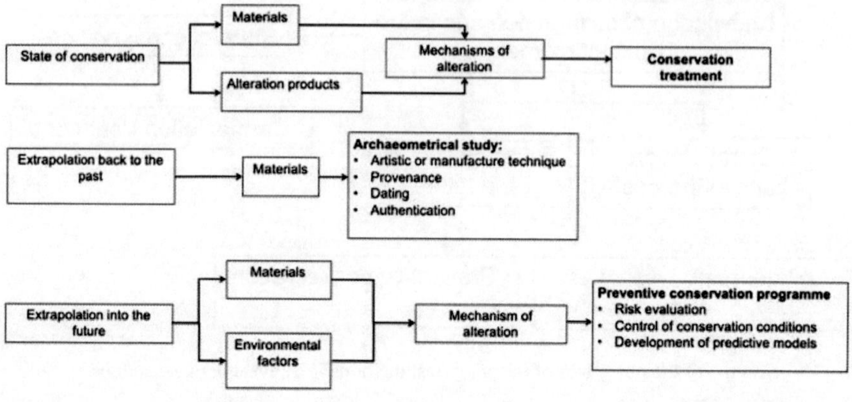

Fig. 1.7 Scheme of the information provided from analytical data

As illustrated in Fig. 1.7, data compiled in the archaeometry and conservation of cultural heritage studies are obtained from three primary sources: the object, the surrounding environment, and the products and methods used in the intervention treatments. This topic is described in more detail in the next section.

1.3.1 Analytical Information Obtained from the Object

The information obtained by examining and analyzing the heritage assets can be classified into four categories depending on the analytical or examination methodology used: physical, chemical, geological, and biological.

1.3.1.1 Physical Information

Mechanical, rheological, electrical, thermodynamic or optical, or surface features, among other properties of the material, determine the behavior of the object. Characterization of these properties in heritage assets is often carried out using standardized procedures proper to materials science. The National Standards Policy Advisory Committee [31] defines a standard as "a prescribed set of rules, conditions, or requirements concerning definitions of terms; classification of components; specification of materials, performance, or operations; delineation of procedures; or measurement of quantity and quality in describing materials, products, systems, services or practices."

Several national and international institutions are devoted to developing standards for guaranteeing the quality of the products manufactured in all kinds of industries and human activities. Among the important organisms that are directly devoted to the standardization of materials and procedures involved in the conservation of cultural heritage, special mention should be made of the Italian Committee NORMAL, the

Technical Committee 346 for conservation of the cultural heritage of the European Standard Committee (CEN) and the International Union of Laboratories and Experts in Construction Materials, Systems, and Structures (RILEM). Most common standard tests applied in studies in conservation of cultural heritage are designed for characterizing the original materials of the object and their alterations and for testing new products and conservation methods. RILEM establishes several categories of standards, namely tests for characterizing the structure of the material (i.e., porosity accessible to water, air permeability, thermal conductivity), tests for characterizing mechanical and rheological properties of the material (i.e., mechanical strength, mechanical compression test, shearing test), tests for characterizing surface properties of the material (i.e., abrasion resistance test, adhesion test, hardness test), tests for characterizing the hydric properties of the material (i.e., water absorption coefficient, water drop penetration test), test for characterizing corrosion processes and other alterations of the material (i.e., neutral salt spray, Kesternich), durability tests (i.e., salt crystallization, freeze–thaw cycle stability). Other frequently applied standard tests are specific for adhesive joints or impact resistance. Technical examination of the object with instruments that record images in 2D or 3D obtained from electromagnetic radiation, electrons, or sound provides varied information on size, shape, and superficial structure (debris, dust, superficial deposits, crusts, cracks, pores, fissures, fractures, laminations, lixiviations, spots, or efflorescences) and internal structure (presence of foreign bodies, anomalies, or damages).

Examination with optical microscopes of microsamples prepared as thin-sections allows the identification of the histological features of the type of organic material (i.e., wood, parchment, textile, paper, ivory, horn, leather). Examination of paint microsamples prepared as cross-sections allows for the characterization of the pictorial strata distribution that is essential for establishing the artistic technique used by the painter.

1.3.1.2 Chemical Information

The information on chemical types from the heritage assets is abundant and diverse. Depending on the analytical method, the data are qualitative or quantitative. Qualitative information includes the identification of elements and electroactive species present in the material and molecular and structural information that involves the identification of functional groups, crystallinity, and cell parameters. Occasionally, the identification of organic compounds requires the quantitation of species and the application of multiparametric methods for data processing. Advanced instrumentation simultaneously yields chemical and morphological information registered as compositional linear, 2D, or 3D mappings.

In parallel to the data from the object, aging studies performed in the laboratory on specimens that mimic the original materials that compose the objects are carried out for assessing theoretical models developed for describing the behavior of the object in the short, medium, and long-term. A variety of alteration products formed, such as oxidation, reduction, polymerization, scission, hydration, dehydration, or dehydrogenation, are the target species.

1.3.1.3 Geological Information

Characterizing of rocks used in sculptures and monuments and their alterations is made from the examination of thin-sections of these materials with a petrographic microscope and the study of optical properties of minerals and rocks such as color, pleochroism, refractive index, birefringence, extinction, interference color and figures, optic sign and sign of elongation, etc. This methodology also enables the characterization of pigments, ceramics, glass, glazes, plasters, metals, and slags.

Geosciences that investigate the past, present, and future behavior of our planet by developing theoretical models based on experimental data are also important contributors to the study of heritage. Dating methods are essential tools in archaeometrical research and theoretical models of climate change are currently being used for predicting the effects of this phenomenon on cultural property in the future.

1.3.1.4 Biological Information

The main targets of heritage conservation studies are the identification and quantitation of the microorganism species (fungi, algae, or bacteria) and the analysis of their metabolic products as a result of their activity. Another target of heritage conservation studies is the reaction pathways of metabolic activity of all living beings that can interact with heritage assets. Changes in the object due to the biological activity of morphological and chemical types are determined as described above with physical and chemical methods.

1.3.2 Analytical Information Obtained from the Environment

The term "environment" refers to the physical and chemical characteristics and weather conditions of the surroundings in which the object stays. Characterizing the environment allows for a complete picture of the conservation conditions of the object. Three types of environment can be distinguished: aerial, terrestrial, and underwater. Most cultural property is exposed to the atmosphere in outdoor or indoor conditions. In such instances, the determination of chemical, weather, and other physical parameters of the environment enables the identification of the causes of the alterations identified in the object. Monitoring a broader range of weather parameters is required in conservation studies of outdoor monuments. Weather parameters related to the water in the atmosphere (humidity, condensation, frost, precipitation regime: rain and snow), temperature, wind characteristics, and sky radiation are of great interest. In parallel, the concentration of atmospheric pollutants, such as suspended particulate matter, marine aerosols, SO_2, O_3, NO_2, and CO_2, are also of interest. Other environmental factors to be controlled are solarization and vibrations caused by road and rail traffic.

Determination of the pollutants content in the air (CO_2, NO_x, SO_2, O_3, organic compounds, marine aerosols, or suspended matter), temperature, and humidity is also commonly required for characterizing the aerial environment in which the object is exposed. The environment conditions exert a differential influence on the behavior of the object depending on its internal structure. In particular, water circulation throughout the micropore network is responsible for the internal erosion of the porous network due to frost. The growth of microcrystals of insoluble salts transported by the circulating water inside the pores is also frequent.

Monitoring parameters such as light intensity, temperature, and humidity of the indoor environment surrounding the object is of great interest for properly planning preventive conservation programs in museums and historical buildings. Volatile organic compounds (VOC) emanations from furniture and exhibition cases and suspended particulate matter are also monitored in indoor environments.

Characterization of burial and underwater environments in which the object stayed for many centuries yields crucial information. Determination of physical and chemical properties of the soil (i.e., pH, conductivity, chemical composition, ionic strength, etc.) is of interest in studying archaeological artifacts stored in burial conditions. In these ambient, the water can dissolve and carry ionic species from the soil that can promote the acid or alkaline attack and further lixiviation of ionic species from the object materials. Similarly, studies that include the characterization of the content of ionic species, chemical parameters such as pH, ionic strength, and physical parameters including conductivity, temperature, or density of the aqueous solution in equilibrium with waterlogged and underwater objects are also of interest.

1.3.3 Analytical Information Obtained from the Conservation Processes

The modern conception of the restoration and conservation tasks as activities that use scientific methodologies has led to increased demand for the analytical control of the intervention by the personnel in charge of these activities. The analytical information varies depending on the specific intervention and the type of heritage asset. Cleaning operations on ceramics or metals require monitoring the conductivity of the cleaning water bath until the content of soluble salts is reduced to a particular threshold value. Chemical, mechanical, and laser cleaning are usually controlled. In this case, changes in the surface morphology of the object are controlled by comparing data obtained before and during the intervention using microscopy techniques. Changes in the visual appearance of the object, in particular, color are controlled by colorimetric methods. Original materials lixiviated and extracted by the cleaner and the latter remaining in the microporous structure of the object are identified and determined by spectroscopic or chromatographic techniques. The cleaning products and formulations are also subjected to analytical studies to improve their efficiency. Chemical and physical properties such as pH, ionic strength of aqueous solutions, solubility parameters of solvents, vapor pressure, boiling temperature, viscosity, and surface tension, among others, are controlled.

Control of consolidation and adhesion treatments requires the monitoring of morphological, chemical, and physical properties of the consolidant-object system. Changes in mechanical strength, color, gloss, distribution in the microstructure of the object, and penetration depth of the consolidant are commonly assessed. Preliminary studies of the stability and compatibility of consolidants can be performed with suitable natural and accelerated aging trials.

Finishes and protective coatings of paintings, corrosion inhibitors in metals, or water-repellents in stone are also analytically controlled. Likewise, for the consolidants and adhesives, accelerated aging trials are made in preliminary studies to check the suitability of the proposed coatings and fillers. Environmental conditions during application, especially temperature and humidity, are routinely measured as these weather parameters significantly influence the final result of the treatment.

The efficiency of biocides is another preliminary experiment carried out before a disinfection treatment of a cultural good. For this purpose, laboratory specimens that mimic the original are subjected to trials with different products to select the more effective. The stability of the biocides is also tested using accelerated aging trials. These experiments are aimed at characterizing their resistance to aging agents.

The conditions of transport and storage of cultural goods are strictly controlled to guarantee the preservation of the integrity of the object. Parameters such as the inert character of the packaging materials, temperature, humidity, or vibrations, among others, are considered in such instances.

1.4 Analytical Methodologies Applied to Archaeometry and Cultural Heritage Research

1.4.1 Requirements of the Analytical Methodologies

The analysis of a cultural asset is conditioned by its singular and inimitable character and the requirement to preserve its integrity as much as possible. For this reason, the analytical method selected must mandatorily accomplish a series of requirements. As illustrated in Fig. 1.8, the analytical method selected by the analyst must provide the information required to achieve the aims fixed for the case study. When applying a single analytical technique does not offer the required information on the heritage asset under study, the analytical method can be planned with the inclusion of several techniques that are appropriately combined to provide complementary data. The use of multitechnique approaches is currently an increasing trend in cultural

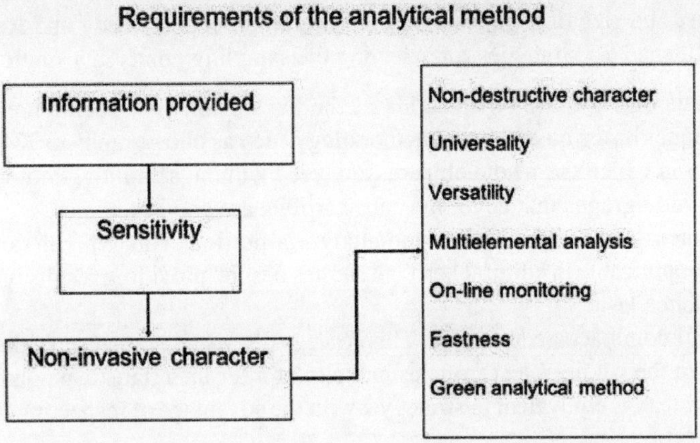

Fig. 1.8 Requirements of an analytical method applied to the study of cultural heritage

heritage. The analyst must be sure that the techniques selected have suitable sensitivity for detecting the target species. Once the analyst has selected all the analytical techniques that accomplish the requirements of suitability and sensitivity, a second refinement in the design of the analytical method is carried out by choosing preferably non-invasive techniques that do not require sampling. If it is not possible to perform an analytical method with non-invasive techniques, invasive ones requiring sampling are included. Other desirable requirements are the non-destructive character of the analytical technique for which the sample is not destroyed during the analysis and, therefore, can be analyzed by a second technique. The universality of the technique enables the analysis of objects with varied compositions, shapes, and sizes. Versatility for providing data on the entire object or any part. The capacity for performing multielemental analysis is preferred over techniques that cannot simultaneously analyze elements. Sometimes, the ability for the online monitoring of a physical or chemical magnitude is essential. Fastness is another requirement that can be necessary when the object is highly damaged and the intervention must be carried out as soon as possible. In the last decades, the requirement of sustainability for any human activity has become even more indispensable and has promoted the emergence of green analytical chemistry. This new trend has also arrived in the field of cultural heritage.

1.4.2 Sampling Strategy

Regardless of whether the analytical technique used is invasive or non-invasive, most studies on cultural heritage require sampling. For planning a complete sampling strategy, the analyst should make decisions about the location of sampling points, the number and size of samples, and the sampling method. Reedy and Reedy [32] have established six strategies for selecting the sampling points in a single object:

1. Analysis of the entire object. This method is suitable for specific non-invasive techniques based on a holistic methodology such as photography or X-ray radiography in which the whole object is analyzed without sampling. Photographs or X-ray radiographs that cover the entire artifact are good examples.
2. Homogenate the entire object and analyze a portion. This type of sampling is hardly applicable in cultural heritage studies despite providing results with a high confidence level.
3. Take randomly located samples. This strategy allows an estimation of the composition of the whole object from a small part of it. From a statistical point of view, this strategy is equivalent to strategy 2 with the advantage of respecting the object analyzed.
4. Choose regularly patterned samples. In this strategy, samples are taken at regular intervals across the object. The level of confidence in the results is similar to strategy 3. Nevertheless, there is the risk of biasing results if the object has a spatial heterogeneities pattern at the same scale that the sampling interval.
5. Haphazardly or arbitrarily select points. In this strategy, the analyst imposes restrictions on particular positions of a randomly or regularly patterned sampling for aesthetic or preservation reasons. Thus, there is a high risk of biasing results.
6. Intentionally select or sample components not yet examined. This is the strategy most frequently applied in cultural heritage research despite its high risk of biasing results. Good examples are the identification of alterations or the characterization of the artist's palette.

The number of samples depends upon the level of precision required in the analysis, the aims of the study, and the type of object. The larger the number of samples, the higher repeatability is reached in the analysis. The larger the number of aims (i.e., identification of materials, identification of alteration products, identification of both materials and alteration products), the larger the number of samples. Multicomponent objects (i.e., sculptures with a metallic internal skeleton, wood body, many colors, and five or more superimposed paint layers) require more samples. The respect to the integrity of the object always limits the number of samples. Therefore, a compromise between analysis quality and object preservation must be achieved.

Two main factors condition the size of the sample. The requisites of respecting the integrity of the object and the degree of heterogeneity of the material analyzed impose the higher and lower sample size limit, respectively. Therefore, a compromise between these two factors must be achieved in each particular case study. Samples below 1 mm^3 are taken for preparing cross-sections in canvas paintings, but larger samples are needed in wall paintings composed of thicker grounds for assuring that the sample includes all strata.

The sampling method depends on the analytical technique selected for performing the analysis. As far as possible, the sampling method should not be invasive; for instance, smudges of pigment transferred onto the opposite page can be used as samples of a miniature decorating an illuminated manuscript. Nevertheless, the most common sampling method is the mechanical taking of samples on the micro to nanoscale using needles, lancets, scalpels, and, less frequently, instruments specially designed. The materials retained in the swab are directly analyzed in the swab or, most frequently, are redissolved in the laboratory, and the solution obtained is analyzed. Soluble materials can be extracted with cotton swabs impregnated with an appropriate solvent that is rolled on the surface of the object. Abrasive transferring to a small SiC disk of a few grains of the solid material is occasionally used in Fourier transform infrared (FTIR) spectroscopy. Sample grains can be abrasively transferred to appropriate graphite electrodes of diameter 0.25–1 mm in voltammetry of immobilized microparticles (VIMP) [33].

More complex sampling strategies are applied when the research is aimed at studying a collection of objects. A frequent case study is the characterization of the artist's palette from the analysis of several paintings. The sampling should be stopped when all the samples of the same color yield duplicated results.

1.4.3 Preparation of Samples

Most of the analytical techniques applied in cultural heritage studies include the pretreatment of the sample as a step before the measurement step. The pretreatment can include, in turn, one or more steps. Figure 1.9 schematizes some of the most frequent pretreatments applied in the analysis of cultural goods [25].

Powdering or grinding of samples is the simplest preparation method used in instrumental techniques such as X-ray diffraction (XRD), nuclear magnetic resonance (NMR), differential thermal analysis (DTA), and thermogravimetric analysis

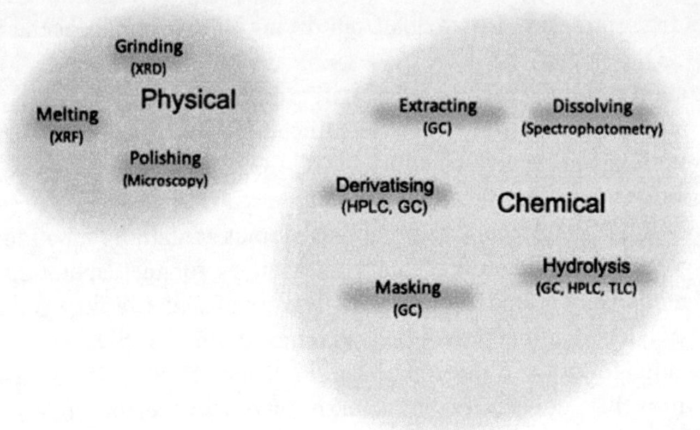

Fig. 1.9 Scheme of the most frequent pretreatment methods used in cultural heritage analysis and examination

(TGA). In some cases, a second step of mixing with an inert binder, i.e., KBr for FTIR spectroscopy, and pressing to form a disk, is required.

Melting a mixture of the sample and a fluxing substance (i.e., lithium tetraborate) that lowers the melting temperature and allows the formation of a solid solution with a disk shape. This pretreatment in X-ray fluorescence (XRF) is necessary to analyze ceramics and glass materials.

For the microscopic examination, samples are prepared as cross-sections, thin-sections, and by mounting the sample on a glass slide using a mounting medium. Thin and cross-sections are prepared by embedding the sample in a fluid polyester, acrylic, or epoxy polymer, and the block is led to cure so that a cross or thin-section of the sample can be obtained by mechanical polishing the block with SiC abrasive disks and, eventually, with alumina suspensions or diamond paste. Removal of atoms, ions, molecules, and clusters of particles from the surface of the sample can be accomplished by bombardment with a thin and intense focused beam of ions (i.e., Ga^+). This method is used in focused ion beam (FIB) combined with scanning electron microscopy (SEM) for obtaining micrometric cross- and thin-sections.

Dissolution of the sample in a suitable solvent is a pretreatment applied in several spectroscopic and chromatographic techniques [i.e., UV–Vis spectrophotometry, high-performance liquid chromatography (HPLC), thin-layer chromatography (TLC), atomic absorption spectroscopy (AAS)]. Sometimes an acidic or alkaline attack is required to dissolve the sample.

Liquid–liquid extraction is carried out as separation and concentration pretreatments in techniques such as gas chromatography (GC). More recently, and following the trend of green analytical chemistry, solid-phase microextraction (SPME) techniques have been successfully applied in chromatographic techniques.

Binding media contain polymerized molecules that cannot be directly analyzed by most chromatographic techniques. In such instances, acidic or alkaline hydrolysis is carried out for releasing the monomeric constituents of the polymer molecules: amino acids from proteins, fatty acids from drying oils, or monosaccharides from plant gums [33].

Derivatization reagents react with the analyte molecules to modify their structure and confer properties that allow their identification with fluorescent detectors in HPLC or provide the necessary volatility in GC [33].

Suppression of interfering species is another pretreatment necessary in some types of samples. C8, C12, or C18 chromatographic columns containing functional ligands are specially designed as complementary alternatives for the separation of polar, neutral, and moderately nonpolar molecules. The use of sequestering agents of metal ions such as ethylenediaminetetraacetic acid tetrasodium salt (EDTA) has also been successfully applied to avoid the complexation of the released amino acids by the metal ions from the pigments in the analysis of proteinaceous binding media of paintings [33].

1.4.4 *Data Measurement and Processing*

Like in other fields of analytical chemistry applications, signal processing methods are commonly used. Filtering is used in spectroscopic techniques for improving the signal-to-noise ratio and data compressing. Fourier transform (FT) and wavelet transform (WT) are applied for pattern recognition, quantification, background removal, multiscale regression, or derivative calculations.

In X-ray spectrometric techniques (XRF, coupled with X-ray microanalysis SEM–EDX) interelement or matrix effect is corrected by theoretical methods such as the *k*-ratio (ratio of the element X-ray intensity in the specimen to the pure element intensity) or by applying corrections to the *k*-ratio using the standard ZAF method (atomic number, self-absorption and fluorescence effects) or by relying on fundamental parameters method.

Descriptive statistics and estimation methods are used in the field of cultural heritage. Linear regression methods and multivariate techniques such as cluster analysis and discriminant analysis are abundantly employed.

1.5 An Overview of Scientific Methods Applied in Archaeometry and Cultural Heritage Research

The variety and complexity of the materials that are part of the cultural assets and the different aims of the research on heritage demand the application of all types of scientific methods that can provide information with little or no damage to the object. The methods currently available can be classified into two main categories: (1) analytical methods and (2) dating methods. The following gives a brief overview of state of the art in the examination and analysis of cultural heritage.

1.5.1 *Analytical Methods*

The categorization of analytical methods can be made according to different criteria. In this section, the analytical methods have been classified according to the chemical or physico-chemical phenomenon on which is based the method. Table 1.1 summarizes the methodologies most commonly applied, which have been categorized by the physico-chemical fundamentals of the method.

1.5.1.1 Chemical Methods

Chemical or classical methods include all those based on chemical reactions carried out with no or simple instrumentation. Microchemical or spot tests can be considered

Table 1.1 Classification of analytical methods

Class	Subclass
Chemical	Spot tests
	Staining tests
	Gravimetric, volumetric methods
Physico-chemical	Microscopy methods
	Surface analysis methods
	Electroanalytical methods
	Spectroscopic methods
	Diffraction methods
	Thermoanalytic methods
	Mass spectrometric methods
	Separation methods
	Methods based on the effects on the atomic nucleus
	Non-invasive methods

the first analytical methodology applied to the study of cultural goods, and they are still used in specialized laboratories. The high sensitivity of the reactions involved allows the use of this method in samples at the microscale. Other advantages are their easiness, fastness, and feasibility for identifying the analyte within the paint layer in samples prepared as cross-sections. The development of the reaction can be observed with stereomicroscopes at low magnification. The method can be used for identifying inorganic and organic pigments, binding media, varnishes, and adhesives. Reactions that result in the formation of a precipitate, volatile product, colored or fluorescent complex, and redox product, among others, are used in spot trials [29]. In parallel to the spot tests, staining tests developed from the classical histological tests for analyzing biological tissues are applied for detecting and discriminating binding media of natural origin: proteinaceous media composed of egg, animal glue, casein, drying oils, or plant gums.

Immunological methodologies based on immunoassays are also successfully applied to identify proteinaceous media. The high specificity of the antigen–antibody reaction allows discriminating between different animal species of the protein or multiple antigens. These techniques are characterized by their high sensitivity. Two types of immunological techniques have been applied in the analysis of proteinaceous media: enzyme-linked immunosorbent assays (ELISA) [29, 34] and immunofluorescence microscopy (IFM) [35].

Despite their simplicity, quantitative classical methods such as volumetric and gravimetric analysis have been displaced by instrumental techniques. Elemental analysis (EA) of organic matter in archaeological remains, also named CHNX, allows for the determination of the fractions of the mass of carbon, hydrogen, nitrogen, and heteroatoms (X: S, halogens) present in the sample. This type of study, carried out by the gravimetric method, is currently performed with automatized elemental analyzers [21].

1.5.1.2 Microscopy Methods

Microscopy methods probably are those first applied to the study of cultural heritage. Mention of the use of magnification lens is reported in manuscripts as that by R. Bacon in the thirteenth century. Simplest optical microscopes can offer useful morphological information in various applications: identification of anatomic elements of organic materials such as wood, paper, textile, or bonds; surface alterations such as pits, cracks, or fissures; strata distribution in paint samples prepared as cross-sections. Petrographic or metallographic microscopes with sophisticated configurations can measure the optical properties of minerals and rocks in samples prepared as thin-sections and metallic phases, respectively. These instruments have given rise to petrographic microscopy and metallography. The IFM mentioned above uses a fluorescence microscope to identify and discriminate the origin species of the proteinaceous media in paintings and polychromed sculptures.

Electron microscopy, confocal microscopy, and atomic force microscopy (AFM) were successively introduced in the field of study of cultural heritage after the development of the instrumentation in the first half of the twentieth century, the former, and in the second half, the two latter. Nowadays, electron microscopy is abundantly used in cultural heritage research. Among the two classes of electron microscopes, transmission (TEM) and scanning (SEM), the latter is undoubtedly the instrument with the most extended use. The resolution provided by electron microscopes ranges from 1.9 to 20 nm for SEM until 0.1 nm in TEM. Technological advances have progressively have given rise to new models with improved electron sources (i.e., field emission, FESEM), the gaseous environment in the specimen chamber (environmental, ESEM), or coupled equipment that dehydrates samples (cryo-SEM) that allows the examination of organic and inorganic materials with high water content. A focused ion beam can be coupled to the SEM (FIB–SEM). Nevertheless, the main advantage of the SEM and TEM is the ability of coupling with X-ray microanalysis systems (energy-dispersive (SEM–EDX, TEM–EDX, STEM–EDX, scanning transmission electron microscopy) and wavelength-dispersive SEM–WDX) that enable the elemental analysis of the sample. TEM instrumentation is mainly used for characterizing microorganisms and the shape, dimensions, and composition of particles at the micro- and nanoscale in all kinds of heritage materials (rocks, ceramics, bones, fossils, fibers, wood, or microorganisms). SEM–EDX is used for characterizing the morphology and elemental composition of varied materials, pigments, rocks, ceramics, glass, metals, ivory, or bonds. Examining samples prepared as cross-sections and, more recently, trenches made with FIB [36] allows for establishing the distribution and composition of a sample's different original and corrosion layers.

Atomic force microscopy is a scanning methodology with a resolution below the nanometer. The topographic (3D) image is formed by the raster scanning of a probe that interacts with the forces exerted by the sample surface. AFM can be used to perform force spectroscopy, particularly for measuring the sample's mechanical properties, such as the elastic or Young's modulus. The topographic imaging technique is used in research that requires a fine examination of the object's surface, as in obsidian hydration dating [37], and for controlling the morphological and mechanical changes of paintings subjected to cleaning or consolidation treatments [38].

Confocal laser microscopy enables a 3D reconstruction of the sample and multidimensional (2D and 3D) images over time as well as spectrophotometric analysis. Although the use of this technique is lesser extended, this technique has been applied in studies on various cultural goods such as historical photographs, glass windows biofilms, or cleaning treatments [39].

1.5.1.3 Surface Analysis Methods

Surface analysis techniques provide information on the composition of the surface of a solid or liquid. Here the surface is defined as the outer layer of atoms or molecules of an object, with thickness in the range of 5–20 nm, that is in contact with a second phase (liquid, air, vacuum, etc.). The composition of the object varies in depth until reaching that of the bulk, with a transition zone thickness ranging up to a few micrometers. Although nowadays there is a wide range of surface analysis techniques, their application to the study of cultural heritage is still limited. Nevertheless, some of them have already some use [i.e., fast atom bombardment (FAB), electron energy loss spectrometry (EELS), Rutherford backscattering spectroscopy (RBS), spectroscopic ellipsometry (SE)]. Prominent among these are X-ray photoelectron spectroscopy (XPS) and secondary ion mass spectrometry (SIMS) [40]. Emission energy spectra yielded from low-energy electrons emitted from samples irradiated with an intense monochromatic X-ray microbeam are provided by XPS instruments. This technique has the advantage of discerning elements in different states of oxidation and quantifying them. SIMS identify atoms, molecules, ions, and clusters of species at trace level in a punctual or mapping mode that allows obtaining 2D compositional maps with a resolution of 10 nm. Identification of corrosion processes and inorganic and organic materials in paintings, sculptures, monuments, ceramic or glass, are the applications of these techniques.

1.5.1.4 Electroanalytical Methods

Several electrochemical methods are used in cultural heritage research, as shown in Fig. 1.10.

Conductometry is a simple technique widely applied for monitoring the removal of salts in water baths for cleaning archaeological ceramics by measuring the decrease in the conductivity of successive baths [41]. Ion-sensitive electrodes (ISEs) used in potentiometry enable the measurement of a wide range of ions. Among them, the hydrogen-ion selective glass electrodes are abundantly applied to measure the pH of heritage materials, which are prone to corrosion. The corrosion is significant in

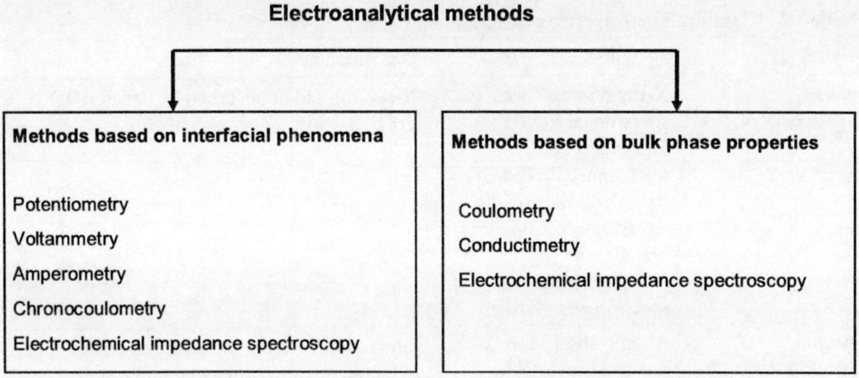

Fig. 1.10 Scheme of electrochemical techniques applied to the study of cultural heritage

the presence of acidic compounds and high levels of humidity, such as cellulosic fibers used in the manufacture of paper and textiles. Another common application is the control of the pH of the formulations employed for cleaning paintings (canvas, panel, or wall) [42]. Coulometry, combined with quartz crystal microbalances, is applied to determine the number of corrosion products in metal objects. The quartz crystal microbalance is employed in the ASTM test for studying the corrosion of metal objects [ASTM B 808-97 Standard Test Method for Monitoring of Atmospheric Corrosion Chambers by quartz crystal microbalance, ASTM G96-90 Standard Guide for On-line Monitoring of corrosion in plant Equipment (Electrical and Electrochemical Methods)].

1.5.1.5 Spectrometric Methods

Table 1.2 shows a classification of these spectrometric techniques commonly applied in the study of cultural heritage according to the phenomenon on which they are based.

X-ray spectrometry encompasses a wide range of instrumental techniques based on the emission of characteristic X-rays or γ-rays emitted by a micro spot of the material when it is excited by high-energy particles (electrons, protons) or an intense X-rays beam [43]. The depth of the analyzed material is in the order of a few to hundreds of μm under the surface, depending on the technique and the working conditions. Advanced instrumentation can achieve detection limits in the range of ppb with synchrotron sources to ppm with conventional X-ray sources. Three main classes of methodologies can be distinguished depending on the excitation source, as summarized in Table 1.3.

Conventional X-ray tubes and (poly) capillary X-ray focusing optics used in conventional XRF instruments operate with X-ray beams in the range of Ø ~ 10–50 μm diameter. The more advanced benchtop μ-XRF spectrometers, especially synchrotron-based μ-XRF, employ radiation sources with compound refractive lens

Table 1.2 Classification of spectrometric techniques

Class	Techniques	Fundamentals
X-ray spectrometry	X-ray fluorescence spectrometry (XRF) Electron probe microanalysis (EPMA) Particle-induced X-ray emission spectrometry (PIXE) X-ray absorption spectrometry (XAS)	X-ray emitted or absorbed by a material irradiated with X-ray or particles
Atomic spectrometry	Atomic absorption spectroscopy (AAS) Atomic emission spectrometry (AES)	Interaction between the electromagnetic radiation and the atoms in the free state
Molecular spectrometry	UV–vis spectrometry Infrared spectrometry Raman spectrometry	Interaction between the electromagnetic radiation and the mater in condensed phase
Magnetic effect spectrometry	Nuclear magnetic resonance spectrometry (NMR) Electron spin resonance spectrometry (ESR)	Effect of the magnetic field on atoms
Colorimetry	Colorimetry	Determination of the color perception

Table 1.3 X-ray spectrometry techniques

Excitation source	Techniques
X-rays beam	X-ray fluorescence (XRF) spectrometry Micro-X-ray fluorescence (μ-XRF) spectrometry Synchrotron radiation X-ray fluorescence spectrometry (SXRF) X-ray absorption spectrometry (XAS) X-ray absorption near-edge structure spectrometry (XANES)
High-energy electron beam	Scanning electron microscopy–X-ray microanalysis (SEM–EDX) Electron probe microanalysis (EPMA)
High-energy proton beam	Proton-induced X-ray emission (PIXE) spectrometry Micro particle-induced X-ray emission (μ-PIXE) spectrometry
	Proton-induced γ-ray emission (PIGE) spectrometry

systems, which yield beams of 0.5–2 μm diameter spot. Portable/in situ μ-XRF has been recently developed, offering beams of 50–200 μm cross-section.

XAS and μ-XAS allow for characterizing the local geometric and electronic structure of the sample. Synchrotron radiation is used as a radiation source as it provides intense and tunable X-ray beams. XANES, also called near-edge X-ray absorption fine structure (NEXAFS), is an elemental analysis technique local bonding-sensitive that enables the determination of the partial density of the empty states of a molecule.

μ-XANES can yield data on the speciation (oxidation state), and the coordination environment of the studied element is obtained. These data are presented as elemental and chemical maps with areas at ca. (40–50 × 20) μm^2 and a step size of ~ 0.25 μm. These techniques have improved mineralogical characterization and chemical imaging of samples at the submicrometer scale.

Like SEM coupled with EDS or WDS detectors, EPMA (or EPXMA and EMPA) uses high-energy electrons as the exciting source of the atoms of the compounds present in the sample. These techniques, especially SEM–EDX, are, by far, the most extensively used in art and art conservation. They have become routine methods of analyzing art and archaeological objects and monitoring conservation treatments [43–45].

A proton beam generated from a Van de Graaff accelerator is used in proton-induced X-ray emission (PIXE) to promote the emission of characteristic X-rays from the sample. In PIGE, a fast protons beam (in the order of MeV) impacts the sample and excites the nuclei of the atoms such that gamma rays are emitted. The gamma-ray spectrum obtained also exhibits characteristic emission lines that enable the identification of the different elements present in the sample.

Applications of these techniques in the analysis of cultural heritage are abundant. Elemental composition allows the identification of historical pigments, mineral phases of rocks, ceramic, glass, and glaze composition, metals and alloys composition, mineral phases of bonds, ivory, and related materials gemstones, jewelry, inks, stamps, and their corrosion products. More complex studies of characterization of the artistic or manufacturing technique of artifacts, authentication, or provenance can be made with the qualitative and quantitative data on the elemental composition of objects.

Although atomic absorption spectrometric techniques were frequently applied in the second part of the twentieth century for determining the elemental chemical composition of archaeological ceramics or stones, they have been progressively replaced by emission spectroscopic techniques. Currently, two sources for exciting the samples are preferably used: inductively coupled plasma (ICP) and laser beams. ICP devices can be coupled to spectrometric detectors (ICP–ES) or mass spectrometers (ICP–MS) [46]. Solid samples can be analyzed by coupling a laser ablation device (LA–ICP–MS). This technique is used for quantitative analyses of inorganic materials such as ceramics, rocks, or glass. The laser-induced breakdown spectrometry (LIBS) also uses laser sources of different wavelengths. A laser beam impacts a small area of the surface of the solid object yielding a plume of atomized material that emits characteristic radiations. This technique provides sensitive identification; nevertheless, the lack of reproducibility of the plume prevents the determination of the elemental composition of the studied material [47].

FTIR spectrometry, Raman spectrometry, and UV–vis spectrometry are the three molecular spectrometric techniques that have abundant applications in studying cultural heritage. Whereas the two former techniques operate in the IR region of the electromagnetic spectrum and provide information on both inorganic and organic compounds, the latter operates in the UV–visible region and is mainly applied for characterizing organic compounds. All techniques require little pretreatment of the samples, usually analyzed in the solid state, apart from dyes that must be dissolved when analyzed with UV–vis spectrometry. The notable improvement in the instrumentation has extended the scope of applications to the analysis of cultural goods, monitoring conservation treatments, and assessing new conservation products and methods [48–53]. FTIR methods in the middle IR region embrace attenuated total reflectance (ATR), diffuse reflection Fourier-transform infrared spectrometry (DRIFT), and FTIR microspectrometry (μFTIR). Fiber optics reflectance spectrometry (FORS) in the middle and near-infrared region can be used for in situ analyses. Raman spectrometry has also improved instrumentation with fiber optic devices in mobile Raman equipment, micro-Raman instrument (μRaman), and surface-enhanced Raman scattering (SERS) in which the signal is notably enhanced. UV–vis spectrometers have increased sensitivity and can be used in absorbance or reflectance mode. Additionally, UV–vis spectrometers can be used as detectors coupled to high-performance liquid chromatography (HPLC–DAD). Applications of FTIR spectrometry embraces the identification of inorganic and organic materials present in all kind of objects, aging studies, control of cleaning, and consolidation treatments. Similarly, Raman spectrometry is a handy tool for analyzing inorganic and organic materials and control of conservation treatments. In particular, SERS is a sensitive technique for the analysis of colorants. UV–vis spectrometry has been very useful in the characterization of dyes, although it is currently being displaced by liquid chromatographs coupled with mass spectrometer detectors (HPLC–MS). Photo- and thermal aging studies of conservation products are other main applications of this technique [54].

Electron spin resonance (ESR) spectrometry and nuclear magnetic resonance (NMR) spectrometry are applied to study different characteristics of the heritage objects. NMR has been applied in complementing the study of the structure of organic compounds present in heritage objects complementary to other spectrometric, chromatographic, or X-ray diffraction (XRD) techniques. Magnetic resonance imaging (MRI) has eventually been used to map water distribution or consolidant fluids into the porous network of stone monuments [55]. In contrast, ESR spectrometry is mainly used as a dating method (vide infra) and, less frequently, for identifying pigments with paramagnetic properties and the influence of certain pigments for altering supports such as stone, leather, or paper.

Colorimetric studies devoted to the psychophysical characterization of the object's color currently use the three-chromatic coordinates of the CIELAB colorimetric space (CIE 1976 L*a*b*). Advanced spectrometric colorimeters include several standard illuminating sources (A, C, D65), and modes of measuring the reflecting radiation (45°/0°, diffuse/8°). The spectral factors of radiance, reflection, and diffuse reflection are used to measure the spectral characteristics of the object's color. Micro-fading testers use a thin beam of light to measure the changes undergone by materials highly sensitive to light exposure with a spectrometric colorimeter. The low invasiveness of the test enables the application of this method directly to the studied objects.

Colorimetry has varied applications in cultural heritage research and practice: measuring the color of artworks, control of photoaging in preventive conservation programs, control of cleaning and consolidation processes, selecting paints for reintegration processes, and accelerated aging studies of new restoration products and treatments [56].

1.5.1.6 Diffraction Methods

X-ray diffraction initially applied for identifying the crystalline structure of materials has become an excellent tool in studying cultural heritage and provides much information on the mineralogical composition of varied materials. The pattern charts obtained using X-ray diffractometers are fingerprints that allow the identification of unknown compounds present in the sample or alteration products. Applications of this technique are diverse: historical pigments, stones, ceramics, and metals from archaeological sites, monuments, and sculptures and their alterations (i.e., efflorescences, corrosion products), and, to a lesser extent, organic materials (i.e., ivory, cellulosic materials). Advanced instruments such as micro XRD (μXRD) or synchrotron-XRD (SRXRD) are used to the detriment of the conventional powder-XRD. μ-SRXRD uses beam sizes of $0.5 \times 0.9\ \mu m^2$ and can analyze small areas with a sample spot size of $30 \times 40\ \mu m^2$. Portable XRD instruments for in situ analysis have also been successfully used [57].

1.5.1.7 Thermoanalytical Methods

Thermoanalytical methods are of interest in cultural heritage research because they can provide information on the effect that temperature changes induce on the materials and that, in turn, informs on the composition of the material. In cultural heritage there are three techniques preferably applied: thermogravimetric analysis (TGA), differential thermal analysis (DTA), and differential scanning calorimetry (DSC). These techniques have a wide range of applications: characterization of some properties of synthetic polymers used as binders, adhesives, or consolidants such as glass transition temperature, crystallinity or curing efficiency, determination of the content of plasticizer, identification of binders, the characterization of alterations of mortars from monuments and mural paintings, in archaeometrical studies of ceramics, and studies of the effect of the humidity on sensitive materials of cellulosic or proteinaceous nature [58].

1.5.1.8 Mass Spectrometric Methods

Mass spectrometry (MS) is a powerful tool in identifying organic materials due to its great potential to resolve molecular structures without or with simple sample pretreatments. Although the complexity of art materials often restricts the application of this technique in cultural heritage studies, coupling MS with chromatographic methods has expanded the scope of their applications (vide infra). Different

MS techniques are used in the analysis of cultural goods that mainly differ in the method for introducing the sample in an ionized state in the MS. Depending on the technique, the applications are different. Direct infusion MS, direct pyrolysis MS (DPMS), and direct temperature-resolved MS (DTMS) are used for analyzing substances that are complex mixtures of molecules that can be thermally separated, such as natural resins and their aging products. Fast-atom bombardment (FAB) mass spectrometry and, in particular, SIMS is used for studying the strata composition of paintings because this technique allows the identification of pigments and binders and can provide mappings of distribution of elements and molecular species in samples prepared as cross-sections. ICP-MS, LA-ICP-MS are valuable techniques for identifying and determining the elemental composition of various materials (ceramic, glass, stone, paints, etc.). Glow discharge [ion source]-mass spectrometry (GD-MS) and graphite-assisted laser-induced desorption/ionization (LDI) mass spectrometry have been employed in the study of oxidation products formed in aged varnishes prepared with terpenoid resins. Matrix-assisted laser-induced desorption/ionization (MALDI) mass spectrometry is used for characterizing colorants and proteinaceous media in paintings. In such instances, a preliminary step of enzymatic cleavage of the macromolecule fragments with a characteristic mass-to-charge ratio (m/z) value turns the mass spectrum into a "fingerprint" of the analyzed substance [33].

The combined technique of tandem mass spectrometry MS/MS has started to be used more recently for resolving molecular structures of aged varnishes and discriminating the isotopic composition of lead pigments in provenance studies [59].

1.5.1.9 Separation Methods

Chromatographic and electrophoretic methods have been applied in the field of cultural heritage. These techniques are particularly useful owing to their ability to separate the components that compose most materials constituted of complex mixtures of organic compounds. Chromatographic methods comprise a variety of techniques that cover a wide range of analytical heritage casuistries. The complex and time-consuming sample preparation is their main weakness; despite that, these techniques are abundantly applied in specialized laboratories due to their ability to identify and quantify the components of mixtures of organic nature, natural and synthetic small molecules, macromolecular or polymerized substances used by artists as binding media, varnishes and coatings, adhesives or consolidants.

Technological advances have marked the development and application of chromatographic techniques. More straightforward techniques, such as paper chromatography (PC) and thin-layer chromatography (TLC) have been progressively replaced by gas chromatography (GC) or liquid chromatography (LC). The development of new injection and detection systems has resulted in instrumentation enhancements and has also extended the scope of applications of these techniques.

Pretreatment of samples, previously described in Sect. 1.4.3, is applied depending on the sample and the chromatographic technique. TLC, which replaces PC since the 1960s, is used for identifying proteinaceous, lipid and polysaccharide binding media and, to a lesser extent, natural resins. A preliminary step of hydrolysis is required in the analysis of the former. A few methods for analyzing proteinaceous media using ion chromatography (IC) with cationic exchange columns have been proposed. IC with anionic exchange columns is frequently used to determine the concentration of typical anions forming the soluble salts responsible for altering ceramics, stone, or mortars of monuments. This method is also used for controlling the dechlorination of archaeological underwater metal objects. Size exclusion chromatography (SEC), which provides mass distributions of high molecular mass species, has been applied in the study of the composition of natural resins, such as Manila copal with a high degree of polymerization, or the degree of cross-linking of acrylic resins. Studies of the fragmentation of natural and synthetic polymers used as binding media or coatings due to aging and the characterization of the materials extracted by cleaning treatments with organic solvents have also been interesting applications of SEC [33].

HPLC with sensitive reverse-phase columns as a stationary phase and a gradient elution with two solvents is frequently employed in analyzing several organic materials present in cultural goods. HPLC coupled with UV–vis spectrometers using a photodiode-array (HPLC–PDA) has been used mainly in the analysis of dyes. HPLC with a spectrofluorimeter detector (HPLC–FD) is used to identify drying oils and proteinaceous media [33, 60]. This type of detector requires a preliminary derivatization step to confer the analytes fluorescent properties. Mass spectrometers used as detectors coupled to an HPLC have been abundantly applied with the development in the last decades of ionization systems by electrospray (ESI) and atmospheric-pressure chemical ionization (APCI). These systems have solved the problem of introducing significant volumes of liquid samples from the HPLC column to the MS. In most laboratories, HPLC coupled in series with DAD and MS detectors is used. This configuration has been applied to analyze organic materials: dyes, proteins, lipids, and natural resins. Less frequently, HPLC has been coupled with tandem mass spectrometry for the high-performance liquid chromatography–atmospheric pressure chemical ionization (tandem) mass spectrometry (nanoLC-nanoESI-Q-qTOF-MS-MS) in proteomic studies. These instruments have enabled the identification of the animal species of proteinaceous binders used in paintings. Cryoprobes suitable for LC-NMR have been proposed for studying several aged binding media: linseed oil, egg tempera, and acrylic resin [33].

GC coupled with a mass spectrometer as the detector (GC–MS) has become one of the most extended techniques for analyzing all kinds of organic substances in cultural heritage in recent decades. The ability of the MS detectors for characterizing the molecular structure of the compounds separated in the column of the chromatograph has displaced the flame ionization detectors (FID) that required a prior calibration of the system with reference and standard compounds. Nowadays, the applications of this technique are broad due to the availability of standards of amino acids, fatty acids, monosaccharides, and hydrocarbons with the appropriate purity and the development of improved pretreatments. The latter involves two-step procedures of hydrolysis and derivatization according to the type of material, namely proteinaceous media, plant gums, drying oils, and waxes. Specific one-step pretreatment procedures that enable the breakdown and derivatization of macromolecules, such as methanolysis, have also succeeded in the analysis of polysaccharides and highly polymerized natural resins [33, 61]. The difficulty of analyzing synthetic polymers used in modern art paintings and sculptures due to their low volatility has been circumvented by using pyrolysis-GC-MS (Py-GC-MS). Three pyrolyzer techniques are available: Curie-point, furnace, and filament pyrolyzers, the latter the more widespread. Besides direct pyrolysis, online and offline derivatization procedures are used to improve results. Methylation with hydroxide of tetramethylammonium (TMAH), the so-called thermally assisted hydrolysis and methylation (THM), is the most common method of online derivatization [33].

Capillary electrophoresis (CE) is the primary electrophoretic method that has application in the analysis of cultural heritage. Although this methodology is less extended than the chromatographic methods, several procedures for analyzing resins, drying oils, waxes, animal glues, and plant gums have been proposed [33, 62].

1.5.1.10 Methods Based on Effects Induced in the Atomic Nucleus

Two techniques based on the changes induced in the nucleus of the atoms when irradiated with accelerated particles or γ radiation have been applied to the study of cultural goods: neutron activation analysis (NAA) and Mössbauer spectrometry. NAA uses a research nuclear reactor as a source of accelerated neutrons that interact with the nucleus of the atoms in the sample that, in turn, emit characteristic γ rays that allow the identification of the elements present in the sample at a trace level. Although the technique is not invasive, hence enabling the analysis of the entire object without sampling, the exposition of the object to intense irradiation with high-energy particles restricts its application. Thus, NAA has been used in provenance studies of metal, ceramics, and stone objects [63]. Mössbauer spectrometry applications in the study of cultural heritage are mainly focused on Fe-containing materials such as pigments, minerals, rocks, and ceramics. The interest in this technique is based on its ability to discriminate different states of oxidation and coordination of iron ions associated with changes in the mineralogical composition of objects subjected to heating processes, which enables its use as an archaeometrical thermometer for ceramic and rock objects [64].

1.5.1.11 Miscellaneous Non-invasive Methods

This section describes the applications to the study of the cultural heritage of various instrumental techniques that are characterized for providing images. Here, the image is defined as a 2D or 3D graphical representation of the punctual values of a physical or chemical property measured in a specific part or the entire object when exposed to a wave source. Electromagnetic and sound waves of specific wavelengths are the two main phenomena used. On the other hand, the imaging methods can be broadly grouped according to the procedure employed for obtaining the image. If the image is formed from a source that irradiates the entire or a selected part of the object, the method is considered "full-field imaging." If the image is formed with point-by-point scanning of the entire object or the selected portion with a thin beam of radiation emitted by the source, the method is considered "scanning imaging" (mapping) [65]. A subclassification can be made in the case of methods that use electromagnetic radiation for imaging the object according to the region of the electromagnetic field in which the technique operates. Table 1.4 summarizes the principal instrumental techniques classified according to the imaging method and the type of wave phenomenon.

The ground-penetrating radar (GPR) is an instrument that operates in the micro- and radiowaves region (100 MHz–1 GHz). This technique is mainly used for examining buildings and archaeological sites. The pulse of radiation emitted by a source that penetrates a heterogeneous material is sensitive to the differences in electromagnetic properties of its different components. By measuring the response with punctual scanning and further processing of the data, graphs are obtained where structural anomalies are recognizable such as buried objects, interphases among layers or strata, presence of water or cavities.

In the IR region, IR reflectography (IRR) provides an image by the contrast between the intensity of absorbance of the different zones of the object when irradiated with IR light. The method provides information about canvas and panel painting

Table 1.4 Instrumental techniques based on the images produced by electromagnetic and sound waves

Wave phenomenon	Region	Technique
Full-field		
Electromagnetic radiation	Infrared (IR)	IR photography IR reflectography Thermography
	Ultraviolet–visible (UV–Vis)	Photography Photogrammetry Interferometric techniques
	γ-rays, X-rays	Radiography Computer tomography Gammagraphy
Scanning		
Electromagnetic radiation	Micro- and radiowaves	Georadar
	IR-terahertz	Scanning IR reflectography Tetrahertz imaging Optical coherence tomography
	Visible	Multi- and hyperspectral imaging[a] 3D laser scanner Laser imaging detection and ranging
	X-rays	Scanning X-ray beam imaging
Sound waves	–	Ultrasound testing

[a] The electromagnetic signal ranges between the visible to the near-IR region

underpainting and underdrawings made with carbon black sticks or pens on reflective chalk or gypsum ground. Thermography provides temperature mappings from the IR radiation that emits/reflects the objects. This technique helps detect structural defects and moisture infiltrating the walls and grounds of monuments. Tetrahertz time-domain spectroscopic imaging (THz–TDS) operates with radiation sources in the 0.1–10 THz interval of frequencies, which corresponds to wavelengths in the range 0.03–3 mm. The instrument operates like an "acoustic echo" forming the image by the contrast between the intensity of absorbance of the different zones of the object. This technique is mainly used to obtain mappings of the interfaces inside a paint multilayer. Optical coherence tomography (OCT) operates with a source in the near-IR. It yields interference measurements from which are mapped the distribution of materials and interfaces between the multiple layers of the painting. The technique can accumulate 2D maps that combined yield 3D images with a resolution of 1 μm until variable depths depending on the opacity and complexity of the object.

Photography in the visible region is probably the most extended full-field technique in cultural heritage studies in its two versions of macro- and microphotography, the latter being performed with microscopes. Nowadays, the instrumentation has notably improved, and images, which are formed by contrasting the intensity of absorbance of the different zones of the object, can be obtained not only in the visible region but also in the near-IR regions by using special filters or films sensitive to these radiations. Using Wood lamps, photographs can be obtained in the near-UV region as the films are sensitive to this radiation. If a UV filter is used, a fluorescence photograph of the object is obtained. Conventional photography provides mainly morphological information, and IR- photographs inform on subsurface morphology, whereas UV and fluorescence photographs provide mainly compositional data.

Photogrammetry yields 2D and 3D mathematical models from digital imaging capture of objects from afar and their further processing of the images. Passive optical sensors have been used that consist of digital cameras built for metric purposes. Off-the-shelf digital cameras can be used in close-range photogrammetry. The processing procedures combine fundamentals of optical and projective geometry. This technique is used in architectural measurements of buildings and archaeological site cartography. Electronic speckle pattern interferometry (ESPI) and holographic interferometry (HI) are based on the interference patterns formed with beams of coherent light that contain data on the amplitude and the phase wave. The anomalies observed in the interference pattern allow the identification of structural defects such as pores or fissures of paintings.

Multi- and hyperspectral imaging are imaging techniques that map reflectance values at pixel resolution by scanning the entire object or a selected part. The term hyperspectral is used when the detector used provides a continuous reflectance spectrum/pixel, whereas in multispectral instruments the reflectance values are measured in a few spectral bands (3–20) in the visible and IR regions. By processing the acquired set of data, a vast number of images of the object at specific wavelengths are obtained. These techniques provide information about the spatial distribution of pigments and binding media, underdrawings, and repaintings. These techniques are also used to digitalize documents for their safe storage.

3D scanning operates as an active optical sensor that scans the object with a thin and intense light beam, preferably using laser technology. A record of an object requires performing scans of the object from some different reference positions. The mathematical processing of the data recovered forms a 3D cloud from which can be built a three-dimensional image that accurately reproduces the 3D shape of the object. The appearance of the object (i.e., color) can also be recorded and reproduced in a 3D model. 3D laser scanners are mounted on a tripod (terrestrial laser scanner, TLS), mobile laser scanners (MLS) are mounted on a vehicle (handheld, boat), whereas airborne laser scanners (ALS) are operated from an aerial vehicle. 3D scanners can broadly be classified into two categories: (i) time-of-flight (ToF) systems that operate similarly to conventional RADAR and SONAR by measuring the time interval between the emission and the detection of a light pulse reflected by the object; and (ii) triangulation systems that project a light beam in a geometric configuration so that the direction of the light from a known position can be determined. Although 3D laser is mainly used in the architectural examination of buildings and archaeological sites, its use is increasingly extended to all kinds of movable cultural goods for virtually reconstructing missing parts that can be built with 3D printers. Laser Imaging Detection and Ranging (LIDAR) instruments used in archaeology and topography for studying the topology of interest areas of the Earth's surface are operated from planes. They are composed of the ALS, which provides the 3D cloud, a Global Positioning System (GPS), an inertial navigation system (INS), which enables knowing the plane's position and trajectory, and a video camera for recording images of the land area measured.

Radiography, gammagraphy, and electron emission radiography are full-field methods that provide images by contrasting the absorbance intensity of the parts of the object with different compositions. Due to the higher degree of penetration of γ-rays and the requirement of special sources of excitation, this technique has restricted use in objects composed of materials with high mass absorption coefficients such as heavy metals or dense stone. In contrast, X-ray radiographs, also called macroscopic X-ray radiographs (XRR), are increasingly used due to the advanced instrumentation that currently provides digital images. Applications of this technique are abundant: spatial discrimination of pigments, underdrawings, underpaintings, presence of hidden structures or objects of metal or other materials in sculptures and panels. By exposing paintings and jewels to intense X-ray sources operating at 300 kV a radiograph formed by the emitted γ-radiation is obtained that can provide an image of the internal structure of these objects made with dense materials that cannot be studied with conventional radiographic instrumentation. X-ray computerized tomography (CT) is an advanced method that provides 3D radiographs of an object by acquiring a series of radiographs recorded under different angles of incidence of the X-ray beam, whereas the object rotates around an axis perpendicular to the X-ray beam. The mathematical processing of the acquired data yields a virtual 3D image of the entire object formed by the different intensities of absorption of the materials composing the object in its inner parts. CT can be performed with an X-ray synchrotron source that emits thin monochromatic X-ray beams of tunable energy. Laminography is a variant of X-ray CT that allows the study of multilayered paintings by introducing modifications in the instrument so that the rotation takes place around an axis tilted relative to the X-ray beam direction.

Scanning techniques in the X-ray region, already described in prior sections, include μ-XRF, which provides information about the distribution of elements in the object, μ-XRD which informs about the distribution of crystallographic phases, XAS, and μ-XANES which provide chemical state and/or electronic environment contrast. Recently, confocal μ-XRF (CXRF) has been developed by improving SEM–EDX. This instrumentation provides depth-selective imaging of paintings. Macroscopic scanning X-ray fluorescence (MA-XRF) uses a high-energy X-ray millibeam to scan large areas of the object. These techniques have been mainly used for analyzing inner layers in paintings, underdrawings, and underpaintings.

Ultrasound testing scans the object with a sound waves beam > 20 kHz and measures the attenuation and the time spent by the wave emitted by a sensor or transducer to arrive at the receiver. There are two main modes of operating: pulse-echo, in which a unique sensor acts as the emitter and receiver, and trough-transmission, in which the beam emitted by an emitter is detected by a transducer acting as a receiver. This technique is used mainly for imaging hidden structures in archaeological sites. Less frequently, this technique is applied for monitoring the cleaning of archaeological objects.

1.5.2 Dating Methods

Dating methods aim to provide the basic chronological framework of a specific event, the manufacture of an artifact, or the death of a living being. That leads to establishing a method for measuring time, in other words, to have a clock that enables one to know the zero-setting or "time zero" in which the surveyed event took place and thus serves as a reference or starting point for designing a standardized time scale. A second requirement for defining the dating method is the calibration of the time scale. The calibration is carried out by correlating a measurable time-dependent quantity or parameter with a calendrical time scale. The calendrical time scale is made via a series of events that took place at well-known dates and exhibited well-known values of the time-dependent parameter selected. Undoubtedly, this procedure has limitations in applicability mainly due to the type of materials studied, the state of preservation of the sample, and the age range [66–68].

Dating methods can be broadly classified into absolute and relative methods. Absolute methods provide the time passed since an event took place on a defined time scale, and relative methods infer serial or sequential relationships. They are non-chronometric methods in which specific dates are not provided; instead, the event is compared to the date of other events provided by stylistic or stratigraphic studies. Not always the dating methods are absolute or relative sensu stricto. Sometimes the dating method needs a time scale provided by another chronology method as archaeomagnetism. Radiocarbon analysis is an absolute dating method within certain limits. Obsidian hydration dating can be applied in either absolute or relative mode.

Another classification of dating techniques is based on the phenomenon or the measurement technique employed for establishing the date of the event. Table 1.5 illustrates this classification.

It is worthy to note that the selection of the dating method in a certain case study depends on the characteristics of the object and the timing range of application of the dating method. Figure 1.11 shows a scheme of the chronological ranges provided for some of the most applied dating methods.

Table 1.5 Dating methods according to the phenomenon in which they are based

Type of phenomenon	Dating method
Effects of the atomic nuclei	Radiometric methods Side effects of the internal radioactivity
Magnetic	Paleomagnetism–archaeomagnetism
Chemical	Obsidian hydration Amino acid racemization Metal corrosion FUN (fluorine, uranium, nitrogen)
Biochemical	Dating based on molecular markers/mutation rate
Biological	Dendrochronology Valve Palinology
Geological	Varve
Geophysical	Terrestrial orbital variations

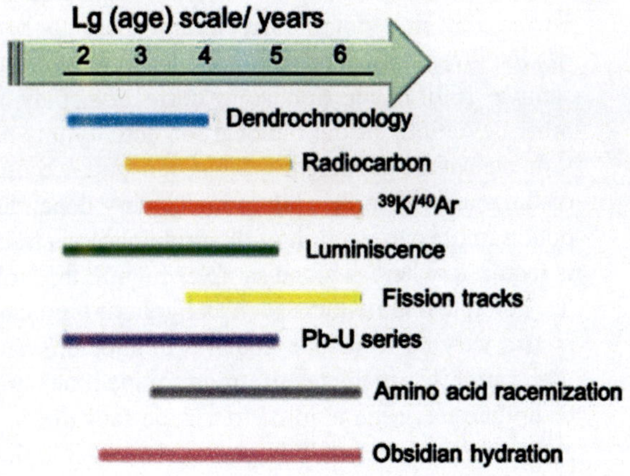

Fig. 1.11 Chronological ranges provided from some of the most applied dating methods

1.5.2.1 Methods Based on Effects of the Atomic Nucleus

The first type of dating methods included in this group directly measures the internal radioactivity of the object. A second group is composed of those others that study phenomena associated with the changes that radioactivity has induced in the object.

Radiometric dating

Radiometric dating is based on determining the concentration of daughter isotopes (stable or radioactive) or parent isotopes formed as a product of a radioactive chain from a parent isotope. As examples, six main methods are discussed here: (a) radiocarbon, (b) potassium–argon, (c) uranium-series dating, (d) lead isotopic composition, (e) oxygen isotopic composition, and (f) some other than ^{14}C cosmogenic radioisotopes.

(a) *Radiocarbon dating*: since its development by Willard Libby (1908–1980) in 1946, this method has progressively extended its applications from archaeology and paleontology to geosciences and environmental sciences in specialized laboratories that routinely apply this technique [69]. Atmosphere, oceans biosphere (living and dead biomass), and fossil fuels form the global terrestrial ^{14}C reservoir in which an equilibrium is established between the production of ^{14}C by cosmic rays from ^{14}N and its loss by radioactive decay. This dynamic equilibrium is characterized by a constant value of the $^{14}C/^{12}C$ ratio. This value decreases when the biomaterial (i.e., living matter) of an object is no longer exchanging carbon with the reservoir so that this ratio can be correlated with the date at which the exchange was stopped. The $^{14}C/^{12}C$ ratio is determined by measuring the emission rate of β particles, and alternatively by measuring the ^{14}C content in a suitable accelerator mass spectrometer (AMS) able to quantify the low amounts of radionuclide contents currently found in archaeological samples [69]. In addition, some corrections may be required to compensate for other deviations of the radiocarbon activity due to external factors such as the influence of the geomagnetic field, sunspots activity, reservoir size, or man-made disturbances. Sample pretreatments vary depending on the type of sample. In bone samples, extraction of the proteinaceous fraction from the carbonate bulk is required, whereas wood samples require the isolation of cellulose from other less reliable components such as lignin or humic acids. The degree of accuracy is also varying from one material to another. While soils and sediments have high variability in the results, mortars and lime burials are accurately dated. This technique has been applied to a wide range of art and archaeological objects, and human remains of living beings from geological ages, geological materials, constructive materials such as lime mortars, textiles, and all kinds of crafts and artworks (paintings, ceramics, etc.) [66–68, 70–73].

(b) *Potassium–argon dating*: this method is based on measuring the content of ^{40}Ar as the resulting product of the spontaneous radioactivity of the ^{40}K isotope with a half-life of 1250 million years contained in K-bearing minerals. The clock starts when, during the cooling down of the lava, the temperature of the minerals sinks below a certain and mineral-specific closing temperature, as then the exit of argon by diffusion is not anymore possible. The age of archaeological and geological remains trapped between two lava flows can be determined in this way. The ^{40}Ar gas is released by fusion of the sample and the $^{40}Ar/^{40}K$ ratio is determined. The content of ^{40}Ar is measured by MS and ^{40}K is quantified by AAS or flame photometry. In the single crystal laser fusion (SCLF) the sample is irradiated with an argon laser. The argon-argon method is a variant of the potassium–argon method. When the sample is irradiated with neutrons in a nuclear reactor, the unstable isotope ^{39}Ar (half-life of 269 years) is formed and decays by β emission to ^{40}Ar. The dating is based on comparing the $^{40}Ar/^{39}Ar$ ratio in the sample with that of a mineral with a known age, which is irradiated under identical conditions A thermal ionization mass spectrometer (TIMS) is employed [66–68, 70–73].

(c) *Uranium series dating*: it is based on the determination of the $^{234}U/^{230}Th$ or $^{235}U/^{231}Pa$ ratios. The isotopic ratios are determined by TIMS, multicollector inductively coupled plasma mass spectrometry (MCICPMS) or LA-ICPMS. This technique is mainly applied to rock art, fossil teeth, coral and mollusk shells, and stalagmitic calcite [66–68, 71–75].

(d) *Lead isotopic composition analysis*: while ^{204}Pb has no long-lived radioactive parent, and thus, its content in geologic formations remains unchanged, the rest of the lead isotopes have progenitor parents with different radioactive decay $^{238}U \rightarrow {}^{206}Pb$, $^{235}U \rightarrow {}^{207}Pb$, and $^{232}Th \rightarrow {}^{208}Pb$, respectively. Therefore, lead ores from different geological sources can be discriminated by the ratios of the four stable Pb isotopes according to their geologic age, parent–daughter isotope ratios, and weathering. Determination of lead isotope content is carried out by different methods: TIMS, quadrupole inductively coupled plasma mass spectrometry (Q-ICPMS), inductively coupled plasma sector field mass spectrometry (ICP-SMS), multiple collector inductively coupled plasma mass spectrometry (MC-ICP-MS), and inductively coupled plasma time-of-flight mass spectrometry (ICP-TOF-MS). This technique has enabled sourcing various archaeological metal objects made with lead, silver, copper, bronze, and also faience, glass, glazes, and pigments [66–68].

(e) Oxygen-isotope ratio ($^{18}O/^{16}O$) varies because $H_2{}^{18}O$ has a slightly lower vapour pressure than $H_2{}^{16}O$ so that the ocean water is enriched in ^{18}O, whereas the rainwater is enriched in ^{16}O. In glaciations, ocean water was significantly enriched in ^{18}O as large amounts of rainwater were trapped in the glacial ice, whereas in warmer periods the $^{18}O/^{16}O$ ratio was lower. The variations of the $^{18}O/^{16}O$ ratio associated with the seasonal or global terrestrial climatic periods can be determined in calcareous tests, in long cores drilled in polar ice-caps or in shells of foraminifera forming deep-sea sediments using a MS. Seasonal layers can be established with a standard deviation of ± 10 years. Another application of this technique is the dating of bone and teeth [66–68, 76].

(f) *Other cosmogenic radioisotopes*: other radioisotopes, which are produced by the nuclear reaction yield by the action of cosmic rays on certain stable atoms present in the atmosphere or exposed solid materials on earth, can be used for dating purposes. Table 1.6 summarizes their main characteristics [66, 77].

Dating methods based on the side effects of the internal radioactivity

In this subsection are described several dating methods based on the cumulative effect of nuclear radiation on the crystal structure: (a) fission-track dating; (b) dating methods by stimulated luminescence, namely thermoluminescence and optical stimulated luminescence; and (c) ESR.

(a) *Fission-track dating*: several minerals present in rocks (i.e., zircon, mica, apatite, etc.) and natural glasses such as obsidian contain uranium. In such instances, this technique can be applied. The method is based on the determination of the ratio of tracks per surface unit due to the spontaneous fission of the ^{238}U and the extra tracks per surface unit induced by exposure of the sample in a nuclear reactor. If the mineral is heated, the tracks are annealed; therefore measuring the tracks formed after heating provides a dating method of the event. The tracks density measured before exposition to the nuclear reactor quantifies the spontaneous fission, and tracks formed in the sample after reactor exposure measure the uranium present in the sample. Zircon and obsidian are commonly analyzed with this method. Tracks, usually with a length of ca. 0.01 mm, are identified and counted in samples prepared as cross-sections with a high-magnification microscope. Aqueous hydrogen fluoride or sodium hydroxide solutions are employed for etching the tracks to make them visible under a microscope. Step-heating plateau correction method and isothermal plateau fission-track (ITPFT) techniques allow for overcoming the effect of thermal annealing. The fission-track method has been mainly used for dating pottery in a time range of 700–2300 years, obsidian knife-blades and flakes have been dated in the range 400–600 years, and bones and teeth have also been dated [66–68, 71, 72, 78, 79].

(b) *Stimulated luminescence dating*: under this term, two techniques are included in which the material emits a luminescence signal that increases with age. Natural radionuclides belonging to the ^{238}U, ^{232}Th, or ^{40}K decay chains present in the material at trace level can emit α, β, and γ rays. The latter, together with cosmic radiation, transfer energy that is accumulated by electrons in trap states with a long mean life. By suitable stimulation, this accumulated energy can be emitted

Table 1.6 Cosmogenic radioisotopes used for dating and in geological studies (adapted from [66])

Isotope	Half-life (thousands of years)	Formation process	Applications
^{129}I	15,700	Spallation (Xe)	Groundwater tracer
^{10}Be	1600	Spallation (N and O)	Sedimentation processes
^{26}Al	720	Spallation (Ar)	Rocks and meteorites
^{36}Cl	308	$^{35}Cl\ (n,\gamma)^{36}Cl$	Rocks, groundwater tracer
^{41}Ca	103	$^{40}Ca\ (n,\gamma)^{41}Ca$	Carbonate rocks

as a luminescence signal. Although these methods have poorer precision than radiocarbon dating, they can provide dating on a broader age range. In thermoluminescence (TL), the mineral or ceramics sample is heated up to 500 °C, and the emitted photons, in the wavelength range 100–400 nm, are measured using a photomultiplier detector. Age is estimated as the ratio between the paleodose or accumulated thermoluminescent dose and the annual dose rate. Portable spectrometers have been used for in situ analysis. Pottery has been characterized in an age range between 30,000 and 50,000 years [66–68, 71, 72, 80]. Optically stimulated luminescence (OSL) uses light emission from luminescence centers of certain minerals with trapped electrons due to defects or impurities of the crystalline lattice as an analytical signal. Phototransferred thermoluminescence (PTTL) measures the luminescent signal emitted by the electrons in shallow traps, which is proportional to the electron population of the deep traps. Applications of these techniques include anthropological remains such as teeth and bones, metallurgical slags, and diverse geological materials such as polymineral sediments, zircon, calcite, stalagmitic calcite, meteorites, flint, quartz, loess, and volcanic materials (obsidian, tephra) [66–68, 71, 72, 81].

(c) *Electron spin resonance (ESR)*: similarly to TL in ESR, the accumulated dose of nuclear radiation defects is measured. The application of a resonance high-frequency electromagnetic radiation generated from a strong steady magnetic field, which is appropriately modulated, is used for determining the paramagnetic centers (trapped unpaired electrons) present in the material. This method often requires a pretreatment of the sample consisting of crushing followed by grains etching with a weak acid. Sometimes heating is also applied in order to remove unstable signals. The main applications of this method are secondary carbonates precipitated such as stalactites, stalagmites, and flowstones (speleothems) found in caves or hydroxyapatite composing teeth enamel, bones, mollusks, and shells. The age range usually determined with this technique is between 1 and 10,000 million years [66–68, 71, 72, 82].

1.5.2.2 Methods Based on Magnetic Effects

Dating methods based on magnetic effect embrace experimental procedures aimed at determining the remanent magnetization of certain materials that recorded the direction or orientation changes in the Earth's magnetic field in the past.

Paleomagnetism and magnetostratigraphy: study the natural remanent magnetization (NRM), which is a permanent magnetization of rocks due to the alignment in the same direction as the Earth's field of fine grains of iron oxide dispersed in the rock matrix. This magnetization occurs when the temperature is raised (i.e., lava from a Vulcan eruption or heating during the formation of igneous rocks) and is maintained after the subsequent cooling. Measurement of this remanent magnetism is experimentally performed employing spinner magnetometers. Reference curves need data from a previously established time scale based on other dating methods, such as dendrochronology. This method has provided accurate data on Paleolithic time scales.

Archaeomagnetism: is a branch of paleomagnetism for the dating of archaeological materials. Here the recording mechanism is the thermoremanence (TRM), which is the remanent magnetization of iron oxide impurities of clays during the pottery firing in the kiln. Dating of sunbaked bricks or discrimination of cast coins from struck ones, or dating firing temperature of pottery have also been other interesting applications in archaeomagnetism. This method is generally used for dating archaeological materials within the last 10,000 years [66–68, 71, 83]. Eventually, the method has been applied for dating older remains ascribed to *Homo erectus* (1.64 $\pm$ 0.03 million years) [83].

1.5.2.3 Chemical Methods

In this section, several dating methods are described that are based on measuring the result of a chemical reaction that has induced surface alterations, isomorphic or chemical changes in the composition of the original material: (a) Obsidian hydration, (b) amino acid racemization, (c) substitution of fluorine, uranium, nitrogen (FUN) dating, and (d) metal aging via corrosion monitoring.

Obsidian hydration: this volcanic glass was used for manufacturing diverse hunting or domestic tools by prehistorical men by adapting the form of the fragment with varied shapes and sizes for each specific application. From the moment of manufacture, a growing hydration layer started to be formed on the surface of the artifact so that its thickness could be used for dating it. This method has the drawback of a notable dependence on burial temperature. Analysis of the hydration layer is carried out by several methods, from the simplest optical microscopy until AFM, SIMS, secondary ion mass spectrometry–surface saturation (SIMS–SS), or sputter-induced optical spectrometry (SIPS) mass spectrometry–surface saturation (SIMS–SS) or sputter-induced optical spectrometry (SIPS) [66–68, 71, 72, 84].

Amino acid racemization: the degree of racemization acquired by the proteinaceous material that remains in bones and related biological materials expressed as the ratio “dextro amino acid (D) enantiomeric form/levo amino acid (L) enantiomeric form” can be used as an indicator of the age of the object. Several factors influence the reliability of this method: the racemization curve is temperature-dependent (both during burial and analysis). A second factor is the degradation of the proteinaceous material during burial, particularly, the breaking down of the protein macromolecules in lower molecular weight fragments by hydrolysis reactions. The age range covered by this method is 200–100,000 years. Different amino acids can be analyzed depending on the archaeological material and age range. Aspartic acid provides dating data until 10,000–50,000 years, whereas isoleucine provides data until several million years in teeth samples. This method has been applied to fossil shells, bones, teeth, wood, plant remains, and coral [66–68, 71, 72, 85].

FUN dating: is a relative dating method applied to the study of bones found in an archaeological site or deposit that was well-dated with an absolute dating method. The FUN method is based on the comparison between the content increase undergone by fluorine (F) and uranium (U) when these elements are incorporated in the hydroxyapatite mineral composing the bones and the nitrogen (N) content decrease due to the progressive protein decomposition. GC and reverse-phase HPLC techniques are used [68, 86].

Corrosion methods: several models for dating metal objects have been proposed in the last decade based on the measurement of electrochemical characteristics (thickness, composition) of the corrosion layer formed on the surface of the object. These models are described in following chapters of this book.

1.5.2.4 Biological and Biochemical Methods

Dendrochronology: studies the morphology of the seasonal tree rings, particularly the width, which varies between a few millimeters to several tenths. The tree rings can be observed in samples cut as a transversal cross-section of a fallen tree. Each tree ring is correlated to seasonal variations so that this method can be used for direct dating of archaeological sites and provides information about the climate. For enlarging the time scale cross-dating between trees of different ages, archaeological timbers and fossil trees can be carried out. Currently, dendrochronology is considered one of the more accurate dating methods. Among others, it has demonstrated a distortion in the time scales provided by other dating methods, and it has established a relationship between Europe and the Near East during the Neolithic and Bronze Ages [66–68, 71, 87].

Analysis of pollen and spores: the study of these biological materials preserved in acid soils and peat bogs is the aim of palynology. The type of botanical species growing in a particular terrestrial region is an indicator of the climate, and the latter can estimate the site's age. The soil samples are sieved, followed by chemical treatments to isolate the biological material that is stained with the suitable dye for enhancing the morphological characteristics and examined with an optical or scanning electron microscope [67, 68, 88].

1.5.2.5 Genetic Methods

Dating based on molecular markers/mutation rate: this method is based on the identification of the molecular changes in the nucleotide sequences of the desoxyribonucleic acid (DNA) associated with mutations between two species or taxonomical groups [66, 89].

1.5.2.6 Geological Methods

This dating techniques are based on the correlation between the composition of archaeological and geological materials and the climatic variations in the Earth during the recent (roughly the last 2 million years) geological time.

Varve or rhythmites chronology is based on examining the distribution of the layered sediments formed in the beds of lakes during the seasonal melting of glaciers and the subsequent deposition of particles of different sizes supplied by streams. This stratified structure can be correlated with seasonal climate changes. This dating method has improved the radiocarbon time scale back to late-glacial times. Like varve, the study of the loess deposits of windblown sediments provides almost continuous registers of climate changes, particularly the glacial stages. These geological formations are found in the middle latitudes of the Northern hemisphere, South America and New Zeeland. The study of fossils and their relative abundance can indicate the water temperature in which these organisms lived. This information is obtained from terrigenous sediments of the ocean bed and biogenic ooze formed with skeletons of marine microfauna [66, 67, 90].

1.5.2.7 Geophysical Methods

The classical work of Milankovitch [91, 92] on the control of the climate by the orbital variations, further developed by Imbrie and Imbrie [93], is based on the elaboration of theoretical models aimed at adjusting the insolation curve to precessional cycles, the variations in the ecliptic obliquity and the Earth's orbit eccentricity cycles. In turn, the model can be correlated to climate change, particularly the build-up and decay of glaciations.

References

1. The Oxford Encyclopedia of Mesoamerican Cultures (2001) Oxford University Press
2. Beaulieu P-A (1989) Reign of Nabonidus, King of Babylon (556–539 BC). Yale University Press
3. Silverberg R (1997) Great adventures in archaeology. University of Nebraska Press
4. Kelly, RL, Thomas, DH (2013) Archaeology: down to earth. Cengage Learning
5. Leute U (1987) Archaeometry. An introduction to physical methods in archaeology and the history of art. VCH ED, New York
6. Aitken MJ (1961) Physics and archaeology. Interscience, New York
7. Olin JS (1982) Future directions in archaeometry. A round table. Smithsonian Institution Press, Washington DC
8. ICOM-CC (2008) Resolution adopted by the ICOM-CC membership at the 15th Triennial conference, New Delhi, 22–26 Sept
9. Plenderheit HJ, Werner AEA (1967) The conservation of antiquities and works of art. Oxford University Press, London
10. Brandi C (1963) Teoría del restauro, Edizioni di Storia e Letteratura, Einaudi, Rome
11. Riegl A (2008) Modern cult of monuments: its character and its origin. MIT Press

12. Carta del Restauro (1972) Italian restoration chart of 1972. Available in https://sangem.s3.amazonaws.com›2015/11. Accessed on 21 Dec 2022
13. Hague Convention (1954) Convention for the protection of cultural property in the event of armed conflict with regulations for the execution of the convention. Available in https://en.unesco.org›1954_Convention_EN_2020. Accessed on 21 Dec 2022
14. Australia ICOMOS (1999) The Burra Charter: the Australia ICOMOS charter for places of cultural significance. 1979; rev. 1999. Burwood. Available in http://australia.icomos.org/wp-content/uploads/BURRA-CHARTER-1999_charter-only.pdf. Accessed on 10 Mar 2022
15. Nadolny J (2003) The first century of published scientific analyses of the materials of historical painting and polychromy, circa 1780–1880. Rev Conserv 4:39–51
16. Sease C (1996) A short history of archaeological conservation. Stud Conserv 41:157–161
17. Fernandes Vieira G, Dias Coelho LJ (2011) Arqueometría: Mirada histórica de una ciencia en desarrollo. Revista CPC 13:107–133
18. Caley E (1949) Early history and literature of archaeological chemistry. J Chem Educ 26:242–268
19. Caley E (1951) Klaproth as a pioneer in the chemical investigation of antiquities. J Chem Educ 28:64–66
20. Caley ER (1967) The early history of chemistry in the service of archaeology. J Chem Educ 44:120–123
21. Doménech-Carbó MT (2018) Análisis químico y examen científico de patrimonio cultural. Sintesis, Madrid
22. Plenderleith HJ (1998) A history of conservation. Stud Conserv 43:29–143
23. Bridgman CF (1964) The amazing patent on the radiography of paintings. Stud Conserv 9:135–139
24. Plesters J (1956) Cross-sections and chemical analysis of paint samples. Stud Conserv 2:110–157 (and references therein)
25. Doménech-Carbó MT, Osete-Cortina L (2016) Another beauty of analytical chemistry: chemical analysis of inorganic pigments of art and archaeological objects. ChemTexts 2:14 (and references therein)
26. Olin JS, Salmon ME, Olin CH (1969) Investigations of historical objects utilizing spectroscopy and other optical methods. Appl Optics 8:29–39
27. Feller RL (1954) Dammar and mastic infrared analysis. Science 120:1069–1070
28. Hall ET (1963) Methods of analysis (physical and microchemical) applied to paintings and antiquities. In: Thomson G (ed) Recent advances in conservation. Butterworths, London, pp 29–32
29. Doménech-Carbó MT, Doménech-Carbó A (2022) Spot test: past and present. ChemTexts 8:4 (and references therein)
30. Ward P (1986) The nature of conservation a race against time. The Getty Conservation Institute, Marina del Rey
31. National Standards Policy Advisory Committee (1978) National policy on standards for the United States and a recommended implementation plan. National Standards Policy Advisory Committee, Washington D.C., p 6
32. Reedy TJ, Reedy CL (1988) Statistical analysis in art conservation research. The Getty Conservation Institute, Los Angeles
33. Doménech-Carbó MT (2008) Novel analytical methods for characterising binding media and protective coatings in artworks. Anal Chim Acta 621:109–139 (and references therein)
34. Zevgiti S, Sakarellos C, Sakarellos-Daitsiotis M, Ioakimoglou E, Panou-Pomoniss E (2007) Collagen models as a probe in the decay of works of art: synthesis, conformation and immunological studies J Pept Sci 13:121–127
35. Tamburini D, Vermeulen M, Ortiz Miranda AS, Walton MS (2022) Technical steps towards enhanced localization of proteins in cultural heritage samples by immunofluorescence microscopy and micro-reflectance imaging spectroscopy. Microchem J 176:107243

36. Doménech-Carbó MT, Álvarez-Romero C, Doménech-Carbó A, Osete-Cortina L, Martínez-Bazán ML (2019) Microchemical surface analysis of historic copper-based coins by the combined use of FIB-FESEM-EDX, OM, FTIR spectroscopy and solid-state electrochemical techniques. Microchem J 148:573–581
37. Laskaris N, Liritzis I, Bonini M, Ridi F, Kersting R, Al-Otaibi F (2017) AFM and SIMS surface and cation profile investigation of archaeological obsidians: new data. J Cult Herit 25:101–112
38. Lee Y, Martín-Rey S, Osete-Cortina L, Martín-Sánchez I, Bolívar-Galiano F, Doménech-Carbó MT (2018) Evaluation of a gelatin-based adhesive for historic paintings that incorporates citronella oil as an eco-friendly biocide. J Adhes Sci Technol 32:2320–2349
39. Nieto-Villena A, Martínez Mendoza JR, Ramírez-Saito MA, Arauz-Lara JL, de la Cruz-Mendoza JA, Guerrero-Serrano AL, Ortega-Zarzosa G, Solbes-García A (2019) Exploring confocal microscopy to analyze ancient photography. J Cult Herit 36:191–199
40. Adriaens A, Dowsett MG (2006) Applications of SIMS to cultural heritage studies. Appl Surf Sci 252:7096–7101 (and references therein)
41. Koob SP, Ng WY (2000) The desalination of ceramics using a semi-automated continuous washing station. Stud Conserv 45:265–273
42. Niven B, Campbell L (2006) Effects of selected aqueous treatments on the properties of two papers. Stud Conserv 51:189–198
43. Janssens K (2005) A survey of the recent use of X-ray beam methods for non-destructive investigations in the cultural heritage sector. In: Van Grieken R, Janssens K (eds) Cultural heritage conservation and environment impact assessment by non-destructive testing and micro-analysis. AA Balkema Publishers, London, pp 265–308 (and references therein)
44. Demortier G (1991) Review of the recent applications of high energy microprobes in art and archaeology. Nucl Inst Meth Phys Res Section B 54:334–345 (and references therein)
45. Schreiner M, Melcher M, Uhlir K (2007) Scanning electron microscopy and energy dispersive analysis: applications in the field of cultural heritage. Anal Bioanal Chem 387:737–747 (and references therein)
46. Marengo E, Aceto M, Robotti E, Liparot MC, Bobba M, Pantò G (2005) Archaeometric characterisation of ancient pottery belonging to the archaeological site of Novalesa Abbey (Piedmont, Italy) by ICP–MS and spectroscopic techniques coupled to multivariate statistical tools. Anal Chim Acta 537:359–375
47. Díaz Pace DM, Mugueta MA, Molina J (2021) Characterization of archaeological remains from a military settlement in the Argentinian Pampas (19th century): new insights by laser-induced breakdown spectroscopy. Spectrochim Acta B 186:106322
48. Bitossi G, Giorgi R, Mauro M, Salvadori B, Dei L (2005) Spectroscopic techniques in cultural heritage conservation: a survey. Appl Spectrosc Rev 40:187–228 (and references therein)
49. Derrick MR, Stulik D, Landry MJ (1999) Infrared spectroscopy in conservation science. Getty Conservation Institute, Los Angeles (and references therein)
50. Edwards HGM, Chalmers JM (eds) (2005) Raman spectroscopy in archaeology and art history. The Royal Society of Chemistry, Cambridge (and references therein)
51. Vandenabeele P, Edwards HGM, Moens L (2007) A decade of raman spectroscopy in art and archaeology. Chem Rev 107:675–686
52. Sessa C, Weiss R, Niessner R, Ivleva NP, Stege H (2018) Towards a surface enhanced raman scattering (SERS) spectra database for synthetic organic colourants in cultural heritage. The effect of using different metal substrates on the spectra. Microchem J 138:209–225
53. Ding L, Gong T, Wang B, Yang Q, Liu W, Pemo R, Metok T (2021) Original article non-invasive study of natural dyes in textiles of the Qing dynasty using fiber optic reflectance spectroscopy. J Cult Herit 47:69–78
54. Ahmed HE, Tahoun IF, Elkholy I, Shehata AB, Ziddan Y (2017) Identification of natural dyes in rare Coptic textile using HPLC-DAD and mass spectroscopy in museum of Faculty of Arts, Alexandria University, Egypt. Dyes Pigm 145:486–492
55. Camaiti M, Casieri C, De Luca F, Fantazzini P, Terenzi C (2007) The use of portable single-sided relaxometry and laboratory imaging NMR devices in stone conservation. Stud Conserv 52:37–49

56. Lorusso S, Natali A, Matteucci C (2007) Colorimetry applied to the field of cultural heritage: examples of study cases, conservation science in cultural heritage. Quaderni di Scienza della Conservazione 7:187–220
57. Bertrand L (2007) Synchrotron imaging for archaeology, art history, conservation, and palaeontology. In: Creagh D, Bradley D (eds) Physical techniques in the study of art, archaeology and cultural heritage, Chapter 2. Elsevier
58. Odlyha M (2003) The applications of thermoanalytical techniques to the preservation of art and archaeological objects. In: Brown ME, Gallagher PK (eds) Handbook of thermal analysis and calorimetry, Chapter 2. Elsevier
59. Napolitano MP, Prieto Conaway MC, Kuo P-C, Arslanoglu J, Yost RA (2019) High resolution/accurate mass tandem MS of isotopically complex cluster ions from the artists' pigment lead white. Int J Mass Spectrom 439:19–26
60. Quillen S, Learner T (2006) A review of the classes, structures, and methods of analysis of synthetic organic pigments. J Am Inst Conserv 45:107–125 (and references therein)
61. Colombini MP, Modugno FC (2004) Characterisation of proteinaceous binders in artistic paintings by chromatographic techniques. J Sep Sci 27:147–160 (and references therein)
62. Kaml I, Kenndler E (2007) Characterization of natural organic binding media in museum objects by capillary electrophoresis. Curr Anal Chem 3:33–40
63. Glascock MD (2000) The Status of activation analysis in archaeology and geochemistry. J Radioanal Nucl Chem 244:537–541
64. Casellato U, Vigato PA, Russo U, Matteini M (2000) A Mössbauer approach to the physico-chemical characterization of iron-containing pigments for historical wall paintings. J Cult Herit 1:217–232
65. Janssens K, Dik J, Cotte M, Susini J (2010) Photon-based techniques for nondestructive subsurface analysis of painted cultural heritage artifacts. Acc Chem Res 43:814–825
66. Doménech-Carbó A (2015) Dating: an analytical task: ChemTexts 1:5 (and references therein)
67. Geyh MA, Schleicher H (1990) Absolute age determination—physical and chemical dating methods and their application. Springer, Berlin, Heidelberg
68. Aitken MJ (1990) Science-based dating in archaeology Longman archaeology series. New York pp 1−13
69. Taylor RE (1987) Radiocarbon dating: an archaeological perspective. Academic Press, Orlando
70. Tuniz C, Norton G (2008) Accelerator mass spectrometry: new trends and applications. Nucl Inst Meth Phys Res B 266:1837–1845 (and references therein)
71. Wintle AG (1996) Archaeologically-relevant dating techniques for the next century small, hot and identified by acronyms. J Archaeol Sci 23:123–138 (and references therein)
72. Taylor RE (1987) Dating techniques in archaeology and paleoanthropology. Anal Chem 59:317–331
73. Rodríguez-López A, Khandekar N, Gates G, Newman R (2007) Materials and techniques of a Spanish renaissance panel painting. Stud Conserv 52:81–100
74. Hames WE (2021) K/Ar and $^{40}Ar/^{39}Ar$ dating methods, 2nd edn. In: Alderton D, Elias SA (eds) Encyclopedia of geology. Elsevier, pp 50−65
75. Van Calsteren P, Thomas L (2006) Uranium-series dating applications in natural environmental science. Earth-Sci Rev 75:155–175
76. Huth TE, Passey BH, Cole JE, Lachniet MS, McGee D, Denniston FR, Truebe S, Levin NE (2022) A framework for triple oxygen isotopes in speleothem paleoclimatology. Geochim Cosmochim Acta 319:191–219
77. Dunai T (2010) Cosmogenic nuclides. Cambridge Univeristy Press, Cambridge
78. Bigazzi G, Hadler Neto JC, Iunes PJ, Osório Araya AM (2005) Fission-track dating of South American natural glasses: an overview. Rad Meas 39:585–594
79. Jolivet M, Lebatard AE, Reyss JL, Bourlès D, Mackaye HT, Lihoreau F, Vignaud P, Brunet M (2008) Can fossil bones and teeth be dated using fission track analysis? Chem Geol 247:81–99
80. Zacharias N, Schwedt A, Buxeda i Garrigós J, Michael CT, Mommsen H, Kilikoglou V (2007) A contribution to the study of post-depositional alterations of pottery using TL dating analysis. J Archaeol Sci 34:1804–1809

81. Meriç N, Koşşal M, Atlıhan MA, Yüce UR (2008) OSL properties of anthropological bone and tooth radiation. Phys Chem 77:685–689
82. Eighmy J, Sternberg R (1990) Archaeomagnetic dating. The University of Arizona Press, Tucson
83. Hus J, Ech-Chakrouni S, Jordanova D (2002) Origin of magnetic fabric in bricks: its implications in archaeomagnetism. Phys Chem Earth 27:1319–1331
84. Liritzis I, Laskaris N (2011) Fifty years of obsidian hydration dating in archaeology. J Non-Cryst Solids 357:2011–2023
85. Arany S, Ohtani S, Yamamoto T, Sugiyama T (2007) Comparison of aspartic acid racemization between mammoth and human dentinal tissues. Arch Oral Biol 52:20–25
86. Haddy A, Hanson A (1982) Nitrongen and fluorine dating of Moundville skeletal samples. Archaeometry 24:37–44
87. Martinelli N (2004) Climate from dendrochronology: latest developments and results. Global Planet Change 40:129–139 (and references therein)
88. Ellison JC (2008) Long-term retrospection on mangrove development using sediment cores and pollen analysis: a review. Aq Bot 89:93–104 (and references therein)
89. Zuckerkandl E, Pauling LB (1962) Molecular disease, evolution, and genic heterogeneity. In: Kasha M, Pullman B (Eds) Horizons in biochemistry. Academic Press, New York, pp 189–225
90. Fisher TG (2007) Abandonment chronology of glacial Lake Agassiz's Northwestern outlet. Palaeogeogr Palaeoclim Palaeoecol 246:31–44
91. Milankovitch MM (1941) Canon of Insolation and the Ice-age Problem, Königlich Serbische Akademie, Beograd (English translation by the Israel Program for Scientific Translations, published for the US Department of Commerce and the National Science Foundation, Washington DC, 1969)
92. Dimitrijević MS (2020) Milutin Milanković and climate changes leading to ice ages. ChemTexts 6:23
93. Imbrie J, Imbrie JZ (1980) Modeling the climatic response to orbital variations. Science 207:943–953

Additional Literature

94. Adriaens A, Dowsett M (eds) (2021) Spectroscopy, diffraction and tomography in art and heritage science. Elsevier, Amsterdam
95. Ambrose SH, Katzenberg MA (eds) (2010) Biogeochemical approaches to paleodietary analysis. Springer, Berlin-Heidelberg
96. Artioli G (2010) Scientific methods and cultural heritage. Oxford University Press
97. Buckley E (ed) (2020) Archaeometry: theory and practice. Willford Press
98. Colombini MP, Degano I, Nevin A (eds) (2022) Analytical chemistry for the study of paintings and the detection of forgeries. Springer, Berlin
99. Edwards HGM, Vandenabeele P, Colomban P (2022) Raman spectroscopy in cultural heritage preservation. Cultural heritage science series. Springer, Berlin
100. Fukunaga K (2016) THz technology applied to cultural heritage in practice. Cultural heritage science series. Springer, Berlin
101. Gmelin JF (1781) Experimenta Nonnulla cum Mumiis Instituta. Commentationes Societatis regiae scientiarum Gottinggensis historicae et philosophicae classis 4:3–25
102. Göbel F (1842) Ueber den Einfluss der Chemie auf die Ermittelung der Völker der Vorzeit oder Resultate der chemischen Untersuchung metallischer Altertümer inbesondere der in den Ostseegouvernements vorkommenden, Behufs der Ermittelung der Völker, von welchen sie abstammen, Erlangen

103. Klaproth MH (1798) Mémoires de I'académie royale des sciences el belles-lettres, Berlin, Classe de philosophic experimentale, 97–113. A German version of this same paper "Beitrag zur numismatische Docimasie," was later published in Sammlung der deutseken Abhandlungen welche in der honigliche Akademie der Wissenschaflen zu Berlin vorgelesen warden in den Jahren 1792–1797, Experimental-Philosophic, 1799:3–14, and still later under the same title a modified German version was published in (1801) in Allgemeines Journal der Chemie 6:227–244
104. Leute U (1987) Archaeometry: an introduction to physical methods in archaeology and the history of art. Wiley-VCH
105. Liritzis I, Singhvi AK, Feathers JK, Wagner GA, Kadereit A, Zacharias N, Li S-H (2013) Luminescence dating in archaeology, anthropology, and geoarchaeology. An overview. Springer
106. Martini M, Galli A (2022) Application of materials science in the study of cultural heritage. MDPI, Basel
107. Mulholland SC, Rapp Jr G (eds) (2013) Phytolith systematics. Springer
108. Rink WJ, Thompson JW (eds) (2013) Encyclopedia of scientific dating methods. Springer
109. Romig A, Lindblom K (2016) Discovery of Radiocarbon Dating The University of Chicago. American Chemical Society. Available on https://www.acs.org/content/dam/acsorg/education/whatischemistry/landmarks/willard-libbyradiocarbon-dating.pdf. Accessed 6th June 2023.
110. Shackley MS (ed) (2013) Archaeological obsidian studies. Method and theory. Springer
111. Taylor RE, Aitken MJ (eds) (2013) Chronometric dating in archaeology. Springer
112. Taylor RE, Bar-Yosef O (2020) Radiocarbon dating. An archaeological perspective. Routledge
113. Varella EA (2013) Conservation science for the cultural heritage. Lecture notes in chemistry series 79. Springer

Chapter 2
Electrochemical Processes and Techniques

2.1 Introduction

Generically speaking, electrochemistry is a branch of science that deals with the interrelation of electrical and chemical phenomena. Central topics are (i) the transfer of charges (electrons, ions) across interfaces (mostly interfaces between metals and ion conductors, in some cases also interfaces between immiscible ion solutions), and (ii) the study of ion and electron conduction in bulk phases (solids, solutions, conducting films). These processes are normally accompanied by a number of phenomena like gas evolution, deposition or dissolution of solids, adsorption/desorption at the electrode surface, chemical reactions in the electrolyte, double-layer charging, etc. [1–3]. In this chapter, we will focus the attention on the electrochemical techniques predominantly applied in the study of cultural heritage.

There is a variety of electrochemical techniques potentially applicable to the study and preservation of cultural heritage. From a historical perspective, during the twentieth century, the focus was on electrolysis techniques whose application for stabilizing metallic objects was widely studied for the first two decades of the century [4, 5]. These techniques were considerably expanded [6–8], but their use further decayed upon recognition of the complexity of the corrosion/stabilization processes [9]. Currently, there is a renewed interest in electrolysis procedures, which will be treated in the chapters devoted to metallic heritage.

The use of electrochemical techniques as analytical tools to achieve information of value for archaeologists, conservators, and restorers was largely concentrated in two fields: (i) conventional potentiometric, voltammetric, or amperometric analysis aimed to determine selected analytes; for instance, chloride in metal stabilization baths; (ii) studies of metal corrosion processes based on polarization curves and impedance techniques.

The analytical use of electrochemistry in the cultural heritage field has been notably expanded in the last decades by the development of solid-state methodologies [10]. The roots of these techniques can be associated with mineral analysis

A. Doménech-Carbó and M. T. Doménech-Carbó, *Electrochemistry for Cultural Heritage*, Monographs in Electrochemistry,
https://doi.org/10.1007/978-3-031-31945-7_2

based on Glazunov's electrography and related tests developed in the first half of the twentieth century [11, 12]. In the 1960s, the use of carbon paste electrodes provided a (mainly) voltammetric approach to the analysis of solids [13–15]. Here, the solid is embedded into a paste formed by graphite powder with a non-conducting (nujol oil, paraffin oil) or conducting (aqueous solutions, ionic liquids) binder, the resulting electrode being put into contact with a given electrolyte.

The consolidation of electrochemical techniques as suitable for the analysis of solids comes to a great extent from the development of the voltammetry of immobilized particles (VIMP) proposed by Scholz et al. in the late 1980s [16–21]. Initially, the technique has been called abrasive stripping voltammetry. The VIMP consists of recording the voltammetric response of a microparticulate deposit of a solid material, abrasively transferred to the surface of an inert electrode. Mostly paraffin-impregnated graphite electrodes are used which are in contact with a suitable electrolyte. The essential condition to be accomplished in order to obtain a genuine solid-state analysis is that the solid is insoluble in the selected electrolyte solution.

Voltammetric measurements consist of the measurement of the current passing through an electrochemical cell when a time-dependent potential difference is established between a working electrode and a reference electrode. Experimental arrangements involve a third electrode (auxiliary electrode or counter electrode) so that the cell current passes through the working and auxiliary electrode, and the reference electrode is almost completely kept current free. The reference electrode only provides the reference point for the potential control. The potential difference between the working and reference electrodes equals the potential difference between the working electrode and the solution. The resulting current/potential graph is called the voltammogram. There are different types of potential inputs to be applied. In the simplest cases, the potential (difference) linearly varies with time; these are the linear potential scan and cyclic voltammetries (LSV and CV, respectively). Pulsed techniques add a series of potential pulses to the linear potential scan or to a staircase ramp. The different modes (normal pulse voltammetry, differential pulse voltammetry, square wave voltammetry, ...) permit to discriminate against capacitive currents [20].

The VIMP permits obtaining well-defined voltammetric responses using amounts of sample in the nanogram to microgram range, thus making it of obvious interest for the analysis of cultural goods. The VIMP has been applied to different materials, organic and inorganic pigments, metals, woods, ceramic pastes, glass, and glazed materials, etc. [21–23].

In this chapter, the fundamentals of the electrochemical techniques more extensively used in the cultural heritage domain will be described assuming that the reader has a reasonable knowledge of general electrochemistry. Accordingly, in the first part of the chapter, the essential aspects of the theoretical description of the electrochemical processes occurring in VIMP analysis will be treated. In the second part of the same, the fundamentals of impedance techniques (electrochemical impedance spectroscopy, EIS) and scanning electron microscopy (SECM), both used in the heritage domain, will be treated. Although these techniques do not exhaust the list of available ones, they define the basic background for the intersection between electrochemistry and cultural heritage.

2.2 Voltammetry of Immobilized Microparticles

2.2.1 *Voltammetry, General Aspects and Conventions*

The term voltammetry designates a family of electrochemical techniques, where the current passing through an electrochemical cell is recorded when a potential difference, continuously varying with time according to a predefined pattern, is applied between a working electrode and a counter electrode and the potential of the working electrode is set versus a reference electrode. The resulting current/potential graph is the so-called voltammogram. This graph reflects all electrochemical phenomena associated to the pass of current, including faradaic processes (i.e., those involving changes in the chemical composition of the electrolyte solution and/or electrode, including gas evolution, metal deposition, etc.) and non-faradaic ones. The latter are caused by double-layer charging/discharging. They are usually seen as undesired perturbations to the faradaic response, and will be further treated in Chap. 3 within a more detailed description of voltammetry.

There is a variety of voltammetric techniques depending on the characteristics of the potential/time input applied to the electrochemical cell [1–3]. In principle, these are mainly aimed to study the faradaic processes occurring as a result of the application of the potential inputs. The processes in which there is a net transfer of electrons from the working electrode to species in the electrolyte solution (or adhered to the electrode surface) are termed as cathodic. These are electrochemical reductions. The opposite processes, in which there is electron transfer from such species to the working electrode, are called anodic and they are oxidations.

In the voltammograms, the (difference of) potential applied to a working electrode is measured relative to a given reference electrode (typically Ag/AgCl and Hg/Hg_2Cl_2—calomel electrode—). According to the sign criteria established by the International Union of Pure and Applied Chemistry (IUPAC), anodic (oxidation) currents are taken as positive, whereas cathodic (reduction) currents are taken as negative. Historically, however, the inverse criterion was frequently used to draw the voltammograms. The result is that much literature data were depicted contrary to the IUPAC recommendations. Due to the impossibility of recovering much of such 'old' data to rearrange them according to the IUPAC rules, we have reproduced in this book the original figures, which are in several cases in disagreement with the current IUPAC rule.

The simplest voltammetric technique is linear potential scan voltammetry in which the potential is varied linearly with time at a given potential scan rate v. Usually, the scan rate ranges between 1 and 10^4 mV s^{-1}, but in several techniques potential scan rates of 10^{-3} mV s^{-1} or even lower are used. This technique is routinely expanded by cycling the potential between two limits; this is called cyclic voltammetry. In principle, there is no limit for the number of potential cycles to be applied, thus defining the repetitive voltammetry.

In a second group of voltammetric techniques, potential pulses are applied. In these techniques, the pulses are superimposed to a linear or staircase ramp. The pulse techniques have in common that they minimize the measurements of undesired capacitive currents. Among the variety of pulse techniques, differential pulse voltammetry and square wave voltammetry are probably the most widely utilized. In the former, the pulses, characterized by their amplitude (usually 5–25 mV) and duration, are periodically applied (with a frequency typically varying between 1 and 10^5 s^{-1}) superimposed to the ramp function. The current is measured before and after each pulse and the difference is amplified. This is common to square wave voltammetry, where a square wave potential waveform (characterized by potential step increment, square wave amplitude and frequency) is superimposed to a staircase ramp. The voltammograms can be displayed as forward and backward currents and the difference between them. This last is the quantity usually displayed as square-wave voltammogram. These techniques offer an inherently high sensitivity, but the interpretation of the voltammograms may involve difficulties due to the eventual appearance of 'artifacts' needing to be accurately described [20].

2.2.2 *Solid-State Transformations*

As previously noted, the VIMP is a sensitive technique for the electrochemical analysis of solids [24–26]. The relevant aspect to underline is that, in contrast to traditional solution-phase electrochemistry, the voltammetric response is characteristic of the solid phases attached to the electrode surface. There are three main types of electrochemical processes occurring in VIMP analysis; the first one consists of the topotactic transformation of one solid to another. An idealized representation of the process is depicted in Fig. 2.1. Here, a solid particle is in contact with an electrochemically silent base electrode and an electrolyte solution defining a three-phase system. According to the theoretical model developed by Lovric, Scholz, Oldham, and others [27–31], the redox reaction proceeds via electron transfer through the electrode/particle interface and ion transfer through the particle/electrolyte interface. For reasons of charge conservation, the exchange of electrons between the solid and the inert electrode material has to be coupled with the exchange of charge-balancing cations or anions between the active solid and the surrounding solution. The electrochemical process starts at the particle/base electrode/electrolyte three-phase junction line and advances through the solid particle via electron hopping between immobile redox centers and intercalation of ions into the lattice or their exit to the solution.

Whereas many solids will simply dissolve when an oxidation or reduction is proceeding (e.g. reductive dissolution of metal oxides; Sect. 2.2.2), there are many solids which can remain stable (solid) because insertion-type electrochemical reactions happen. Insertion electrochemical systems can be treated as follows, taking, for

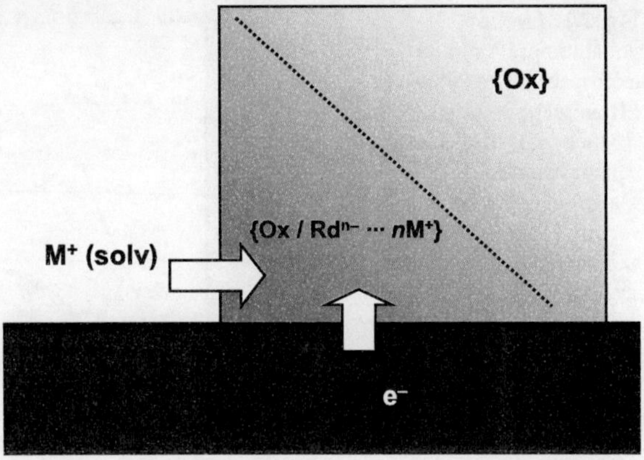

Fig. 2.1 Scheme for the cation-assisted electrochemical reduction of a cation-permeable solid attached to an inert electrode in contact with an electrolyte in which both the parent solid and its reduced form are insoluble. The dotted line represents the boundary of the reaction front at a given reaction time

simplicity, monovalent ions: The overall process for the reduction of a neutral solid (oxidized form), {Ox}, assisted by the entrance of monovalent cations, M^+, to form a reduced, ion-permeating solid, $\{Rd^{n-} \cdots nM^+\}$, can be represented as:

$$\{Ox\}_{solid} + nM^{+}{}_{solv} + ne^{-} \rightarrow \{Rd^{n-} \cdots nM^{+}\}_{solid} \tag{2.1}$$

where the subscripts solid and solv denote solid and solvated species, respectively. The reaction (2.1) can be driven from the right side to the left side when the potential is reversed. Alternatively, the reduction of an $\{Ox^{n+} \cdots nX^-\}_{solid}$ can be coupled with the exit of anions, and the oxidation of the reduced form with the ingress of anions. Again, taking, for simplicity, monovalent ions, one can write:

$$\{Ox^{n+} \cdots nX^{-}\}_{solid} + ne^{-} \rightarrow \{Rd\}_{solid} + nX^{-}{}_{solv} \tag{2.2}$$

Reaction (2.2) can proceed from right to left, when the potential is reversed. What reaction mechanism is operative depends on the chemical composition of the solid and its ability to host mobile cations or anions and to provide their transport through the solid phase. Under conditions of thermodynamic control (electrochemical reversibility), this type of process can be treated as a reversible, diffusion-controlled three-phase process [21, 37] with two coupled electrode reactions, the transfer of electrons and the transfer of ions. Then, in negative-going potential scans applied to an {Ox} solid, a cathodic peak corresponding to the process described by Eq. (2.1) appears. Upon reversal of the potential scan, an anodic peak corresponding to the back reaction appears. Similar considerations can be applied to the oxidation of a solid {Rd}.

For a simplified two-dimensional scheme such as depicted in Fig. 2.1, charge transfer through the solid occurs via electron transport in the vertical direction and cation diffusion in the horizontal one. Depending on the values of the respective diffusion coefficients (D_e, D_M, respectively) the reaction front (dotted line in Fig. 2.1) will be more or less inclined. Although the transport of electrons is not a physical

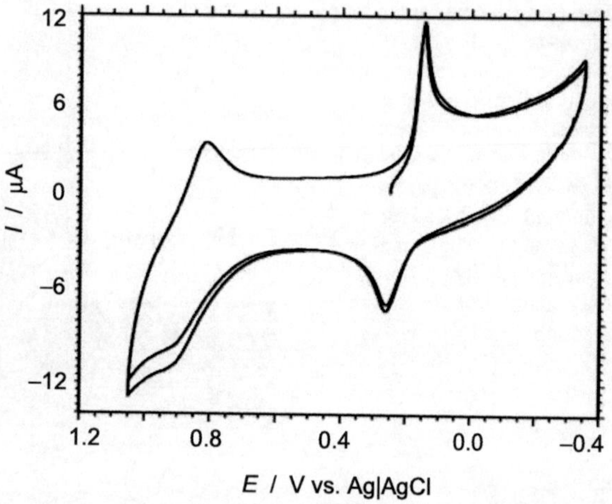

Fig. 2.2 Cyclic voltammogram of a microparticulate deposit of PB on graphite electrode in contact with 0.10 M KNO_3 aqueous solution. Potential scan initiated at 0.20 V versus Ag/AgCl in the negative direction; potential scan rate 50 mV s^{-1}. Note that the axes are opposite to the IUPAC convention

diffusion in the strict sense (it may, e.g., occur via electron hopping between the redox centers), it can formally be given a diffusion coefficient. Two limiting cases can be postulated: (i) when $D_e \ll D_M$, the reaction zone spreads along the solid particle/base electrode interface much faster than along the solid particle/electrolyte solution interface and slow electron diffusion determines the progress of the redox process; (ii) when $D_e \gg D_M$, the reaction front spreads along the solid particle/electrolyte interface much faster than along the solid particle/electrode interface and the reaction progress is governed by slow cation diffusion [27, 30, 31].

With regard to the scheme in Fig. 2.1, it has to be noted that the reduced solid can form a solid solution with the parent oxidized solid or segregate as a new phase [32]. Then, complications associated with miscibility gaps have to be accounted for [33].

The electrochemistry of Prussian blue (PB) in contact with aqueous electrolyte solutions constitutes a paradigmatic example of this kind of process. Figure 2.2 shows a cyclic voltammogram recorded for a microparticulate deposit of PB on a graphite electrode in contact with 0.10 M KNO_3 aqueous solution.

The voltammogram shows two essentially reversible couples of cathodic/anodic peaks at equilibrium potentials (defined by the half-sum of the cathodic and anodic peak potentials) of 0.85 and 0.15 V versus Ag/AgCl. These processes can be described in terms of the K^+-assisted reduction of PB to Berlin white or Everitt's salt [34]:

$$KFe^{III}\left[Fe^{II}(CN)_6\right]_{solid} + K^+{}_{aq} + e^- \rightarrow K_2Fe^{II}\left[Fe^{II}(CN)_6\right]_{solid} \tag{2.3}$$

In turn, the oxidation of PB which produces Prussian yellow ($Fe^{III}[Fe^{III}(CN)_6]$) or Berlin green can be represented as,

$$KFe^{III}\left[Fe^{II}(CN)_6\right]_{solid} \rightarrow Fe^{III}\left[Fe^{III}(CN)_6\right]_{solid} + K^+{}_{aq} + e^- \tag{2.4}$$

Berlin green has been described as a solid solution of PB and Prussian yellow $\{KFe^{III}[Fe^{II}(CN)_6]\}_x$ $\{Fe^{III}[Fe^{III}(CN)_6]\}_{1-x}$, [34–36] and probably Prussian yellow contains some minor amounts of Prussian blue.

In general, the charge-balancing cations/anions are inserted into the solid lattice as intercalation ions. Similar reaction schemes can be formulated in case of the proton-assisted reduction/oxidation of organic compounds in contact with aqueous electrolytes. Here, the ingress/exit of charge-balancing protons is associated with modifications in the system of bonds of the solid phases. Examples of these processes will be discussed in Chaps. 4 and 6.

Interestingly, the study of these ion-insertion/ion-release processes permits the determination of thermochemical properties of individual ions [37] and treats the problem of correlating potential scales in different solvents [38].

2.2.3 *Reductive/Oxidative Dissolution Processes*

As mentioned before some solids are dissolved by reduction or oxidation, i.e., the solid is reduced or oxidized to species in the solution phase leading to a progressive disintegration of the lattice. Then, the voltammograms consist of a unique cathodic or anodic wave without coupled anodic or cathodic counterpart when the potential scan is reversed. A typical case is the reduction of iron oxides, other iron minerals, and iron pigments in contact with acidic electrolytes. In the case of hematite, the net electrochemical process is [39]:

$$Fe_2O_{3solid} + 6H^+{}_{aq} + 2e^- \rightarrow 2Fe^{2+}{}_{aq} + 3H_2O \qquad (2.5)$$

The description of these processes can be made based on considerations of solid-state chemical reaction kinetics [40]. In principle, processes such as described by Eq. (2.5) involve proton insertion into the solid lattice (i.e. protonation of oxide anions and exit of water or hydroxide anions) and/or formation of surface complex species.

Figure 2.3 shows two replicate square wave voltammograms recorded for several commercial iron pigments [41] in contact with 0.10 M HCl aqueous solution. The voltammograms show a series of more or less overlapped cathodic signals between 0.6 and − 0.8 V versus Ag/AgCl. These are attributable to the reduction of iron oxide (hematite) and oxide-hydroxide (goethite) particles with different degrees of hydration and crystallinity. In the case of goethite, the reduction process can formally be described as:

$$FeO(OH)_{solid} + 3H^+{}_{aq} + e^- \rightarrow Fe^{2+}{}_{aq} + 2H_2O \qquad (2.6)$$

The voltammetric response permits the characterization of the mineral component but is also dependent on the particle shape and size distribution, as well as the purity, crystallinity, and degree of hydration of the particles [25]. For instance, the

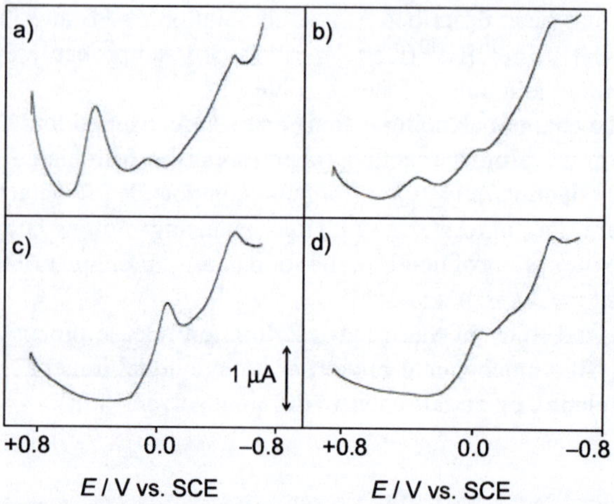

Fig. 2.3 Square wave voltammograms of **a** Bohemian green, **b** Umber, **c** Sienna, and **d** French ochre (Kremer pigments) on Alpino lead in contact with 0.10 M HCl. Potential scan initiated at 0.85 V in the negative direction. Potential step increment 4 mV; square wave amplitude 25 mV; frequency 5 Hz. Reproduced from Ref. [41], with permission. Note that the axes are opposite to the IUPAC convention

electrochemistry of hematite and iron oxide pigments is influenced by the degree of Al^{3+} by Fe^{3+} substitution in the lattice [42].

2.2.4 *Redox Processes with Phase Changes*

This type of process is of importance in the analysis of many inorganic pigments and corrosion products. The most frequent process consists of the reduction of a metal compound to a deposit of the corresponding metal. In the case of litharge (yellow PbO), the overall process is [43, 44]:

$$PbO_{solid} + 2H^{+}{}_{aq} + 2e^{-} \rightarrow Pb_{solid} + H_2O \tag{2.7}$$

Figure 2.4 shows the cyclic voltammograms (three successive potential cycles) recorded for a microparticulate deposit of litharge onto a paraffin-impregnated graphite electrode in contact with acetate buffer at pH 4.75. When the potential scan is initiated at 0.0 V in the negative direction, a prominent cathodic peak appears at ca. − 0.70 V versus Ag/AgCl corresponding to the process described by Eq. (2.7). In the subsequent positive-going potential scan, a tall anodic peak is recorded at − 0.50 V. This last presents the typical symmetric profile of the signals for the oxidative dissolution (stripping) of metal deposits, in this case, represented as:

$$Pb_{solid} \rightarrow Pb^{2+}{}_{aq} + 2e^{-} \tag{2.8}$$

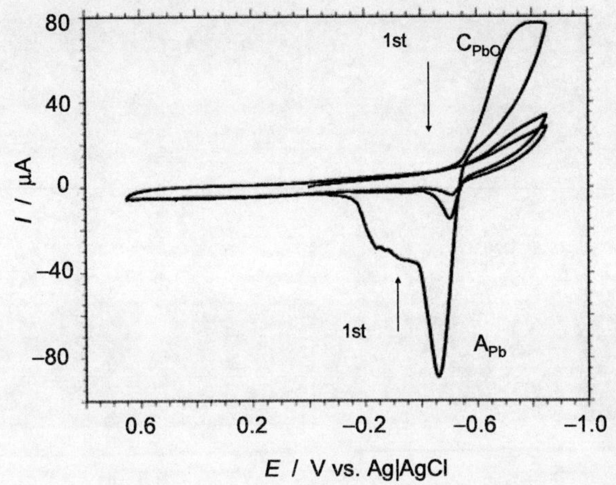

Fig. 2.4 Cyclic voltammetry of a microparticulate deposit of litharge on a paraffin-impregnated graphite electrode immersed into 0.25 M HAc/NaAc aqueous buffer at pH 4.75. Potential scan initiated at 0.0 V in the negative direction; potential scan rate 50 mV s^{-1}. Three successive potential cycles are recorded. Note that the axes are opposite to the IUPAC convention

The decreasing of peak heights during continuous cycling is due to the loss of material from the surface due to the release of Pb^{2+} ions by dissolution. The Pb(II) to Pb(IV) oxidation takes place at more positive potentials (see Chap. 12).

The description of these electrochemical reactions is complicated by the complexity of the involved processes: proton insertion and surface complexation, lattice deconstruction, and reconstruction of the new metal phase with nucleation and growth processes, among others. In the case of litharge in contact with 0.10 M KCl, the atomic force microscopy (AFM) study coupled with the application of reductive potential inputs by Hasse and Scholz [45] revealed the epitactic growth of the metal phase on one side of the PbO crystals. The process can be described in terms of the formation of an intermediate ionomeric layer interposed between the two solid (PbO and Pb) phases (Fig. 2.5).

This view is consistent with AFM images of a microparticulate deposit of litharge onto a graphite plate in contact with aqueous acetate buffer before and after application of a cathodic potential input resulting in the partial PbO to Pb reduction (Fig. 2.5). As can be seen in the images, the formation of metallic grains occurs in the vicinity of the PbO particles. Depending on the experimental conditions, the cathodic process can proceed via intermediate species in solution and the metal deposition can occur in regions of the base electrode separated from the particles of the parent solid. This second, competing pathway can be represented as [44]:

$$PbO_{solid} + 2H^{+}{}_{aq} + 2xe^{-} \rightarrow xPb_{solid} + (1-x)Pb^{2+}{}_{aq} + H_2O \qquad (2.9)$$

Again, the voltammetric response depends not only on the chemical and mineralogical composition of the sample but also on the crystallinity, degree of hydration,

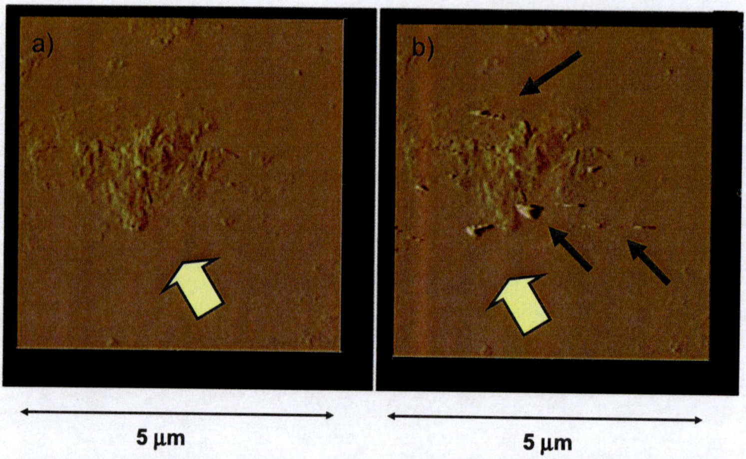

Fig. 2.5 Topographic AFM images of a microparticulate deposit of PbO (gross arrow) onto a graphite plate in contact with 0.25 M HAc/NaAc aqueous solution at pH 4.75 **a** before and **b** after application of a potential step of – 0.70 V versus Ag/AgCl for 2 min. Deposits of electrochemically formed Pb metal are marked with black arrows [54]

etc. of the parent solid. For our purposes, the relevant point to emphasize is that the VIMP methodology provides a sensitive tool to extract information on solid materials like pigments and their alteration products, corrosion products, etc. In particular, the voltammetric response is sensitive to, for instance, pigment–binding interactions, thus prompting the achievement of archaeometric information. These aspects will be treated in Chap. 4.

2.3 Electrochemical Impedance Spectroscopy

2.3.1 *Impedance Measurements and Impedance Spectroscopy*

Voltammetry and electrolysis, as well as other electrochemical techniques (chronoamperometry, chronocoulometry, etc.), are sometimes termed *faradaic techniques* because are chemical changes happen in the electrochemical cell (electrode and/or electrolyte), which are caused by faradaic currents. Faradaic current accounts for charge flow through electrodes which causes chemical changes that are proportional to the charge passing through the electrochemical cell. Other techniques (potentiometry, conductometry, …) are conceived as *non-faradaic*, i.e., proceeding (at least ideally) without chemical changes in the cell. These measurements can be performed at the equilibrium potential of the system and also at any other potential, so that faradaic currents play a more or less significant role. Typically, these are based on the determination of the impedance parameters assigned to an electrochemical cell when a sinusoidal potential input is applied. Here, the electrochemical cell is treated as an electric circuit in which in general there is a phase difference between the applied potential (difference) and the measured current flowing through the cell [46]. In these circumstances, the impedance is expressed as a complex of the form:

$$Z = Z_{real} + jZ_{imag} \quad (j = \sqrt{-1}) \tag{2.10}$$

The circuitry of conventional types of equipment permits the determination of the absolute value of the total impedance, |Z|, and the phase angle, φ, or, equivalently, the real and imaginary components, Z_{real}, Z_{imag}, of the impedance. The essential idea is that the application of a sinusoidal potential of low amplitude (few mV) allows for performing impedance measurements at a given frequency (f) without significant faradaic effects in the cell.

Impedance measurements are of importance in corrosion studies and semiconductor electrochemistry, where they are applied to determine electrochemical band gaps [32]. Remarkably, impedance measurements can be carried out in a large interval of frequencies, typically between 10^5 and 10^{-3} Hz, thus acquiring information on the studied system in a wide time constant scale (in other words, from a wide rate constant scale). Electrochemical impedance spectroscopy (EIS) simply combines impedance measurements in a given interval of frequencies. The resulting variation of impedance parameters with frequency (or angular frequency, $w = 2\pi f$) constitutes the impedance spectrum of the system.

Figure 2.6 shows two of the more frequently representations of impedance spectra for a eurocent coin immersed into mineral water (black circles) and 0.10 M NaCl (red circles). These are the so-called Bode plots corresponding to the |Z| versus log$f$, and $-\varphi$ versus log$f$ representations. One can see that the total impedance increases on decreasing frequency defining, apparently, two steps. In turn, the (minus) phase angle reaches two maxima at frequencies of ca. $10^{4.5}$ and 10 Hz. All phase angles fall in the 0 to $-50°$ range; i.e., the impedance spectra reveal a capacitive behavior intermediate between the purely ohmic response ($\varphi = 0°$) and the purely capacitive one ($\varphi = 90°$).

Figure 2.7 depicts the plots of $-Z_{imag}$ versus Z_{real} (Nyquist representation) for the same spectra. Here, two depressed loops (notice the different scale of the horizontal and vertical axes) at high and low frequencies appear. As will be seen in the next subsection, this impedance response can be described in terms of the existence of (at least) two different relaxation phenomena resulting in the appearance of two time constants.

Two aspects have to be underlined:

(a) Strictly, impedance spectra can only be defined for a stationary system fulfilling the constraints imposed by the theory of linear systems (LST): the system has to be time-invariant during the time of acquisition of impedance parameters, a condition that can be derived from the application of Kramers–Kronig (K–K)

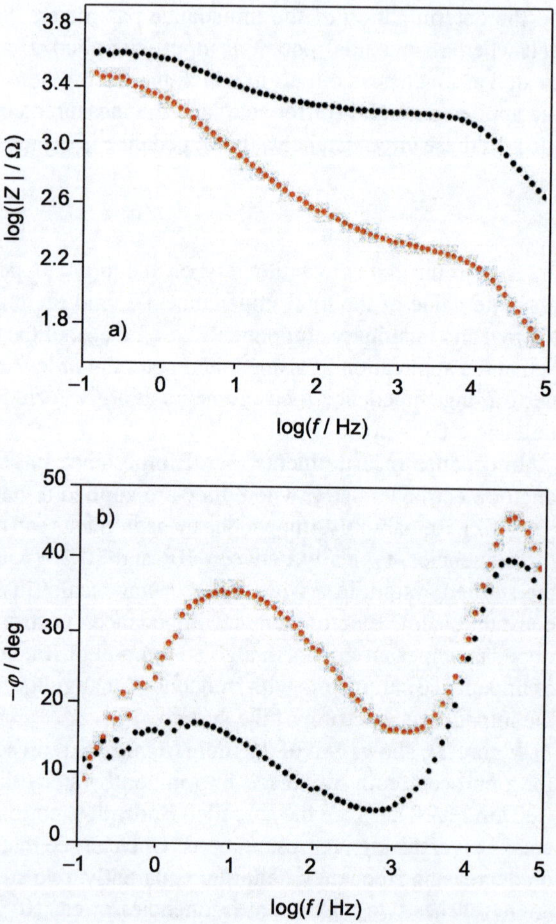

Fig. 2.6 Bode plots of: **a** |Z| versus log*f*, and **b** and $-\varphi$ versus log*f* (c) of the impedance spectra recorded for a eurocent coin immersed into mineral water (black circles) and 0.10 M NaCl (red circles). Frequency range: 0.10–10^5 Hz, peak-to-peak amplitude 10 mV; bias potential − 0.60 V versus Ag/AgCl

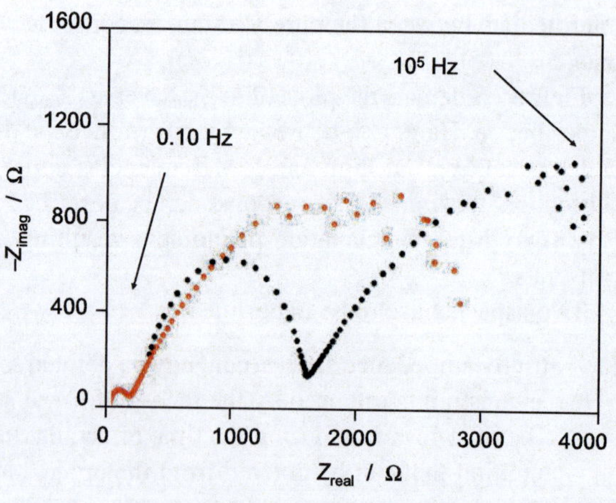

Fig. 2.7 Nyquist representations of $-Z_{imag}$ versus Z_{real} of the impedance spectra recorded for a eurocent coin immersed into mineral water (black circles) and 0.10 M NaCl (red cicles). Frequency range: 0.10–10^5 Hz, peak-to-peak amplitude 10 mV; bias potential − 0.60 V versus Ag/AgCl

transforms [47]. In summary, the essential idea is that EIS experiments should be performed under stationary conditions.

(b) Impedance spectra are in general taken by superimposing the sinusoidal input to a constant potential (bias potential). Frequently, this is the open circuit potential (OCP) of the cell, easily measured with the conventional types of equipment, or the equilibrium potential of a redox probe in the electrolyte, typically, the $[Fe(CN)_6]^{3-}/[Fe(CN)_6]^{4-}$ Under these conditions, the system behaves as an essentially non-faradaic system. However, there is the possibility of performing impedance measurements upon application of a bias potential ensuring the occurrence of faradaic processes. Accordingly, the EIS technique can also take characteristics of a faradaic one.

2.3.2 Circuit Elements and Equivalent Circuits

To acquire information of physicochemical significance, impedance spectra have to be modeled using a given equivalent circuit that: (i) is able to reproduce experimental data in a statistically valid form, and, (ii) is constituted by elements that have a definite physical meaning.

There are five basic circuit elements routinely used in EIS:

(a) Ohmic resistance (R): this element has only a real impedance component equal to the ohmic resistance R. Accordingly, the phase angle is 0.

(b) Capacitor (C): this element possesses only an imaginary impedance component equal to $-j/Cw$, C being the capacity (F) of the capacitor. The phase angle is -90°.

(c) (Auto)induction (L): having only an imaginary impedance component equal to jLw, L being the coefficient of induction (H) of the element. The phase angle is $+90^\circ$. This element, however, rarely appears in electrochemical studies.

(d) Constant phase element (CPE): this element can be interpreted as representative of a non-ideal capacitor. The impedance is expressed as:

$$Z(\omega) = Q_o(j\omega)^{-h} \tag{2.11}$$

where $0 < h < 1$. The ideal capacitor can be viewed as a particular case of CPE when $h = 1$.

(e) Warburg element (W): representative of diffusive effects. Its impedance can be expressed by:

$$Z(\omega) = W(j\omega)^{-1/2} \tag{2.12}$$

expression which is equivalent to Eq. (2.11) taking $h = 1/2$.

The simplest equivalent circuit used for modeling electrochemical cells is illustrated in Fig. 2.8. This is the Randles circuit which is constituted by a resistance R_s

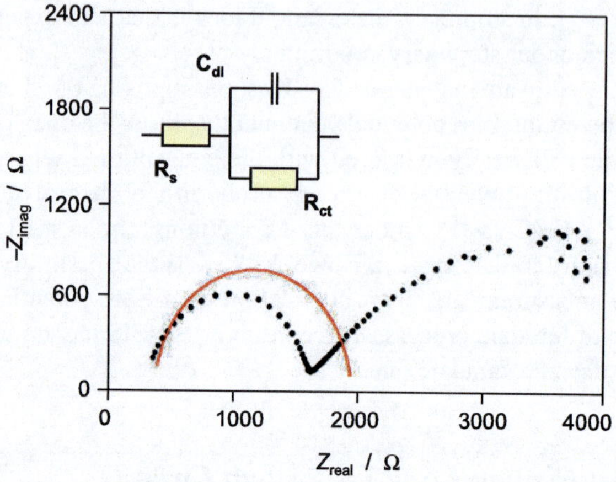

Fig. 2.8 Experimental (black circles) Nyquist plot of the impedance spectra recorded for a eurocent coin immersed into mineral water. The red line corresponds to the 'best' theoretical one using the Randles circuit (inset) using the entire set of data (see text). Frequency range: 0.10–10^5 Hz, peak-to-peak amplitude 10 mV; bias potential − 0.60 V versus Ag/AgCl

in series with a parallel combination of a second resistance R_{ct} and a capacitor C_{dl}. The first element, R_s, is termed the solution resistance and is representative of the ohmic resistance due to the electrolyte solution and the system of metallic conductors and connections in the cell. R_{ct} is the charge transfer resistance (often called electronic resistance), being representative of the potential barrier associated with the transfer of electrons through the electrode/electrolyte interface. This potential barrier is expressed, in terms of circuit elements, as an ohmic resistance. Finally, C_{dl} is the double-layer capacitance, representative of the charge separation produced in the vicinity of the electrode surface associated with the space distribution of ionic species.

The real and imaginary components of the total impedance are:

$$Z_{real} = R_s + \frac{R_{ct}}{1 + R_{ct}^2 C_{dl}^2 \omega^2} \tag{2.13}$$

$$Z_{imag} = -\frac{jR_{ct}^2 C_{dl} \omega}{1 + R_{ct}^2 C_{dl}^2 \omega^2} \tag{2.14}$$

The Nyquist representation consists of a depressed loop whose diameter can be taken as the charge transfer resistance. In turn, the Bode plot of |Z| versus log*f* approximates an inverted "*s*" where the limiting value of the total impedance at high frequencies, Z_{high}, is a measure of the solution resistance. At intermediate frequencies, this representation approximates a straight line of slope − 1.

This equivalent circuit solely contains ohmic and capacitive contributions, so that the phase angle will range between – 90° and 0°. Then, to facilitate the reading, the Nyquist and Bode plots use the sign convention previously indicated.

As can be seen in Fig. 2.8, the experimental data for a eurocent coin immersed into mineral water cannot be satisfactorily reproduced by the Randles equivalent circuit whose 'best' fitting (derived from iterations minimizing the square root of the sum of the squares of the differences between experimental and calculated total impedances at all frequencies) clearly differs from the experimental spectrum, being unable to reproduce the double loop appearing in the Nyquist plot. The Randles circuit, however, can be taken as the basis for subsequent models including new circuit elements. The simplest modification consists of the substitution of the ideal capacitor by a CPE. In this case, the Nyquist and Bode representations are similar to those previously described, the slope of the almost linear region at intermediate frequencies in the |Z| versus log*f* plot being equal to – *h*.

A second frequent modification consists of the addition of a Warburg element in series with R_{ct}. This operates when the diffusive effects in the cell significantly influence the rate of the overall process. The presence of this element yields linear branches in the Nyquist plot, as can be seen in Fig. 2.9.

Although without entirely satisfactory fitting with the experimental impedance spectrum, the introduction of this new circuit element produces a two-segment response. Roughly, this can be summarized as saying that the model circuit contains two time constants associated to the capacitor and the Warburg element. Further improvement can be obtained using the equivalent circuit depicted in Fig. 2.10. The presence of two capacitors is reflected in the two loops in the Nyquist representation, the two inflections in the |Z| versus log*f* plot and the two maxima in the – φ versus

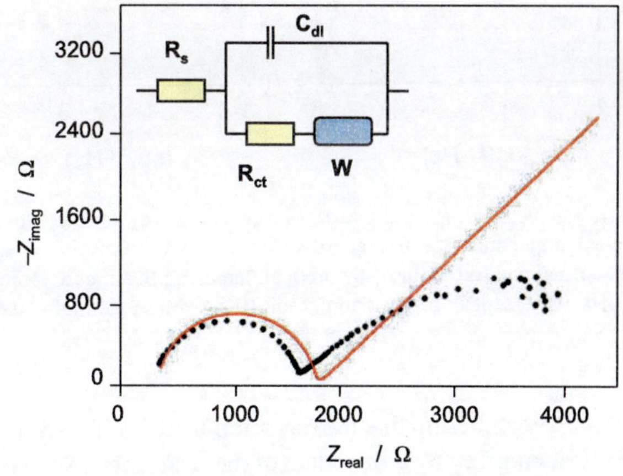

Fig. 2.9 Experimental (black circles) Nyquist plot of the impedance spectra recorded for a eurocent coin immersed into mineral water. The red line corresponds to the 'best' theoretical one using the Warburg-modified Randles circuit (inset). Frequency range: 0.10–10^5 Hz, peak-to-peak amplitude 10 mV; bias potential – 0.60 V versus Ag/AgCl

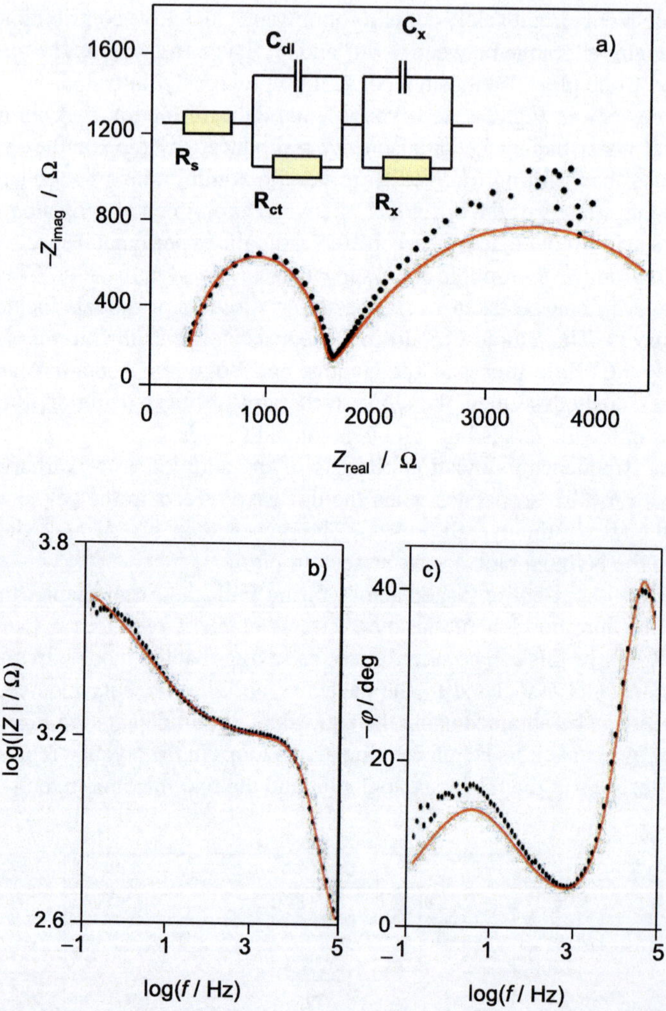

Fig. 2.10 **a** Nyquist, **b** |Z| versus log*f*, and **c** and − *φ* versus log*f* (c) representations of the experimental (black circles) impedance spectra recorded for a eurocent coin immersed into mineral water. The red lines correspond to the 'best' theoretical one using the circuit depicted in the inset. Frequency range: 0.10–10^5 Hz, peak-to-peak amplitude 10 mV; bias potential − 0.60 V versus Ag/AgCl

log*f* representation. Remembering that the resistor plus capacitor circuits are characterized by a time constant given by the product of the resistance by the capacitance (τ = RC), these features are representative of the existence of two time constants which can be associated to two relaxation phenomena occurring at different time scales. Although the 'best' theoretical spectrum does not entirely agree with the experimental one, there is an obvious improvement relative to previous simulations. Examples of more complex impedance spectra modeling will be seen in Chaps. 10 and 11.

2.4 Other Techniques

2.4.1 Combination with Non-electrochemical Techniques

There are numerous electrochemical techniques potentially useful in the analysis of cultural heritage goods. Apart from those briefly mentioned in this chapter, electrochemical quartz crystal microbalance [48], open circuit potential [49, 50], and electrochemical noise measurements [51], among others are of potential interest in the field of cultural heritage. Another important field derives from the combination of hybridization of electrochemical techniques with non-electrochemical ones. Studies involving AFM [40, 41], X-ray diffraction [52], among others [25] have been reported.

2.4.2 Scanning Electrochemical Microscopy

Scanning electrochemical microscopy (SECM) can be seen as a technique of surface analysis directly inspired by scanning electron microscopy although with relevant differences between them. In its usual version, SECM involves the record of the current flowing through a microelectrode or ultramicroelectrode tip (typically between 1 and 25 μm diameter) moving horizontally and vertically relative to the surface of the substrate material under study, both being in contact with a suitable electrolyte. The system is complemented with reference and auxiliary electrodes thus acting as a four-electrode cell.

In the basic operation mode, a redox probe such as $[Fe(CN)_6]^{3-}/[Fe(CN)_6]^{4-}$ is placed in the electrolyte solution while a constant potential difference is established between the tip and the reference electrode. This is the tip potential (E_T) that is fixed at a value at which the redox probe is oxidized or reduced under diffusion control. When the tip is placed at a distance of a few micrometers above the surface, the current passing through the microelectrode depends on the distance to the surface at each point and the conductivity of this zone. Accordingly, displacing horizontally the tip one obtains an electrochemical topography of the surface [53].

Applying a given potential input (E_S) to the substrate, one can modulate its electrochemical response, for instance, promoting reductive or oxidative processes, which in turn affect the tip response. This permits a variety of combinations of E_T and E_S yielding information on the electrochemical reactivity of the substrate.

Figure 2.11 shows the color map (a) and the SECM topographic image (b) obtained for a deposit of microparticles of Naples yellow, a well-known lead pigment (lead(II)

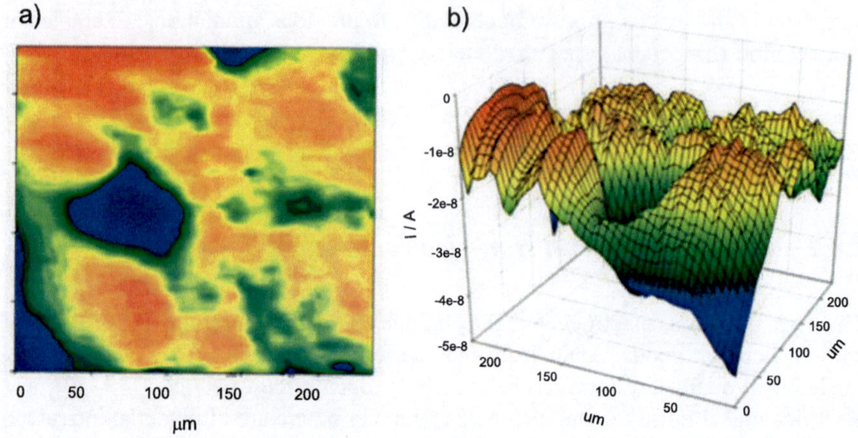

Fig. 2.11 SECM color map (**a**) and topographic image (**b**) of a microparticulate deposit of Naples yellow (Kremer K43100) onto a Pt electrode immersed into 2.0 mM $K_4[Fe(CN)_6]$ plus 0.25 M HAc/NaAc, pH 4.75. $E_T = +0.30$ V; $E_S = 0.00$ V. Adapted from Ref. [54], with permission

antimoniate, $Pb_3(SbO_4)_2$), placed onto a Pt electrode using 2.0 mM $K_4[Fe(CN)_6]$ plus 0.25 M HAc/NaAc, pH 4.75 as the electrolyte solution [54]. Here, no potential was applied to the surface, whereas the tip was submitted to a potential of 0.30 V versus Ag/AgCl. At this potential, $[Fe(CN)_6]^{4-}$ is oxidized under diffusive control so that the tip current shows positive feedback features in the vicinity of conducting regions and negative feedback features when passing near the non-conducting ones.

Accordingly, the blue and green zones in Fig. 2.9 correspond to the regions where the base Pt electrode is directly exposed to the electrolyte solution; the yellow, orange, and red regions represent the regions covered by the pigment particles whose topography is traced by the tip. The *z*-axis in Fig. 2.9b represents the tip current (I_T) increasing from the upper to the lower part of the diagram. To convert I_T values into 'true' distances between the tip and the surface of the pigment particles, there is a need for the previous calibration of the tip current/distance relationship under the used experimental conditions [53] (Fig. 2.11).

2.5 A Note on Thermochemical Calculations

Electrochemical data are the primary source of much thermochemical parameters and, conversely, thermodynamic data are relevant for electrochemistry. For these purposes, electrochemical reversibility is needed. In this section, we will briefly describe several calculations dealing with the topics treated in the book. The starting point can be the following example of an electrochemical process,

$$Cu^{2+}{}_{aq} + 2e^- \rightarrow Cu_{solid} \tag{2.15}$$

The Nernst equations relate the standard electrode potential, $E°$, with the variation of Gibbs free energy involved in this process, $\Delta G°$,

$$\Delta G° = -nFE° \tag{2.16}$$

where n is the number of transferred electrons reflecting the electron stoichiometry of the process ($n = 2$ in our example) and F the Faraday constant. The Nernst equation expresses the equilibrium potential of the redox couple in arbitrary conditions as,

$$E_{eq}(Cu^{2+}/Cu) = E°(Cu^{2+}/Cu) + \frac{RT}{nF}\ln\left(\frac{a_{Cu2+}}{a_{Cu}}\right) \tag{2.17}$$

Here, a_{Cu2+} and a_{Cu} are the thermochemical activities of $Cu^{2+}{}_{aq}$ and Cu_{solid} species. The latter can be taken as unity (activity of a pure phase) and, in general, activities are approximated by concentrations so that, at 1 atm (or, essentially equivalent, 1 bar) and 298 K,

$$E_{eq}(Cu^{2+}/Cu) = E°(Cu^{2+}/Cu) + \frac{RT}{nF}\log[Cu^{2+}] \tag{2.18}$$

Tables of standard potentials provide $E°(Cu^{2+}/Cu) = 0.34$ V versus NHE (normal hydrogen electrode). This value corresponds to the Gibbs free energy for the process described by Eq. (2.15), for which the well-known thermodynamic definitions yield,

$$\Delta G° = -\Delta G_f^{\circ}(Cu^{2+}{}_{aq}) = 65.6\ \text{kJ mol}^{-1} \tag{2.19}$$

$\Delta G_f^{\circ}(Cu^{2+}{}_{aq})$ being the Gibbs free energy of formation of $Cu^{2+}{}_{aq}$.

Now, let us consider an electrochemical process such as,

$$PbO_{solid} + 2H^{+}{}_{aq} + 2e^{-} \rightarrow Pb_{solid} + H_2O \tag{2.20}$$

Remembering that the Gibbs free energies of formation of $H^{+}{}_{aq}$ and Pb_{solid} are zero, the standard potential of the PbO/Pb couple can be calculated from:

$$\Delta G°(PbO/Pb) = \Delta G_f^{\circ}(H_2O) - \Delta G_f^{\circ}(PbO) = -2FE°(PbO/Pb) \tag{2.21}$$

Taking $\Delta G_f^{\circ}(H_2O) = -237.2$ kJ mol^{-1} and ΔG_f° (PbO, litharge) $= -188.6$ kJ mol^{-1}, one obtains, $E°(PbO/Pb) = 0.25$ V versus NHE. Note that this is the equilibrium potential of the system when all species have unity activity; i.e., pH = 0.0 and PbO and Pb forming separate solid phases. For our purposes, a convenient expression is that providing the variation of the equilibrium potential of the couple with the pH. The equilibrium potential in V versus NHE at 298 K is,

$$E_{eq}(PbO/Pb) = E°(PbO/Pb) - (0.059\ \text{V})\,\text{pH} \tag{2.22}$$

These parameters can also be related to the solubility equilibrium of PbO,

$$PbO_{solid} + 2H^{+}{}_{aq} \rightarrow Pb^{2+}{}_{aq} + H_2O \tag{2.23}$$

and the well-known electron transfer process,

$$Pb^{2+}{}_{aq} + 2e^{-} \rightarrow Pb_{solid} \tag{2.24}$$

Combining the above equations, one obtains,

$$\Delta G_s(PbO) = -RT \ln K_s(PbO) = E°(PbO/Pb) - E°(Pb^{2+}/Pb) \tag{2.25}$$

where K_s(PbO) is the product of solubility of PbO. Taking the tabulated value of $E°(Pb^{2+}/Pb) = -0.126$ V, one obtains $K_s(PbO) = 1.5 \times 10^{-4}$. This solubility product is representative of the equilibrium described by Eq. (2.22) occurring at pH 0.0. Naturally, the calculations can be inverted and starting from the solubility product obtaining the equilibrium potentials.

It is interesting to note that for processes such as described by Eq. (2.19), the Nernst equation reflects the dependence of the equilibrium potential on the pH and the thermodynamic activities of the solid phases:

$$E_{eq}(PbO/Pb) = E°(PbO/Pb) + \frac{(0.059\ \text{V})}{2} \log\left(\frac{a_{PbO}}{a_{Pb}}\right) - (0.059\ \text{V})\text{pH} \tag{2.26}$$

Here, two situations are possible: If PbO and Pb segregate as separate, 'pure' solid phases, their activities will be unity and the activities in Eq. (2.25) cancel. However, if PbO and Pb form a unique, mixed solid phase, the ratio between their respective thermochemical activities can be expressed by the ratio of their respective molar fractions in the solid solution [55], x_{PbO}, x_{Pb}, so that,

$$E_{eq}(PbO/Pb) = E°(PbO/Pb) + \frac{(0.059\ \text{V})}{2} \log\left(\frac{x_{PbO}}{x_{Pb}}\right) - (0.059\ \text{V})\text{pH} \tag{2.27}$$

A practical aspect to be accounted for is the conversion of electrode potentials between the normal hydrogen electrode and other scales. The typical case is the Ag/AgCl reference electrode, constituted by a silver wire surrounded by solid AgCl in contact with a KCl aqueous solution. Here, the redox process is,

$$AgCl_{solid} + e^{-} \rightarrow Ag_{solid} + Cl^{-}{}_{aq} \tag{2.28}$$

The standard potential is 0.223 V versus NHE, corresponding to the equilibrium potential at 1 atm and 298 K when the activity of chloride ions is unity. The equilibrium potential will depend on the concentration (activity) of chloride ions:

$$E_{eq}(\mathrm{AgCl/Ag}) = E^{\circ}(\mathrm{AgCl/Ag}) + (0.059\ \mathrm{V})\log[\mathrm{Cl}^{-}] \tag{2.29}$$

Thermochemical calculations can provide information on the reversibility of the processes under study, when a comparison between experimental potentials and those calculated from thermodynamic data is made. For instance, the equilibrium potential for the reduction of litharge to lead in contact with aqueous buffer at pH 5.0, will be (assuming separate solid phases) of – 0.04 V versus NHE; i.e., – 0.27 V versus Ag/AgCl. In contrast, experimental voltammetric data (see Fig. 2.4 and Chap. 3) yield peak potentials of ca. – 0.5 V versus Ag/AgCl. The difference between these values suggests that there are deviations from electrochemical reversibility in the voltammetric response. Further, it is possible that the calculations did not take into account side equilibria, so that the thermodynamic calculations yielded wrong results. In fact, in the case of oxidative dissolution of metals, the formation of complexes of metal ions has to be accounted for. For instance, the oxidation of lead metal in acetate buffer has to be represented as,

$$\mathrm{Pb}_{\mathrm{solid}} + z\mathrm{Ac}^{-}{}_{\mathrm{aq}} \rightarrow \mathrm{Pb(Ac\text{-})}_{z}^{(2-z)+}{}_{\mathrm{aq}} + 2\mathrm{e}^{-} \tag{2.30}$$

rather than by the inverse of Eq. (2.23). Then, the equilibrium potential depends on the complexation equilibria constants, β_z, and the concentration of acetate ion in the electrolyte solution.

In summary, the application of thermochemical data to interpret experimental electrochemistry data has to be made by considering all chemical equilibria, and only then one may judge about effects of irreversibility. Additionally, in the case of slightly soluble compounds, the slow dissolution rate may be influential.

References

1. Bard AJ, Faulkner L (2022) Electrochemical methods: fundamentals and applications, 3rd edn. Wiley, New York
2. Hamann CH, Hamnett A, Vielstich W (2007) Electrochemistry, 2nd edn. Wiley VCH, Weinheim
3. Bagotsky VS (2006) Fundamentals of electrochemistry, 2nd edn. Wiley-Interscience, Hoboken
4. Rosemberg GA (1917) Antiquités en fer et en bronze: leur transformation dans la terre contenant de l'acide carbonique et des chlorures et leur conservation. Copenhagen
5. Rathgen F (1924) Die Konservierung von Altertumsfunden: II und III Teil. Walter de Gruyter, Berlin
6. Plenderleith HJ (1956) The conservation of antiquities and works of art. Oxford University Press, London
7. Stambolov T (1967) The corrosion and conservation of metallic antiquities and works of art. Central Research Laboratory for Objects of Art and Science, Amsterdam

8. Organ RM (1968) Design for scientific conservation of antiquities. IIC/Butterworths, London
9. Barrio J, Chamón J, Pardo AI, Arroyo M (2009) Electrochemical techniques applied to the conservation of archaeological metals from Spain: a historical review. J Solid State Electrochem 13:1767–1776
10. Doménech-Carbó A, Doménech-Carbó MT, Costa V (2009) Electrochemical methods in archaeometry, conservation and restoration (monographs in electrochemistry series, Scholz F, ed). Springer, Berlin-Heidelberg
11. Scholz F, Meyer B (1994) Electrochemical solid state analysis—state of the art. Chem Soc Rev 23:341–347
12. Scholz F (2021) Glazunov's electrography—the first electrochemical imaging and the first solid-state electroanalysis. J Solid State Electrochem 25:2705–2715
13. Bauer D, Gaillochet MP (1974) Etude du comportement de la pate de carbone a compose electroactif incorporé. Electrochim Acta 19:597–606
14. Brainina KhZ, Lesunova RP (1974) Use of a paste electrode for electrochemical phase analysis 1. Theoretical aspects. Formation of soluble products. Zh Anal Khim 29:1302–1308
15. Brainina KhZ, Vidrevich MB (1981) Stripping analysis of solids. J Electroanal Chem 121:1–28
16. Scholz F, Nitschke L, Henrion G (1989) A new procedure for fast electrochemical analysis of solid materials. Naturwiss 76:71–72
17. Nitschke L, Henrion G, Damaschun F, Scholz F (1989) A new technique to study the electrochemistry of minerals. Naturwiss 76:167–168
18. Scholz F, Nitschke L, Henrion G (1989) Identification of solid materials with a new electrochemical technique—the abrasive stripping voltammetry -. Fresenius Z Anal Chem 334:56–58
19. Scholz F, Lange B, Jaworski A, Pelzer J (1991) Analysis of powder mixtures with the help of abrasive stripping voltammetry and coulometry. Fresenius' J Anal Chem 340:140–144
20. Lovrić M (2002) Square-wave voltammetry. In: Scholz F (ed) Electroanalytical methods. Springer, Berlin
21. Doménech-Carbó A (2010) Voltammetric methods applied to identification, speciation, and quantification of analytes from works of art: an overview. J Solid State Electrochem 14:363–379
22. Doménech-Carbó A (2011) Tracing, authentifying and dating archaeological metal using the voltammetry of microparticles. Anal Meth 3:2181–2188
23. Doménech-Carbó A, Doménech-Carbó MT (2018) Electroanalytical techniques in archaeological and art conservation. Pure Appl Chem 90:447–462
24. Scholz F, Meyer B (1998) Voltammetry of solid microparticles immobilized on electrode surfaces. In: Bard AJ, Rubinstein I (eds) Electroanalytical chemistry, A series of advances, vol 20. Marcel Dekker, New York, pp 1−86
25. Scholz F, Schröder U, Gulabowski R, Doménech-Carbó A (2014) Electrochemistry of immobilized particles and droplets, 2nd edn. Springer, Berlin-Heidelberg
26. Doménech-Carbó A, Labuda J, Scholz F (2013) Electroanalytical chemistry for the analysis of solids: characterization and classification (IUPAC technical report). Pure Appl Chem 85:609–631
27. Lovrić M, Scholz F (1997) A model for the propagation of a redox reaction through microcrystals. J Solid State Electrochem 1:108–113
28. Lovrić M, Hermes M, Scholz F (1998) The effect of the electrolyte concentration in the solution on the voltammetric response of insertion electrodes. J Solid State Electrochem 2:401–404
29. Oldham KB (1998) Voltammetry at a three-phase junction. J Solid State Electrochem 2:367–377
30. Lovrić M, Scholz F (1999) A model for the coupled transport of ions and electrons in redox conductive microcrystals. J Solid State Electrochem 3:172–175
31. Schröder U, Oldham KB, Myland JC, Mahon PJ, Scholz F (2000) Modelling of solid state voltammetry of immobilized microcrystals assuming an initiation of the electrochemical reaction at a three-phase junction. J Solid State Electrochem 4:314–324
32. Doménech-Carbó A (2020) Electrochemistry of porous materials, 2nd edn., Chap. 3. Taylor & Francis, Boca Raton

33. Scholz F, Lovrić M, Stojek Z (1997) The role of redox mixed phases $\{ox_x(C_n\ red)_{1-x}\}$in solid state electrochemical reactions and the effect of miscibility gaps in voltammetry. J Solid State Electrochem 1:134–142
34. Dostal A, Meyer B, Scholz F, Schröder U, Bond AM, Marken F, Shaw SJ (1995) Electrochemical study of microcrystalline solid Prussian blue particles mechanically attached to graphite and gold electrodes: electrochemically induced lattice reconstruction. J Phys Chem 99:2096–2103
35. Grandjean F, Long GJ, Samian L (2012) The pivotal role of Mössbauer spectroscopy in the characterization of Prussian blue and related iron cyanide complexes. Mössbauer Effect Ref Data J 35:143–153
36. Doménech-Carbó A, Doménech-Carbó MT, Osete-Cortina L, Donnici M, Guasch-Ferré N, Gasol RM, Iglesias MA (2020) Electrochemical assessment of pigments-binding medium interactions in oil paint deterioration: a case study on indigo and Prussian blue. Herit Sci 8:71
37. Scholz F, Doménech-Carbó A (2019) The thermodynamics of insertion electrochemical electrodes—a team play of electrons and ions across two separate interfaces. Angew Chem Int Ed 58:3279–3284
38. Doménech-Carbó A (2012) Solvent-independent electrode potentials of solids undergoing insertion electrochemical reactions: part I. Theory J Phys Chem C 116:25977–25983
39. Grygar T (1996) The electrochemical dissolution of iron(III) and chromium(III) oxides and ferrites under conditions of abrasive stripping voltammetry. J Electroanal Chem 405:117–125
40. Grygar T (1998) Phonomenological kinetics of irreversible electrochemical dissolution of metal-oxide microparticles. J Solid State Electrochem 2:127–136
41. Doménech-Carbó A, Doménech-Carbó MT, Valle-Algarra FM, Gimeno-Adelantado JV, Osete-Cprtina L, Bosch-Reig F (2016) On-line database of voltammetric data of immobilized particles for identifying pigments and minerals in archaeometry, conservation and restoration (ELCHER database). Anal Chim Acta 927:1–12
42. Grygar T, Bezdicka P, Hradil D, Doménech A, Marken F, Pikna L, Cepriá G (2002) Voltammetric analysis of iron pigments. Analyst 127:1100–1107
43. Zakharchuk N, Meyer S, Lange B, Scholz F (2000) A comparative study of lead oxide modified graphite paste electrodes and solid graphite electrodes with mechanically immobilized lead oxides. Croat Chem Acta 73:667–704
44. Doménech-Carbó A, Doménech-Carbó MT, Mas-Barberá X (2007) Identification of lead pigments in nanosamples from ancient paintings and polychromed sculptures using voltammetry of nanoparticles/atomic force microscopy. Talanta 71:1569–1579
45. Hasse U, Scholz F (2001) In situ atomic force microscopy of the reduction of lead oxide nanocrystals immobilised on an electrode surface. Electrochem Commun 3:429–434
46. Retter H, Lohse H (2005) Electrochemical Impedance Spectroscopy. In: Scholz F (ed) Electroanalytical methods. Springer, Berlin
47. Macdonald DD, Sikora A, Engelhardt G (1998) Characterizing electrochemical systems in the frequency domain. Electrochim Acta 43:87–107
48. Doménech-Carbó A, Sánchez-Ramos S, Yusá-Marco DJ, Moya-Moreno M, Gimeno-Adelantado JV, Bosch-Reig F (2004) Determination of the boron/lead ratio in ceramic materials based on electrochemical quartz crystal microbalance. Electroanalysis 16:1814–1822
49. Sassolini A, Colozza N, Papa E, Hermansson K, Cacciotti I, Arduini F (2019) Screen-printed electrode as a cost-effective and miniaturized analytical tool for corrosion monitoring of reinforced concrete. Electrochem Commun 98:69–72
50. Doménech-Carbó A, Peiró-Ronda MA, Vives-Ferrándiz J, Duffó GS, Farina S (2021) 'Dry' electrochemistry: a non-invasive approach to the characterization of archaeological iron objects. Electrochem Commun 125:106992
51. Martínez-Lázaro I, Doménech-Carbó A, Doménech-Carbó MT, Pastor MT, Amigó V (2010) Electrochemical criteria for evaluating conservative treatments applied to contemporary metallic sculpture. A case study. J Solid State Electrochem 14:437–447
52. Meyer B, Ziemer B, Scholz F (1995) In situ X-ray diffraction study of the electrochemical reduction of tetragonal lead oxide and orthorhombic Pb(OH)Cl mechanically immobilized on a graphite electrode. J Electroanal Chem 392:79–83

53. Bard AJ, Mirkin MV (eds) (2001) Scanning electrochemical microscopy. Marcel Dekker, New York-Basel
54. Doménech-Carbó A, Doménech-Carbó MT, Silva M, Valle-Algarra FM, Gimeno-Adelantado JV, Bosch-Reig F, Mateo-Castro R (2015) Screening and mapping pigments in paintings using scanning electrochemical microscopy (SECM). Analyst 140:1065–1075
55. Jaworski A, Stojek Z, Scholz F (1993) A comparison of simulated and experimental abrasive stripping voltammetric curves of ionic crystals: reversible case. J Electroanal Chem 354:1–9

Additional Literature

56. Blum D, Leyffer W, Holze R (1996) Pencil-leads as new electrodes for abrasive stripping voltammetry. Electroanalysis 8:296–297
57. Castelló-Palacios A, Montoya N, Doménech-Carbó A, Guerola-Blay V, Pérez-Marín E, Doménech-Carbó MT (2022) Valencian red imprimaturs since Francisco Ribalta (1565–1628). J Cult Herit 53:184–189
58. Cisternas R, Kahlert H, Wulff H, Scholz F (2015) The electrode response of a tungsten bronze electrode differ in potentiometry and voltammetry and give access to the individual contributions of electron and proton transfer. Electrochem Commun 56:34–37
59. Diard JP, Montilla C (2014) Non-intuitive features of equivalent circuits for analysis of EIS data. The example of EE reaction. J Electroanal Chem 735:99–110
60. Doménech-Carbó A, Montoya N, Scholz F (2011) Estimation of individual Gibbs energies of cation transfer employing the insertion electrochemistry of solid Prussian blue. J Electroanal Chen 657:117–122
61. Doménech-Carbó A, Koshevoy IO, Montoya N, Pakkanen TA (2011) Estimation of free energies of anion transfer from solid-state electrochemistry of alkynyl-based Au(I) dinuclear and Au(I)-Cu(I) cluster complexes containing ferrocenyl groups. Electrochem Commun 13:96–98
62. Doménech-Carbó A, Koshevoy IO, Montoya N, Karttunen AJ, Pakkanen TA (2011) Determination of individual Gibbs energies of anion transfer and excess Gibbs energies using an electrochemical method based on insertion electrochemistry of solid compounds. J Chem Eng Data 56:4577–4586
63. Doménech-Carbó A, Koshevoy IO, Montoya N (2016) Separation of the Ionic and electronic contributions to the overall thermodynamics of the insertion electrochemistry of some solid Au(I) complexes. J Solid State Electrochem 20:673–681
64. Dostal A, Kauschka G, Reddy SJ, Scholz F (1996) Lattice contractions and expansions which accompany the electrochemical conversion of Prussian blue and the reversible and irreversible insertion of rubidium and thallium ions. J Electroanal Chem 406:155–1563
65. Gozález-Meza OA, Larios-Durán ER, Gutiérrez-Becerra A, Casillas N, Escalante JI, Bárcena-Soto M (2019) Development of a Randles-Ševčík-like equation to predict the peak current of cyclic voltammetry for solid metal hexacyanoferrates. J Solid State Electrochem 23:3123–3133
66. Gritzner G (1998) Single-ion transfer properties: a measure of ion solvation in solvents and solvent mixtures. Electrochim Acta 44:73–83
67. Grygar T (1995) Kinetics of electrochemical reductive dissolution of iron(III) hydroxy-oxides. Collect Czech Chem Commun 60:1261–1273
68. Grygar T, Subrt J, Bohaceck J (1995) Electrochemical dissolution of goethite by abrasive stripping voltammetry. Collect Czech Chem Commun 6:950–959
69. Grygar T (1996) Electrochemical dissolution of iron(III) hydroxy-oxides: more information about the particles. Coll Czech Chem Commun 61:93–106
70. Grygar T, Marken F, Schröder U, Scholz F (2002) Voltammetry of microparticles: a review. Coll Czech Chem Commun 67:163–208

71. Hermes M, Lovrić M, Hartl M, Retter U, Scholz F (2001) On the electrochemically driven formation of bilayered systems of solid Prussian-blue-type metal hexacyanoferrates: a model for Prussian blue cadmium hexacyanoferrate supported by finite difference simulations. J Electroanal Chem 501:193–204
72. Scholz F, Dostal A (1995) The formal potentials of the solid metal hexacyanometalates. Angew Chem Int Ed 34:2685–2687

Chapter 3
Voltammetry: The Essentials

3.1 General Aspects

The major part of electrochemistry is focused on the study of processes involving the transfer of electrons through the interface between an electronic conductor (electrode) and an ionic conductor (electrolyte). The analytical applications of electrochemistry in the field of cultural heritage are mainly based on voltammetric measurements. The theoretical aspects of these signals have been extensively studied in solution electrochemistry [1, 2], and their essential results are summarized here.

Voltammetry is an instrumental technique widely used in analysis and for the characterization of redox active species in solution. It relies on recording of the current passing through a two- or three-electrode cell upon application of a time-dependent potential difference between the working and counterelectrodes (i.e., a potential difference between the working electrode and the solution phases). In two-electrode cells, the counterelectrode acts also as reference electrode. In 3-electrode cells, a third electrode, the reference electrode is added and the potential of the working electrode is set against the constant potential of the reference electrode, through which practically no current flows. The role of this electrode is manifold. It allows to use potentiostats and compensation of solutions resistance, and it provides a constant reference potential, which is independent of the current flowing through the working and counterelectrodes. Practically, no current flows through the reference electrode. In most cases, voltammetry is applied to register faradaic signals, where there are changes in the composition ('electrolysis') of the system, i.e., at least at the working electrode | solution interface. Among others, gas evolution or metal deposition/dissolution processes can occur.

The resulting current–potential graph is called voltammogram. Depending on the time dependence of the potential, one can define different techniques. As briefly commented in Chap. 2, the two basic groups of voltammetric techniques are constituted by linear potential scan techniques (linear and cyclic voltammetry) and pulsed techniques (differential pulse voltammetry, square wave voltammetry, …).

A. Doménech-Carbó and M. T. Doménech-Carbó, *Electrochemistry for Cultural Heritage*, Monographs in Electrochemistry,
https://doi.org/10.1007/978-3-031-31945-7_3

Chronoamperometry can be seen to some extent as a limiting case of the above when a single rectangular pulse is applied to the cell.

3.2 The Electrochemical Reaction

The electrochemical processes taking place during the performance of a voltammetric experiment involve *charge transfer* at the working electrode|solution interface and concomitantly *charge transport* through the solution. The design of the electrochemical cell and the use of a potentiostat makes sure that the current is controlled only by the electrochemical processes at the working electrode. If the metal wiring and the electrode material do not possess relevant ohmic resistance, the current is controlled by the electrochemical reaction at the working electrode | electrolyte interface (coupled to mass transport, possibly also to the rate of chemical reactions in the near-electrode space). Only in unfavorable cases, the ohmic resistance of the electrolyte solution may affect the current–potential response.

When the ohmic resistance of the electrolyte solution is high, the potential gradient at the working electrode|solution interface may reach out into the solution. This will prompt migration of electroactive species, if they are ions (possess a charge). Depending on the charge and the electrode potential, migration can enhance or depress the current through the electrode. Similar to the *Debye length* around ions, at electrodes a so-called *Gouy length* has been defined, which depends on the charge on the electrode surface (i.e., the potential) and the electrolyte concentration [3]. Migration is the working modus of electrophoresis, a technique which is conventionally (and unfounded) not considered as an electroanalytical but a separation technique, as separation of ions is the main goal. In voltammetric measurements, migration is mostly undesired. To avoid it, a high concentration of ions is needed, which are capable of charge transport in the solution phase, but do not undergo oxidation or reduction reactions at the working electrodes: this is the supporting electrolyte. When the mass transport to the electrode surface is controlled by the concentration gradient of the electroactive species, diffusion control is operative. If the electrolyte solution is stirred (or rotating electrodes are used), the transport of electrolyte species is additionally driven by the gradients of momentum, which is then called the convective diffusion regime. Voltammetric experiments as described here are generally performed in quiescent solutions containing a supporting electrolyte in sufficiently large concentration (typically, between 0.1 and 1 mol L^{-1}) to ensure that the transport of the electroactive species is under diffusion control. Note that this does not necessarily mean that the rate of the electrochemical process was governed by diffusion phenomena; other kinetic constraints (vide infra) can act as a rate-determining step of the electrochemical process.

The theoretical description of the voltammetric processes is based on the Nernst equations, the Fick diffusion equations, and the condition of electroneutrality [4]. Let us consider an *n*-electron process represented as,

$$Ox_{solv} + ne^- \rightarrow Rd_{solv} \tag{3.1}$$

where the subscript solv indicates that the oxidized (Ox) and reduced (Rd) species are solvated. It is assumed that this process occurs at the working electrode surface when the electrode is polarized, i.e., when a favorable potential difference is established between the working and reference electrodes (i.e., between the working electrode and solution phases). As a result, the composition of the solution in the vicinity of the electrode changes progressively so that time-dependent concentration gradients of Ox and Rd are formed in the region near the electrode surface. In this region, the so-called diffusion layer, the concentration of the redox species varies during the voltammetric measurements. When the thickness of this layer, typically between 0.01 and 0.1 mm, is clearly smaller than the dimension of the working electrode, planar diffusion of the redox species takes place. Microelectrodes of diameters around 10–50 μm can also be used. At these electrodes, semispherical or cylindrical diffusion occurs.

Under conditions of diffusion control at planar macroelectrodes (typical areas between 0.01 and 1.0 cm^2), the semi-infinite boundary conditions apply. Here, the faradaic current, I (typically expressed in μA), for the reduction or oxidation of the electroactive species whose concentration in the solution bulk is c (mol cm^{-3}), and its diffusion coefficient ($cm^2\ s^{-1}$) is D, at a plane electrode, is then given by:

$$I = \pm nFAD\left(\frac{\partial c}{\partial x}\right)_{x=0} \tag{3.2}$$

where A (cm^2) represents the electrode area, n the number of transferred electrons per mol of electroactive species, and F the Faraday constant (9.64853383(83) × 10^{10} μC mol^{-1}). x is the distance (cm) from the electrode surface perpendicular to the same. The current is proportional to the gradient of concentration of electroactive species at the electrode/electrolyte interface.

The electron transfer through the electrode|electrolyte interface is a heterogeneous chemical reaction where the electron, acting as a reactant, must pass a certain potential barrier. Roughly speaking, when the potential barrier is small, the interfacial charge transfer is fast in both the direct and inverse directions and the process exhibits electrochemical reversibility. If there is a large potential barrier, the kinetics is more or less slow and non-reversible o irreversible conditions are attained.

The kinetics of the heterogeneous reaction rate at planar macroelectrodes under the aforementioned conditions can be described by means of the Butler–Volmer equation:

$$j = -nFk^\circ\left[c^\circ{}_{red}\exp\left[\frac{nF(1-\beta)(E-E^{o\prime})}{RT}\right] - c^\circ{}_{ox}\exp\left[-\frac{\beta nF(E-E^{o\prime})}{RT}\right]\right] \tag{3.3}$$

In this equation, j represents the current density (μA cm^{-2}), $c°_{ox}$ and $c°_{red}$ represent, respectively, the concentrations of the oxidized and reduced forms of the electroactive species at the electrode surface. This equation introduces two kinetic parameters, $k°$, the rate constant of the heterogeneous electron transfer process at the standard potential, (cm s^{-1}), and β, the symmetry factor. This last characterizes the symmetry of the energy barrier that has to be surpassed during charge transfer [5, 6]. In the above equation, E represents the applied potential and $E^{o\prime}$ the formal electrode potential (sometimes, and rather rarely, close to the standard electrode potential). The difference $E - E^{o\prime}$ is called the overvoltage. This quantity is a measure of the excess energy imparted to the electrode beyond the equilibrium potential for the reaction [1]. As discussed in detail by Compton, Molina et al. the Butler–Volmer equation is 'phenomenological', i.e., that it was not derived from thermodynamic, fundamental models such as the Marcus–Hush theory [7, 8].

Classically, the coefficient β of an electrochemical reaction consisting of a single elementary step involving the simultaneous uptake of n electrons from the electrode was taken as the product of the number of electrons involved in the rate-determining step, n_a, and the electron transfer coefficient, α. This is defined as,

$$\alpha = \pm\left(\frac{RT}{F}\right)\left(\frac{\mathrm{d}\ln|j|}{\mathrm{d}E}\right) \tag{3.4}$$

the sign depending on the cathodic/anodic nature of the process. The simultaneous transfer of more than one electron is, however, highly improbable and, currently, it is recommended to write β as the sum of the number of electrons n' transferred before the rate-determining step and the electron transfer coefficient α of the latter [9], i.e., $\beta = \alpha + n'$.

3.3 Reversible Solution-Phase Voltammetry Under Diffusion Control

At first, the case of an electrochemical reversible electrode reaction of dissolved redox species will be considered. The term reversibility here means the compliance with the Nernst equation for the surface concentration of the electroactive species. Under semi-infinite diffusive control, cyclic voltammograms consist of a pair of cathodic (reduction) and anodic (oxidation) peaks as illustrated in Fig. 3.1. Here, the cyclic voltammogram recorded at a gold electrode in a 2.0 mM solution of $K_4[Fe(CN)_6]$ in 0.10 M KCl is depicted. In this experiment, the potential varies linearly with time at a rate v between two extreme (or switching) potentials, starting from a given initial potential. In the initial anodic scan, an oxidation peak appears at a peak potential (E_{pa}) ca. 0.25 V which is coupled, upon reversal of the direction of the potential scan, with a reduction peak at a peak potential (E_{pc}) of 0.18 V.

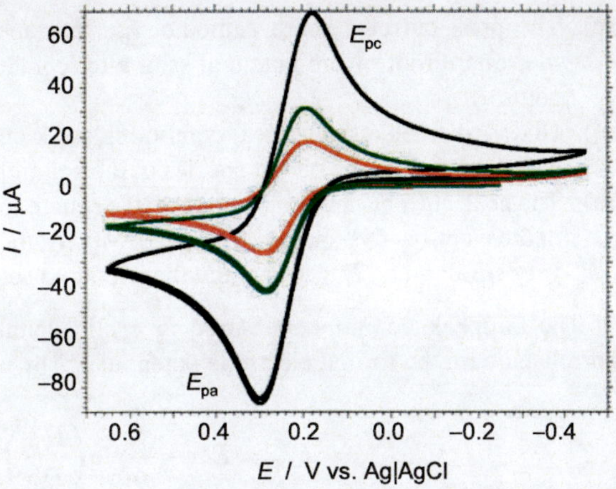

Fig. 3.1 Cyclic voltammogram at gold electrode of a 2.0 mM solution of $K_4[Fe(CN)_6]$ in 0.10 M KCl aqueous solution. Potential scan initiated at − 0.45 V versus Ag/AgCl in the positive direction; potential scan rates of 20 (red), 50 (green), and 100 mV s^{-1} (black). Note that the axes are opposite to the IUPAC convention

The electrochemical process can be described as:

$$[Fe(CN)_6]^{4-}{}_{solv} \leftrightarrows [Fe(CN)_6]^{3-}{}_{solv} + e^- \tag{3.5}$$

The theoretical description of these processes is based on the Nernst equations, the Fick diffusion equations, and the condition of electroneutrality [4]. The resulting current/potential curve requires numerical integration of the constitutive equations. The solutions for planar electrodes take the form [1, 3, 4]:

$$I = nFAcD^{1/2}\left(\frac{nF}{RT}\right)^{1/2} v^{1/2}\Psi(E - E^{\circ\prime}) \tag{3.6}$$

where $\Psi(E - E^{\circ\prime})$ represents a numerical function of the difference between the applied potential and the formal electrode potential of the redox couple. Under the above conditions, the peak current, I_p, is given by the Randles–Ševčik equation [1]:

$$I_p = 0.446nFAc\sqrt{\frac{nFvD}{RT}} \tag{3.7}$$

Several relationships are experimentally observed [1, 10]:

(a) Peak potentials are independent of the potential scan rate (as long as the system behaves electrochemically reversibly). The mid-peak potential ($E_{mp} = (E_{pa} + E_{pc})/2$) can be taken as a formal potential ($E^{\circ\prime}$) close to (but not coincident with) the standard potential of the redox couple.

(b) The peak potential separation ($E_{pa} - E_{pc}$) and the peak to half peak potential separation ($E_p - E_{p/2}$) are 59/n mV regardless of the potential scan rate, v (as long as the system behaves electrochemically reversible).

(c) The peak currents (both cathodic, I_{pc}, and anodic, I_{pa}) are proportional to the square root of the potential scan rate (condition of semi-infinite diffusion control).
(d) Under fixed electrochemical conditions, peak currents are proportional to the concentration of the parent species in the solution.
(e) The peak currents are proportional to the square root of the diffusion coefficients of the electroactive species (D_{ox}, D_{rd}). Mostly, the two diffusion coefficients are very similar, i.e., $D_{ox} \approx D_{rd}$ and the anodic to cathodic peak ratio is ≈ 1.

The midpeak potential measured in cyclic voltammetric experiments can be correlated with the formal electrode potential $E^{o\prime}$ by means of the relationship:

$$E_{mp} = E^{o\prime} - \frac{RT}{nF} \ln\left(\frac{D_{ox}}{D_{rd}}\right)^{1/2} \tag{3.8}$$

In turn, $E^{o\prime}$ can be correlated with the standard electrode potential E° of the redox couple. In favorable cases, E° can reasonably be approximated by E_{mp}.

As previously noted, the concentration of the oxidized and reduced species varies with time and the distance to the electrode. As also previously noted, the concentration gradients are practically established in the diffusion layer [3] so that experiments at disk macroelectrodes occur under the aforementioned semi-infinite planar diffusion conditions. The direct estimate of diffusion coefficients can also be made from chronoamperometric experiments in which the potential is stepped from an initial value where no oxidation or reduction of the electroactive species takes place (i.e., far from $E^{o\prime}$), to a second value cathodic or anodic enough to promote its diffusion-controlled oxidation or reduction. The resulting chronoamperogram is a current–time graph that, for a reversible n-electron transfer process, is then given by the Cottrell equation:

$$I = \frac{nFAcD^{1/2}}{\pi^{1/2}t^{1/2}} \tag{3.9}$$

The thickness of the diffusive layer at a time t, $\delta(t)$, can be approached by the product $(Dt)^{1/2}$.

It has to be underlined that the above expressions correspond to the faradaic currents without consideration of other effects, which will be discussed below. Some of which uncompensated ohmic drops and capacitive double-layer effects always accompany faradaic processes (see below). Of course, there are possibility of successive charge transfer processes [12, 13].

Diffusive effects are conditioned by the geometry and size of the electrode. In the case of microelectrodes, the diffusion occurs under radial hemispherical conditions and the above equations should be modified accordingly [3, 12].

Again, with regard to the cyclic voltammetric experiment in Fig. 3.1, one can state that, roughly speaking, in the region at the foot of the voltammetric peaks the current increases exponentially with the potential. In the region of potentials

150–200 mV after the peak, the reaction rate is controlled by the diffusion of the electroactive species. This last region can be described in terms similar to the familiar chronoamperometric experiment in which a constant potential sufficiently cathodic or anodic to promote the diffusion-controlled reduction or oxidation is applied. Here, the Cottrell equation states the proportionality of the current with the inverse of the square root of time [11]. This same type of relationship applies in the diffusion region of cyclic voltammograms, resulting in the proportionality between currents and the inverse of the square root of the difference between the applied potential and a given threshold potential [14].

3.4 Resistive and Capacitive Effects

The above voltammetric response corresponds exclusively to the faradaic current–potential response in the ideal case in which there is strict reversibility. This treatment, however, ignores the fact that the metal|electrolyte interface acts as a capacitor. Here, the charge can be accumulated via excess or deficit of surface electrons in the metal and excess or deficit of solvated ions in the electrolyte. This last effect is accompanied by the orientation of dipole molecules of the solvent [2].

The polarized metal|electrolyte interface forms the so-called electrochemical double layer which comprises an inner layer of adsorbed and oriented solvent molecules and specifically adsorbed cations (anions), a layer of solvated ions approaching the interface. The anions (cations) in this region approximate the metal surface until a minimum distance, thus defining the so-called outer Helmholtz plane. In turn, the plane passing by the centers of the adsorbed ions defines the inner Helmholtz plane. Accordingly, the charge excess in the diffuse layer decays toward the solution bulk where ionic charges are compensated.

Obviously, the above description is 'static' and should necessarily be refined to adopt a more realistic 'dynamic' view taking into account molecular motion. For our purposes, the aspects to remark are: (i) the potential varies with the distance at the electrode surface through the electrochemical double layer; (ii) the thickness of the layer depends on the ionic charge and concentration, the dielectric constant of the electrolyte solution and the charge of the electrode on the metal surface. The potential of zero charge (EPZ) also depends on the above factors (excluding the surface charge), as this is the potential at which there is no excess charge on either side of the electrochemical double layer.

The formation of the electrochemical double layer causes the appearance of capacitive currents. These currents are associated with its charging/discharging, the double layer being characterized by the so-called differential double layer capacitance, C_{dl} [2]:

$$C_{\mathrm{dl}} = \frac{\mathrm{d}q}{\mathrm{d}E} = \frac{\mathrm{d}q}{\mathrm{d}t}\frac{\mathrm{d}t}{\mathrm{d}E} \tag{3.10}$$

Then, the capacitive current corresponds to the variation of the accumulated charge q with time:

$$I = \frac{dq}{dt} = C_{dl}\frac{dE}{dt} \tag{3.11}$$

Introducing the charge density at the electrode surface, σ, and the electrode area, A, one can write:

$$I = \frac{dq}{dt} = \frac{d(\sigma A)}{dt} = \sigma\frac{dA}{dt} + A\frac{d\sigma}{dt} \tag{3.12}$$

Then, combining Eqs. (3.10) and (3.11):

$$I = \sigma\frac{dA}{dt} + AC_{dl}\frac{dE}{dt} \tag{3.13}$$

Equation (3.13) tells us that in a voltammetric experiment (where the applied potential E always varies with time) a capacitive current appears regardless of the possible time variation of the electrode area.

There are other possible capacitive effects associated with other charge separation phenomena occurring in electrochemical cells. Of particular interest are those associated with the peculiar electrochemistry of semiconductor electrodes, where space charging processes occur [2].

Apart from capacitive effects, the electrochemical cells possess resistive effects due to the ohmic resistance to charge transport through the different phases, especially through the electrolyte solution. This implies that the electrochemical cell behaves in general as an electric circuit with a certain resistance R and a certain capacity C (this last at least being the C_{dl}). In a voltammetric experiment where the potential scan begins at a potential E_{start}, the variation of applied potential E with time can be expressed as [1]:

$$E = E_{start} + vt = R(dq/dt) + q/C \tag{3.14}$$

In this equation, q is the charge passed at a time t and v the potential sweep rate. Accordingly, in the absence of faradaic effects, the net current I flowing at a time t is:

$$I = vC + (E_{start}/R - vC)e^{-\frac{t}{RC}} \tag{3.15}$$

The corresponding current–potential expression is:

$$I = vC + (E_{start}/R - vC)e^{-\frac{E-E_{start}}{vRC}} \tag{3.16}$$

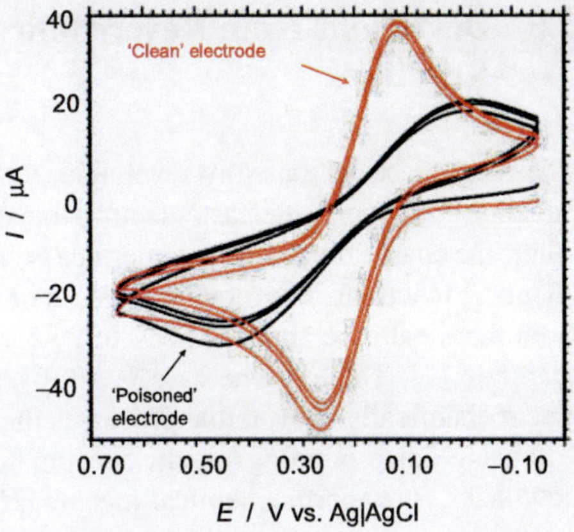

Fig. 3.2 Cyclic voltammograms of a 1.0 mM solution of $K_4[Fe(CN)_6]$ in 0.10 M KCl aqueous solution at: **a** 'clean' and **b** 'poisoned' glassy carbon electrodes. Potential scan initiated at − 0.10 V versus Ag/AgCl in the positive direction; potential scan rate 50 mV s^{-1}. Note that the axes are opposite to the IUPAC convention

Excluding double-layer effects, Eq. (3.16) describes the background current present in voltammetric experiments associated with uncompensated ohmic drops and other than double-layer capacitance effects. This current is superimposed on those associated with faradaic and electrochemical double-layer effects, thus distorting the purely faradaic current/potential response described by Eq. (3.5) for the case of electrochemically reversible, diffusion-controlled processes.

For practical purposes, resistive and capacitive effects result in the increase of peak separation, wave broadening, and peak current decrease. An example is shown in Fig. 3.2 where the cyclic voltammograms recorded on a $[Fe(CN)_6]^{4-}$ solution at a glassy carbon submitted to the usual cleaning and polishing procedures ('clean' electrode), and at a glassy carbon having a previous 'history' resulting in the so-called poisoning of the electrode surface, associated with adsorbates, impurities and/or inclusions. Here, the increased peak separation may also be a sign of some electrochemical irreversibility caused by surface poisoning. The difference between the voltammetric records in Figs. 3.1 and 3.2 is obvious.

Due to the general appearance of disturbing resistive and capacitive effects, for practical purposes, the criteria of electrochemical reversibility can be reformulated as follows:

(a) At small potential scan rates ($v \to 0$), the peak-to-peak potential separation tends to $59/n$ mV.
(b) The mid-peak potential is independent of the potential scan rate.

There is a need, however, to discern between resistive and capacitive effects and those associated with the degree of electrochemical reversibility and coupled chemical reactions.

3.5 Deviations from Reversibility and Coupled Chemical Reactions

The ideal electrochemically reversible, diffusion-controlled behavior described in Sect. 3.1 corresponds to 'fast' electron transfer processes, i.e., for electrode reactions where the charge transfer at the interface is faster than the mass transport or associated chemical reactions. For moderate values of the rate constant, the behavior separates from the ideal case, and the peak-to-peak separation increases and varies with the scan rate [1]. The extreme case is a fully irreversible electrochemical process. In electrochemically irreversible processes there is only one (cathodic or anodic) peak without its corresponding (anodic or cathodic) counterpart. The peak potential, which now lacks pure thermochemical meaning, is proportional to the logarithm of scan rate:

$$E_{\mathrm{p}} = E^{\circ\prime} - \frac{RT}{\beta F}\left[0.780 - \ln k^{\circ} + \ln\sqrt{\frac{\beta F D v}{RT}}\right] \tag{3.17}$$

where β is the symmetry factor of the electron transfer process and k° the electrochemical rate constant whose meaning was discussed in Sect. 3.2. As in the reversible case, assuming diffusive control, the peak current can be expressed as:

$$I_{\mathrm{p}} = 0.227nFAck^{\circ}\exp\left[-\frac{\beta F(E_{\mathrm{p}} - E^{\circ\prime})}{RT}\right] \tag{3.18}$$

and the peak to half peak potential difference at 298 K is $48/\beta$ (mV).

Apart from the above, it has to be emphasized that the electrochemical processes are in general multistep and that there is an enormous variety of possibilities, among others:

(a) Those involving chemical reactions coupled to the electron transfer process. These reactions can precede, be parallel, or subsequent to the electrons transfer. Different cases have been theoretically described, the variation of peak current function, $i_{\mathrm{p}}/v^{1/2}$, with v being one of the parameters used to discriminate between them [1–3, 15–18]. Of particular interest are the electrocatalytic processes [1–4, 19].

(b) Those involving adsorption of reactants and/or products onto the electrode surface. Adsorption-controlled processes are in general characterized by the appearance of symmetric peaks and the proportionality of peak currents with v [1, 2].

There are a variety of electrochemical processes in which the electron transfer is accompanied by chemical reactions involving species in the solution phase or attached to the electrode surface. Figure 3.3 shows one of the best-known examples, the oxidation of dopamine in aqueous media. The electrochemical oxidation of dopamine proceeds via an initial two-electron, two-proton oxidation yielding the

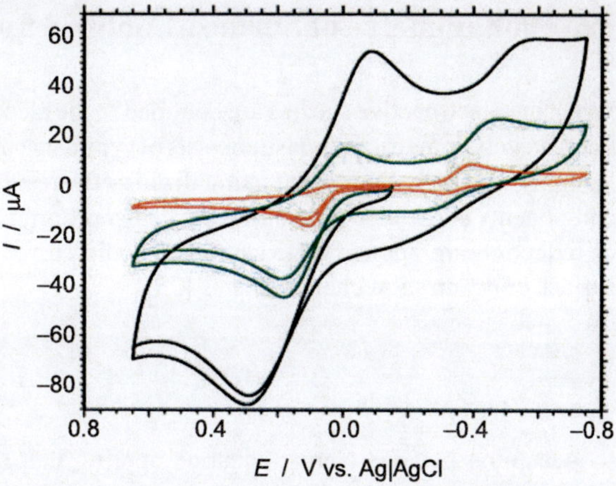

Fig. 3.3 Cyclic voltammogram at glassy carbon electrode of a 2.0 mM solution of dopamine in 0.10 M potassium phosphate buffer at pH 7.0. Potential scan initiated at − 0.10 V versus Ag/AgCl in the positive direction; potential scan rate 10, 20, and 100 mV s^{-1}. Note that the axes are opposite to the IUPAC convention

o-quinone which is followed by a cyclization reaction yielding leuko-aminochrome. This species is more easily oxidized than the parent catecholamine and can experience further oxidation to aminochrome, this process competing with a disproportionation reaction [20, 21].

Since the cyclization reaction following the initial two-electron, two-proton oxidation of the *o*-catechol group is fast, the initial anodic wave at 0.10 V does not exhibit a coupled cathodic peak. The rapidly formed leuco-aminochrome possesses an *o*-quinone unit which is reduced to the corresponding *o*-catechol form in an essentially reversible process at a mid-peak potential of − 0.35 V.

The variation of the sweep rate offers detailed information on this kind of process. In the case of dopamine electrochemistry, if this parameter is increased, the opportunity for completing the cyclization reaction decreases, and the cathodic peak coupled to the initial anodic step should appear. The processes involving proton transfer steps can usually be modeled in terms of successive protonation (chemical reaction, abbreviated by C) and electron transfer (abbreviated by E) steps. Thus, CE, ECE, ECE, ECEC, CECE, … reaction pathways have been described. In cases of reversible behavior, the peak potentials of reactions

$$\mathrm{Ox}_{\mathrm{aq}} + q\mathrm{H^+}_{\mathrm{aq}} + ne^- \rightarrow \mathrm{H_qRd}^{(q-n)+}{}_{\mathrm{aq}} \tag{3.19}$$

vary linearly with the pH of the electrolyte, the slope (absolute value) of the corresponding graph is 59(q/n) mV at 298 K. In the case of irreversible electron transfer, the slope of the same representations can be approximated by 59(q/β).

The precedent considerations can be extended to pulsed voltammetric techniques [22–24]. These are characterized by strongly reduced capacitive effects and the capability to increase the resolution of multiple signals compared with linear potential scan voltammetry. There are, however, problems with the appearance of 'artifacts' under several conditions so accurate data analysis is needed [22–24].

3.6 Voltammetry of Surface-Confined Species

When the electroactive species are confined to the electrode surface, the linear potential scan voltammetric response consists of symmetrical or almost symmetrical peaks. Roughly speaking, this can be rationalized as the result of the rapid exhaustion of the components at the electrode surface when appropriate potentials for their oxidation or reduction are applied. The cathodic/anodic current is expressed in terms of the surface concentration change rate as [25]:

$$i(t) = \pm nFA\left(\frac{\mathrm{d}\Gamma_{\mathrm{ox}}}{\mathrm{d}t}\right) \tag{3.20}$$

Assuming that the Nernst equation applies, that a unique electroactive species with a surface concentration Γ (=$\Gamma_{\mathrm{ox}} + \Gamma_{\mathrm{rd}}$) is reversibly reduced or oxidized, and assuming that the adsorption follows the Langmuir isotherm, the theoretical current-potential curve at a potential scan rate v must satisfy the relationship [1]:

$$I = \frac{n^2F^2vA\Gamma_{\mathrm{ox}}(b_{\mathrm{ox}}/b_{\mathrm{rd}})\mathrm{e}^{\frac{nF(E-E^{\circ\prime})}{RT}}}{RT\left[1 + (b_{\mathrm{ox}}/b_{\mathrm{rd}})\mathrm{e}^{\frac{nF(E-E^{\circ\prime})}{RT}}\right]^2} \tag{3.21}$$

In this equation, the coefficients b_i (i = Ox, Rd) are equal to $\Gamma_i\exp(-\Delta G^{\circ}_i/RT)$, ΔG°_i being the standard free energy for surface attachment (adsorption). The peak current is then given by:

$$I_{\mathrm{p}} = \frac{n^2F^2vA\Gamma}{4RT} \tag{3.22}$$

In the general, non-reversible case, the Butler–Volmer kinetics for an n-electron transfer between surface-confined species can be expressed as [26, 27]:

$$I = nFAk^{\circ}\left[\Gamma_{\mathrm{ox}}\mathrm{e}^{\frac{\beta F(E-E^{\circ\prime})}{RT}} - \Gamma_{\mathrm{rd}}\mathrm{e}^{-\frac{(1-\beta)nF(E-E^{\circ\prime})}{RT}}\right] \tag{3.23}$$

where k is the rate constant of the electrochemical reaction (s^{-1}), and Γ_{ox}, Γ_{rd}, are the surface concentrations of the oxidized and reduced forms of the electroactive species (mol cm^{-2}). Here, the formal electrode potential of the redox couple is related to the standard potential and the Gibbs free energies of surface attachment of the oxidized and reduced forms, $\Delta G^{\circ}_{\mathrm{ox}}$, $\Delta G^{\circ}_{\mathrm{rd}}$ by means of the relationship [28],

$$E^{\circ\prime} = E^{\circ} + \left(\frac{\Delta G^{\circ}_{\mathrm{ox}} - \Delta G^{\circ}_{\mathrm{rd}}}{nF}\right) \tag{3.24}$$

whereas the peak potential becomes independent on potential scan rate. In the case of irreversible electron transfer, however, the peak, when the concentration of surface-confined species is small, can be approximated by:

$$E_p = E^{o\prime} \pm \frac{RT}{\beta F} \ln[\beta F/RT] \tag{3.25}$$

whereas the current-potential curve can be approximated by [26]:

$$I = nFAk_o \left[-\frac{\beta F(E - E^{o\prime})}{RT} \right] \exp\left[\frac{k_o RT}{\beta F v} \exp\left[-\frac{\beta F(E - E^{o\prime})}{RT} \right] \right] \tag{3.26}$$

3.7 Voltammetry of Oxidation/Reduction of Ion-Permeable Solids

At first glance, one may assume that the voltammetry of microparticulate deposits in VIMP experiments could be derived from that of surface-confined species described in the previous section. There are, however, two significant differences appearing in VIMP experiments described in Chap. 2: (i) the solid reactants form a discontinuous, thick layer on the base electrode surface rather than a thin film on the same; (ii) the VIMP processes involve charge transfer through the particle/electrode/electrolyte solution three-phase junction.

In the case of the reversible topotactic solid-to-solid redox transformations described in Sect. 2.2.1, the voltammetric response of a reaction such as:

$$\{\mathrm{Ox}\}_{\mathrm{solid}} + \mathrm{M^+}_{\mathrm{solv}} + \mathrm{e^-} \rightarrow \{\mathrm{Rd^-} \cdots \mathrm{M^+}\}_{\mathrm{solid}} \tag{3.27}$$

can be modeled in terms of the parameter z defined as [28]:

$$z = \left(\frac{D_{\mathrm{M+,solid}}}{D_{\mathrm{M+,electrolyte}}} \right)^{1/2} \left(\frac{\rho_{\mathrm{solid}}}{c_{\mathrm{M+}}} \right) \tag{3.28}$$

Here, $D_{\mathrm{M+,solid}}$ and $D_{\mathrm{M+,electrolyte}}$ are the diffusion coefficients of the cation M^+ in the solid particle and in the electrolyte solution, respectively. ρ_{solid} is the molar density of the solid and $c_{\mathrm{M+}}$ the concentration of the charge-balancing cation in the electrolyte bulk. The mid-peak potential for $z < 0.1$ can be approximated by the equation:

$$E_{\mathrm{mp}} = E^{o\prime}_{\mathrm{app}} + \frac{RT}{F} \ln c_{\mathrm{M+}} \tag{3.29}$$

where $E^{o\prime}{}_{\mathrm{app}}$ represents an apparent electrode potential characterizing the redox couple. Under these conditions, the electrochemical process is controlled by the diffusion of M^+ through the solid particle. When $z > 10$, the mid-peak potential becomes independent on the cation concentration in the electrolyte solution [28]:

$$E_{mp} = E^{o\prime}_{app} + \frac{RT}{2F}\ln\left(\frac{D_{M+.solid}}{D_{M+,electrolyte}}\right) + \frac{RT}{F}\ln\rho_{solid} \tag{3.30}$$

and the voltammetric peak currents decrease very much; for larger z values the peak currents tend to zero.

This approach can be refined considering the effects associated with the limited size of the particles. The model of Schröder et al. [29] yields a relatively simple expression for the chronoamperometric current/time curve of the type:

$$I = A + Bt^{-1/2} - Ct^{1/2} \tag{3.31}$$

where the parameters A, B, C are related to the molar density of the solid and the diffusion coefficients of electrons and mobile ions into the solid. This relationship applies at short times for reversible electron transfer when the ionic concentration in the solution phase was high enough to assume that the diffusion of mobile ions through the electrolyte solution is not rate-determining. The above result is particularly interesting because the chronoamperometric curve displays a peculiar profile characteristic of this type of solid-state process [29].

At long times, however, the chronoamperometric current tends to have functional relationships between the current and the time of the type,

$$I = Ge^{-Ht} \tag{3.32}$$

where, again, G and H are coefficients depending on the diffusion coefficients of electrons and/or ions, molar volume, and shape and size of the particles. Under favorable conditions, experimental chronoamperometric data allow us to calculate the values of diffusion coefficients in ion-permeable solids [30, 31].

Another approach on the voltammetry of ion-permeable solids was developed around studies on ion intercalation electrochemistry in lithium, sodium, etc. batteries. Here it is assumed that the general process can be represented by a general intercalation isotherm of form [32],

$$a(M^+)_{solid} = Ga(M^+)_{solut} \tag{3.33}$$

where $a(M^+)_{solid}$ and $a(M^+)_{solut}$ represent the thermodynamic activity of the ion M^+ in the solid host and the electrolyte solution, respectively. In Eq. (3.34), G is a constant, function of the electrode potential, whose meaning is analogue to the partition constant described in ion-permeable redox polymers. In ion intercalation models, it is assumed that under equilibrium conditions the intercalation level of M^+ into the solid, x, defined as the fraction of interstitial sites occupied by the intercalated ion in the oxidized form of the solid, can be expressed as [33],

$$a(\mathrm{M}^+)_{\mathrm{solid}} = \left(\frac{x}{1-x}\right)\exp[-gx] = Ga(\mathrm{M}^+)_{\mathrm{solut}} \tag{3.34}$$

where g is a dimensionless constant representative of the ion-guest interaction. In the presence of a large concentration of M^+ in the electrolyte solution and in the absence of ion–solid interaction, the above equation reduces to the Langmuir isotherm,

$$\frac{1-x}{x} = \exp\left[\frac{nF(E-E^\circ)}{RT}\right] \tag{3.35}$$

This model assumes that the solid lattice contains a series of empty interstitial sites that can be free or occupied by the cation and that no interaction between cells of the lattice is assumed to take place. Then, the potential difference across the solid/electrolyte solution interface during the intercalation of a single cation species can be expressed by equalizing thermodynamic potentials of the species transported across the crystal/electrolyte interface. To model the guest ion–solid lattice interaction, the Langmuir isotherm was replaced by the Frumkin isotherm [32, 33],

$$\frac{1-x}{x} = \exp\left[\frac{gxnF(E-E^\circ)}{RT}\right] \tag{3.36}$$

as before referring to equilibrium conditions. These formulations can be expanded to describe voltammetric measurements including the consideration of phase changes and different diffusional and kinetic regimes [34–36].

3.8 Voltammetry of Oxidative/Reductive Dissolution Processes

These processes can be treated in terms of solid-state kinetic models. In most practical cases, the kinetics of solid-state reactions can be modeled using differential equations of the form:

$$\frac{\mathrm{d}a}{\mathrm{d}t} \propto f(a) = kf(a) \tag{3.37}$$

where a ($0 < a < 1$) represents the degree of conversion of the parent solid into a new component. Grygar [37, 38] applied this formalism to the reductive and oxidative dissolution of metal compounds assuming that, if the reaction is promoted by the application of a potential E, a function characterizing the influence of potential, $k(E)$, can be inserted into the kinetic equation:

$$\frac{\mathrm{d}a}{\mathrm{d}t} \propto k(E)f(a) \tag{3.38}$$

The electrochemical reaction can be considered as an irreversible process following Butler–Volmer kinetics, so that:

$$k(E) = k_o \exp\left[-\frac{\beta F}{RT}(E - E^{o\prime})\right] \tag{3.39}$$

In electrochemical experiments, *a* can be identified as the ratio between the charge passed at a given time *q*, and the maximum charge able to be transferred, q_o. Applying this formalism leads to a variety of kinetic models with more or less complicated voltammetric curves. In a chronoamperometric experiment in which a fixed potential *E* is applied, Eq. (3.36) can be written as [37]:

$$\frac{dq}{dt} = kq_o\left(\frac{q}{q_o}\right)^{\eta} \tag{3.40}$$

whereη is an exponent whose value will be decided by experimental data. Integration of this equation between $t = 0$ and t yields:

$$q/q_o = 1 - [1 + (1 - \eta)\,kt]^{\frac{1}{1-\eta}} \tag{3.41}$$

Then, the chronoamperometric current is:

$$I = q_o(1 - \eta)k\{1 + [(1 - \eta)\,kt]\}^{\frac{\eta}{1-\eta}} \tag{3.42}$$

In this case, the I/q ratio is particularly appropriate for comparing the model with experimental data [37, 38]:

$$\frac{I}{q} = \frac{(1 - \eta)k}{1 - (1 - \eta)kt} \tag{3.43}$$

3.9 Voltammetry of Solid-to-Solid Redox Transformations

Typically, this type of process appears in the electrochemical reduction of metal salts to metal deposits. These processes, of importance in inorganic pigment electrochemistry, offer considerable complexity. In general, the cathodic process involves phase changes and several other processes. For instance, in the case of the reduction of litharge to lead, the overall process:

$$PbO_{solid} + 2H^{+}{}_{aq} + 2e^{-} \rightarrow Pb_{solid} + H_2O_{liq} \tag{3.44}$$

is, according to the Hasse and Scholz model [39] a topotactic transformation of PbO to Pb accomplished by a proton insertion to form an ionophoric intermediate layer as schematized in Fig. 3.4.

The overall process can be understood as a sequence of processes, each one of which is, in principle, potentially rate-determining [40]:

Proton diffusion toward the particle surface and insertion and transport into the solid forming an ionophoric layer in the solid:

$$H^+{}_{aq}(\text{bulk solution}) \rightarrow H^+{}_{aq\ PbO\ surface} \rightarrow H^+{}_{aq\ PbO} \quad (3.45)$$

Electron injection and transport into the solid:

$$PbO_{solid} + e^-{}_{electrode} \rightarrow PbO\left(e^-\right)_{solid} \quad (3.46)$$

Formation of metal nuclei:

$$H^+{}_{aq\ PbO} + PbO\left(e^-\right)_{solid} \rightarrow Pb_{nucleus} + H_2O_{ionophoric\ layer} \quad (3.47)$$

Growth and coalescence of metallic nuclei:

$$Pb_{nucleus} \rightarrow Pb_{solid} \quad (3.48)$$

Water release from the ionophoric layer:

$$H_2O_{ionophoric\ layer} \rightarrow H_2O_{liq} \quad (3.49)$$

This pathway can be complicated by, for instance, the presence of complexing species such as chloride ions with their corresponding diffusion through the solution, insertion into the solid, and release from the ionophoric layer.

Additionally, one can consider a second competing pathway involving the formation of Pb^{2+} ions in the ionophoric layer:

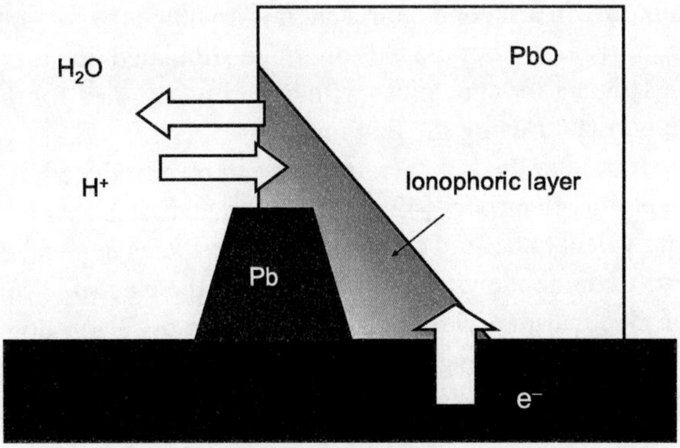

Fig. 3.4 Schematics for the topotactic reduction of litharge to lead in VIMP experiments based on [39]

$$H^{+}{}_{aq\,PbO} + PbO\,(e^{-})_{solid} \rightarrow Pb^{2+}{}_{ionophoric\ layer} + H_2O_{ionophoric\ layer} \quad (3.50)$$

Followed by their release into the solution:

$$Pb^{2+}{}_{ionophoric\ layer} + H_2O_{ionophoric\ layer} \rightarrow Pb^{2+}{}_{aq} + H_2O_{liq} \quad (3.51)$$

and their electrochemical reduction either from the solution or in the ionophoric layer:

$$Pb^{2+}{}_{ionophoric\ layer} + 2e^{-} \rightarrow Pb_{solid} \quad (3.52)$$

$$Pb^{2+}{}_{aq} + 2e^{-} \rightarrow Pb_{solid} \quad (3.53)$$

This complicated situation could give rise to different electrochemical responses depending on the composition of the electrolyte solution, the shape and size distribution of the parent solid on the electrode surface, and the characteristic time of the experiment.

Jaworski et al. [41] proposed a theoretical model for the *n*-electron reduction process of a MA_m salt to M,

$$\{MX_m\}_{solid} + nme^{-} \rightarrow M_{solid} + mX^{n-}{}_{solv} \quad (3.54)$$

assuming that Nernstian equilibrium was established at the electrode surface. Then, the thermodynamic activities of the metal salt (a_{MXm}), the metal (a_M), and the X^{m-} anions (a_{Xm-}) in the solution should verify [41]:

$$E = E^{\circ} + \left(\frac{RT}{nmF}\right)\ln\frac{a_{MXm}}{a_M\,a_{Xn-}} \quad (3.55)$$

In a first approach, it is assumed that there is the formation of a mixed salt–metal crystal during the voltammetric experiments. Then, the $a_{MX}m/a_M$ activity ratio is assumed to be equal to the ratio between their respective mole fractions in the solid mixture. In a second approach, it is assumed that the metal crystal segregates so that $a_M = 1$. These approaches provided simulated linear potential scan voltammograms in satisfactory agreement with experimental data for the reduction of silver halides, in particular using the first model.

A second theoretical approach can be introduced by accounting for the modeling of porous electrodes [42–44] because, at least in several aspects, they can be formally equivalent to a set of nonconducting particles deposited onto a base electrode. Then, two extreme views are possible. First, the electrode surface can be seen as an array of identical particles each one having a diffusion domain where there is a significant concentration gradient of charge-balancing ions. Second, the diffusion domains merge so that the system behaves like the surface-confined species treated in Sect. 3.5.

The voltammetric behavior will depend on the size of the particles and the potential scan rate.

A phenomenological approach has also been proposed to describe reductive/oxidative dissolution processes. Let us first consider the general expression of the current in a voltammetric experiment involving the reversible interconversion between surface-confined species. In the simplest case, Eq. (3.21) can be expressed as:

$$I = \frac{n^2F^2A\Gamma v}{RT}\frac{e^{nF(E-E^{\circ\prime})/RT}}{\left[1+e^{nF(E-E^{\circ\prime})}/RT\right]^2} \tag{3.56}$$

This equation is formally equivalent to that obtained assuming that the transferred charge varies with the applied potential following a logistic growth of the form [41]:

$$\frac{d(q/q_o)}{dE} = g\left(\frac{q}{q_o}\right)\left(1-\frac{q}{q_o}\right) \tag{3.57}$$

where q is the charge passed when a potential E is applied, q_o is the maximum charge able to be transferred, and g represents a constant parameter representative of the rate of variation of q with E. This type of equation was introduced by Verhulst to describe the time variation of a population tending to a limiting value or carrying capacity [45, 46]. In principle, the logistic functions can be applied to quantities reflecting physical, geographical, biological, etc. systems whose growth was limited by a certain carrying capacity [47]. The application of logistic approximations to electrochemical systems has been treated in literature [48, 49], but its physical meaning is controversial [50]. Accordingly, the approach presented here has to be taken as an operational approximation. Integration of Eq. (3.57) yields:

$$\ln\left(\frac{q/q_o}{1-q/q_o}\right) = gE + H \tag{3.58}$$

H being the integration constant. If this constant is equaled to $-\,gE^{\circ\prime}$, one obtains:

$$q = q_o\frac{e^{g(E-E^{\circ})}}{1+e^{g(E-E^{\circ})}} \tag{3.59}$$

The expression for the current in a voltammetric experiment carried out at a potential scan rate v can be derived from,

$$I = \frac{dq}{dt} = \frac{dq}{dE}\frac{dE}{dt} = \pm v\frac{dq}{dE} \tag{3.60}$$

so that the current-potential curve will be:

$$I = q_o g v \frac{e^{g(E-E^\circ)}}{\left[1 + e^{g(E-E^\circ)}\right]^2} \tag{3.61}$$

This equation is equivalent to Eq. (3.56) taking $g = nF/RT$ and $q_o = nFA\Gamma$.

The above equations correspond to the reversible case. To account for deviations from reversibility in the electron transfer and the effects associated with the solid-state transformations (changes in the volume of the system under study, kinetic of the deconstruction/reconstruction of solid lattices, …), one can use different logistic equations. The logistic equation developed by Richards [51] is particularly useful. This equation can be written as [40]:

$$\frac{d\left(\frac{q}{q_o}\right)}{dE} = g\left(\frac{q}{q_o}\right)\left[1 - \left(\frac{q}{q_o}\right)^{\gamma}\right] \tag{3.62}$$

where γ is an exponent whose value is representative of the above factors determining the deviation of the system from reversibility. Integration yields for the variation of the charge passed and the current with the potential in a linear potential scan voltammetric experiment:

$$q = q_o\left[\frac{e^{\gamma g(E-E^{o\prime})}}{1 + e^{\gamma g(E-E^{o\prime})}}\right]^{\frac{1}{\gamma}} \tag{3.63}$$

$$I = q_o g v \frac{e^{g(E-E^{o\prime})}}{\left[1 + e^{\gamma g(E-E^{o\prime})}\right]^{\frac{1}{\gamma}+1}} \tag{3.64}$$

Now, the peak potential and peak current are, respectively,

$$E_p = E^{o\prime} + \frac{\ln\gamma}{\gamma g} \tag{3.65}$$

$$I_p = \frac{q_o g v e^{-\frac{\ln\gamma}{\gamma}}}{\left[1 + e^{-\ln\gamma}\right]^{\frac{1}{\gamma}+1}} \tag{3.66}$$

This model has been tested for lead pigments showing a satisfactory agreement between theory and voltammetric data [40]. This can be seen in Fig. 3.5 where the linear potential scan voltammograms recorded at two different sweep rates for litharge deposits attached to graphite electrodes in contact with 0.10 M H_2SO_4 are compared with theoretical voltammograms, both expressed in terms of the I/I_p ratio.

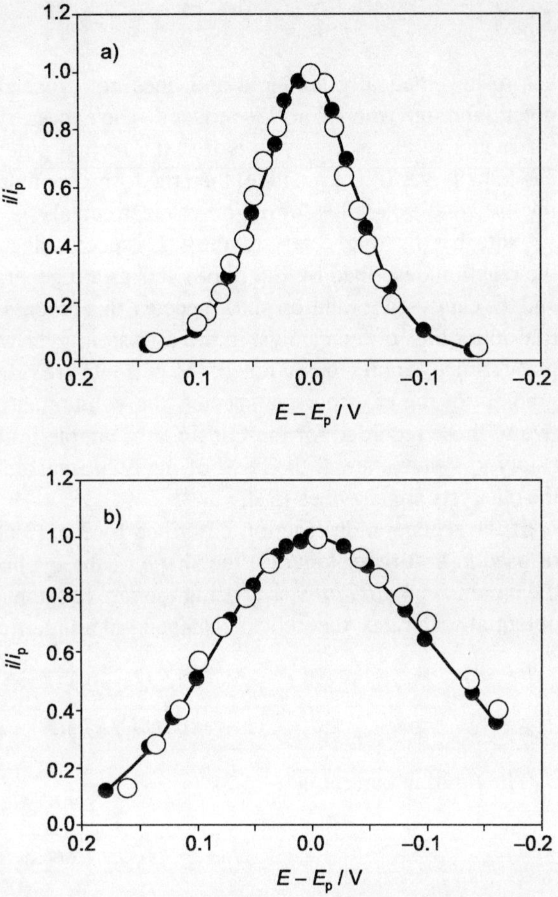

Fig. 3.5 Superimposed experimental (circles) and theoretical (solid circles and continuous lines) linear potential scan voltammograms for litharge-modified graphite electrode in contact with 0.10 M H_2SO_4 aqueous solution. Potential scan rates of **a** 2 and **b** 8 mV s^{-1}. Theory from Eq. (3.64) taking **a** $g = -34.9\ V^{-1}$, $\gamma = 1.05$; **b** $g = -31.5\ V^{-1}$, $\gamma = 0.40$. Reproduced from Ref. [40], with permission. Note that the axes are opposite to the IUPAC convention

3.10 Electrocatalysis

An important class of electrochemical processes is that in which there is a catalytic promotion of the oxidation or reduction of a species. Classically, a catalyst species whose electrochemical reduction or oxidation is facile under determined conditions, promotes the reduction or oxidation of a substrate species S through a regenerative pathway [1, 3, 52]. Then, the following sequence is operative (charges are omitted for simplicity):

$$\text{Cat}^{\text{Ox}} + e^- \rightarrow \text{Cat}^{\text{Rd}} \tag{3.67}$$

$$\text{Cat}^{\text{Rd}} + \text{S}^{\text{Ox}} \rightarrow \text{Cat}^{\text{Ox}} + \text{S}^{\text{Rd}} \tag{3.68}$$

It has to be emphasized that the direct electrochemical reduction of the substrate,

$$S^{Ox} + e^- \rightarrow S^{Rd} \tag{3.69}$$

has to be kinetically hindered and does not proceed in appreciable extent in the potential range where Cat^{Ox} is reduced. The necessary (but not sufficient) condition for reductive electrocatalysis is that the formal potential of the catalyst reduction has to be larger (more positive) than the formal potential of the S^{Ox}/S^{Rd} couple. The inverse condition holds for oxidative electrocatalysis.

Note that the 'true' catalyst is the Cat^{Rd} species electrochemically generated so that the reaction described by Eq. (3.68) acts as a regenerative step. When the substrate and the catalyst are solution phase species, this process leads to much steeper concentration profiles of the catalyst in the diffusion layer where the reaction (3.69) takes place. Since, as previously noted, the currents are proportional to the concentration gradient of the electroactive species, the voltammetric currents are enhanced relative to those recorded for the Cat^{Ox}/Cat^{Rd} couple in the absence of substrate. Such catalytic systems are at the heart of many biosensors; e.g., glucose sensors, where the catalysts are enzymes [53].

If the regeneration reaction described by Eq. (3.68) proceeds under conditions of pseudo-first-order kinetics, the shape of the cyclic voltammograms depends on the parameter $k_{reg}RT/nFv$, k_{reg} being the rate constant for the reaction [1, 3]. At low potential scan rates, theoretical s-shaped voltammetric curves are expressed as,

$$I \approx \frac{nFAc(Dk_f)^{1/2}}{1 + \exp[(nF/RT)(E - E^{o\prime})]} \tag{3.70}$$

The limiting current is:

$$I_{lim} = nFAc\sqrt{Dk_f c_{cat}} \tag{3.71}$$

At large sweep rates, the catalytic reaction has no effect on the voltammetric response. Then the voltammogram is equivalent to that recorded for a reversible n-electron, diffusion-controlled couple.

Interestingly, there is the possibility of electrocatalytic processes in which either the solid or any species in solution act as the catalyst or substrate [13]. Accordingly, the sensitivity in the detection of a solid analyte can be enhanced significantly. Among different possibilities, the most interesting case for our purposes is that where the voltammetric signal of an electroactive solid is enhanced by the presence of a substrate in the electrolyte solution. Assuming that the catalyst behaves as an ion-insertion solid, the catalytic reaction can be represented as [15]:

$$\{Cat^{Ox}\}_{solid} + nM^+{}_{sol} + ne^- \rightarrow \{Cat^{Rdn-} \cdots nM^+\}_{solid} \tag{3.72}$$

This reaction is accompanied by the regeneration reaction:

$$S^{Ox}{}_{sol} + \{Cat^{Rdn-} \cdots nM^+\}_{solid} \rightarrow \{Cat^{Ox}\}_{solid} + S^{Rd}{}_{sol} + nM^+{}_{sol} \tag{3.73}$$

As a result, the voltammetric signal for the catalyst reduction is enhanced. Different examples of solid-state catalytic processes have been reported in the context of the electrochemistry of porous materials [54].

References

1. Bard AJ, Faulkner LR (2001) Electrochemical methods, 2nd edn. Wiley, New York
2. Scholz F (2015) Voltammetric techniques of analysis: the essentials. ChemTexts 1:17
3. Oldham KB, Myland JC (1994) Fundamentals of electrochemical science. Academic Press, San Diego
4. Nicholson RS, Shain I (1964) Theory of stationary electrode polarography. Anal Chem 36:706–723
5. Inzelt G (2014) Crossing the bridge between thermodynamics and electrochemistry. From the potential of the cell reaction to the electrode potential. Chem Texts 1:1–11
6. Seeber R, Zanardi C, Inzelt G (2015) Links between electrochemical thermodynamics and kinetics. ChemTexts 1:18
7. Batchelor-McAuley C, Kätelhön E, Barnes EO, Compton RG, Laborda E, Molina A (2015) Recent advances in voltammetry. ChemistryOpen 4:224–260
8. Tanner EEL, Barnes EO, Tickell CB, Goodrich P, Hardacre C, Compton RG (2015) Application of asymmetric Marcus–Hush theory to voltammetry in room-temperature ionic liquids. J Phys Chem C 119:7360–7370
9. Guidelli R, Compton RG, Feliu JM, Gileadi E, Lipkowski J, Schmickler W, Trasatti S (2014) Defining the transfer coefficient in electrochemistry: an assessment (IUPAC technical report). Pure Appl Chem 86:245–258
10. Bard AJ, Inzelt G, Scholz F (eds) (2008) Electrochemical dictionary. Springer, Berlin-Heidelberg
11. Bockris JO'M, Reddy AKN (1977) Modern electrochemistry. Plenum Press, New York
12. Štulík K, Amatore C, Holub K, Mareček V, Kutner W (2000) Microeletrodes. Definitions, characterization and applications (IUPAC Technical Report). Pure Appl Chem 72:1483–1492
13. Polcyn DS, Shain I (1966) Multistep charge transfers in stationary electrode polarography. Anal Chem 38:370–375
14. Ginzburg G (1978) Data treatment in cyclic voltammetry. Anal Chem 50:375–376
15. Doménech-Carbó A (2020) Electrochemistry of porous materials, 2nd edn. Chap. 15. Taylor & Francis, Boca Raton
16. Nicholson RS (1969) Some examples of the numerical solution of nonlinear integral equations. Anal Chem 37:667–671
17. Nicholson RS, Shain I (1965) Theory of stationary electrode polarography for a chemical reaction coupled between two charge transfers. Anal Chem 37:178–190
18. Kudirka PJ, Nicholson RS (1972) Experimental evaluation of cyclic voltammetry theory for disproportionation reactions. Anal Chem 44:1786–1794
19. Polcyn DS, Shain I (1966) Theory of stationary electrode polarography for a multistep charge transfer with catalytic (cyclic) regeneration of the reactant. Anal Chem 38:376–382
20. Amatore C, Savéant J-M (1978) Do ECE mechanisms occur in conditions where they could be characterized by electrochemical techniques? J Electroanal Chem 86:227–232
21. Ciolowski EL, Maness KM, Cahill PS, Wightman RM, Evans DH, Fosset B, Amatore C (1994) Disproportionation during electrooxidation of catecholamines at carbon-fiber microelectrodes. Anal Chem 66:3611–3617
22. Stojek Z (2010) Pulse Voltammetry. In: Scholz F (ed) Electroanalytical methods. Guide to experiments and applications. Springer, Berlin
23. Molina A, Gonzalez J (2015) Pulse voltammetry in physical electrochemistry and electroanalysis. In: Scholz F (ed) Monographs in electrochemistry. Springer, Berlin

24. Mircevski V, Komorsky-Lovrić S, Lovrić M (2007) Square wave voltammetry. monographs in electrochemistry series. In: Scholz F (ed). Springer, Berlin-Heidelber
25. Laviron E (1979) General expression of the linear potential sweep voltammogram in the case of diffusionless electrochemical systems. J Electroanal Chem 101:19–28
26. Laviron E (1979) The use of linear potential sweep voltammetry and of ac voltammetry for the study of the surface electrochemical reaction of strongly adsorbed systems and of redox modified electrodes. J Electroanal Chem 100:263–270
27. Myland JC, Oldham KB (2005) Quasireversible cyclic voltammetry of a surface confined redox system: a mathematical treatment. Electrochem Commun 7:282–287
28. Lovrić M, Hermes M, Scholz F (1998) The effect of the electrolyte concentration in the solution on the voltammetric response of insertion electrodes. J Solid State Electrochem 2:401–404
29. Schröder U, Oldham KB, Myland JC, Mahon PJ, Scholz F (2000) Modelling of solid state voltammetry of immobilized microcrystals assuming an initiation of the electrochemical reaction at a three-phase junction. J Solid State Electrochem 4:314–324
30. Doménech-Carbó A (2004) A model for solid-state voltammetry of zeolite-associated species. J Phys Chem B 108:20471–20478
31. Doménech-Carbó A (2015) Theoretical scenarios for the electrochemistry of porous silicate-based materials: an overview. J Solid State Electrochem 19:1887–1903
32. Levi MD, Aurbach D (2008) Kinetics of electrochemically induced phase transitions in ion-insertion electrodes and the chemical diffusion coefficient. J Solid State Electrochem 12:409–420
33. Levi MB, Aurbach D (1999) Electrochim Acta Frumkin intercalation isotherm Ð a tool for the description of lithium insertion into host materials: a review 45:167−185
34. Gavilán-Arriazu EM, Barraco DE, Leiva EPM (2021) Fast charging of alkali-ion batteries at the single-particle level: the impact of particle geometry on diffusional and kinetic bottlenecks in voltammetry. J Solid State Electrochem 25:2793–2806
35. Malaie K, Scholz F, Schröder U, Wulff H, Kahlert H (2022) ChemPhysChem: e202200364
36. Doménech-Carbó A, López S, Chandía B, Cáceres G, Muñoz E (2023) Cation con-intercalation in copper(II) hexacyanoferrates. ChemPhysChem 24:02200853
37. Grygar T (1995) The electrochemical dissolution of iron(III) and chromium(III) oxides and ferrites under conditions of abrasive stripping voltammetry. J Electroanal Chem 405:117–125
38. Grygar T (1998) Phenomenological kinetics of irreversible electrochemical dissolution of metal-oxide microparticles. J Solid State Electrochem 2:127–136
39. Hasse U, Scholz F (2001) In situ atomic force microscopy of the reduction of lead oxide nanocrystals immobilised on an electrode surface. Electrochem Commun 3:429–434
40. Doménech-Carbó A (2022) Description of solid-to-solid redox processes based on the voltammetry of immobilized particles methodology: a logistic approximation. J Phys Chem C 126:11822–11832
41. Jaworski A, Stojek Z, Scholz F (1993) A comparison of simulated and experimental abrasive stripping voltammetric curves of ionic crystals: reversible case. J Electroanal Chem 354:1–9
42. Barnes EO, Chen X, Li P, Compton RG (2014) Voltammetry at porous electrodes: a theoretical study. J Electroanal Chem 720–721:92–100
43. Chan HTH, Kätelhön E, Compton RG (2017) Voltammetry at electrodes decorated with an insulating porous film: Understanding the effects of adsorption. J Electroanal Chem 801:135–140
44. Ward KR, Compton RG (2014) Quantifying the apparent 'catalytic' effect of porous electrode surfaces. J Electroanal Chem 724:43–47
45. Verhulst P-F (1838) Notice sur la loi que la population poursuit dans son accroissement. Corresp Math Phys 10:113–121
46. Ausloos M (2014) Gompertz and verhulst frameworks for growth and decay description. Int J Comput Anticip Syst 30:15–36
47. Doménech-Carbó A (2019) Rise and fall of historic tram networks: logistic approximation and discontinuous events. Physica A 522:315–323

48. Torresi RM, Córdoba de Torresi SI, Gonzalez ER (1999) On the use of the quadratic logistic differential equation for the interpretation of electrointercalation processes. J Electroanal Chem 461:161–166
49. Diard J-P, Le Gorrec B, Montella C (1999) Logistic differential equation A general equation for electrointercalation processes? J Electroanal Chem 475:190–192
50. Sadkowski A (2000) On the application of the logistic differential equation in electrochemical dynamics. J Electroanal Chem 486:92–94
51. Richards FJA (1959) Flexible growth function for empirical use. J Exp Bot 10:290–300
52. Molina A, González J, Laborda E, Wang Y, Compton RG (2011) Catalytic mechanism in cyclic voltammetry at disc electrodes: an analytical solution. PhysChemChemPhys 13:14694–14704
53. Wang J (2008) Electrochemical glucose biosensors. Chem Rev 108:814–825
54. Doménech-Carbó A, García H, Marquet J, Bourdelande JL, Herance JR (2006) Modelling electrocatalysis of hydroquinone oxidation by nicotinamide dinucleotide coenzyme encapsulated within SBA-15 and MCM-41 mesoporous aluminosilicates. Electrochim Acta 51:4897–4908

Additional Literature

55. Adán-Mas A, Silva TM, Guerlou-Demourgues L, Montemor M-F (2018) Application of the Mott-Schottky model to select potentials for EIS studies on electrodes for electrochemical charge storage. Electrochim Acta 289:47–55
56. Amatore C, Savéant JM, Tessier D (1983) Charge transfer at partially blocked surfaces. A model for the case of microscopic active and inactive sites. J Electroanal Chem 147:39–51
57. Bisquert J, Garcia-Belmonte G, Fabregat-Santiago F (1999) Modelling the electric potential distribution in the dark in nanoporous semiconductor electrodes. J Solid State Electrochem 3:337–347
58. Bott AW (1998) Electrochemistry of semiconductors. Curr Sep 17:87–91
59. Chan HTH, Kätelhön E, Compton RG (2017) Voltammetry at electrodes decorated with an insulating porous film: understanding the effects of adsorption. J Electroanal Chem 801:135–140
60. de Levie R, Vogt A (1990) On the electrochemical response of rough electrodes : Part II. The transient response in the presence of slow faradaic processes. J Electroanal Chem 281:23–28
61. Doménech-Carbó A, Doménech-Carbó MT, García H, Galletero MS (1999) Electrocatalysis of neureotransmitter catecholamines by 2,4,6-triphenylpyrylium ion immbilized inside zeolite Y supercages. Chem Commun: 2173–2174
62. Doménech-Carbó A, Aucejo R, Alarcón J, Navarro P (2004) Electrocatalysis of the oxidation of methylenedioxyamphetamines at electrodes modified by cerium-doped zirconias. Electrochem Commun 6:719–723
63. Doménech-Carbó A (2012) Solvent-independent electrode potentials of solids undergoing insertion electrochemical reactions: part I. Theory J Phys Chem C 116:25977–25983
64. Doménech-Carbó A, Scholz F, Montoya N (2012) Solvent-independent electrode potentials of solids undergoing insertion electrochemical reactions: part III. Experimental data for prussian blue undergoing electron exchange coupled to cation exchange. J Phys Chem C 116:25993–25999
65. Doménech-Carbó A, Koshevoy IO, Montoya N, Pakkanen TA, Doménech-Carbó MT (2012) Solvent-independent electrode potentials of solids undergoing insertion electrochemical reactions: part II. Experimental data for alkynyl-diphosphine dinuclear Au(I) complexes undergoing electron exchange coupled to anion exchange. J Phys Chem C 116:25984–32599
66. Fabregat-Santiago F, Mora-Seró I, Garcia-Belmonte G, Bisquert J (2003) Cyclic voltammetry studies of nanoporous semiconductors. capacitive and reactive properties of nanocrystalline TiO_2 electrodes in aqueous electrolyte. J Phys Chem B 107:58–768
67. Gerischer H (1969) Charge transfer processes at semiconductor-electrolyte interfaces in connection with problems of catalysis. Surf Sci 18:97–122

68. Gonzalez J, Lopez-Tenes M, Molina A (2013) Non-nernstian two-electron transfer reactions for immobilized molecules: a theoretical study in cyclic voltammetry. J Phys Chem C 117:5208–5220
69. Gozález-Meza OA, Larios-Durán ER, Gutiérrez-Becerra A, Casillas N, Escalante JI, Bárcena-Soto M (2019) Development of a Randles-Ševčík-like equation to predict the peak current of cyclic voltammetry for solid metal hexacyanoferrates. J Solid State Electrochem 23:3123–3133
70. Gavilán-Arriazu EM, Barraco DE, Ein-Eli Y, Leiva EPM (2022) J Solid State Electrochem 26:1995–2003
71. Grygar T (1997) Dissolution of pure and substituted goethites controlled by the surface reaction under conditions of abrasive stripping voltammetry. J Solid State Electrochem 1:77–82
72. Grygar T, Bezdicka P (1998) Electrochemical dissolution of Cr^{III} and Cr^{IV} oxides. J Solid State Electrochem 2:31–38
73. Krulic D, Fatouros N (2011) Peak heights and peak widths at half-height in square wave voltammetry without and with ohmic potential drop for reversible and irreversible systems. J Electroanal Chem 652:26–31
74. Levi MD, Aurbach D (1997) The mechanism of lithium intercalation in graphite film electrodes in aprotic media. Part 1. High resolution slow scan rate cyclic voltammetric studies and modeling. J Electroanal Chem 421:79–88
75. Lovrić M (2002) Square-wave voltammetry. In: Scholz F (ed) Electroanalytical methods. Springer, Berlin
76. Lovrić M, Scholz F (1997) A model for the propagation of a redox reaction through microcrystals. J Solid State Electrochem 1:108–113
77. Lovrić M, Scholz F (1999) A model for the coupled transport of ions and electrons in redox conductive microcrystals. J Solid State Electrochem 3:172–175
78. Malaie K, Scholz F, Schröder U (2023) A thermodynamic model for the insertion electrochemistry of battery cathodes. ChemElectroChem in press. https://doi.org/10.1002/celc.202201118
79. Marcus RA (1956) On the theory of oxidation-reduction reactions involving electron transfer. I. J Chem Phys 24:966–978
80. Molina A, Gonzalez J, Henstridge M, Compton RG (2011) Voltammetry of electrochemically reversible systems at electrodes of any geometry: a general, explicit analytical characterization. J Phys Chem C 115:4054–4062
81. O'Dea JJ, Osteryoung JG (1993) Characterization of quasi-reversible surface processes by square-wave voltammetry. Anal Chem 65:3090–3097
82. Oldham KB (1998) Voltammetry at a three-phase junction. J Solid State Electrochem 2:367–377
83. Mirčeski V, Lovrić V (2001) Ohmic drop effects in square-wave voltammetry. J Electroanal Chem 497:114–124
84. Myland JC, Oldham KB (1983) An analytical expression for the current-voltage relationship during reversible cyclic voltammetry. J Electroanal Chem 153:43–54
85. Schmalzried H (1995) Chemical kinetics of solids. VCH Wenheim
86. Ward KR, Gara M, Xiong L, Lawrence NS, Hartshorne RS, Compton RG (2013) Thin-layer vs. semi-infinite diffusion in cylindrical pores: a basis for delineating Fickian transport to identify nano-confinement effects in voltammetry. J Electroanal Chem 695:15–24

Chapter 4
Analytical Issues

4.1 Generalities

For practical purposes, we can distinguish between qualitative analysis, aimed to identify the presence of selected chemical species in the work of art samples, and quantitative ones, aimed to determine the quantitative composition of that sample. This is in general limited to a few selected components and is often obtained in terms of relative proportions of different components (for instance, the Sn/Cu ratio in a bronze alloy). In this case, there is relative quantification, because there are other components present in an unknown proportion of the sample. When the proportion of a given component with respect to the total sample is obtained, the quantification is qualified as absolute.

In this chapter, quantification methods based on the application of electrochemical techniques will be described. These will be limited, for obvious reasons, to solid-state techniques. Interestingly, electrochemical techniques can also be applied to the discrimination—but also quantification—between the oxidation states of a given species. This case will be termed speciation analysis and will be treated separately from identification and quantification methods. It should be noted, however, that the term speciation can be applied to, for instance, the discrimination between different minerals of the same chemical composition.

A. Doménech-Carbó and M. T. Doménech-Carbó, *Electrochemistry for Cultural Heritage*, Monographs in Electrochemistry,
https://doi.org/10.1007/978-3-031-31945-7_4

4.2 Identification of Components

4.2.1 Voltammetric Parameters

The identification of the components of a sample from a cultural good is an obvious analytical target in heritage studies. The term 'component' here does not necessarily mean a chemical element but possibly also a chemical compound or a specific mineral phase. This task, however, can involve considerable complexity because many samples exhibit different degrees of heterogeneity, frequently incorporating alteration products, and/or are distributed in different stratified layers. In the case of elemental analysis, the classical analytical approach involves the dissolution of the sample, logically leading to a partial loss of information.

The VIMP offers, as described in Chap. 2, a sensitive methodology for the analysis of solid samples at the nanogram level. This methodology has a drawback/advantage: not all components of a given archaeological or work of art sample are electroactive under the selected experimental conditions. This limits the range of components able to be directly determined via VIMP; however, this limitation can be an advantage for a number of applications because the analytical response is focused on one or a few components of interest [1].

A clear example of the above considerations is the analysis of pigments in frescoes. Since the pigments are more representative of authorship than preparative layers and determine the chromatic effect of the painting, their identification is necessary for historic studies and conservation and restoration. This aim is complicated by the fact that pigments are highly diluted in the pictorial layers. In this context, pigments can be identified from SEM/EDX data, infrared, and Raman spectroscopies, but the last analytical techniques are conditioned by the large interference produced by the gross spectral bands due to the preparative layers.

The identification of a unique component using VIMP in principle requires the comparison of the voltammetric response of the sample with that of a standard under the same experimental conditions. An example was provided in Fig. 3.1 where the voltammetry of minium blank and a sample extracted from the red pictorial layer of wall paintings of a rock-hewn church of Lalibela (Ethiopia), dated back to the twelfth century, can be compared [2]. Here, the VIMP methodology was performed by immobilizing the solid particles onto a glassy carbon electrode embedded into a porous polymer, a strategy used in the study of zeolites and clays [3, 4].

Figure 4.1 reveals a clear similarity between the general voltammetric profiles of the pure pigment and the wall painting sample, but there is no total coincidence between them. This feature reflects the difference in the size and shape of pigment particles and the effect of the accompanying components of the pictorial sample, aspects that will be treated in detail in the next chapter.

The identification criteria are based on peak potentials and other related parameters: onset potentials, peak-to-half peak potential difference, and half-peak width, marked in Fig. 4.2 over the square wave voltammogram recorded for cuprite (Cu_2O), a ubiquitous corrosion product of copper and bronze objects, in contact

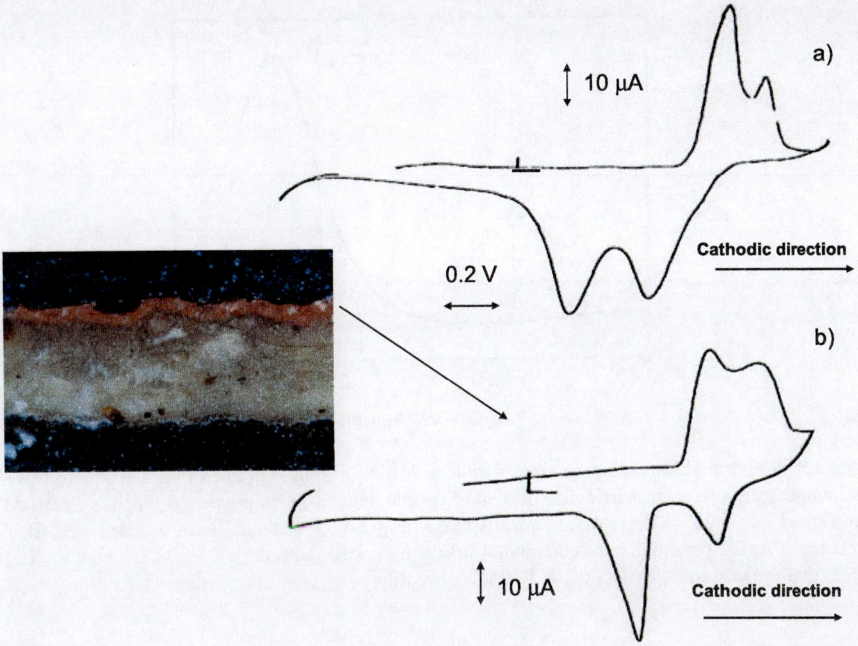

Fig. 4.1 Cyclic voltammograms of submicroparticulate deposits of **a** minium, **b** sample extracted from the red paint layer of a fresco from the 'Biete Maryan' (LA-9–7/98) wall paintings of the rock-hewn church in Lalibela (Ethiopia) embedded into Paraloid B72 ® film on glassy carbon electrode in contact with 1 M NaCl aqueous solution. Potential scan initiated at 0.45 V versus Ag/AgCl; potential scan rate 100 mV s^{-1}. Inset: optical microscopy image of a cross-section of the sample showing the red pictorial layer. Adapted from ref. [2], with permission. Note that the axes are opposite to the IUPAC convention

with acetate buffer using the VIMP technique. The voltammogram shows a tall peak corresponding to the reduction process:

$$Cu_2O_{solid} + 2H^+_{aq} + 2e^- \rightarrow 2Cu_{solid} + H_2O \quad (4.1)$$

In conventional VIMP experiments, the net amount of solid sample attached to the working electrode cannot be accurately controlled. Then, the peak currents—which in principle will be proportional to the amount of electroactive solid—vary from one to another replicate experiment. Accordingly, parameters associated with potential measurements –which are much less sensitive to the amount of sample–are used for identification purposes. The main parameters are indicated in Fig. 3.2: the peak potential (E_p) corresponds to the maximum current (peak current, I_p) taken from a suitable baseline whereas the half-peak potential ($E_{p/2}$) is the potential at a current half of the peak current ($I_{p/2}$). The onset potential (E_{onset}) can be defined as the potential of zero current which is obtained as the intersection between the baseline and the prolongation of the almost-linear (near inflection) region of the

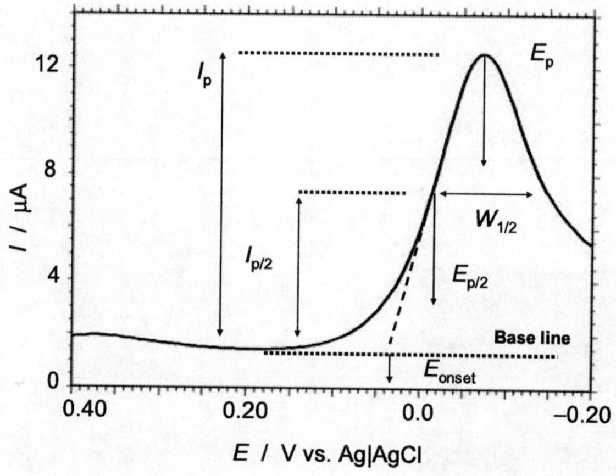

Fig. 4.2 Voltammetric parameters used for identification purposes marked over the square wave voltammogram of a microparticulate deposit of cuprite on paraffin-impregnated graphite electrode immersed into 0.25 M HAc/NaAc aqueous buffer at pH 4.85. Potential scan initiated at 0.40 V versus Ag/AgCl; potential step increment 4 mV; square wave amplitude 25 mV; frequency 5 Hz. Note that the axes are opposite to the IUPAC convention

central region of the voltammetric wave. The potential difference between the points at the half-peak current in the ascending and descending branches of the peak defines the half-peak width, $W_{1/2}$.

Of course, all the above parameters are dependent on the electrolyte solution and other factors (temperature, ...) and the electrochemical conditions (starting potential, potential scan rates, etc.) of the voltammetric experiment. Under fixed conditions, the aforementioned parameters can be used for identification purposes. Additional shape-characterizing parameters are the peak-to-half peak difference ($E_p-E_{p/2}$), and the onset potential to peak potential difference ($E_{onset}-E_p$), but often other easily defined parameters such as the ($E_{3/4}-E_{1/4}$) potential difference are used. Additionally, the voltammetric response can be influenced by the shape and size distribution of the particles of the solid.

In the case of multi-component systems where the individual components appear as separate particles, the voltammetric response of the different species will in principle be superimposed. Data treatment using semi-derivative and semi-integrative techniques offers increasing capabilities of signal resolution [5, 6]. Then, the identification of individual components can in principle be achieved using the above parameters, but they should be examined with caution: the interaction of the electroactive components of the sample with the matrix of the same and also the mutual interaction between electroactive components can distort the individual voltammetric responses [7]. Additionally, the above interactions, as well as the net amount of solid transferred onto the surface of the base electrode [8], may determine an increase in the uncompensated ohmic drops and capacitive effects increasing distortions of the voltammetric signals [9].

Figure 4.3 shows an example of the application of the VIMP methodology to a complex multi-component system. In this figure, the square wave voltammograms of a microparticulate deposit of the corrosion products recovered from buried leaded bronze artifacts of the Roman archaeological site of Gadara (Jordan) are successively recorded in the negative and positive directions shown. In the negative-going potential scan (a), a series of overlapping cathodic signals between −0.10 and −0.75 V versus Ag/AgCl is recorded. These signals can be attributed to the superposition of reduction processes of copper corrosion products (C_{Cu}), mainly cuprite and malachite (peak at ca. −0.15 V) and tenorite (−0.40 V) and lead corrosion products (C_{Pb}) occurring between −0.45 and −0.75 V. In the subsequent positive-going potential scan, two well-defined signals appear at −0.55 (A_{Pb}) and 0.05 V (A_{Cu}). These correspond to the oxidative dissolution of the metallic deposits of Pb and Cu electro-generated by the previous reduction of copper and lead corrosion products. Possible Sn-based voltammetric signals are usually absent due to the general phenomenon of destannification affecting bronze artifacts [10] and their masking by lead-based ones, in general much more intense.

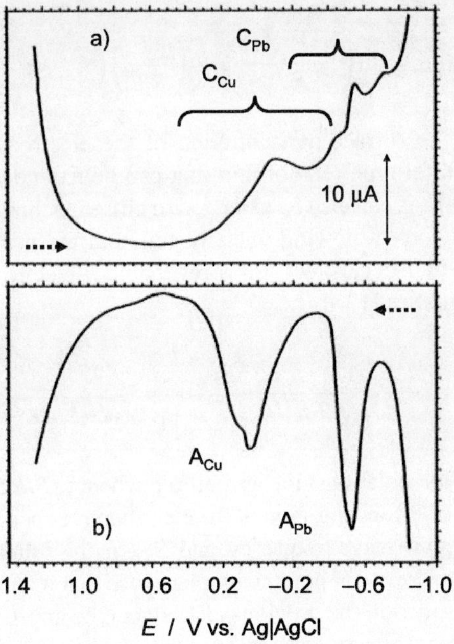

Fig. 4.3 Square wave voltammograms of a microparticulate deposit of the corrosion products from leaded bronze artifacts of the Roman archaeological site of Gadara (Jordan), dated back to the fourth century CE, attached to a graphite electrode. Electrolyte solution: 0.25 M HAc/NaAc, pH 4.75. Potential scan initiated at **a** 1.25 V in the negative direction, **b** −0.85 V in the positive direction. Potential step increment 4 mV; square wave amplitude 25 mV; frequency 5 Hz. Samples courtesy of Wassef Al Sekheneh, Yarmouk University Irbid-Jordan. The dotted arrows mark the direction of the potential scan. Note that the axes are opposite to the IUPAC convention

4.2.2 Tafel-Type Analysis

The voltammetric parameters described in the preceding section can be complemented with those determined by means of the so-called Tafel-type analysis of voltammetric curves. This derives from the relationship between the current (I) and the applied potential (E) recorded under steady-state, stationary conditions. This can be written as [11]:

$$\ln I = a + b(E - E^{\circ}) \tag{4.2}$$

where a and b are constants. Although in general these conditions do not strictly apply in voltammetric experiments, in most cases this type of relationship applies in the rising region of the voltammetric peaks. This is valid for irreversible, diffusion-controlled charge transfer processes involving species in the solution phase [12] but also for the oxidation/reduction of surface-confined species studied in Chap. 3. In this second case, under conditions of electrochemical reversibility, at the foot of the voltammetric wave, Eq. (3.52) tends to,

$$\ln I = \ln\left(\frac{n^2F^2A\Gamma}{RT}\right) + \left(\frac{nF}{RT}\right)\left(E - E^{\circ'}\right) \tag{4.3}$$

where Γ represents the surface concentration of the electroactive species. Equation (4.3) defines a Tafel-type relationship that can be extended to the irreversible case. This kind of analysis can also be extended to pulsed techniques. In square wave voltammetry, the net current flowing (usually called difference current) during the cathodic and anodic half-cycles, ΔI, for a reversible n-electron transfer involving species in solution satisfies [13, 14],

$$\Delta I = \frac{n^2F^2AD^{1/2}cf^{1/2}E_{SW}}{RT\pi^{1/2}} \frac{e^{nF(E-E^{\circ'})/RT}}{\left[1 + e^{nF(E-E^{\circ'})/RT}\right]^2} \tag{4.4}$$

tending to a Tafel-type relationship, as before, when ($nF/RT)E-E^{\circ\prime}$) << 1. In Eq. (4.4), c represents the concentration of the electroactive species, D is its diffusion coefficient, f is the square wave frequency, and E_{SW} is the square wave amplitude.

For our purposes, the relevant point to emphasize is that many solid-state voltammetric processes described in the previous chapter can be approached by Tafel-type current/potential relationships operating at the foot of the cathodic or anodic peaks. These can be expressed as:

$$I = hq_o e^{\frac{\beta F}{RT}(E-E^{\circ'})}; \quad \text{or} \quad \ln I = \ln h + \frac{\beta F}{RT}(E - E^{\circ'}) \tag{4.5}$$

where q_o represents the maximum charge passed, in principle corresponding to the exhaustion of the electroactive substance attached to the base electrode surface and

h is an electrochemical constant characterizing the process and the voltammetric parameters. This equation, however, has the drawback that in VIMP experiments the value of q_o cannot in general be exactly fixed and that $E^{o'}$ is also in general unknown. For these reasons, generalized Tafel-type plots consisting of linear relationships between the logarithm of the current to peak current ratio (I/I_p) and the applied potential minus peak potential difference can be used [15, 16]. For instance, in the case of the solid-to-solid transformations studied in Sect. 3.8, one can obtain generalized current/potential curves in linear potential scan voltammograms as [17]:

$$\frac{I}{I_p} = \frac{e^{\frac{\ln\gamma}{\gamma}}\left(1 + e^{-\ln\gamma}\right)^{\frac{1}{\gamma}+1} e^{g(E-E^{o'})}}{\left[1 + e^{\gamma g(E-E^{o'})}\right]^{\frac{1}{\gamma}+1}} \tag{4.6}$$

From Eqs. (3.64) and (3.66). Replacing the formal electrode potential with the peak potential given by Eq. (3.65) leads to the limiting expression for the foot of the voltammetric wave [17]:

$$\ln\left(\frac{I}{I_p}\right) = \ln\left(1 + e^{-\ln\gamma}\right) + g(E - E_p) \tag{4.7}$$

This corresponds to a generalized Tafel-type relationship predicting a linear dependence of $\ln(I/I_p)$ on E–E_p. This relationship is independent of the amount of sample transferred onto the electrode surface so that the slope and the ordinate at the origin of the corresponding representation can be used to characterize different materials.

The precedent relationship has a drawback with respect to the sensitivity of the current measurements to the choice of a proper baseline. A less sensitive relationship can be used taking the central region of the voltammetric wave. For the case of solid-to-solid transformations, one can take [17]:

$$\frac{I_p - I}{I_p} \approx \left(\frac{e^{\frac{\ln\gamma}{\gamma}}\left(1 + e^{-\ln\gamma}\right)}{2^{\frac{1}{\gamma}+1}}\right)\left[2^{\frac{1}{\gamma}+1} + \frac{\ln\gamma}{\gamma} - g(E - E_p)\right] \tag{4.8}$$

Equation (4.8) corresponds to a linear variation of the $(I_p–I)/I_p$ ratio with $E–E_p$. This is roughly equivalent to the empirical logarithmic relationship previously introduced as a modified Tafel-type analysis [18].

4.3 Quantification

4.3.1 General Aspects

The quantification of a selected component in cultural heritage samples is an obvious analytical target in the fields of archaeometry, conservation, and restoration. In classic solution electroanalysis, quantification of analytes can be obtained via exhaustive electrolysis, potentiometric, amperometric, and voltammetric techniques, among others. The charge passed during a certain Faradaic process permits determining the mass of the analyte in the sample using Faraday's electrolysis laws. Usually, the analytical problem can be reduced to the determination of the concentration of a given electroactive species in a problem solution.

In the case of linear potential scan voltammetric measurements, charges can be determined by transforming current/potential into current/time curves and their integration, so that the area under a voltammetric peak is representative of the charge passed. This method is complicated, however, by the need of subtracting the non-faradaic contributions. Another problem is that modern digital instruments never measure over the complete time span, but only during certain time increments. This leads to more or less significant losses of charge. Accordingly, the most frequent approach is based on the determination of peak currents. Under fixed electrochemical conditions, diffusion-controlled processes (both reversible and irreversible) display linear variations of the peak current with the concentration of the electroactive species, and external calibration strategies can be used. In the case of metals such as Ag, Cd, Cu, Pb, Sn, and Zn, the oxidative dissolution (stripping analysis) analytical strategy can be used to determine the concentration of their ionic species in solution [19]. Naturally, the performance of these analytical methods is conditioned by: i) the occurrence of interference and matrix effects; ii) the existence of coupled chemical reactions, and adsorption phenomena, ... producing deviations from the diffusion-controlled regime.

4.3.2 Voltammetric and Coulometric Quantification Strategies

The application of solution-phase electrochemistry methods in the field of cultural heritage, in general, would involve a preceding dissolution step of a solid sample using a suitable reagent. This requires handling highly diluted solutions of the analytes increasing the opportunities for sample loss and/or contamination and, in summary, the loss of analytical information. Most importantly, any dissolution leads to a loss of information about the solid as a specific phase. For these reasons, solid-state electrochemical methods of quantification can be adopted. These include methods of relative quantification (i.e., determining the relative quantities of two selected components of the sample) and methods of absolute quantification (i.e., determining the net concentration or abundance of a selected species in the sample).

In the following, we will refer to a solid sample containing one or several electroactive components. Generically, in both relative and absolute methods of quantification, there is a need for: (i) obtaining well-defined signals of the analyte(s) and eventually other 'auxiliary' (vide infra) species and (ii) the knowledge of the relationship between the 'intensity' of the signal and the net amount of the species responsible for this signal. This generalized 'intensity' can be the charge in chronocoulometric experiments, peak currents, or peak areas in voltammograms, but also peak potentials or other electrochemical parameters.

In many electrochemical experiments, it is possible to assume that the generalized intensity of the voltammetric signal associated with a component J, Y_J, is proportional to the net amount (mass or mol) of this component, m_J, attached to the electrode surface, i.e.:

$$Y_J = g_J m_J \tag{4.9}$$

where g_J is an electrochemical constant depending on the molecular mass of the analyte and the number of transferred electrons in the redox process. g_J depends on physico-chemical conditions (electrolyte, temperature, ...) and the electrochemical conditions of operation (initial potential, potential scan rate in linear techniques, pulse parameters in pulsed techniques, ...). Unfortunately, the determination of g_J is problematic because (i) it is difficult to control the amount of sample transferred onto the electrode surface; (ii) even if this quantity were known with reasonable accuracy, it is difficult to estimate the exact effective amount of sample attached to the electrode. This last can be rationalized by taking into account that the solid sample forms a more or less irregular deposit of particles onto a base electrode as shown in Fig. 4.4 [20]. In typical VIMP experiments at paraffin-impregnated graphite electrodes, the base electrode usually consists of a set of more or less irregular laminas. As schematized in Fig. 4.5, since the electroactive region of the solid particles is initially limited to the surroundings of the particle/base electrode/electrolyte three-phase junction, the effective amount of sample cannot be known prior to voltammetric runs.

Additionally, note that the abrasive transference of the solid particles will produce a more or less intensive scratching of the electrode surface in each individual experiment. Finally, the intrinsic electroactivity of graphite edge planes is, in principle, the much more active sites than the basal plane sites [21, 22]. Accordingly, the nature and surface conditioning of the basal electrode eventually forming different types of surface oxygen functionalities [23, 24] can notably influence the voltammetric response.

Accordingly, replicate VIMP experiments on 'fresh' microparticulate deposits taken from the same sample, in general, do not reproduce the values of Y_J. However, if two components A and B displaying well-resolved signals are considered, their relative amount can be calculated on the basis of the relationship:

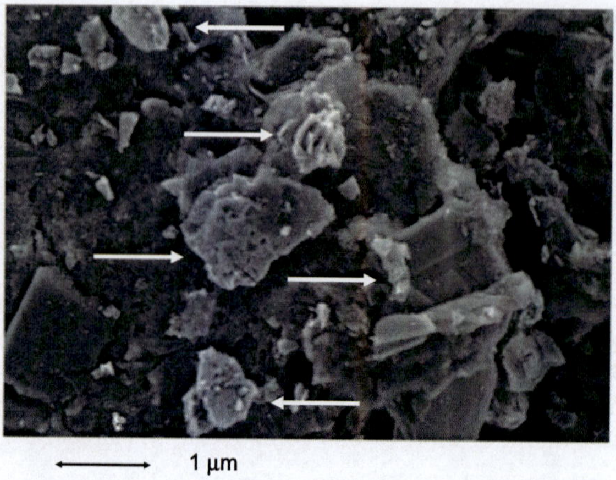

Fig. 4.4 SEM image of a microparticulate deposit of bronze corrosion products over graphite bar. The grains of corrosion products were extracted from an artificial patinated bronze probe by pressing a graphite cylinder (2 mm diameter) over the metallic lamina. The arrows mark several grains of corrosion products. Taken from ref. [20], with permission

Fig. 4.5 Schematics for a deposit of microparticles of two solid components transferred onto the surface of a graphite plate assumed to be formed by irregularly distributed laminas

$$\frac{Y_B}{Y_A} = \left(\frac{g_B}{g_A}\right)\left(\frac{m_B}{m_A}\right) \tag{4.10}$$

providing that the g_B/g_A ratio is known. This quantity can reasonably be estimated from experiments using standards prepared with known m_B/m_A ratios.

One way to directly access quantitative information is through the performance of chronocoulometric experiments. In these experiments, a constant potential is applied to the working electrode recording the resulting current/time and charge/time curves. When two electroactive species A and B produce electrochemical processes at well-separated potentials, the application of successive potential steps permits determining the corresponding consumed charges Q_A and Q_B. Then, the number of mol of A and

B existing in the sample, x_A and x_B, verify:

$$x_A/x_B = (Q_A/Q_B)(n_B/n_A) \tag{4.11}$$

n_A and n_B being the electron stoichiometry of the respective A- and B-localized electron transfer processes. This methodology, combined with VIMP, has been systematically applied by Scholz et al. [25, 26] for determining the composition of solid phases. The advantage of coulometric experiments is their intrinsic 'absolute' quantitative character; the drawback is that they offer a relatively large opportunity for the appearance of systematic deviations.

Remarkably, electrochemical quantification of solid samples allows determining the composition of mixtures of solids [27] but also for the determination of the composition of non-stoichiometric phases [28]. These methodologies can be combined with the application of electrochemical pretreatments. This is the case of the quantification of CdS in cadmium pigments (cadmium sulfide and cadmium sulfoselenide) reported by Cepriá et al. [29] from pigment plus AgCl mixtures, the latter being used as an internal standard. Here, a reductive pretreatment followed by an oxidative scan was applied. In this last step, stripping peaks for Ag and Cu oxidative dissolution processes were recorded. The height of such peaks was found to be linearly dependent on the Cd:Ag molar ratio in the mixture. The high inherent sensitivity of oxidative dissolution processes of metal deposits generated electrochemically has also been exploited for the relative quantification of metal ions in archaeological glass and glazes [30]. A similar philosophy was applied by Grygar and van Oorschot for determining the limit of voltammetric detection of ferric oxides in soil samples [31]. Here, quantification resulted from peak area measurements using an internal standard of MnO_2. A potentially interesting complement to voltammetric and coulometric methods is the use of an electrochemical quartz crystal microbalance, whose application for determining boron in ceramic materials has been described [32].

4.3.3 Standard Addition Methods

In solution-phase electrochemistry, it has to be accounted for the presence of interference and matrix effects. These effects derive from the presence of other species displaying signals overlapped with those of the analyte and when such species modify (via complexation, for instance) the response of the analyte. A well-known example of the matrix effect is the determination of Cu^{2+} in natural waters via stripping voltammetry. In 'clean' river water, usually, there are no important complexing effects and the analytical process, which involves a previous Cu^{2+} to Cu metal reduction, is completed by the measurement of the signal for the Cu to Cu^{2+} aq oxidation. In seawater, however, the high concentration of chloride ions stabilizes the Cu(I) oxidation state and the stripping process consists of the oxidation of Cu metal to Cu(I)-chloride complexes.

In VIMP experiments, there is the possibility of complex interference and matrix effects which will be treated individually for different types of samples in subsequent chapters. For quantification purposes, generically, we can compensate–or minimize–matrix effects using, as in solution-phase electroanalysis, the strategy of standard additions. This consists of the addition of known amounts of analyte to the sample and determining the values of intensity of the analytical signal.

This application of the extended solution-phase strategy to solid-state electrochemistry requires the combination of another well-known methodology: the use of an internal standard [33, 34]. This consists of the addition of known amounts of a new solid substance to the sample. Two essential requirements must be fulfilled: (i) the standard (S) will provide a voltammetric signal clearly separated from that of the analyte (A); ii) the intensity of the S-signal, as well as that of the analyte, has to be linearly dependent on the net amount of each component in the sample attached to the electrode.

The general problem is to determine the mass fraction f_A of a given component in a solid sample. For this purpose, let us assume that we dispose of sample aliquots of mass m, each one containing a mass m_A of A (so that $f_A = m_A/m$). Now, we spike these aliquots with a known m_S mass of an internal standard S so that its mass fraction f_S $(= m_S/(m + m_S))$ is the same. Performing a replicate voltammetric measurement of such S-enriched samples will provide a constant ratio between the intensities of the A- and S-localized signals, Y_A and Y_S, satisfying:

$$\frac{Y_A}{Y_S} = \left(\frac{g_A}{g_S}\right)\left(\frac{m_A}{m_S}\right) = \left(\frac{g_A}{g_S}\right)\left(\frac{f_A}{f_S}\right)\left(\frac{m}{m + m_S}\right) \tag{4.12}$$

Here, g_A and g_S are the respective coefficients of electrochemical response whose ratio, g_A/g_S, can be determined in blank experiments using synthetic A + S mixtures of known composition. Naturally, the problem is that the matrix of the problem sample, in general unknown, may differ from that used in the reference mixtures.

Accordingly, it is possible to combine the above internal standard methodology with standard additions of analyte to obtain 'absolute' quantifications. For this purpose, we prepare a series of A-enriched and S-enriched samples. Each solid specimen is prepared by adding weighted masses of the original sample (m), the internal standard S (m_S), and the standard of the analyte A ($m_A{}^*$) so that the m_S/m ratio remains constant. Now, the mass fraction of A will be:

$$f_A + f_A^* = \frac{m_A + m_A^*}{m} \tag{4.13}$$

So that the intensities of the A- and S-localized signals, Y_A and Y_S, will verify:

$$\left(\frac{Y_A}{Y_S}\right)^* = \left(\frac{g_A}{g_S}\right)\left(\frac{f_A + f_A^*}{f_S}\right)\left(\frac{m}{m + m_S}\right) \tag{4.14}$$

or, equivalently:

$$\left(\frac{Y_A}{Y_S}\right)^* = \left(\frac{Y_A}{Y_S}\right) + \left(\frac{g_A}{g_S}\right)\left(\frac{f_A^*}{f_S(1+f_S)}\right) \tag{4.15}$$

Since the ratio f_A^*/f_S ($= m_A^*/m_S$) of each specimen is known, the plot of $(Y_A/Y_S)^*$ versus $f_A^*/f_S(1+f_S)$ will provide a straight line of slope g_A/g_S, ordinate at the origin Y_A/Y_S, and abscissa at the origin $-f_A/(f_S(1+f_S))$. Since f_S is known, the value of the mass fraction of A in the problem sample can easily be determined from the above parameter.

Analysis of uncertainties by means of the usual model of error propagation [33, 34] indicates that the optimal conditions for standard addition experiments correspond to values of g_A/g_S and m/m_S equal to unity. As the main limitation, the requirement of sample amounts of about 1–2 mg (as required for weighting in conventional analytical balances). This is not important, however, in several cases in which pigments and ceramic materials are involved.

This methodology can in principle be applied to any type of voltammetric signal, in particular to stripping peaks corresponding to the oxidative dissolution of metal deposits. In this case, no formation of intermetallic compounds of A and S must occur. As previously noted, it is a requirement that the signals of the analyte and the internal standard have to be totally independent. For practical purposes, the voltammetric peaks must be separated by at least 150–200 mV. The above conditions define the 'electrochemical compatibility' between A and S and apply to the determination of the content of boron [33] and zirconium [34] in ceramic materials using ZnO as the internal reference compound and H_3BO_3 and $Li_2B_4O_7$, and ZrO_2, as the respective standards.

Figure 4.6 shows the cyclic voltammetric responses of ZnO and $Li_2B_4O_7$ attached to paraffin-impregnated graphite electrodes in contact with 0.10 M NaCl + 0.25 M mannitol. This electrolyte was used due to the ability of mannitol (as well as other polyalcohols) for complexing boron. The respective cathodic processes for the reduction of Zn (C_1) and $Li_2B_4O_7$ (C_2) can be represented as:

$$ZnO_{solid} + 2H^+_{aq} + 2e^- \rightarrow Zn_{solid} + H_2O \tag{4.16}$$

$$Li_2B_4O_{7solid} + 14H^+_{aq} + 12e^- \rightarrow 4B_{solid} + 2Li^+_{aq} + 7H_2O \tag{4.17}$$

Upon cycling the potential scan in the positive direction, the deposits of Zn and B are oxidatively dissolved giving rise to the anodic signals A_1 and A_2, respectively:

$$Zn_{solid} \rightarrow Zn^{2+}_{aq} + 2e^- \tag{4.18}$$

$$B_{solid} + 4H_2O + (mannitol)_{aq} \rightarrow [(H_4BO_4^-)(mannitol)]_{aq} + 4H^+_{aq} + 3e^- \tag{4.19}$$

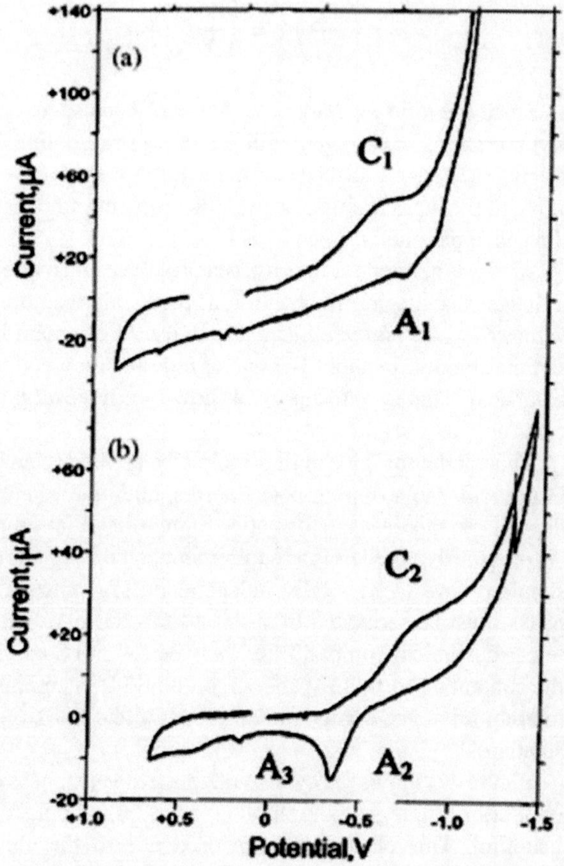

Fig. 4.6 Cyclic voltammograms of paraffin-impregnated graphite electrodes modified by: **a** ZnO and **b** $Li_2B_4O_7$, immersed into 0.10 M NaCl + 0.25 M mannitol. Potential scan initiated at 0.0 V versus Ag/AgCl in the negative direction; potential scan rate 100 mV s^{-1}. Adapted from ref. [33], with permission. Note that the axes are opposite to the IUPAC convention

Figure 4.7 shows a series of square wave voltammograms recorded for different ZnO + $Li_2B_4O_7$ mixtures attached to graphite bars in contact with 0.10 M NaCl + 0.25 M mannitol aqueous solution. Note that, given the insulating/semiconducting nature of elemental boron, the electrochemical reaction can take place only at the solid particle/electrolyte solution/base electrode three-phase junction. One can see how the intensities of the B and Zn stripping signals vary systematically with the proportion of both components in the solid mixture. Application of the standard addition methodology previously described led to satisfactory quantification of boron in minerals and ceramic samples [33].

The voltammetric profile depends on the potential and time of application of the previous reductive electrodeposition step and is, of course, also dependent on the pH and the presence of complexing agents in the electrolyte solution. Apart from the possible formation of intermetallic compounds, the stripping response is influenced by the modification of the electrode surface due to the sequential deposition of the metals. In the case illustrated in Fig. 4.7, the B deposition occurs at potentials less negative than those of Zn deposition. This means that the nucleation and growth of

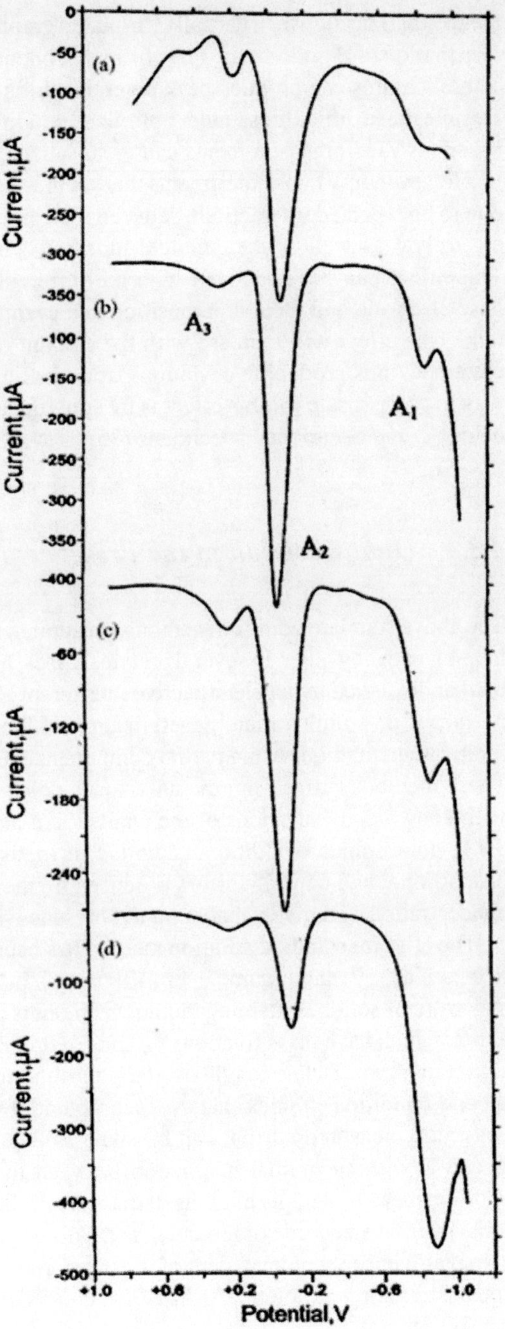

Fig. 4.7 Square wave voltammograms of paraffin-impregnated graphite electrodes modified with different ZnO + $Li_2B_4O_7$ mixtures. Electrolyte solution was 0.10 M NaCl + 0.25 M mannitol. $Li_2B_4O_7$/ZnO mass ratios of: **a** 2.110; **b** 0.650; **c** 0.312; **d** 0.078. Potential scan initiated at −1.05 V versus Ag/AgCl in the positive direction after an electrodeposition step of 60 s at that potential. Potential step increment 4 mV; square wave amplitude 25 mV; frequency 15 Hz. Adapted from ref. [33], with permission. Note that the axes are opposite to the IUPAC convention

elemental B occur over a partially Zn-plated graphite surface. Then, the shape and size distribution of B grains may differ from that obtained without previous Zn deposition. These features can produce peak potential shifts and distorted background currents. Despite these difficulties, under optimized conditions, standard addition experiments based on VIMP measurements can yield satisfactory results.

However, in VIMP experiments there is the possibility of deviations from linearity due to unexpected interactions between the components of the solid mixture and the electrolyte as well as their mutual blocking when attached to the base electrode. This effect can be associated with the different adherence of the particles to the base electrode and their superposition and eventually, other inter-component cross-talk. This effect will increase with the amount of sample transferred onto the base electrode, thus producing deviations from the linearity predicted by Eq. (4.15). One way to compensate for this effect is by applying a standard addition dilution method using A- and S-enriched specimens [35].

4.3.4 *Quantification in the Presence of Interferents*

The above standard addition methods are applied when the analyte signal is not overlapping with any other. In several practical cases, however, this signal is superimposed on that displayed by an electroactive interferent Z. This problem has been addressed by means of a multivariate generalization of the standard addition method [36] or multivariate methods using partial component and partial least squares analysis [37]. These methods suffer from the need for the disposal of a set of calibration samples with known concentrations of the analyte. An alternative to be applied to VIMP is the H-point standard addition method. This method allows for a direct correction of proportional and constant errors produced by the matrix in the determination of the concentration of a given analyte in the presence of a direct interferent [38].

The H-point standard addition method has been used for determining lead and tin in ceramics [39] and organic dyes [40] using VIMP. For this purpose, let us consider a mixture of solids containing unknown amounts of two electroactive compounds, A and Z, which have mass fractions f_A and f_Z. It will be assumed that the voltammetric signals of A and Z are strongly overlapping although they behave independently. This second condition implies that, at each potential within the common voltammetric wave, the measured current can be taken as the sum of the individual currents for A and Z without mutual interaction between them. As in the precedent standard addition method, sample aliquots of mass m will be enriched with a weighted amount of an internal standard compound S, m_S. This compound will display a voltammetric signal also entirely independent of those recorded for A and Z, appearing at a clearly separated region of potentials. Again, of course, it will be assumed that all materials can be accurately powdered and thoroughly mixed giving rise to a homogeneous distribution of the different solid components within the S-spiked specimens.

Under the above favorable circumstances, the H-point standard addition method can be applied by measuring the peak current of S, I_S, and the current at two potentials,

E_1 and E_2, of the common A + Z voltammetric wave, I_1 and I_2. For a given specimen, the I_1/I_S and I_2/I_S current ratios will satisfy:

$$\frac{I_1}{I_S} = \left(\frac{g_{1A}}{g_S}\right)\left(\frac{f_A m}{m + m_S}\right) + \left(\frac{g_{1Z}}{g_S}\right)\left(\frac{f_Z m}{m + m_S}\right) \tag{4.20}$$

$$\frac{I_2}{I_S} = \left(\frac{g_{2A}}{g_S}\right)\left(\frac{f_A m}{m + m_S}\right) + \left(\frac{g_{2Z}}{g_S}\right)\left(\frac{f_Z m}{m + m_S}\right) \tag{4.21}$$

As before, g_S represents the electrochemical coefficient of the response of S, while g_{1A}, g_{1Z}, g_{2A}, and g_{2Z} represent the coefficients of the response of A and Z at the potentials E_1 and E_2. These coefficients of response are dependent on the molecular mass of each component and characterize the redox properties of the respective analyte (number of electrons transferred, reversibility, diffusion coefficients for electrons and ions), the electrochemical conditions (potential scan rate, etc.), and other experimental parameters (electrolyte, pH).

The standard additions experiment can be performed by preparing a series of S-enriched specimens containing weighted amounts of a standard of A, m_A*. Then, the I_1/I_S and I_2/I_S ratios become:

$$\left(\frac{I_1}{I_S}\right)^* = \left(\frac{g_{1A}}{g_S}\right)\left(\frac{f_A + f_A^*}{f_S}\right)\left(\frac{m}{m + m_S}\right) + \left(\frac{g_{1Z}}{g_S}\right)\left(\frac{f_Z m}{m + m_S}\right) \tag{4.22}$$

$$\left(\frac{I_2}{I_S}\right)^* = \left(\frac{g_{2A}}{g_S}\right)\left(\frac{f_A + f_A^*}{f_S}\right)\left(\frac{m}{m + m_S}\right) + \left(\frac{g_{2Z}}{g_S}\right)\left(\frac{f_Z m}{m + m_S}\right) \tag{4.23}$$

These equations predict that plots of $(I_1/I_S)^*$ and $(I_2/I_S)^*$ versus $f_A{}^*m/f_S(m + m_S)$ $(= m_A{}^*/f_S(m + m_S))$ must give straight lines of slopes g_{1A}/g_S and g_{2A}/g_S, respectively. Such straight lines intersect at a point (H-point) where:

$$\left(\frac{f_A^*}{f_S}\right)\left(\frac{m}{m + m_S}\right)_{\text{H point}} = \left(\frac{m}{m + m_S}\right)\frac{\left(\frac{g_{2A} - g_{1A}}{g_S}\right)\left(\frac{f_A}{f_S}\right) - \left(\frac{g_{2Z} - g_{1Z}}{g_S}\right)\left(\frac{f_Z}{f_S}\right)}{\left(\frac{g_{1A}}{g_S}\right) - \left(\frac{g_{2A}}{g_S}\right)} \tag{4.24}$$

If the potentials E_1 and E_2 are taken so that the corresponding intensities of the interferent are equal; i.e., if $g_{1Z} = g_{2Z}$, the above equation reduces to:

$$\left(\frac{f_A^*}{f_S}\right)_{\text{H point}} = -\frac{f_A}{f_S} \tag{4.25}$$

Then, the f_A/f_S ratio, and hence, the value of f_A (i.e., the concentration of A in the original sample) can be determined from the abscissa of the H-point. Naturally, this requires previous knowledge of the voltammetric responses of the analyte and the interferent.

As previously noted, the uncertainty associated with the m/m_S ratio can be minimized by weighting relatively high amounts of sample and the reference material. Assuming that all mass measurements have the same uncertainty, the optimal situation occurs for m/m_S equal to unity. On the other hand, there is a need to account for uncertainties associated with the measure of the currents (I_1, I_2, and I_S). In practice, reasonable measurements can be obtained when $0.05 < I_1/I_2$ and $I_2/I_S < 20$, the uncertainty in the current ratios being close to the unity. Application of the H-point method to the determination of lead and tin in archaeological glazes [39] and alizarin and purpurin in pictorial samples [40] has been reported.

In general, the composition of the matrix and possible interfering species is unknown. Additionally, there exists the possibility of deviations from linearity in the voltammetric responses due to the interactions between components and/or the influence of other effects (shape and size distribution of the particles, for instance). Accordingly, the linear dependences are given by Eqs. (4.12) to (4.15) and (4.20) to (4.24) and can be seen as the asymptotic tendencies attained at small transferred masses. Accordingly, relatively large dilutions of the electroactive components into an electrochemically silent matrix such as silica or clays are desirable. These asymptotic relationships are of interest in other analytical contexts and contribute to solving a general problem in analytical chemistry: the determination of an analyte in the presence of unknown interferents and unknown matrix effects [41].

4.4 Speciation

The term speciation can be applied in general to any analytical technique aimed to discriminate between different directly related species differing in structure, oxidation state, etc. Here, we will focus the attention on species existing in two oxidation states. The essential idea is that voltammetric data can be used to identify the prevailing oxidation state and even to quantify the proportion between the involved states. One case of coexistence of different oxidation states is the Prussian blue/Berlin green/Everett's salt, but an equivalent situation can be found in two organic pigments, Brazilwood and Campeche wood (logwood or litmus Campeche). The former is mainly composed of the reduced from brazilin, tetracyclic with two aromatic rings, one pyrone, and one five-membered unsaturated ring, which is frequently accompanied by its oxidized form, brazilein. In the second case, the main coloring agents are hematoxylin and hematein, also reduced and oxidized forms with a common benzopyrone structure [42]. Given the close structural similarity between the oxidized and reduced redox forms, it is in general difficult to discriminate between such forms.

A first electrochemical approach to this analytical problem was formulated by Scholz and Hermes [43], who proposed a simple method based on cyclic voltammetric experiments of application to reversible redox processes involving species in solution. Here, the entire voltammogram is shifted along the current axis depending on the redox state. Then, the difference between the absolute values of the extreme currents measured in the voltammogram at potentials symmetric to the mid-peak

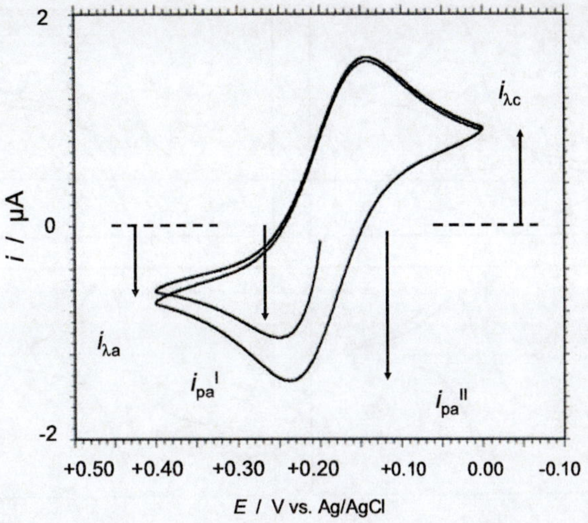

Fig. 4.8 Peak currents and extreme (switching) currents used to determine the oxidation state of the electroactive species displaying a reversible behavior superimposed to an experimental cyclic voltammogram at Pt electrode recorded for a 0.50 mM $K_4[Fe(CN)_6]$ + 0.50 mM $K_3[Fe(CN)_6]$ solution in 0.10 M $NaClO_4$. Potential scan rate 10 mV s^{-1}. Reproduced from ref. [44], with permission. Note that the axes are opposite to the IUPAC convention

potential ($|I_{\lambda a}| - |I_{\lambda c}|$) see Fig. 4.8) in repetitive voltammetry was found to be linearly dependent on the molar ratio of the reduced (or oxidized) form of the reactant.

A second methodology was proposed based on the measurement of the ratio between the anodic (or cathodic) peak current in the first (I_{pa}^{I}) and second (I_{pa}^{II}) scans in cyclic voltammograms initiated just at the mid-peak potential of the redox couple [44]. This method is based on the common observation that, in the case of electrochemically reversible systems, when the potential scan is initiated at a potential clearly more cathodic or more anodic than the formal electrode potential of the couple, the cyclic voltammogram remains essentially indistinguishable regardless of the oxidation state of the electroactive species (see Fig. 4.9). This is due to the efficient oxidation of the reduced form and/or the reduction of the oxidized form before the system reaches the potentials near the mid-peak potential. However, when the potential scan is initiated just (or very close to) at the formal electrode potential, the initial current will be cathodic if the oxidized form prevails or anodic if the reduced form predominates. As a result, the peak current in the first scan will clearly differ from the peak current in the second and successive scans. This is reflected in Fig. 4.9, where cyclic voltammograms initiated at different potentials for $K_3[Fe(CN)_6]$ solutions in 0.10 M KCl are depicted. This method permits to distinguish between extreme redox states and, after calibration, leads to quantify their proportion in a mixture of states.

The application of the above methods is illustrated in Fig. 4.10 for $[Fe(CN)_6]^{3-}$/$[Fe(CN)_6]^{4-}$ solutions at different potential scan rates. One can see that both the ($|I_{\lambda a}| - |I_{\lambda c}|$) difference and the I_{pa}^{I}/I_{pa}^{II} ratio vary linearly with the molar fraction of

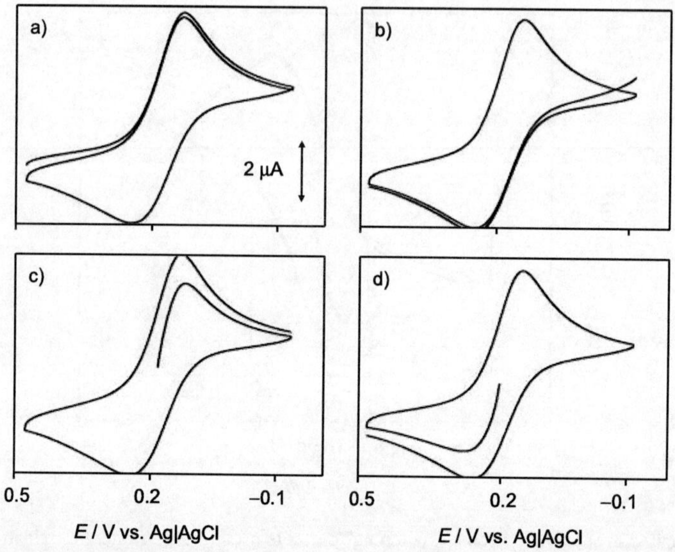

Fig. 4.9 Cyclic voltammograms at gold electrode of 1.67 mM $K_3[Fe(CN)_6]$ solution in 0.10 M KCl. Potential scan initiated at **a** 0.48 V versus Ag/AgCl in the negative direction; **b** −0.12 V in the positive direction; **c** 0.18 V in the negative direction; d) 0.18 V in the positive direction. Potential scan rate 50 mV s^{-1}

oxidized (or reduced) form, f_{ox}. Although the practical application of these methods can be disturbed by the presence of uncompensated ohmic and capacitive effects in the cell, several applications to reversible solid-state processes in the cultural heritage domain will be commented on in further chapters.

4.5 Electrochemical Data Processing

4.5.1 *Handling Electrochemical Signals*

Regardless of the concrete electrochemical technique used for analytical purposes (coulometry, voltammetry, impedance spectroscopy, ...), the physical quantities measured in the experiments can be submitted to processes of background subtraction, deconvolution, derivation, integration, ..., common to other instrumental techniques [6]. The reader will find several examples of the above data treatments along the text. These methods are in general available in commercial types of equipment, including portable devices, and permit the refinement of the raw data. Of course, all these operations should be employed after a judicious interpretation of the base results because several manipulations may produce artifacts distorting their analytical significance.

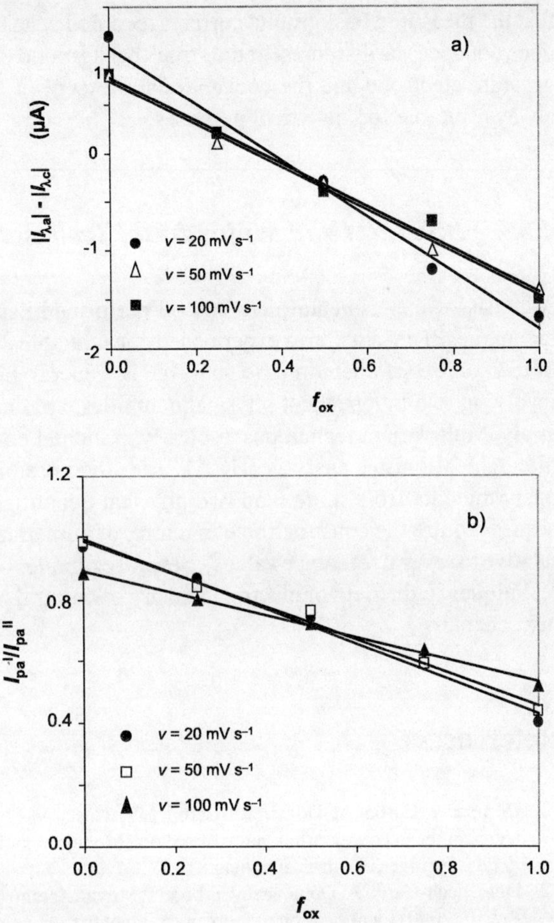

Fig. 4.10 Plots of: **a** $|I_{\lambda a}| - |I_{\lambda c}|$ and **b** $I_{pa}{}^{I}/I_{pa}{}^{II}$ versus the molar fraction of oxidized form, f_{ox} for different $K_4[Fe(CN)_6]$ plus $K_3[Fe(CN)_6]$ mixtures (total concentration 1.0 mM) in 0.10 M $NaClO_4$. Experimental data from voltammetric data in conditions such as in Fig. 4.9. Adapted from ref. [44], with permission. Note that the axes are opposite to the IUPAC convention

A case of particular interest is the use of background subtraction in VIMP. Typically, this consists of the subtraction from the voltammogram recorded at a sample-modified graphite electrode of the blank voltammogram recorded for the bare graphite electrode. This operation can produce erratic results because of the abrasive nature of the process of electrode conditioning. This process leads not only to a partial occupation of the graphite surface area but also produces a scratching of the surface. As a result, the exposed area varies and, more importantly, there is a modification of the graphite active sites. As previously indicated, it is believed that the edge planes are more active than basal plane sites [21, 22]. Additionally, different types of surface oxygen functionalities can be generated also influencing the voltammetric response [23, 24]. Ideally, we should dispose of a blank scratching material standardly used for electrode conditioning and reproducing the mechanical effect without any chemical modification. Unfortunately, the variety of electrode modifiers makes it enormously difficult to find a 'universal' scratching material. The result is that the measured background current (recorded at an unmodified graphite electrode) does not necessarily represent the "true" background existing in the sample-modified graphite electrode and the conventional protocol of background subtraction is not always suitable for analytical purposes.

4.5.2 *Bivariant and Multivariant Techniques*

Analysis of experimental data can be performed using bivariant and multivariant techniques. The latter have experienced considerable development in the last decades in the context of chemometric analysis. In general, bivariant techniques, i.e., those involving two independent physical quantities, are much more intuitive and easy to apply. Multivariant techniques, typically, principal components analysis (PCA) and hierarchical cluster analysis (HCA), offer the possibility of combining simultaneously the data from more than two physical quantities [45]. In general, these techniques require the construction of a matrix of numerical data (eventually normalized relative to several reference values). Several examples of applications of chemometric techniques in the field of electrochemistry of cultural heritage will be exposed in the next chapters.

References

1. Doménech-Carbó A, Doménech-Carbó MT, Costa V (2009) Electrochemical methods in archaeometry, conservation and restoration (Monographs in Electrochemistry series, Scholz F, Ed). Springer, Berlin-Heidelberg
2. Doménech-Carbó A, Doménech-Carbó MT, Moya-Moreno M, Gimeno-Adelantado JV, Bosch-Reig F (2000) Identification of inorganic pigments from paintings and polychromed sculptures immobilized into polymer film electrodes by stripping differential pulse voltammetry. Anal Chim Acta 407:275–289
3. Ghosh PK, Mau AW-H, Bard AJ (1984) Clay-modified electrodes: Part II. Electrocatalysis at bis(2,2'-bipyridyl) (4,4'-dicarboxy-2,2'-bipyridyl)Ru(II)-dispersed ruthenium dioxide—hectorite layers. J Electroanal Chem 169:315–317
4. Liu H-Y, Anson FC (1985) Electrochemical behavior of cationic complexes incorporated in clay coatings on graphite electrodes. J Electroanal Chem 184:411–417
5. Dalrymple-Alford P, Goto M, Oldham KB (1977) Peak shapes in semidifferential electroanalysis. Anal Chem 49:1390–1394
6. Grdeń M (2017) Semi-differential analysis of irreversible voltammetric peaks. J Solid State Electrochem 21:1045–1058
7. Doménech-Carbó A, Doménech-Carbó MT, López F, Valle-Algarra FM, Osete-Cortina L, Arcos E (2013) Electrochemical characterization of Egyptian blue pigment in wall paintings using the voltammetry of microparticles methodology. Electroanalysis 25:2621–2630
8. Meyer B, Zhang S, Scholz F (1996) The quantitative analysis of mixed crystals CuS_xSe_{1-x} with abrasive stripping voltammetry and a redetermination of the solubility product of CuSe and the standard potential of the Cu/CuSe electrode. Fresenius J Anal Chem 356:267–270

9. Doménech-Carbó A, Doménech-Carbó MT, Redondo-Marugán J, Osete-Cortina L, Vivancos-Ramón MV (2016) Electrochemical characterization of corrosion products in leaded bronze sculptures considering ohmic drop effects on Tafel analysis. Electroanalysis 28:833–845
10. Bernard MC, Joiret S (2009) Understanding corrosion of ancient metals for the conservation of cultural heritage. Electrochim Acta 54:5199–5205
11. Petrii OA, Nazmutdinov RR, Bronschtein MD, Tsirlina GA (2007) Life of the Tafel equation: current understanding and prospects for the second century. J Electroanal Chem 52:3493–3504
12. Reinmuth WH (1960) Irreversible systems in stationary electrode polarography. Anal Chem 32:1891–1892
13. Ramaley L, Krause MS (1969) Theory of square wave voltammetry. Anal Chem 41:1362–1365
14. Krause MS, Ramaley L (1969) Analytical application of square wave voltammetry. Anal Chem 41:1365–1369
15. Doménech-Carbó A, Doménech-Carbó MT, Martínez-Lázaro I (2008) Electrochemical identification of bronze corrosion products in archaeological artefacta. Case study. Microchim Acta 162:351–359
16. Doménech-Carbó A, Doménech-Carbó MT, Edwards HGM (2008) Quantitation from Tafel analysis in solid-state voltammetry. Application to the study of Cobalt and Copper pigments in severely damaged frescoes. Anal Chem 80:2704–2716
17. Doménech-Carbó A (2022) Description of solid-to-solid redox processes based on the voltammetry of immobilized particles methodology: a logistic approximation. J Phys Chem C 126:11822–11832
18. Doménech-Carbó A, Doménech-Carbó MT, Pasíes-Oviedo T, Bouzas-Bello MC (2011) Application of modified Tafel analysis to the identification of corrosion products on archaeological metals using voltammetry of microparticles. Electroanalysis 23:2803–2812
19. Wang J (1985) Stripping analysis. principles, instrumentation and applications. VCH, Weinheim-New York
20. Doménech-Carbó A, Ramírez-Barat B, Petiti C, Goidanich S, Doménech-Carbó MT, Cano E (2020) Characterization of traditional artificial patinas on copper using the voltammetry of immobilized particles. J Electroanal Chem 877:114494
21. Chen P, McCreery RL (1996) Control of electron transfer kinetics at glassy carbon electrodes by specific surface modification. Anal Chem 68:3958–3965
22. Davies TJ, Hyde ME, Compton RG (2005) Nanotrench arrays reveal insight into graphite electrochemistry. Angew Chem Int Ed 44:5121–5126
23. Barrière F, Downard AJ (2008) Covalent modification of graphitic carbon substrates by non-electrochemical methods. J Solid State Electrochem 12:231–1244
24. Noked M, Soffer A, Aurbach D (2011) The electrochemistry of activated carbonaceous materials: past, present, and future. J Solid State Electrochem 15:1563–1578
25. Scholz F, Müller WD, Nitschke L, Rabi F, Livanova L, Fleischfresser C, Thierfelder C (1990) Fast and nondestructive identification of dental alloys by abrasive stripping voltammetry. Fresenius J Anal Chem 338:37–40
26. Scholz F, Rabi F, Müller WD (1992) The anodic dissolution of dental amalgams studied by abrasive stripping voltammetry. Electroanalysis 4:339–346
27. Scholz F, Lange B, Jaworski A, Pelzer J (1991) Analysis of powder mixtures with the help of abrasive stripping voltammetry and coulometry. Fresenius' J Anal Chem 340:140–144
28. Meyer B, Zhang S, Scholz F (1996) The quantitative analysis of mixed crystals CuS_xSe_{1-x} with abrasive stripping voltammetry and a redetermination of the solubility product of CuSe and the standard potential of the Cu/CuSe electrode. Fresenius' J Anal Chem 356:267–270
29. Cepriá G, García-Gareta E, Pérez-Arantegui J (2005) Cadmium yellow detection and quantification by voltammetry of immobilized microparticles. Electroanalysis 17:1078–1084
30. Doménech-Carbó A, Doménech-Carbó MT, Osete L, Gimeno JV, Sánchez S, Bosch F (2003) Quantitation of metal ions in archaeological glass by abrasive stripping square wave voltammetry using graphite/polyester composite electrodes. Electroanalysis 15:1465–1475
31. Grygar T, van Oorschot IHM (2002) Voltammetric identification of pedogenic iron oxides in paleosol and loess. Electroanalysis 14:339–344

32. Doménech-Carbó A, Sánchez-Ramos S, Yusá-Marco DJ, Moyá-Moreno M, Gimeno-Adelantado JV, Bosch-Reig F (2004) Determination of the boron/lead ratio in ceramic materials based on electrochemical quartz crystal microbalance. Electroanalysis 16:1814–1822
33. Doménech-Carbó A, Sánchez-Ramos S, Yusá-Marco DJ, Moyá-Moreno M, Gimeno-Adelantado JV, Bosch-Reig F (2004) Electrochemical determination of boron in minerals and ceramic materials. Anal Chim Acta 501:103–111
34. Doménech-Carbó A, Moyá-Moreno M, Doménech-Carbó MT (2004) Standard addition method applied to solid-state striping voltammetry. Determination of zirconium in minerals and ceramic materials. Anal Bioanal Chem 380:146–156
35. Doménech-Carbó A, Martini M, de Carvalho LM, Viana C, Doménech-Carbó MT, Silva M (2013) Standard additions-dilution method for absolute quantification in voltammetry of microparticles. Application for determining psychoactive 1,4-benzodiazepine and antidepressants drugs as adulterants in phytotherapeutic formulations. J Pharm Biomed Anal 80:159–163
36. Saxberg BEH, Kowalski BR (1979) Generalized standard addition method. Anal Chem 51:1031–1038
37. Vadeginste BGM, Masart DL, Budens LMC, De Jong S, Lewi PJ, Smeyers-Verbeke J (1998) Handbook of chemometrics and qualimetrics. Elesvier, Amsterdam
38. Bosch-Reig F, Campins-Falcó P (1988) H-point standard additions method. Part 1. Fundamentals and application to analytical spectroscopy. Analyst 113:1011–1016
39. Doménech-Carbó A, Doménech-Carbó MT, Gimeno-Adelantado JV, Bosch-Reig F (2006) H-point standard addition method applied to solid state stripping voltammetry. Application to the determination of lead and tin in archaeological glazes. Anal Bioanal Chem 385:1552–1561
40. Bosch-Reig F, Doménech-Carbó A, Doménech-Carbó MT, Gimeno-Adelantado JV (2007) H-point standard addition method applied to voltammetry of microparticles. Quantitation of dyes in pictorial samples. Electroanalysis 19:1575–1584
41. Doménech-Carbó A, Bosch-Reig F, Montoya N (2020) ATR-FTIR and XRD quantification of solid mixtures using the asymptotic constant ratio (ACR) methods. application to geological samples of sodium and potassium feldspars. Spectrochim Acta A 236:118328.
42. Doménech-Carbó A, Doménech-Carbó MT, Saurí-Peris MC (2005) Electrochemical identification of flavonoid dyes in solid work of art samples by abrasive voltammetry at paraffin-impregnated graphite electrodes. Talanta 66:769–782
43. Scholz F, Hermes M (1999) The determination of the redox state of a dissolved depolariser by cyclic voltammetry in the case of electrochemically reversible systems. Electrochem Commun 1: 345–348. See corrigendum (2000) in Electrochem Commun 2: 814.
44. Doménech-Carbó A, Martini M, de Carvalho LM, Doménech-Carbó MT (2012) Square wave voltammetric determination of the redox state of a reversible oxidized/reduced depolarizer in solution and in solid state. J Electroanal Chem 684:13–19
45. Díaz-Cruz JM, Esteban M, Ariño C (2019) Chemometrics in electroanalysis, monographs in electrochemistry series, Scholz F, Edit. Springer, Berlin-Heidelberg

Additional Literature

46. Barisci JN, Wallace GG, Baughman RH (2000) Electrochemical studies of single-wall carbon nanotubes in aqueous solutions. J Electtroanal Chem 488:92–98
47. Brainina KhZ (1974) Stripping voltammetry in chemical analysis. Halsted, New York
48. Campins P, Bosch F, Verdú J, Molins C (1994) Study of the behaviour of the absorbent blanks in analytical procedures by using the H-point standard additions method (HPSAM). Talanta 41:39–52
49. Campins P, Verdú J, Bosch F (1994) Development of the H-point standard additions method for the use of spectrofluorimetry and synchronous spectrofluorimetry. Analyst 119:2123–2128

50. Caster DM, Toman JJ, Brown SD (1983) Curve fitting of semiderivative linear scan voltammetric responses: effect of reaction reversibility. Anal Chem 55:2143–2147
51. Costa V, Leyssens K, Adriaens A, Richard N, Scholz F (2010) Electrochemistry reveals archaeological materials. J Solid State Electrochem 14:449–451
52. Cura D'Ars de Figueiredo Junior J, de Freitas Cunha Lins V, De Bellis VM (2007) Surface characterization of a corroded bronze-leaded alloy in a salt spray cabinet. Appl Surf Sci 253:7104–7107
53. Huang W, Henderson TLE, Bond AM, Oldham KB (1995) Curve fitting to resolve overlapping voltammetric peaks: model and examples. Anal Chim Acta 304:1–15
54. Jakubowska M (2011) Signal processing in electrochemistry. Electroanalysis 23:553–572
55. Shams E, Abdollahi H, Yekehtaz M, Hajian R (2004) H-point standard additions method in the analysis by differential pulse anodic stripping voltammetry: Simultaneous determination of lead and thin. Talanta 63:359–364
56. Toman JJ, Brown SD (1981) Peak resolution by semiderivative voltammetry. Anal Chem 53:1497–1504
57. Trijueque J, García-Jareño JJ, Navarro-Laboulais J, Sanmatías A, Vicente F (1999) Ohmic drop of Prussian-blue: graphite-epoxy electrodes. Electrochim Acta 45:789–795

Chapter 5
Pigments and Paintings I

5.1 Introduction

Pigments are essential constituents of polychromed artworks and paintings [1]. There is a variety of painting techniques (oil, tempera, fresco, encaustic, etc.) in which a ground layer is applied on the artwork support (wall, canvas, panels or wood body, metal or stone body, etc.) and over that one or several colored paint layer(s). Pigments are defined as substances consisting of small particles that are practically insoluble in their application system and applied as colorants [1]. They differ from dyes, which also belong to the class of colorants, but they are highly soluble in their application medium. In many cases, the artwork is covered by a protective layer of varnish or coating (Fig. 5.1). Eventually, an imprimatur is applied between the ground and the paint layers. For obtaining a paint, the pigment is mixed with a binding medium for forming a dispersion of the pigment particles in the fluid binder, a natural or synthetic organic product with drying properties. The paint applied in fluid state finally results in a more or less flexible solid paint film. The binding medium can include a number of additives that improve the properties of the paint, avoiding aggregation, flocculation of the pigment, avoiding photodeterioration or microbial attack of the binding media, accelerating the drying process, etc. The number and proportion of additives can become high sometimes like in modern waterborne paints. In parallel to the stratified structure of the polychromed artworks, painters often mix several pigments in the same paint layer for obtaining the desired color. In some artistic techniques like oil painting, the painter applies a final glaze paint layer in which the pigments are in considerably low concentration in comparison with the binding medium. The stronger the color (i.e., the higher the absorption coefficient at a wavelength), the smaller the sufficient proportion of pigment. These features underline the analytical difficulty for the identification of pigments in real samples: apart from the deontological requirement of using so minimal amount of sample as possible, [1].

A. Doménech-Carbó and M. T. Doménech-Carbó, *Electrochemistry for Cultural Heritage*, Monographs in Electrochemistry,
https://doi.org/10.1007/978-3-031-31945-7_5

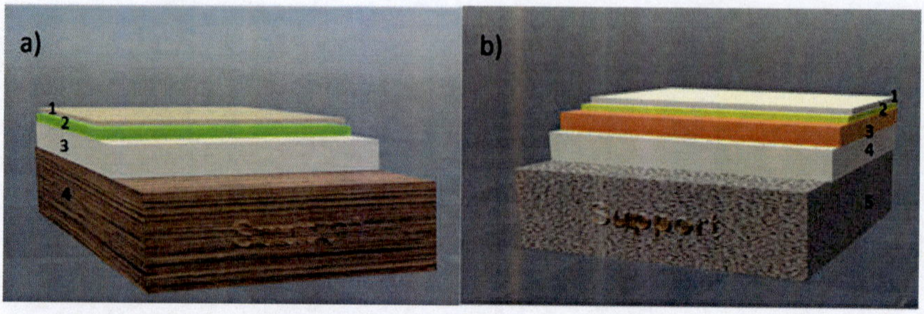

Fig. 5.1 **a** Strata distribution in a panel painting from the top to the bottom: varnish (1); paint layer (2); ground (3); wood support (4). **b** Strata distribution in a gilded decoration of walls: varnish (1); gold lamina (2); red bole (3); ground (4); wall support (5)

Conventional analysis of pictorial samples proceeds via the study of cross-sections extracted from a painting or polychromed artwork with the help of a scalpel. These samples are usually at the microgram level and their examination is made by optical and electron microscopies. In this chapter, the application of solid-state voltammetric techniques to pigment identification in artworks will be described. These techniques can be extended to systems such as textiles where pigments are used dissolved in a proper solvent, generally water, for dying the support.

The application of the VIMP methodology provides a direct way for the identification of pigments in paintings. Since the required net amount of sample can be of only a few nanograms, a plausible strategy is the reuse of the samples previously studied by microscopy.

5.2 Identification of Inorganic Pigments

The voltammetric identification of inorganic pigments was directly inspired by the VIMP mineralogical analysis developed by Scholz et al. [2] (see Chap. 2). The identification of cadmium, copper, lead, mercury, and zinc pigments can be carried out on the basis of the cathodic signals recorded at pigment-modified graphite electrodes in contact with aqueous electrolytes. Here, as illustrated in Fig. 5.2, significantly different voltammetric profiles can be recorded [3]. The discrimination between pigments of different metals can be confirmed by recording the oxidative dissolution signals recorded on scanning the potential in the positive direction from a sufficiently cathodic initial value.

In many cases, however, the identification of individual pigments having similar composition (for instance, azurite and malachite, both copper(II) hydroxycarbonates) is not easy. To optimize the discrimination between them, variations in the electrolyte composition and/or the electrochemical technique and the parameters can be performed. The response of the pigment can be quite sensitive to the electrochemical parameters. This can be seen in Fig. 5.3, where the linear potential scan voltammetry

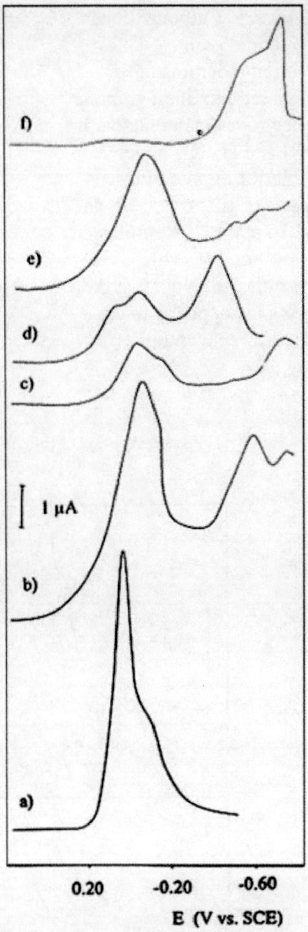

Fig. 5.2 Differential pulse voltammograms of **a** malachite, **b** copper resinate, **c** azurite, **d** verdigris, **e** zinc white, **f** litharge attached to glassy carbon electrodes through a Paraloid B72 polymeric film in contact with 0.50 M HAc/NaAc solution at pH 4.75. Potential scan initiated at 0.60 V versus SCE in the negative direction; potential scan rate 20 mV s^{-1}; pulse amplitude 20 mV. Reproduced from ref. [3], with permission. Note that the axes are opposite to the IUPAC convention

of litharge in contact with 0.10 M H_2SO_4 at different potential scan rates is depicted [4]. Under optimized fixed conditions, the voltammetric curves can be successfully used to discriminate different minerals and pigments for which a database has been assembled [5]. Note that the voltammetric response is also sensitive to the type of graphite used as a base electrode [5].

Figure 5.4 compares the linear potential scan voltammograms of graphite electrodes modified with massicot (PbO), lead-tin yellow (type II), lead white, and $PbCl_2{\cdot}2H_2O$ in contact with 0.10 M H_2SO_4 under fixed conditions [4]. Although the particle size distribution can influence the voltammetric response, one can see that, under fixed electrochemical conditions, the profile of the current/potential curve significantly differs between the different lead compounds. This clearly illustrates the discriminating capability of the VIMP.

In general, a given pigment can be identified using a series of voltammetric parameters as described in Sect. 3.1, as reported for lead [6] and iron [7, 8] pigments. Pulsed

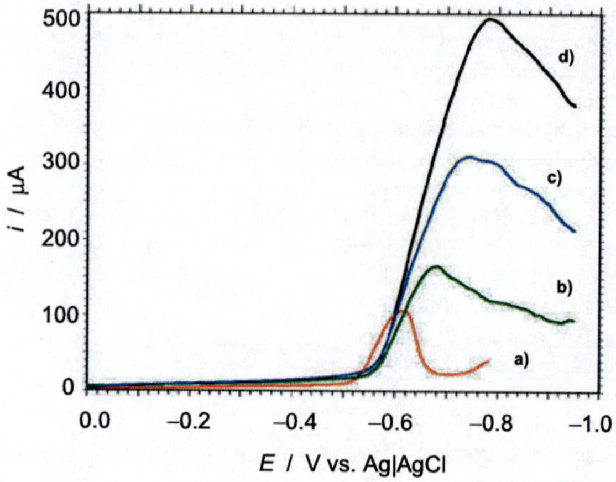

Fig. 5.3 Cathodic linear potential scan voltammograms of litharge-modified graphite electrodes in contact with 0.10 M H_2SO_4 aqueous solution at potential scan rates of **a** 1, **b** 5, **c** 10, and **d** 20 mV s^{-1}. Reproduced from ref. [4], with permission. Note that the axes are opposite to the IUPAC convention

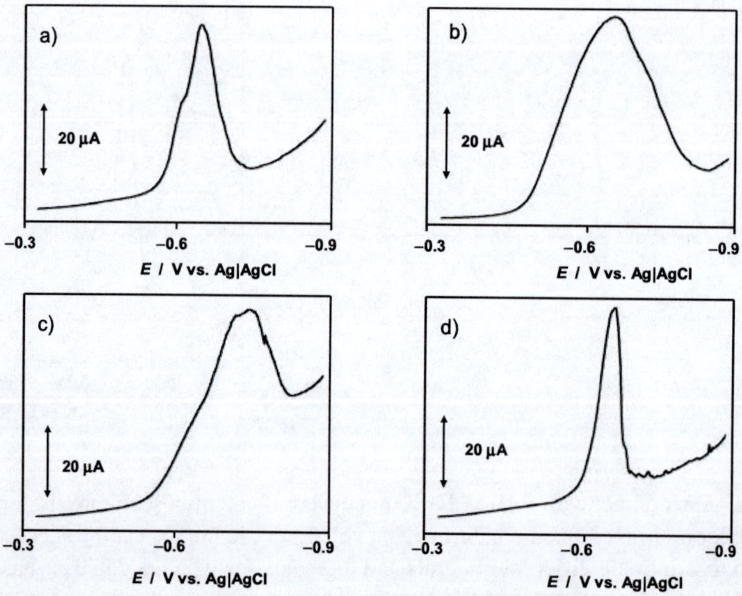

Fig. 5.4 Negative-going linear potential scan voltammograms of graphite electrodes modified with **a** massicot, **b** lead-tin yellow (type iI), **c** lead white, **d** $PbCl_2{\cdot}2H_2O$, in contact with 0.10 M H_2SO_4 at $v = 2$ mV s^{-1}. Reproduced from ref. [4], with permission. Note that the axes are opposite to the IUPAC convention

techniques, in particular square wave voltammetry, have been used because it can discriminate capacitive currents.

In several cases, theoretical modeling such as described in Sect. 3.8 is available. This allows us to identify different minerals and pigments using a short set of numerical parameters. In this context, the application of generalized Tafel-type analysis

Table 5.1 Peak potentials (in mV versus Ag/AgCl), and logistic parameters g, γ, and slope (T_S) and intercept (T_{OO}) of the Tafel-type plots derived from linear potential scan voltammograms of different lead compounds attached to graphite electrodes in contact with 0.10 M H_2SO_4. Potential scan rate of 2 mV s^{-1}. Averaged data from three independent voltammetric measurements at freshly modified electrodes. Reproduced from ref. [4], with permission

Solid	E_p	g (V^{-1})	γ	T_S (V^{-1})	T_{OO}
Litharge	−640 ± 5	−35 ± 1	1.0 ± 0.1	−34 ± 2	1.4 ± 0.2
Massicot	−645 ± 5	−59 ± 1	1.0 ± 0.1	−59 ± 2	2.0 ± 0.3
Lead-tin yellow II	−660 ± 5	−25 ± 1	0.8 ± 0.2	−29 ± 2	2.7 ± 0.3
Lead white	−750 ± 10	−35 ± 1	0.6 ± 0.1	−24 ± 2	1.4 ± 0.2
Naples yellow	−520 ± 10	−75 ± 3	0.9 ± 0.2	−68 ± 3	0.2 ± 0.2
$PbCl_2{\cdot}2H_2O$	−660 ± 5	−60 ± 2	1.5 ± 0.2	−54 ± 2	1.0 ± 0.2

(and its modified version) described in Sect. 4.2.2 can provide a simple approach for identification. Table 5.1 shows an example for the use of these criteria in the electrochemical identification of different lead species [4].

In several cases, the identification of the pigments from paint samples was straightforward. As an example, Fig. 5.5 shows the oil painting *Desposorios místicos de Santa Margarita* painted by Girolamo Francesco Maria Mazzola (Il Parmigianino) (1503 – 1540) accompanied by the cross-sections of one red sample where the paint layer and the ground can clearly be distinguished [5]. Vermilion can be identified as the red pigment in the paint layer, and lead white was identified in the ground as judged by comparing the voltammograms of the samples with those of the reference materials (see Fig. 5.6).

5.3 Pigment Mixtures

The VIMP technique also provides opportunity for identifying or even quantifying pigment mixtures. A case of interest is that of the frescoes of the *Sant Joan del Mercat* in Valencia (Spain) church painted by Antonio Acisclo Palomino in 1707. As studied in more detail in Chap. 6, these frescoes suffered considerable damage and coexisted with other wall paintings by Guilló in the church. VIMP data were permitted to discriminate between the Palomino paintings, where green and blue colors were prepared with azurite and smalt, and Guilló paints, who used malachite as green pigment [9, 10]. These were often accompanied by lead white and Naples yellow.

The discrimination between these pigments was difficult because of the similarity of their respective voltammetric responses, illustrated in Fig. 5.7. Azurite and malachite display epitactic reduction processes to copper metal which can be described as:

Fig. 5.5 **a** *Desposorios místicos de Santa Margarita* painted by the Italian painter Girolano Francesco Maria Mazzola (Il Parmigianino) (1503–1540) exhibited in the Hall town of Malaga, **b** Cross-section of the sample red R: (1) paint layer, (2) ground. Reproduced from ref. [5], with permission

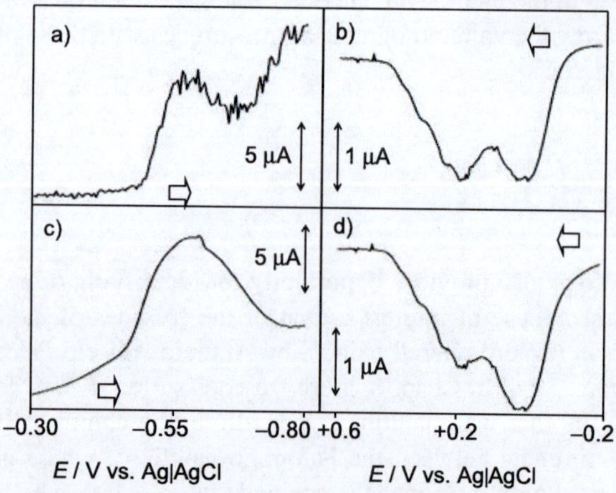

Fig. 5.6 Detail of the square wave voltammograms recorded on a sample **a**, **b** from the *Desposorios místicos de Santa Margarita* painting and lead white **c** and vermilion **d** reference materials attached to graphite lead (Staedtler) in contact with 0.25 M HAc/NaAc, pH 4.75. Potential scan initiated at **a**, **c** 0.65 V in the negative direction and **b**, **d** −1.05 V in the positive direction; potential step increment 4 mV; square wave amplitude 25 mV; frequency 5 Hz. Reproduced from ref. [5], with permission. The gross arrows mark the direction of the voltage scan at approximately the zero (difference) current level. Note that the axes are opposite to the IUPAC convention

Fig. 5.7 Cyclic voltammograms of paraffin-impregnated graphite electrodes modified with: **a** azurite (Kremer cat. n. 10,200), **b** malachite (Kremer cat. n. 10,300), and **c** smalt (Kremer cat. n. 10,000), in contact with 0.50 M potassium phosphate buffer, pH 7.4. Potential scan rate 50 mV s^{-1}. Reproduced from ref. [9], with permission. Note that the axes are opposite to the IUPAC convention

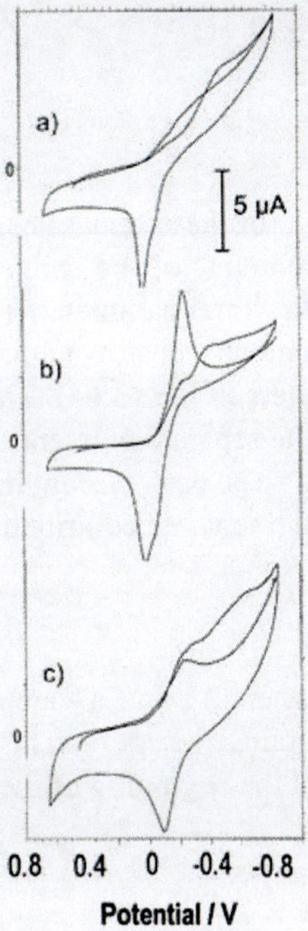

$$2CuCO_3 \cdot Cu(OH)_{2solid} + 6H^+_{aq} + 6e^- \rightarrow 3Cu_{solid} + 2CO_{2aq} + 4H_2O \quad (5.1)$$

$$CuCO_3 \cdot Cu(OH)_{2solid} + 4H^+_{aq} + 4e^- \rightarrow 2Cu_{solid} + CO_{2aq} + 3H_2O \quad (5.2)$$

respectively. In contact with aqueous electrolytes at neutral or moderately acidic pH, these processes appear as a well-defined cathodic wave at ca. −0.10 V vs. Ag/AgCl. In turn, smalt or smalt blue, a glass-type material containing Co^{2+} ions, a cathodic peak at similar potentials corresponding to the Co^{2+} to Co reduction was recorded. This response, as described in the case of cobalt cordierites and cobalt blue (cobalt spinel, Co_3O_4), is sensitive to the tetrahedral/octahedral nature of the coordinative environment of the metal ions [11].

For our purposes, the relevant point to emphasize is that azurite plus cobalt smalt mixtures in Palomino paintings can be quantified using the Tafel-type analysis of the cathodic curve. This requires two basic assumptions: (i) that both pigments display a reduction peak at an identical peak potential and (ii) the pigment particles are separated sufficiently in the paint layers so that their reduction processes are mutually independent. These conditions ensure that the recorded currents at any potential will be the sum of the independent contributions of the pigments. Then, using the model in ref. [4] or any equivalent formulation [9], one can describe the currents of the components X and Y, at the foot of the common voltammetric wave (I_X, I_Y) as (see Sects. 3.9 and 4.2.2):

$$I_X \approx q_X a_X e^{b_X(E-E_p)} \tag{5.3}$$

$$I_Y \approx q_Y a_Y e^{b_Y(E-E_p)} \tag{5.4}$$

In these equations, E_p denotes the common peak potential, q_X and q_Y, the net amounts of the components attached to the electrode, and a_X, b_X, a_Y, and b_Y are the Tafel parameters (slopes and intercepts) of the individual components. In linear potential scan voltammograms, these parameters are function of the g and γ parameters in Eq. (3.64). These can be determined in blank voltammetric experiments of the separate pigments.

The peak current measured in a given voltammetric experiment, I_p, is the sum of the separate contributions of the component X and Y; i.e.:

$$I_P = q_X A_X + q_Y A_Y \tag{5.5}$$

where A_X and A_Y are new parameters that have, in linear potential scan voltammograms, the coefficients g and γ in Eq. (3.66). Accordingly, the measured current (I) at the beginning of the voltammetric wave is:

$$I = q_X a_X e^{b_X(E-E_p)} + q_Y a_Y e^{b_Y(E-E_p)} \tag{5.6}$$

Assuming that the exponents are small, one can write [9]:

$$I \approx (q_X a_X + q_Y a_Y)\exp\left[\frac{q_X a_X b_X + q_Y a_Y b_Y}{q_X a_X + q_Y a_Y}(E - E_p)\right] \tag{5.7}$$

Then, the generalized Tafel-type equation can be expressed as:

$$\ln\left(\frac{I}{I_p}\right) = \ln\left(\frac{q_X a_X + q_Y a_Y}{q_X A_X + q_Y A_Y}\right) + \left(\frac{q_X a_X b_X + q_Y a_Y b_Y}{q_X a_X + q_Y a_Y}\right)(E - E_p) \tag{5.8}$$

and, introducing the Y to X molar ratio, f_{YX} $(= q_Y/q_X)$:

$$\ln\left(\frac{I}{I_p}\right) = \ln\left(\frac{a_X + f_{YX} a_Y}{A_X + f_{YX} A_Y}\right) + \left(\frac{a_X b_X + f_{XY} a_Y b_Y}{a_X + f_{YX} a_Y}\right)(E - E_p) \tag{5.9}$$

This equation reflects a Tafel-type relationship independent of the net amount of sample transferred onto the base electrode. Accordingly, the plots of $\ln(I/I_p)$ vs. $E - E_p$ produce straight lines with a slope and ordinate at the origin, which is intermediate between those recorded for the pure components X and Y separately in their corresponding generalized Tafel-type representations:

$$\ln\left(\frac{I_X}{I_{pX}}\right) = \frac{a_X}{A_X} + b_X(E - E_p) \tag{5.10}$$

$$\ln\left(\frac{I_Y}{I_{pY}}\right) = \frac{a_Y}{A_Y} + b_Y(E - E_p) \tag{5.11}$$

These equations permit the determination of b_X, b_Y, and the a_X/A_X and a_Y/A_Y ratios and, combining it, the determination of the f_{YX} value from Eq. (5.9). Application to pictorial samples from the Palomino's frescoes in the *Sant Joan del Mercat* church allowed us to determine that the smalt/azurite ratios used by the painter varied in a relatively large range between 50 and 86% smalt and that in several points they used pure smalt or pure azurite [9].

5.4 Identification of Organic Pigments

Organic pigments are mainly constituted of anthraquinonic, flavonoid (and analogues, flavonols, flavones, ...), carotenoids and indigoids compounds. Many sorts/types of organic pigments are prepared directly from plants or animals, and they appear as multicomponent systems although generally dominated by a majority compound. The solid-state electrochemistry of organic compounds in contact with aqueous electrolytes predominantly consists of proton-assisted electron transfer processes, i.e., oxidations coupled to proton loss and reductions coupled to proton insertion [12–14]. In fact, the first solid-state electroanalytical studies using carbon paste electrodes were focused on organic compounds (see Additional Literature).

VIMP has been applied to the identification of anthraquinonic [15, 16], flavonoid [16, 17], and other dyes such as curcuma and safflower [18]. Figure 5.8 compares the square wave voltammograms of a series of flavonoid-type dyes in contact with aqueous acetate buffer at pH 4.70. In this figure, semi-derivative convolution was applied to square wave voltammetric data. This procedure, whose physical meaning is well-defined in linear potential scan voltammetry, was applied here to improve peak resolution. It is interesting to remark that the individual compounds frequently display more than one electrochemical pathway so that the voltammetric profile can exhibit several peaks with different degree of reversibility.

In the case of anthraquinonic and flavonoid dyes, typical electrochemical processes involve the oxidative or reductive *o*- or *p*-quinone/*o*- or *p*-catechol interconversion. The proton-assisted nature of such processes results in pH-dependent peak potentials. Figure 5.9 depicts a scheme of typical electrochemical processes of this type for luteolin [17].

Mechanistic studies based on chronoamperometric analysis were consistent with the VIMP model described in Sects. 2.2.1 and 3.6 [19]. The combination of atomic force microscopy (AFM) with VIMP provides evidence, at the nanometric scale, of the solid-state nature of many electrochemical processes involved in dye identification. This is illustrated in Fig. 5.10, corresponding to the application of an oxidative

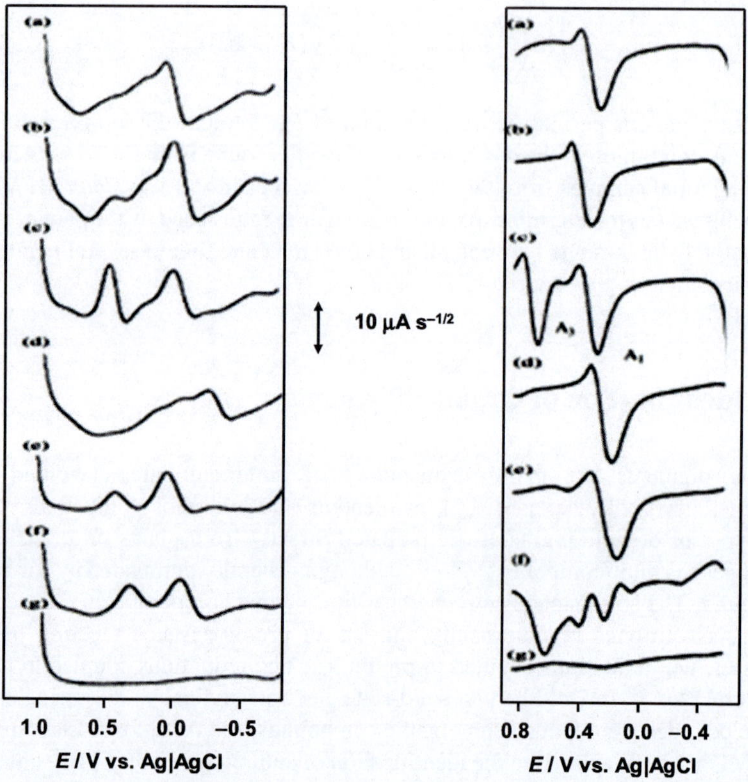

Fig. 5.8 Square wave voltammograms, after semi-derivative convolution, of (from upper to below) **a** Dragon's blood, **b** gamboge, **c** weld, **d** old fustic, **e** Campeche wood, and **f** Brazilwood attached to paraffin-impregnated graphite electrode in contact with 0.25 M HAc plus 0.25 M NaAc solution at pH 4.70. Left side: potential scan initiated at 0.85 V versus Ag/AgCl in the negative direction; right side: potential scan initiated at −0.85 V in the positive direction. Potential step increment 4 mV; square wave amplitude 25 mV; frequency 15 Hz. The voltammogram at the unmodified electrode is depicted in **g**. Reproduced from ref. [17], with permission. Note that the axes are opposite to the IUPAC convention

potential to a microcrystal of indigo anchored onto a graphite plate in contact with aqueous acetate buffer [20]. Under these experimental conditions, indigo is oxidized to dehydroindigo (see Sect. 6.3). The solid-state process is initiated at the solid/base electrode/electrolyte three-phase boundary and expands in the lateral faces of the indigo grain determining significant morphological changes. Note that, while in the electrochemistry of inorganic pigments such as Prussian blue (see Sect. 2.2.1) charge-balancing ions are intercalated/deintercalated into/from the solid lattice where they occupy interstitial positions, the proton-assisted electrochemistry of organic compounds implies severe reorganizations of the system of covalent bonds. For instance, the system of intra- and intermolecular hydrogen bonds of indigo disappears when oxidized to dehydroindigo.

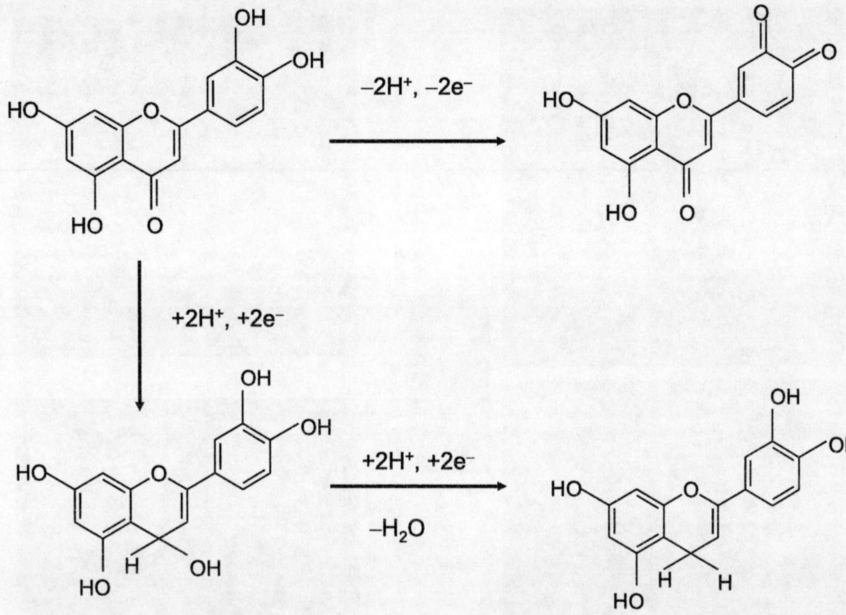

Fig. 5.9 Scheme of the electrochemical processes involved in the solid-state electrochemistry of luteolin in contact with aqueous electrolytes. Reproduced from ref. [17], with permission

In many cases, however, the parent dye and/or its oxidation or reduction products are sufficiently soluble to produce solution-like voltammetric responses, often giving rise to adsorption phenomena in the bare regions of the graphite electrode. This permits to enrich the identification criteria via application of oxidative or reductive potentials [15–17]. Another plausible strategy is the use of complex-forming reactants such as borate and Al^{3+} ions [17]. Voltammetric data can be structured to define a cascade analytical scheme for the identification of different pigments [21, 22].

Figure 5.11 illustrates the application of VIMP to the identification of the dyes present in a cope found in the Church of Castellfort (Valencian Region, Spain). The cope was a sacerdotal garment (also called pluvial) furnished with a hood or a scapulary. This was attributed to the 15th-century production of textile workshops in Morella (Northern Valencian Region, Spain) and dated back ca. 1420. The voltammograms of the red zone were obtained after micro-extraction with EtOH from a 2 mm length fiber and evaporation of the extract onto a graphite electrode. The voltammograms in Fig. 5.10 clearly suggested the use of a mixture of carmine and Brazilwood [17].

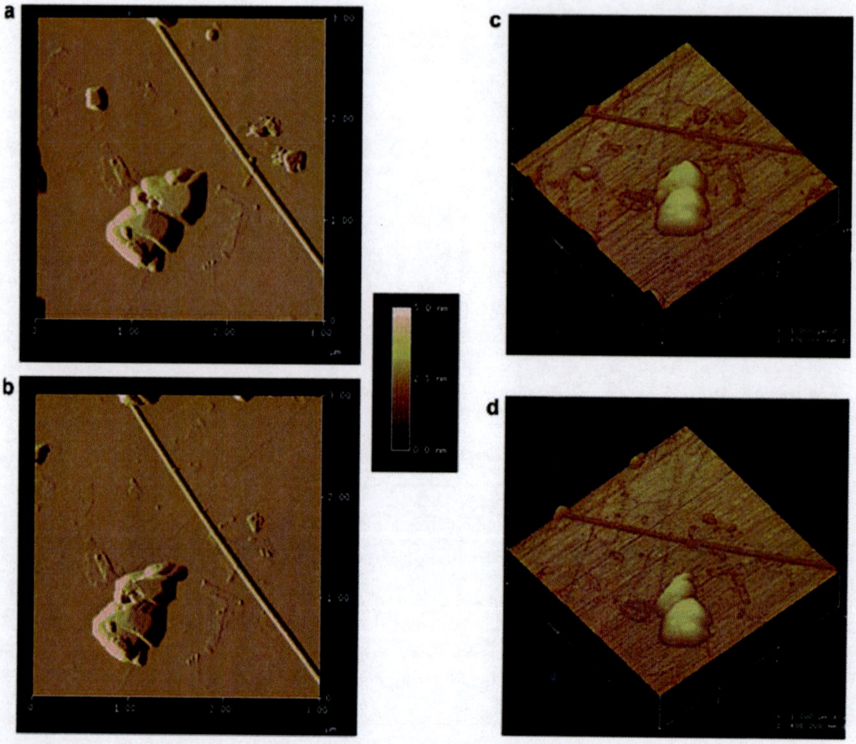

Fig. 5.10 AFM images **a**, **b** and topographic maps **c**, **d** of indigo crystals adhering to a graphite plate in contact with 0.50 M aqueous acetate buffer before **a**, **c**, and after **b**, **d** application of a linear potential scan between 0.00 and 0.75 V vs. Ag/AgCl at a potential scan rate of 10 mV s^{-1}. Reproduced from ref. [20], with permission

5.5 Pigments and Binding Media

As previously commented, pigments appear associated to a more or less complex matrix of components forming the pigmented layer(s) of paintings. In the case of mural paintings, these layers are applied over a plaster support. This is mainly composed of gypsum in the so-called *secco* techniques while in the *fresco* technique, it consists of a mixture of chalk and sand. The pigments usually form a water suspension which can be applied on a wet, fresh, lime mortar (the so-called *buon fresco*), on a dry mortar (*secco fresco*) or on a nearly-dry plaster (*mezzo fresco*). In the case of oil paintings, the pigment particles are embedded into the organic binder, the pictorial layer is usually deposited onto a preparative, *imprimatura*, layer. This layer frequently contains lead white and/or, in most cases, it is deposited over a ground layer made of gypsum, natural red earths, or a combination of Mars colors and red earths. The pigment-binding interactions result in the formation of metal soaps and ionophoric layers (vide infra).

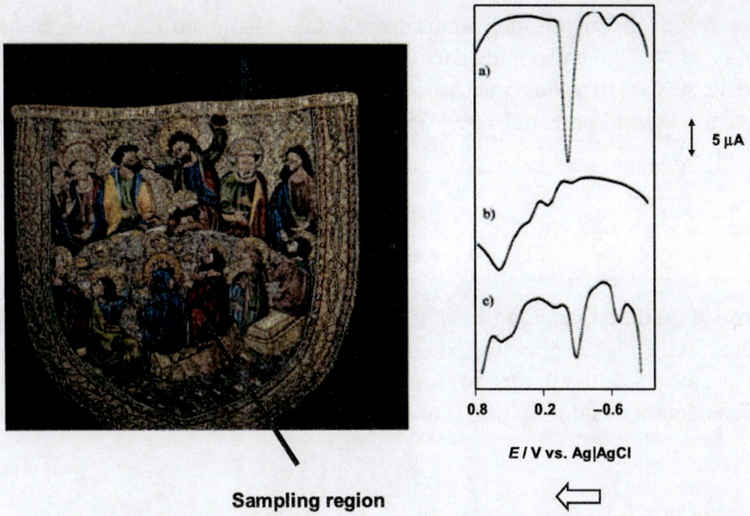

Fig. 5.11 Photographic image of a medieval brocade piece and square wave voltammograms of **a** carmine, **b** Brazilwood, and **c** a sample extracted from one red region of the cope. Electrolyte: HAc plus 0.25 M NaAc solution at pH 4.70. Potential scan initiated at −0.85 V in the positive direction. Potential step increment 4 mV; square wave amplitude 25 mV; frequency 15 Hz. The arrow marks the direction of the potential scan. Adapted from ref. [17], with permission. Note that the axes are opposite to the IUPAC convention

This complex, layered composition of paint strata made it considerably difficult to identify pigments, not only because of their usually low concentration in the paint layers, but also because of the masking of the pigment signals in infrared and diffraction techniques by the much more intense signals of the substrate components.

For our purposes, it is worth to note that the pigment–substrate interaction can distort the voltammetric response, as observed for lead pigments [6]. This was studied in frescoes with regard to Egyptian blue and other copper pigments [23], detecting significant differences in the voltammetric response in $CaCO_3$, SiO_2, and $CaCO_3$ plus SiO_2 mixtures.

These differences are obvious in the shift of the Tafel parameters used for identification. Possible reasons for the alteration of the voltammetric response of the pigments in paint samples relative to that of pure pigments are [23]: (i) the formation of an ionophoric layer (see Sect. 3.8) in the pigment grains whose composition could be modified in the vicinity of other solid material able to compete in such process and/or alter the local pH; (ii) the alteration of the granulometric distribution of the pigment grains during the preparative grinding process; and (iii) the alteration of the surface of the graphite electrode due to the different scratching of the components accompanying the pigment. Additionally, the presence of (generally) insulating components surrounding the pigment particles determines an enhanced ohmic resistance in the electrochemical cell [24].

The effect of uncompensated ohmic drops in the voltammetric response of species in solution phase [25] and confined to electrode surfaces [26, 27] has been repeatedly studied. In the case of strongly adsorbed reactants undergoing a n-electron reversible transfer, the current/potential curve in a linear potential scan experiment can be written as [28]:

$$I = \frac{nFv_{\mathrm{eff}}}{RT}\Gamma \frac{\exp\left[\frac{nF}{RT}(E - E^{\circ} - Ir_{\Omega})\right]}{\left\{1 + \exp\left[\frac{nF}{RT}(E - E^{\circ} - Ir_{\Omega})\right]\right\}} \tag{5.12}$$

This is a modified version of Eq. (3.52). Here, r_{Ω} is the uncompensated ohmic resistance and v_{eff} is the effective potential scan rate defined as $v_{\mathrm{eff}} = v + r_{\Omega}(\mathrm{d}I/\mathrm{d}t)$. In the case of solid-to-solid transformations studied in Sect. 3.8, we can introduce these considerations into Eq. (3.64) to give [4]:

$$I = \frac{v_{\mathrm{eff}}gq_{\mathrm{o}}\mathrm{e}^{g(E-E^{\circ}-Ir_{\Omega})}}{\left[1 + \mathrm{e}^{\gamma g(E-E^{\circ}-Ir_{\Omega})}\right]^{\frac{1}{\gamma}+1}} \tag{5.13}$$

In principle, it is possible to assume that under conditions of slow scan rate voltammetry, both the current and its variation with time can be considered as small so that one can equal v_{eff} with v and adopt an approximate expression for the Tafel-type relationship at the foot of the voltammetric wave as:

$$I \approx \frac{vgq_{\mathrm{o}}\mathrm{e}^{g(E-E^{\circ\prime})}}{1 + vg^{2}q_{\mathrm{o}}r_{\Omega}\mathrm{e}^{g(E-E^{\circ\prime})}} \tag{5.14}$$

In turn, when the applied potential is near to the formal electrode potential:

$$I \approx \frac{vgq_{\mathrm{o}}[1 + g(E - E^{\circ\prime})]}{[1 + vg^{2}q_{\mathrm{o}}r_{\Omega}]} \tag{5.15}$$

For practical purposes, it is convenient to use the Tafel-type relationship derived from the above equation:

$$\ln I \approx \ln\left[\frac{vgq_{\mathrm{o}}}{1 + vg^{2}q_{\mathrm{o}}r_{\Omega}\mathrm{e}^{g(E-E^{\circ\prime})}}\right] + g\left(E - E^{\circ\prime}\right) \tag{5.16}$$

This equation significantly modifies the Tafel-type relationship usable for identification purposes (in the absence of uncompensated ohmic drops) described by Eq. (4.7).

An alternative approach can be made considering separately the faradaic and non-faradaic contributions to the net current I measured in voltammetric experiments. In the electrochemical cells, however, there are uncompensated resistive and capacitive effects resulting in the appearance of a non-faradaic current. In voltammetry, this can be regarded as a background current superimposed to the peaks associated to faradaic processes. The non-faradaic contribution to the net current is mainly associated to the double-layer charging and uncompensated resistance accompanying. In a linear potential scan voltammetric experiment initiated at a given potential, E_s, the non-faradaic current, I_{NF}, can be derived from the differential equation [29]:

$$E = E_s - vt = R\frac{dq}{dt} + \frac{q}{C} \tag{5.17}$$

where R and C represent, respectively, the uncompensated ohmic resistance and double-layer capacitance of the electrochemical cell. Integration of Eq. (5.17) yields:

$$I_{NF} = \left(\frac{E_s}{R} - vC\right)e^{-\frac{t}{vRC}} + vC \tag{5.18}$$

In the case of a solid-to-solid redox process (Sect. 3.8), the net current flowing through the cell is the sum will be [4]:

$$I = q_o g v \frac{e^{g(E-E^{o\prime})}}{\left[1 + e^{\gamma g(E-E^{o\prime})}\right]^{\frac{1}{\gamma}+1}} + vC + \left(\frac{E_s}{R} - vC\right)e^{-\frac{E-E_s}{vRC}} \tag{5.19}$$

Imposing the condition $dI/dE = 0$ leads to the following condition to be accomplished by the peak potential:

$$vq_o g^2 e^{g(E_p-E^{o\prime})}\left[1 - \gamma e^{\gamma g(E_p-E^{o\prime})}\right] - \left(\frac{\frac{E_s}{R} - vC}{vRC}\right)e^{-\frac{E_p-E_s}{vRC}}\left[1 + e^{\gamma g(E_p-E^{o\prime})}\right]^{\frac{1}{\gamma}+2} = 0 \tag{5.20}$$

If the peak potential is near the formal electrode potential, a reasonable assumption for processes with high degree of reversibility at low potential scan rates, one can take,

$$vq_o g^2(1-\gamma) - \left(\frac{\frac{E_s}{R} - vC}{vRC}\right)e^{-\frac{E_p-E_s}{vRC}} 2^{\frac{1}{\gamma}+2} = 0 \tag{5.21}$$

Now, assuming that the potential scan was initiated reasonably near the voltammetric peak (i.e., $(E_p–E_s)/vRC << 1$), it is possible to approximate [4]:

$$E_p = E_s + vRC\left[1 - \frac{v^2RCq_o g^2(1-\gamma)}{2^{\frac{1}{\gamma}+2}\left(\frac{E_s}{R} - vC\right)}\right] \tag{5.22}$$

Equation (5.15) predicts a linear relationship between peak potentials and the charge passed, q_o, just in agreement with data reported by Meyer et al. [30] for the reduction of mixed crystals of CuS_xSe_{1-x} to Cu metal in contact with aqueous media.

A simplified approach to describe the influence of uncompensated ohmic drops in the Tafel-type relationships of voltammetric peaks can be based on the introduction of an additional Ir_Ω term in the current at the foot of the voltammetric wave associated to the reduction/oxidation of an electroactive component A [24]:

$$I \approx h_A q_A \exp[-\beta_A(E - E_p - IR_{\Omega A})] \tag{5.23}$$

where q_A is the maximum charge passed associated to the A-localized process, h_A is the corresponding electrochemical coefficient of response and $R_{\Omega A}$ is the uncompensated ohmic resistance associated to the A-component. Assuming that the voltammetric wave is decayed as a result of the ohmic drop, the peak current can be approximated by [24],

$$I_p = H_A q_A (1 - \chi_A E_p / R_{\Omega A}) \tag{5.24}$$

χ_A and H_A are new electrochemical coefficients whose values will be decided by the experiment. Neglecting capacitive effects and considering that at the foot of the voltammetric wave the exponential term will be in general small, the I/I_p ratio can be approximated by the following generalized Tafel-type relationship [24]:

$$\ln\left(\frac{I}{I_p}\right) \approx \ln\left(\frac{h_A}{H_A}\right) - \ln(1 - h_A q_A \beta_A R_{\Omega A}) - \ln(1 - \chi E_p / R_{\Omega A}) - \beta_A(E - E_p) \tag{5.25}$$

This means that the Tafel representation approaches a straight line of the same slope as in the absence of uncompensated ohmic effects but with different ordinate at the origin.

The effect of electrochemically silent substrates accompanying pigments can be modeled assuming that the electroactive pigment A is diluted within a matrix M forming a homogeneous layer of thickness d and surface area S. The resistance of this layer, R_Ω, is:

$$R_\Omega = \frac{1}{\sigma_{eff}} \frac{d}{S} \tag{5.26}$$

where σ_{eff} is the effective conductivity of the layer (Ω^{-1} cm^{-1}). This quantity can in principle be approximated by [24]:

$$\sigma_{eff} = \frac{\sigma_A m_A + \sigma_M m_M}{m_A + m_M} \tag{5.27}$$

where m_A and m_M are the masses of pigment and matrix existing in the layer and σ_A and σ_M their respective conductivities. Introducing the amount of M relative to A, f (= m_M/m_A), the ohmic resistance of the deposit, R_Ω, is:

$$R_\Omega = \frac{1}{\sigma_{eff}}\frac{d}{S} = \frac{1}{\sigma_A}\frac{d}{S}\left[\frac{1+f}{1+f\frac{\sigma_M}{\sigma_A}}\right] \tag{5.28}$$

Since in general the value of f (as well as q_A) is unknown, it is convenient to consider the product $q_A R_\Omega$ which appears in Eq. (5.15). Assuming that the pigment A experiences a n-electron transfer, and introducing the densities of the pigment and the substrate, ρ_A, ρ_M, the charge passed in the voltammetric experiment is:

$$q_A = nF\frac{m_A}{M_A} = \frac{nF}{M_A}\rho_A dS \tag{5.29}$$

M_A is the molar mass of A. Then, the product $q_A R_\Omega$ follows as:

$$q_A R_\Omega = \frac{nFd^2\rho_A}{M_A\sigma_A}\left[\frac{1+f}{\left(1+f\frac{\sigma_M}{\sigma_A}\right)\left(1+f\frac{\rho_A}{\rho_M}\right)}\right] \tag{5.30}$$

This relationship indicates that the resistance of the pictorial layer will be larger, equal, or smaller than that of a layer of A of identical volume depending on the relative values of the densities and conductivities of A and the matrix M. As a result, the Tafel parameters for a given pigment A will be modified in the presence of an electrochemically inert substrate, a feature in agreement with experimental data for Egyptian blue-containing frescoes [23]. This is a synthetic pigment whose use is documented since the 4th Dynasty of ancient Egypt (ca. 2575–2467 BCE) until the Greco-Roman period (fourth century CE). Based on Vitruvius, it is believed that the pigment was made by firing a mixture of quartz, lime, a copper compound, and an alkali flux at temperatures in the range of 850–1000 °C. The resulting pigment is a multicomponent material whose main constituent is crystalline calcium-copper tetrasilicate, identical to the naturally occurring mineral cuprorivaite of elemental composition $CaCuSi_4O_{10}$. Figure 5.12 shows the square wave voltammograms of azurite and Egyptian blue in contact with aqueous phosphate buffer at pH 7.0 accompanied by those of ploss blue ($CuCH_3COO{\cdot}H_2O$) and Han blue ($BaCuSi_4O_{10}$), two other copper pigments [23]. One can see that all pigments display a cathodic signal at similar potentials. In the case of cuprorivaite, the cathodic process can tentatively be described as:

$$CaCuSi_4O_{10} + 2H_2O + 2e^- \rightarrow Cu(s) + Ca^{2+}(aq) + 4SiO_3^{2-}(aq) + 4H^+(aq) \tag{5.31}$$

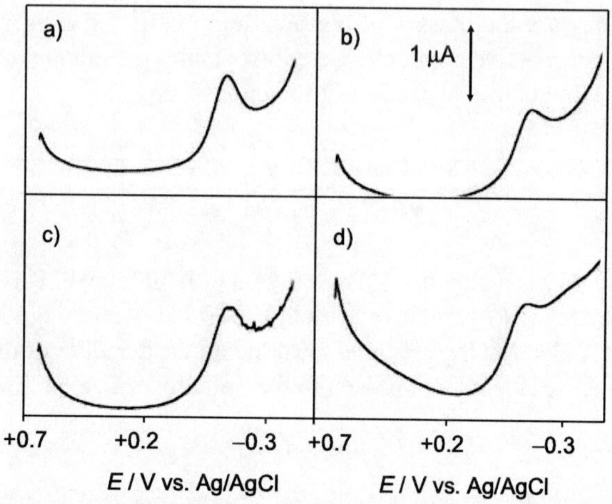

Fig. 5.12 Square wave voltammograms of graphite electrodes modified with: **a** azurite; **b** ploss blue; **c** Han blue; **d** Egyptian blue. Electrolyte: 0.10 M aqueous potassium phosphate buffer at pH 7.0. Potential scan initiated at 0.65 V in the negative direction; potential step increment 4 mV; square wave amplitude 25 mV; frequency 5 Hz. Reproduced from ref. [23], with permission. Note that the axes are opposite to the IUPAC convention

In principle, pigments can be identified with the help of Tafel parameters. However, the voltammetric profile, and hence, such parameters, are dependent on the composition of the pictorial substrate, as denoted by blank experiments on paint specimens prepared from binary and ternary mixtures with SiO_2 and $CaCO_3$ [23]. Figure 5.13 depicts the two-dimensional diagram for pigment identification using the modified Tafel-type analysis based on the linear representations of $\log[(I-I_p)/I_p]$ vs. $\log(E-E_p)$ [23]. Figure 5.13 presents a two-dimensional graph for different pigments plus support specimens prepared from binary and ternary mixtures with SiO_2 and/or $CaCO_3$ using data such as in Fig. 5.12.

In the case of oil paintings, the above resistive effect adds to the formation of metal carboxylates with fatty acids (metal soaps) and ionophoric layers [31], eventually allowing for the identification of different binding media [32]. This matter will be treated in detail in the next section.

The materials accompanying the pigments in paintings can eventually be electroactive. This is the case of binding media produced from porcine blood, which was used in Taiwanese architectural polychromies. Here, voltammetric signals attributable to the Fe(III)/Fe(II) iron couple of hemoglobin can easily be detected. These can be used not only to detect the presence of the binding but also to establish age correlations based on the catalytic enhancement of the Fe(III)/Fe(II) couple in the presence of H_2O_2 [33]. VIMP has been applied to the study and dating of the polychromed architectural decoration existing in different zones of the architectural complex of the Longshan Temple in Lukang (eighteenth century, Taiwan) [33].

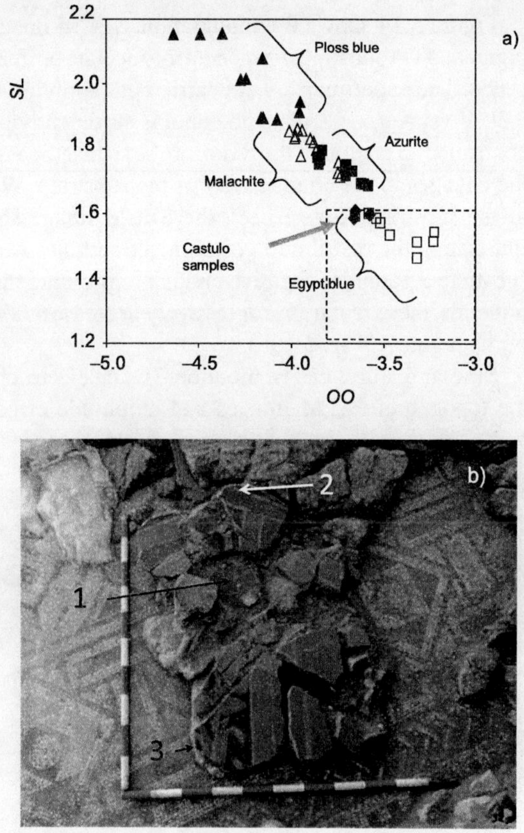

Fig. 5.13 **a** Two-dimensional diagram showing the values of the ordinate at the origin and the slope of the Tafel-type representations obtained from voltammetric records such as in Fig. 5.12 for paint specimens of azurite, malachite, and ploss blue prepared with different SiO_2 and/or $CaCO_3$ supports. The arrow marks the position of samples from the Roman archaeological site of Cástulo (Jaén, Spain). **b** photographic image of the wall paintings showing the position of the blue regions where sampling was made. Adapted from ref. [23], with permission

5.6 Nanoscale Characterization and Mapping of Pictorial Components

Pigment characterization at the nanoscale has been reported combining AFM with VIMP, as described in Sect. 2.2.3. The spatial distribution of pigments [34] and proteins [35] in paint samples at this scale can also be studied by means of the scanning electrochemical microscopy (SECM) technique, briefly introduced in Sect. 2.4.2. Here, the characterization of different pigments can be achieved using the redox competition strategy [36–38], based on the application of different potential inputs to the substrate electrode and, hence, to the sample deposited on it.

Figure 5.14 shows the topographic SECM images of a deposit of Spanish ochre pigment on Pt substrate immersed into acetate buffer containing $Fe(CN)_6^{4-}$ as a redox probe. The experiments were carried out applying to the tip electrode a potential of 0.30 V vs. Ag/AgCl. This potential is sufficiently positive to promote the diffusion-controlled oxidation of the redox probe so that the fluctuations of the tip current reflect the changes in the conductivity of the substrate. When there is no applied potential to the substrate (Fig. 5.12a), the SECM image shows negative feedback features illustrating the distribution of pigment particles over the substrate. Applying to the substrate a potential negative enough to promote the reductive dissolution of Fe(III) minerals, these features progressively decrease as a result of the reductive dissolution of the pigment (Fig. 5.15b).

Similar features can be monitored using electrochemical AFM. Figure 5.15 shows the topographic AFM images and amplitude error channel graphs observed for a

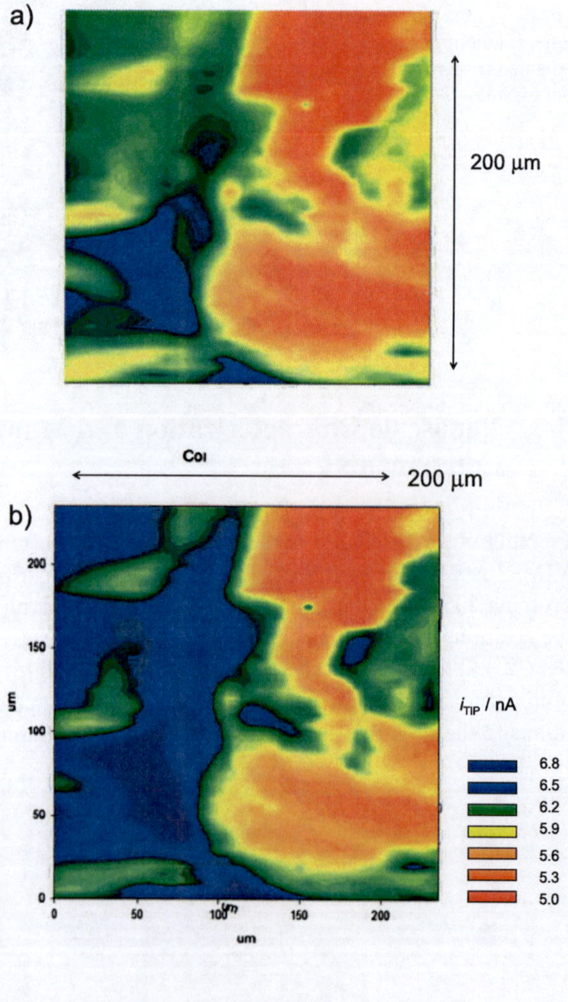

Fig. 5.14 SECM topographic images of a microparticulate deposit of Spanish ochre (K11585 *Kremer pigmente*) on Pt substrate immersed into 2.0 mM $K_4[Fe(CN)_6]$ plus 0.25 M HAc/NaAc, pH 4.75. **a** $E_T = 0.30$ V; $E_S = 0.00$ V; **b** $E_T = 0.30$ V; $E_S = -0.45$ V. Reproduced from ref. [34], with permission

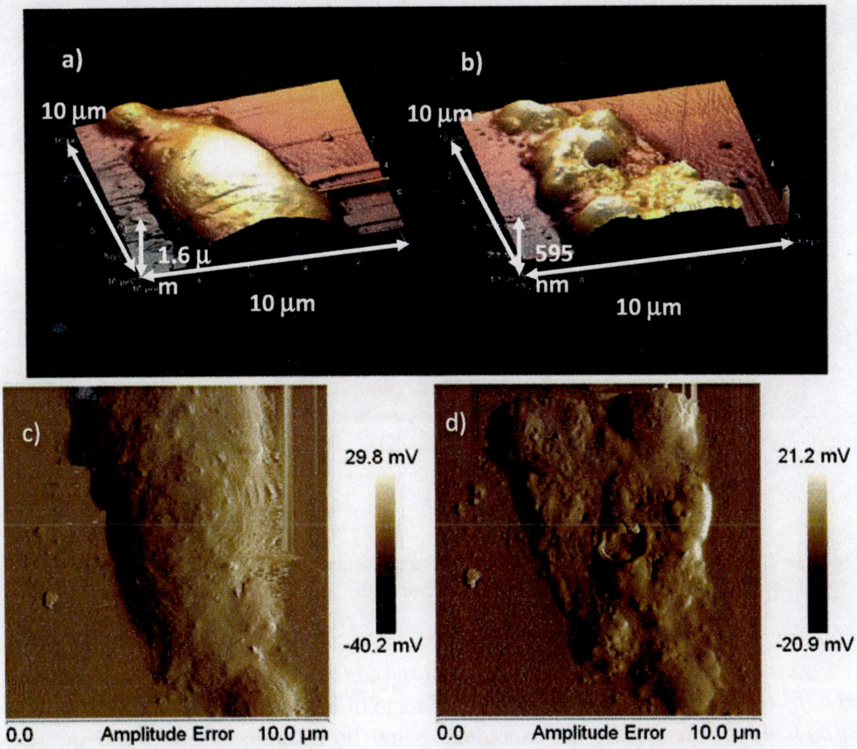

Fig. 5.15 **a**, **b** Topographic AFM image and **c**, **d** amplitude error channel graphs of a deposit of hematite (K48651 *Kremer pigmente*) on a graphite plate in contact with 0.25 M sodium acetate buffer, pH 4,75, **a**, **c** before, and **b**, **d** after application of 10 potential cycles between 0.0 and −0.75 V at a sweep rate of 10 mV s^{-1}. Reproduced from ref. [34], with permission

deposit of hematite on highly oriented pyrolytic graphite plate (HOPG) in contact with aqueous acetate buffer before and after the application of a cathodic input. Under the used experimental conditions, hematite is reductively dissolved so that the size and shape of the pigment grain become significantly modified.

The SECM technique provides information on electrochemical surface topography at the micrometer/submicrometer scale. These characteristics allow us to map distribution of components into stratified paint layers. Figure 5.16 depicts the two basic types of distribution of layers in cross-sections of the paint sample deposited onto a base electrode for SECM measurements.

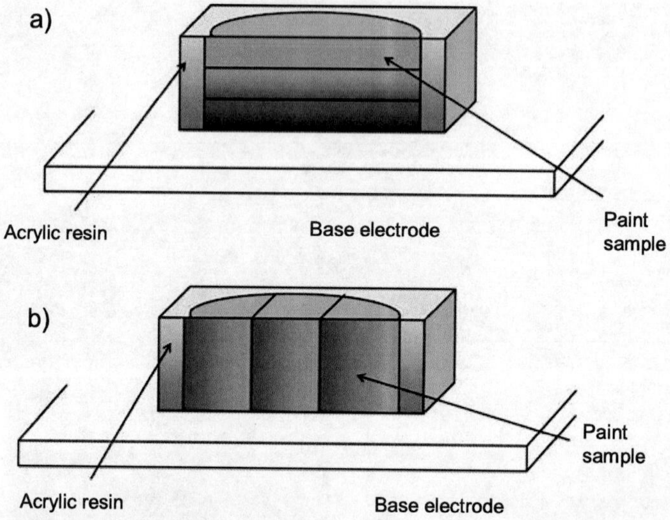

Fig. 5.16 Arrangements for SECM imaging of cross-sections of pictorial samples: **a** parallel; **b** transversal. Reproduced from ref. [34], with permission

Figure 5.17 illustrates the application of the SECM technique for profiling cross-sections of paints. Here, the topographic image of a sample extracted from the *Los Jurados de Valencia ante la Inmaculada* paint (by Jerónimo Jacinto de Espinosa, 1662) can be compared with a microphotograph of the same cross-section. The SECM image, again using $Fe(CN)_6{}^{4-}$ as a redox probe, shows several strata agreeing with the optical microscopy image. The samples studied by SECM (Figs. 5.15, 5.16, and 5.17) were cross-sections embedded in acrylic resin previously studied by conventional SEM/EDX analysis, a strategy also used in VIMP analysis described in the precedent sections.

The mapping and discrimination of different pigments by means of SECM imaging are illustrated in Fig. 5.18, corresponding to a second sample from the painting in Fig. 5.17 again using acetate buffer plus $Fe(CN)_6{}^{4-}$ as the electrolyte. Here, pigment grains appear as negative feedback features and the redox competition strategy permits to discriminate between different pigments. When no external potentials were applied to the substrate (Fig. 5.18a), there is no distinction between different pigments. When a potential of – 0.45 V is applied to the substrate (Fig. 5.18b), some of pigment features vanish, thus suggesting the presence of 'earths' (iron oxide-based pigments) that, in these conditions, experience reductive dissolution processes. If a subsequent potential of – 0.70 V is applied to the substrate, other features become expanded (Fig. 5.18a). This permits to attribute such points to lead pigments that

a)

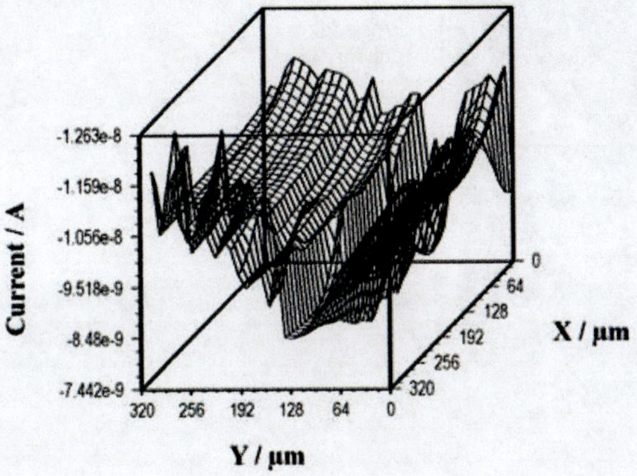

b)

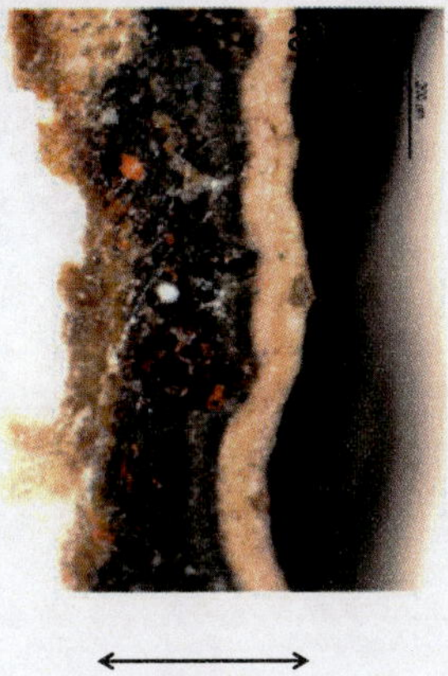

Fig. 5.17 **a** SECM image of the cross-section of a sample extracted from the *Los Jurados de Valencia ante la Inmaculada* paint by Jerónimo Jacinto de Espinosa (1662) and **b** microphotograph (PPL) of the sample. The SECM image recorded upon placing the sample on the substrate electrode in transversal position. Electrolyte: 2.0 mM $K_4[Fe(CN)_6]$ plus 0.25 M HAc/NaAc, pH 4.75. a) $E_T = 0.30$ V; $E_S = 0.00$ V. Adapted from ref. [34], with permission

are reduced to lead metal. This attribution is consistent with topographic images in Fig. 5.19 of a sample from the same painting where the morphology changes in a lead pigment grain (identified in previous SEM/EDX analysis) recorded upon applying a reductive step can be seen. These changes correspond to the growth of a deposit of lead metal in the vicinity of the pigment grain, in agreement with the description of epitactic redox processes discussed in Sects. 2.2.3 and 3.9.

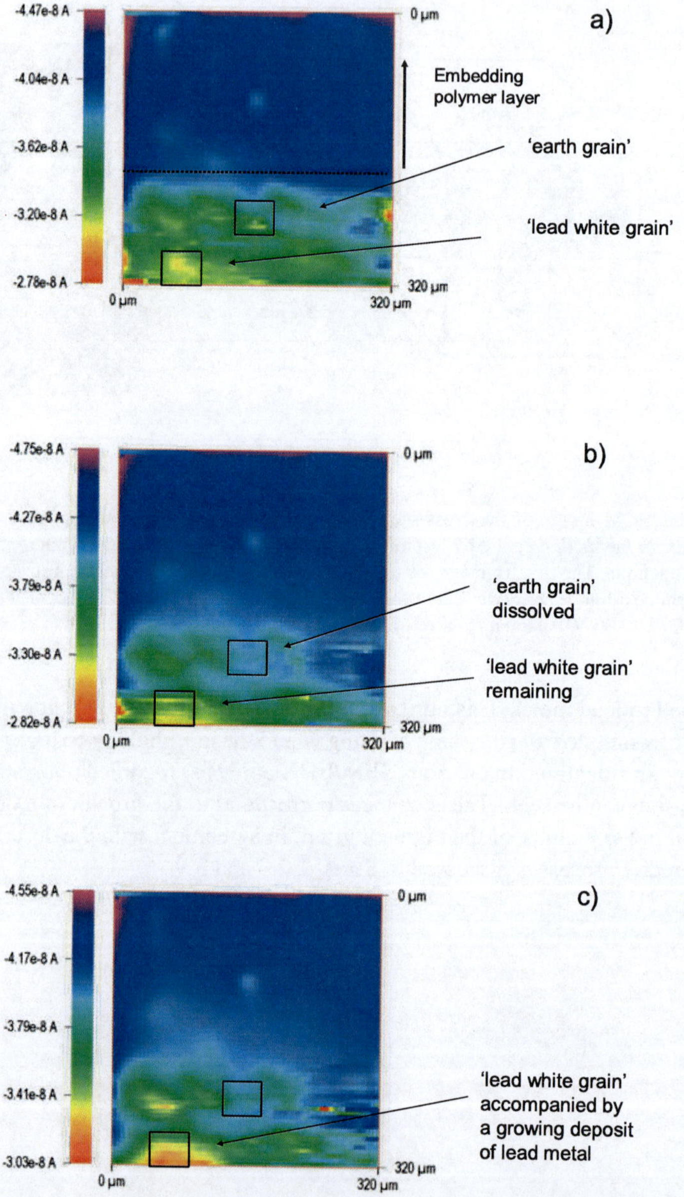

Fig. 5.18 SECM topographic images of a paint sample extracted from the *Los Jurados de Valencia ante la Inmaculada* paint by Jerónimo Jacinto de Espinosa (1662). The sample was arranged transversally on Pt substrate in contact with 2.0 mM $K_4[Fe(CN)_6]$ plus 0.25 M HAc/NaAc, pH 4.75. **a** $E_T = 0.30$ V; $E_S = 0.00$ V; **b** $E_T = +0.30$ V; $E_S = -0.45$ V; **c** $E_T = 0.30$ V; $E_S = -0.70$ V. Reproduced from ref. [34], with permission

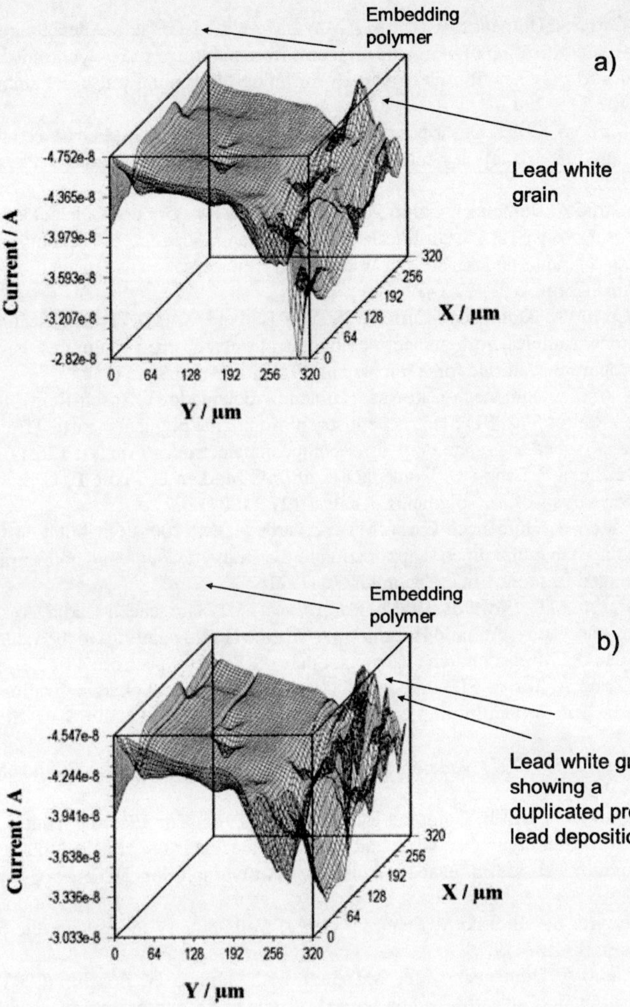

Fig. 5.19 SECM topographic images of a paint sample from the *Los Jurados de Valencia ante la Inmaculada* by Jerónimo Jacinto de Espinosa (1662) arranged transversally on Pt substrate immersed. Electrolyte: 2.0 mM $K_4[Fe(CN)_6]$ plus 0.25 M HAc/NaAc, pH 4.75. **a** $E_T = 0.30$ V; $E_S = 0.00$ V; **b** $E_T = 0.30$ V; $E_S = -0.70$ V. Reproduced from ref. [34], with permission

References

1. Pfaff G (2022) The world of inorganic pigments. ChemTexts 8:15
2. Lange B, Scholz F, Weiss A, Schwedt G, Behnert J, Raezke KP (1993) Abrasive stripping voltammetry—the electrochemical alternative for pigment analysis—Internat Lab 23:23–26

3. Doménech-Carbó A, Doménech-Carbó MT, Moya-Moreno M, Gimeno-Adelantado JV, Bosch-Reig F (2000) Identification of inorganic pigments from paintings and polychromed sculptures immobilized into polymer film electrodes by stripping differential pulse voltammetry. Anal Chim Acta 407:275–289
4. Doménech-Carbó A (2022) Description of solid-to-solid redox processes based on the voltammetry of immobilized particles methodology: a logistic approximation. J Phys Chem C 26:11822–1832
5. Doménech-Carbó A, Doménech-Carbó MT, Valle-Algarra FM, Gimeno-Adelantado JV, Osete-Cprtina L, Bosch-Reig F (2016) On-line database of voltammetric data of immobilized particles for identifying pigments and minerals in archaeometry, conservation and restoration (ELCHER database). Anal Chim Acta 927:1–12
6. Doménech-Carbó A, Doménech-Carbó MT, Mas-Barberá X (2007) Identification of lead pigments in nanosamples from ancient paintings and polychromed sculptures using voltammetry of nanoparticles/atomic force microscopy. Talanta 71:1569–1579
7. Doménech-Carbó A, Doménech-Carbó MT, Gimeno-Adelantado JV, Bosch-Reig F, Saurí-Peris MC, Sánchez-Ramos S (2001) Electrochemistry of iron oxide pigments (earths) from pictorial microsamples attached to graphite/polyester composite electrodes. Analyst 126:1764–1722
8. Grygar T, Bezdicka P, Hradil D, Doménech-Carbó A, Marken L, Pikna L, Cepriá G (2002) Voltammetric analysis of iron pigments. Analyst 127:1100–1107
9. Doménech-Carbó A, Doménech-Carbó MT, Edwards HGM (2008) Quantitation from Tafel analysis in solid-state voltammetry. application to the study of Cobalt and copper pigments in severely damaged frescoes. Anal Chem 80:2704–2716
10. Doménech-Carbó MT, Edwards HGM, Hargreaves MD, Doménech-Carbó A (2012) An authentication case study: Antonio Palomino vs. Vicente Guillo paintings in the vaulted ceiling of the Sant Joan del Mercat church (Valencia, Spain). J Raman Spectrosc 43:1250–1259
11. Doménech-Carbó A, Torres FJ, Alarcón J (2004) Electrochemical characterization of cobalt cordierites attached to paraffin-impregnated graphite electrodes. J Solid State Electrochem 8:127–137
12. Komorsky- Lovrić S (1997) Voltammetry of Azobenzene microcrystal. J Solid State Electrochem 1:94–99
13. Bond AM, Marken F, Hill E, Compton RG, Hügel H (1997) The Electrochemical reduction of Indigo dissolved in organic solvents and as a solid mechanically attached to a basal plane pyrolytic graphite electrode immersed in aqueous electrolyte solution. J Chem Soc Perkin Trans 2:1735–1742
14. Komorsky- Lovrić S, Mirceski V, Scholz F (1999) Voltammetry of organic microparticles. Mikrochim Acta 132:67–77
15. Doménech-Carbó A, Doménech-Carbó MT, Saurí MC, Gimeno JV, Bosch F (2003) Electrochemical identification of anthraquinone-based dyes in solid microsamples by square wave voltammetry using graphite/polyester composite electrodes. Anal Bioanal Chem 375:1169–1175
16. Grygar T, Kuckova S, Hradil D, Hradilova D (2003) Electrochemical analysis of natural solid organic dyes and pigments. J Solid State Electrochem 7:706–713
17. Doménech-Carbó A, Doménech-Carbó MT, Saurí MC (2005) Electrochemical identification of flavonoid dyes in work of art samples by abrasive voltammetry at paraffin-impregnated graphite electrodes. Talanta 66:769–782
18. Doménech-Carbó A, Doménech-Carbó MT, Saurí MC, Gimeno JV, Bosch F (2005) Identification of curcuma and safflower dyes by voltammetry of microparticles using paraffin-impregnated graphite electrodes. Microchim Acta 152:75–84
19. Doménech-Carbó A, Doménech-Carbó MT (2006) Chronoamperometric study of proton transfer/electron transfer in solid state electrochemistry of organic dyes. J Solid State Electrochem 10:949–958
20. Doménech-Carbó A, Doménech-Carbó MT (2008) In situ AFM study of proton-assisted electrochemical oxidation/reduction of microparticules of organic dyes. Electrochem Commun 10:1238–1241

21. Doménech-Carbó A, Doménech-Carbó MT, Calisti M, Maiolo V (2010) Identification of naphtoquinonic and anthraquinonic dyes via sequential potential steps applied to the voltammetry of microparticles methodology. J Solid State Electrochem 14:465–477
22. Doménech-Carbó A, Doménech-Carbó MT, Calisti M, Maiolo V (2010) Sequential identification of organic dyes using the voltammetry of microparticles approach. Talanta 81:404–414
23. Doménech-Carbó A, Doménech-Carbó MT, López F, Valle-Algarra FM, Osete-Cortina L, Arcos E (2013) Electrochemical characterization of Egyptian blue pigment in wall paintings using the voltammetry of microparticles methodology. Electroanalysis 25:2621–2630
24. Doménech-Carbó A, Doménech-Carbó MT, Redondo-Marugán J, Osete-Cortina L, Vivancos-Ramón MV (2016) Electrochemical characterization of corrosion products in leaded bronze sculptures considering ohmic drop effects on Tafel analysis. Electroanalysis 28:833–845
25. Nicholson RS (1965) Some examples of the numerical solution of nonlinear integral equations. Anal Chem 37:667–671
26. Fan RF, Mirkin MV, Bard AJ (1994) Polymer films on electrodes. 25. effect of polymer resistance on the electrochemistry of poly(vinylferrocene): scanning electrochemical microscopic, chronoamperometric, and cyclic voltammetric studies. J Phys Chem 98:1475–1481
27. Trijueque J, García-Jareño JJ, Navarro-Laboulais J, Sanmatías A, Vicente F (1999) Ohmic drop of Prussian-blue:graphite_epoxy electrodes. Electrochim Acta 45:789–795
28. Mircesky V, Lovrić M (2001) Ohmic dorp effects in square-wave voltammetry. J Electroanal Chem 497:114–124
29. Bard AJ, Faulkner LR (1980) Electrochemical methods. Wiley, New York
30. Meyer B, Zhang S, Scholz F (1996) The quantitative analysis of mixed crystals CuS_xSe_{1-x} with abrasive stripping voltammetry and a redetermination of the solubility product of CuSe and the standard potencial of the Cu/CuSe electrode. Fresenius J Anal Chem 356:267–270
31. Hermans JJ, Keune K, van Loon A, Corkery RW, Iedema PD (2016) Ionomer-like structure in mature oil paint binding media. RSC Adv 6:93363–93369
32. Doménech-Carbó A, Doménech-Carbó MT, Mas-Barberá X, Ciarrocci J (2007) Simultaneous identification of lead pigments and binding media in paint samples using voltammetry of microparticles. Arché 2:121–124
33. Doménech-Carbó A, Doménech-Carbó MT, Lee Y, Osete-Cortina L (2012) Potential application of voltammetry of microparticles for dating porcine blood-based binding media used in Taiwanese architectural polychromies. Chem Asian J 7:2268–2273
34. Doménech-Carbó A, Doménech-Carbó MT, Silva M, Valle-Algarra FM, Gimeno-Adelantado JV, Bosch-Reig F, Mateo-Castro R (2015) Screening and mapping pigments in paintings using scanning electrochemical microscopy (SECM). Analyst 140:1065–1075
35. Sciutto G, Prati S, Mazzeo R, Zangheri M, Roda A, Bardini L, Valenti G, Rapino S, Marcaccio M (2014) Localization of proteins in paint cross-sections by scanning electrochemical microscopy as an alternative immunochemical detection technique. Anal Chim Acta 831:31–37
36. Walsh DA, Li LE, Bakare MS, Voisey KT (2009) Visualisation of the local electrochemical activity of thermally-sprayed anti-corrosion coatings using scanning electrochemical microscopy. Electrochim Acta 54:4647–4654
37. Karnicka K, Eckhard K, Guschin DA, Stoica L, Kulesza PJ, Schumann W (2007) Visualisation of the local bio-electrocatalytic activity in biofuel cell cathodes by means of redox competition scanning electrochemical microscopy (RC-SECM). Electrochem Commun 9:1998–2002
38. Calfumán K, Aguirre MJ, Cañete-Rosales P, Bollo S, Llusar R, Isaacs M (2011) Electrocatalytic reduction of nitrite on tetraruthenated metallophorphyrins/Nafion glassy carbon modified electrode. Electrochim Acta 56:8484–8491

Additional Literature

39. Berg JDJ van den, KJ van den, Boon JJ (2002) Identification of non-cross-linked compounds in methanolic extracts of cured and aged linseed oil-based paint films using gas chromatography-mass spectrometry. J Chromatogr A 950:195–211

40. Doménech-Carbó A, Doménech-Carbó MT, Osete-Cortina L (2004) Electrochemistry of archaeological metals: an approach from the voltammetry of microparticles, in Trends in electrochemistry and corrosion at the beginning of the 21st century (dedicated to Professor Dr. Josep M. Costa on the occasion of his 70th birthday), Brillas E, Cabot P-L, Eds..Universitat de Barcelona, Barcelona
41. Doménech-Carbó A, Doménech-Carbó MT, Costa V (2009) Electrochemical methods in archaeometry, conservation and restoration (Monographs in Electrochemistry series, Scholz F, Ed). Springer, Berlin-Heidelberg
42. Doménech-Carbó MT, de Agredos V, Pascaul ML, Osete-Cortina L, Doménech-Carbó A, Guasch-Ferré N, Manzanilla LR, Vidal C (2012) Characterization of Pre-hispanic cosmetics found in a burial of the ancient city of *Teotihuacan* (Mexico). J Archaeol Sci 39:1043–1062
43. Doménech-Carbó MT, Vázquez de Agredos Pascual ML, Osete-Cortina L, Doménech-Carbó A, Guasch-Ferré N, Vidal C (2020) Preparation of body colors in Pre-classical Maya culture: the red pigment in the 19th tomb (Peten, Guatemala). Herit Sci 8: artic 47
44. Kuwana T, French WG (1964) Electrooxidation or reduction of organic compounds into aqueous solutions using carbon paste electrode. Anal Chem 36:241–242
45. Lee Y, Martín-Rey S, Osete-Cortina L, Martín-Sánchez I, Bolívar-Galiano F, Doménech-Carbó MT (2018) Evaluation of a gelatin-based adhesive for historic paintings that incorporates citronella oil as an eco-friendly biocide. J Adhes Sci Technol 32:2320–2349
46. Rehren Th, Pusch EB, Herold A (1998) Glass coloring works within a copper-centered industrial complex in Late Bronze Age Egypt. In McCray P, Ed, The prehistory and history of glassmaking technology. Ceramics and Civilization 8. American Ceramic Society, Westerville
47. Riederer J (1997) Egyptian Blue. In: Fitzhugh EW (ed) Artists' pigments, vol 3. Oxford university Press, Oxford, pp 23–45
48. Robinet L, Corbeil MC (2003) The characterization of metal soaps. Stud Conserv 48:23–40
49. Schultz FA, Kuwana T (1965) Electrochemical studies of organic compounds dissolved in carbon-paste electrodes. J Electroanal Chem 10:95–103

Chapter 6
Pigment and Paintings II

6.1 Electrochemical Characterization of Workshops

In the case of oil paintings, under favorable conditions, there is the possibility of characterizing different authors/workshop techniques based on the differences in the pigment-binding media interactions resulting from the differences in manufacturing. This possibility was studied by comparing the voltammetric response of submicrosamples from 12 oil paintings attributed to Nicolás Borrás (1530–1610), Vicent Castelló (ca. 1587–ca. 1636), Vicente Requena (ca. 1556–1605), Francisco Ribalta (1565–1628), Juan Ribalta (1596/1597–1628), Juan Sariñena (1545–1619), and Jerónimo Jacinto de Espinosa (1600–1667). These painters define the transition from the Renaissance to the Baroque in the Valencian region (Spain). Figure 6.1 shows one of the most representative paintings: *San Cristóbal de Licia* by Vicente Requena, accompanied by the cross-section photographic image of one of the samples. The pictorial layer contains white, yellow, and green pigments appearing as separated grains surrounded by the binder. Based on the EDS analysis of the X-ray spectrum, the yellow grains were identified as composed of lead-tin yellow. In turn, the paint layer was deposited over preparative layers composed of gypsum (g) and red earth accompanied by minium and lead white (r) [1].

When attached to paraffin-impregnated graphite electrodes in contact with an aqueous acetate buffer, these samples displayed square wave voltammetric responses dominated by the cathodic signals due to the reduction of lead white and lead-tin yellow, components of all samples, in the region of potentials between −0.2 and −0.8 V versus Ag/AgCl. As can be seen in Fig. 6.2, these signals present more or less complicated profiles consisting of (at least) three overlapping signals (labeled as I-III). These signals can be interpreted as corresponding to the reduction of the grains of pigment which are in contact with the graphite surface and the electrolyte solution (I), the reduction of Pb^{2+} ions released from the polyelectrolyte layer generated around these grains (II), and the reduction of lead soaps (generically, $Pb^{2+}(RCOO^-)_2$) also accompanying pigment grains (III). The first process is

A. Doménech-Carbó and M. T. Doménech-Carbó, *Electrochemistry for Cultural Heritage*, Monographs in Electrochemistry,
https://doi.org/10.1007/978-3-031-31945-7_6

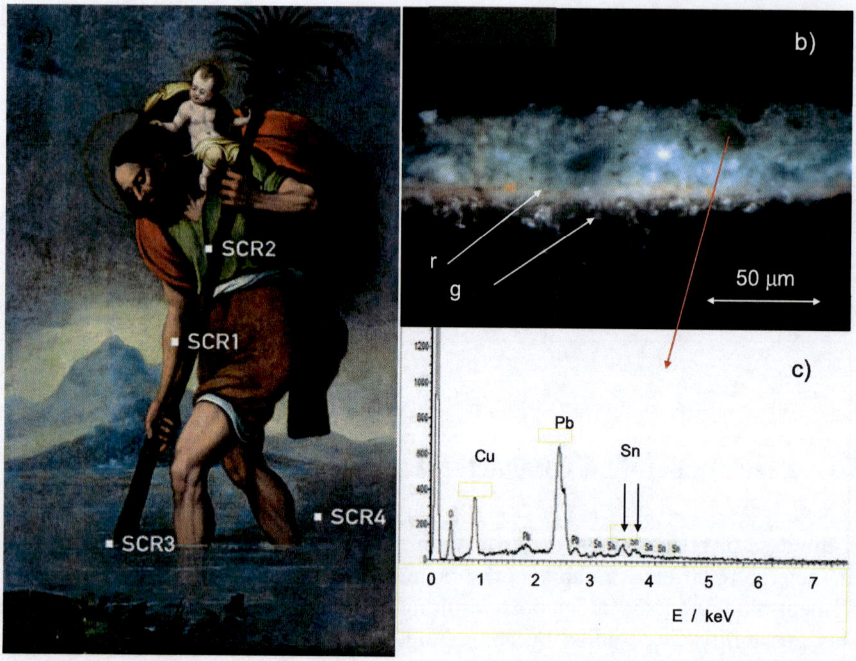

Fig. 6.1 **a** Photographic image of the painting *San Cristóbal de Licia* by Vicente Requena. **b** microphotograph of the cross-section of the SCR2 sample showing the pictorial layer deposited over preparative layers of gypsum (g) and red earth accompanied by minium and lead white (r). **c** X-ray spectrum **c** of a yellow grain of that sample. Reproduced from ref. [1], with permission

supposed to be caused by the following reduction of lead white used in the preparative layers:

$$\{Pb_3(CO_3)_2(OH)_2\}_{solid} + 4H^+_{aq} + 6e^- \rightarrow 3\{Pb\}_{solid} + 2HCO^-_{3\ aq} + 2H_2O \quad (6.1)$$

As in other cases of scarcely soluble compounds, although moderate solubility is attained in acidic media, the rate of dissolution is slow, thus ensuring a solid-state voltammetric response similar to that displayed by highly insoluble compounds (see Additional literature). The reduction of lead soaps (process III) can be formulated as:

$$\left\{Pb^{2+}(RCOO^-)_2\right\}_{soap} + 2H^+_{aq} + 2e^- \rightarrow \{Pb\}_{solid} + 2RCOOH \quad (6.2)$$

Of course, these processes can be described according to the complex scheme drawn in Sect. 3.8 for litharge.

Figure 6.3 shows a backscattered electron image of a cross-section of a sample from a greenish area in the *Ecce Homo* painting by Francisco Ribalta [1]. This image clearly illustrates the existence of an envelope of lead soaps surrounding the pigment grains. Metal soaps can also be identified by their infrared spectral signatures

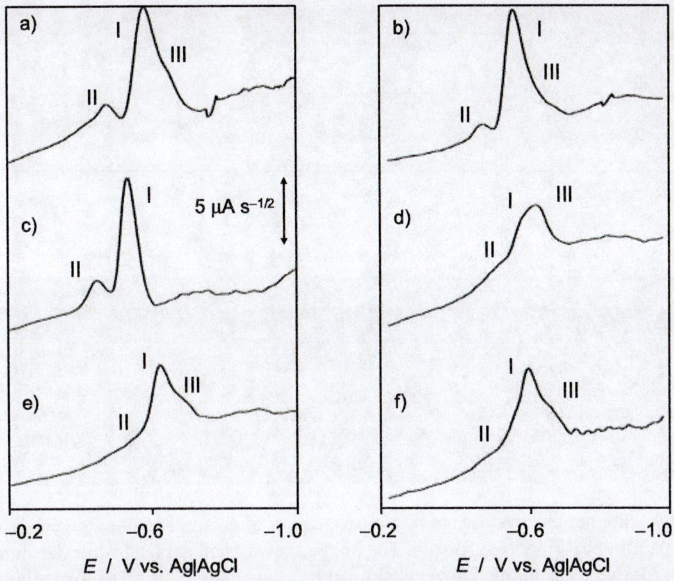

Fig. 6.2 Detail of the region between −0.2 and −0.8 V versus Ag/AgCl of square wave voltammograms, after semi-derivative convolution, recorded for samples from green areas of different paintings by **a**, **b** Juan Sariñena, **c** Jerónimo Jacinto de Espinosa, **d** Vicente Castelló, **e** Francisco Ribalta, and **f** Juan Ribalta, attached to graphite bars in contact with air-saturated 0.25 M HAc/NaAc aqueous solution at pH 4.75. Potential scan initiated at 1.25 V in the negative direction; potential step increment 4 mV; square wave amplitude 25 mV; frequency 5 Hz. Reproduced from ref. [1], with permission. Note that the axes are opposite to the IUPAC convention

[1]. These compounds originate from the complexation of lead ions released from lead pigments by the long-chain saturated fatty acids. This is caused by the partial hydrolysis of the network of polymerized drying oil. It is in turn favored by metal ions which promote the radical generation related to oxidative radical chain reactions [2]. Figure 6.4 shows a scheme of the above processes based on the interpretation of VIMP experiments described in Sects. 3.6–3.8.

The relevant point to emphasize is that the lead-centered voltammetric response reflects not only the composition of the sample (pigment, binder, supporting layers) but also its preparation (mode of application, number, and thickness of preparative and pictorial layers, etc.) and the aging (depending on time, exposure to atmospheric environments, etc.). Naturally, these factors can vary from one to another painting of the same workshop or even between different points of the same painting, but, in principle, one can assume that each painter/workshop maintained over time its peculiar mode of painting.

Then, let us assume that in the paint sample the pigments are in the form of spheroidal grains or radius r surrounded by an envelope of soaps forming a spheroidal ensemble of external radius $r + \delta$. Assuming that in VIMP experiments (see Chap. 3) the peak currents are proportional to the perimeter of the electroactive region, and

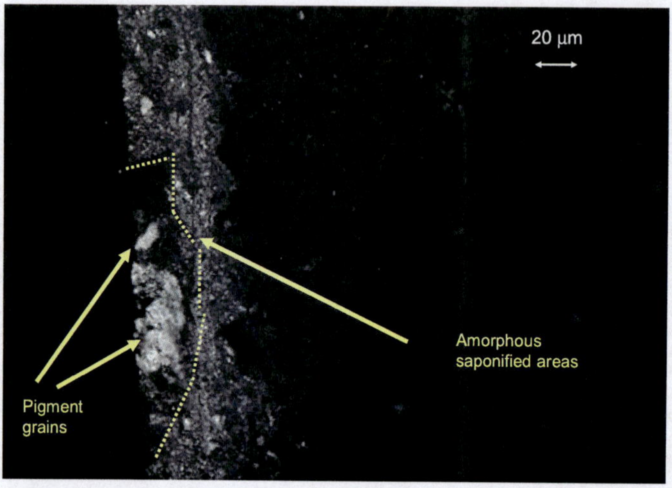

Fig. 6.3 Backscattered electron image of a cross-section of a sample from a greenish area in the *Ecce Homo* painting by Francisco Ribalta. The image shows grains of lead white surrounded by an envelope of amorphous saponified area constituted by lead soaps. This envelope is limited by the dotted line. Reproduced from ref. [1], with permission

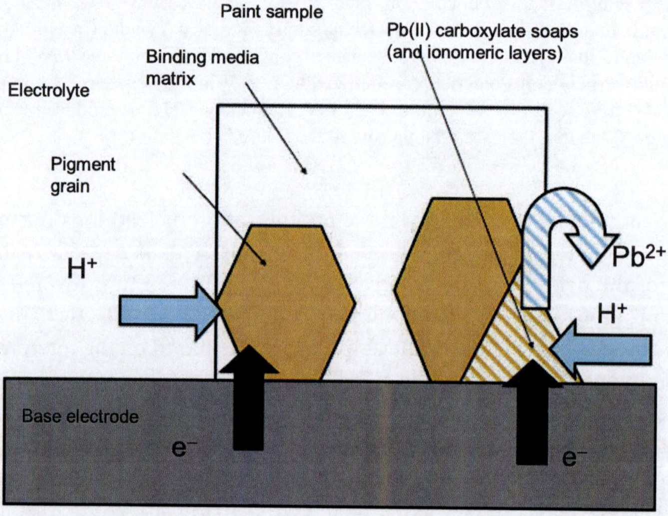

Fig. 6.4 Simplified scheme for interpreting the lead-based electrochemistry of samples from oil paintings. Reproduced from ref. [1], with permission

accepting the aforementioned attributions of peaks I and III to the reduction of pigment grains, released Pb^{2+} ions from the pigment and lead soaps, respectively, the peak currents can be expressed as:

$$I_p(\mathrm{I}) = g_{\mathrm{I}} f_m(2\pi r) \tag{6.3}$$

$$I_p(\mathrm{III}) = g_{\mathrm{III}} f_m[2\pi(r+\delta)] \tag{6.4}$$

where g_{I} and g_{III} are the respective electrochemical coefficients of response (μA g^{-1} cm^{-1}), m the mass of sample put into contact with the base electrode, and f the fraction of grains exposed to the solid particle/base electrode/electrolyte three-phase boundary.

In this frame, both the fraction of pigment grains exposed to the electrolyte solution and the thickness of the soap's layer will depend on the concentration and uniformity of distribution of the pigment grains in the paint specimen. In turn, the thickness of the soap's layer will mirror of the extent of the attack of the binder on the pigment. This can be in principle described in terms of solid-state reaction kinetics (see again Sect. 3.8) so that in the simplest case, the mass of lead soaps, m_{soap}, can be taken as a potential function of the mass of pigment, m_{pigment}:

$$m_{\mathrm{soap}} = k m_{\mathrm{pigment}}^{\eta} \tag{6.5}$$

introducing two constant parameters k, η. Then, the thickness of the soap layer can be related to the radius of the spheroidal particle pigment by,

$$\frac{4}{3}\pi\rho_{\mathrm{soap}}[(r+\delta)^3 - r^3] = k\left(\frac{4}{3}\pi\rho_{\mathrm{pigment}} r^3\right)^{\eta} \tag{6.6}$$

where ρ_{pigment} and ρ_{soap} are the densities of the pigment and the lead pigment soap, respectively. When $\delta \ll r$, one can approximate:

$$\frac{\delta}{r} \approx \frac{k}{3}\left(\frac{\rho_{\mathrm{pigment}}^{\eta}}{3\rho_{\mathrm{soap}}}\right)\left(\frac{4}{3}\pi\right)^{\eta-1} r^{3\eta-3} \tag{6.7}$$

Accordingly, combining this condition with Eqs. (6.5) and (6.6), one can establish a relationship between the peak currents of processes I and III:

$$I_p(\mathrm{III}) \approx \frac{g_{\mathrm{III}}}{g_{\mathrm{I}}}\left(I_p(\mathrm{I}) + \frac{k}{3}\left(\frac{\rho_{\mathrm{pigment}}^{\eta}}{\rho_{\mathrm{soap}}}\right)\frac{I_p(I)^{3\eta-3}}{(2\pi f m g_{\mathrm{I}})^{3\eta-3}\left(\frac{4}{3}\pi\right)^{1-\eta}}\right) \tag{6.8}$$

This relationship is dependent on the mass of the sample transferred onto the electrode surface incorporating a series of parameters (k, f, η) characterizing the preparation of the pigment plus binding system; i.e., of the technical protocol of painting.

An essentially equivalent approach can be derived from Eq. (3.66), assuming that, under fixed electrochemical conditions, the peak currents measured for processes I-III in linear potential scan (or square wave, etc.) voltammograms can be expressed as:

$$I_p(\mathrm{I}) = g_\mathrm{I} m_\mathrm{pigm}; \; I_\mathrm{p}(\mathrm{II}) = g_\mathrm{II} m_\mathrm{rel}; \; I_\mathrm{p}(\mathrm{III}) = g_\mathrm{III} m_\mathrm{soap} \tag{6.9}$$

Now, let us assume that the amounts of lead soaps and released Pb^{2+} ions are:

$$m_\mathrm{soap} = k_\mathrm{sp} m^{\eta}_\mathrm{pigment}; \; m_\mathrm{rel} = k_\mathrm{rp} m^{\chi}_\mathrm{pigment} + k_\mathrm{rs} m^{\lambda}_\mathrm{soap} \tag{6.10}$$

Here k_sp, k_rp, and k_rs are rate constants of the pigment to soap conversion, Pb^{2+} release from pigment particles, and Pb^{2+} release from soap processes, while η, χ, and λ are the respective exponents of the potential functions. Combining the above sets of equations leads to:

$$\frac{I_\mathrm{p}(\mathrm{III})}{I_\mathrm{p}(\mathrm{I})} = k_\mathrm{sp}\left(\frac{g_\mathrm{III}}{g_\mathrm{I}}\right)\left(\frac{I_\mathrm{p}(\mathrm{I})}{g_\mathrm{I}}\right)^{\eta-1} \tag{6.11}$$

$$\frac{I_\mathrm{p}(\mathrm{II})}{I_\mathrm{p}(\mathrm{III})} = \left(\frac{g_\mathrm{II}}{g_\mathrm{III}}\right)\left(\frac{k_\mathrm{rp}}{k_\mathrm{sp}^{\chi/\eta}}\left(\frac{I_p(\mathrm{III})}{g_\mathrm{III}}\right)^{\chi/\eta-1} + k_\mathrm{rs}\left(\frac{I_p(\mathrm{III})}{g_\mathrm{III}}\right)^{\lambda-1}\right) \tag{6.12}$$

again predicting potential dependences between the $I_\mathrm{p}(\mathrm{III})/I_\mathrm{p}(\mathrm{I})$ on $I_\mathrm{p}(\mathrm{III})$, and $I_\mathrm{p}(\mathrm{II})/I_\mathrm{p}(\mathrm{III})$ on $I_\mathrm{p}(\mathrm{III})$. Experimental data such as in Fig. 6.2 permitted to extract $I_\mathrm{p}(\mathrm{II})/I_\mathrm{p}(\mathrm{III})$ versus $I_\mathrm{p}(\mathrm{III})$ plots in Fig. 6.5. Analysis of these data revealed that:

(a) When samples from different paintings by the same author were available, experimental data points could be grouped in two different trend lines, one for samples from flesh and white areas, and the other for yellow-green areas. The former contained lead white, while the second contained lead-tin yellow, often accompanied by lead white and copper pigments.
(b) The trend lines for each painter were consistent with the predictions from Eq. (6.12), thus corresponding to a potential variation of $I_\mathrm{p}(\mathrm{II})/I_\mathrm{p}(\mathrm{III})$ on $I_\mathrm{p}(\mathrm{III})$.
(c) Data points of equivalently colored areas defined a painter/workshop-characteristic trend curve.

Accordingly, there is the possibility of electrochemical discrimination between different painters/workshops using submicrosamples from paints. This requires several favorable conditions (use of common electroactive pigments in particular) and, of course, factors such as the evolution of the workshop techniques, the possibility of repaints, and severe degradation processes can be influential. Despite these difficulties, the described technique can be considered of interest for authentication and event dating of paintings.

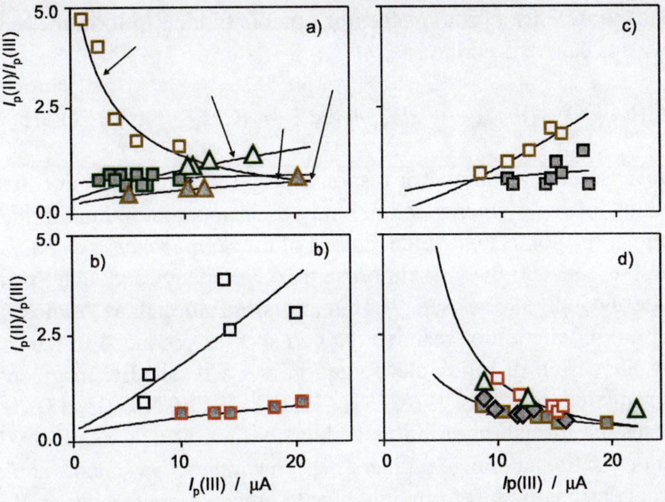

Fig. 6.5 Plots of $I_{\mathrm{p}}(\mathrm{II})/I_{\mathrm{p}}(\mathrm{III})$ versus $I_{\mathrm{p}}(\mathrm{III})$ for samples from yellow and green (empty figures) and white (solid figures) areas of the studied paintings of **a** Jerónimo Jacinto de Espinosa (squares) and Vicente Requena (triangles), **b** Juan Borrás, **c** Juan Sariñena, **d** Francisco Ribalta (squares), Juan Ribalta (triangles), and Vicent Castelló (rhombs). Voltammetric data in conditions such as in Fig. 6.2. Adapted from ref. [1], with permission

6.2 Degradation Processes; Pigment Alteration in Extreme Heritage

Degradation of the components of paintings can be due to mechanical, thermal, hydrous, luminous, chemical, and/or biochemical stress. As previously noted, pigments and binding media can interact and mutually reinforce their degradation processes [3, 4]. In particular, the biodeterioration by fungi and bacteria of reconstructed paint films containing cadmium yellow (CdS) [5] and verdigris (synthetic copper(II) acetate) [6] has been monitored using VIMP and SECM, testing different proteinaceous and oil-binding media. In the case of verdigris, distinctive voltammetric signals were detected corresponding, in agreement with the scenario previously described, to the reduction of pigment grains,

$$\{\mathrm{Cu(Ac)_2}\}_{\mathrm{solid}} + 2\mathrm{H}^+_{\mathrm{aq}} + 2\mathrm{e}^- \rightarrow \{\mathrm{Cu}\}_{\mathrm{solid}} + 2\mathrm{HAc}_{\mathrm{aq}} \tag{6.13}$$

accompanied by the reduction of (mostly protein-complexed in the case of proteinaceous binders) Cu(II) species in the ionomeric layer surrounding the pigment aggregates. This process can be formulated as follows:

$$\{\mathrm{Cu}^{2+}(\mathrm{Bind}^{2-})\}_{\mathrm{Vd\ ionomeric\ layer}} + 2\mathrm{H}^+_{\mathrm{aq}} + 2\mathrm{e}^- \rightarrow \{\mathrm{Cu}\}_{\mathrm{solid}} + (\mathrm{H_2Bind}) \tag{6.14}$$

where (Bind) represents a proton-acceptor organic binder. In turn, the reduction of copper(II) soaps can be described as,

$$\{Cu^{2+}(RCOO^-)_2\}_{Cu\ soap} + 2H^+_{aq} + 2e^- \rightarrow \{Cu\}_{solid} + 2RCOOH \quad (6.15)$$

In several cases, the voltammetric response permits the identification of the degradation products of the pigments and to study their apparent mutual influence. One reported case corresponds to the blue regions of the *Sant Francesc de Paula* painting exhibited in the Tomàs Balvey Arxiu Museum (Cardedeu, Catalonia, Spain), where Prussian blue and indigo were used [7]. Here, Prussian blue acts as a radical scavenger moderating the production of reactive oxygen species produced in the oil-binding medium by the action of ultraviolet radiation. As a result, Berlin green (a solid-solution formulated as $\{KFe^{III}[Fe^{II}(CN)_6]\}_x\{Fe^{III}[Fe^{III}(CN)_6]\}_{1-x}$ [8]) is formed. This promotes the formation of indigo adducts with radicals oxidatively degrading this pigment to isatin, an orange pigment [9]. This interactive scheme could explain the significant chromatic shift from light blue to greenish-orange observed in several parts of the paint.

Solid-state electrochemistry can also be applied to study cases of the so-called *extreme heritage*, i.e., cultural goods which have been submitted to severe deterioration. A case of particular interest was that of the frescoes painted in 1707 by Acisclo Antonio Palomino in the vaulted ceiling of the *Sant Joan del Mercat* church in Valencia (Spain). The frescoes were subjected to gunfire during the Spanish Civil War in 1936 resulting in the destruction of a significant fraction of the wall paint and the severe deterioration of the remaining areas. The restoration works performed since 2001 revealed significant chromatic changes in the paintings; in particular: (i) carnations become brown-reddish and (ii) most of the blue and green robes and vegetal motifs become blackened. VIMP data permitted to explain these chromatic changes as resulting from the thermal stress suffered by the paints during the firing event. The yellow ochre-based pigments are converted into red hematite forms [10] while the blue or greenish blue azurite, identified as the pigment used by Palomino, decomposed to black tenorite (CuO) [11].

Figure 6.6 shows a photographic image recorded in 2004 of a region of ca. 1 m^2 of the damaged frescoes in the vault of the *Sant Joan del Mercat* church in Valencia (Spain) [11]. The image shows the general reddish tint of the images.

Figure 6.7 compares the square wave voltammograms recorded for different iron pigments (left side, generically, iron earth) and three samples extracted from different points of the vault of the church. Hematite displays a cathodic peak at ca. −0.4 V versus Ag/AgCl which becomes broadened on decreasing crystallinity and increasing degree of hydration. In turn, ochres, constituted by hydrated hematite and goethite forms, show cathodic waves between 0.0 and −0.4 V. The relevant point to emphasize is that samples from the vault exhibited the hematite signal clearly prevailing over the ochre signals, thus characterizing the ochre to hematite conversion produced by the firing event and indicating the temperature reached during the same [10]. Dehydroxylation of goethite starts at 200–230 °C and, after the generation of transition

Fig. 6.6 Photographic image of an area of approximately 1 m^2 of the damaged Palomino's frescoes in the vault of the *Sant Joan del Mercat* church in Valencia (Spain). Reproduced from ref. [11], with permission

forms, the formation of typical hematite crystals is completed between 600 and 650 °C (see Additional literature).

The voltammograms of samples from frescoes show additional peaks at ca. 0.0 and −0.58 V. The former can be attributed to the reduction of magnetite and the second to the reduction of lead carbonate and/or lead oxide. Magnetite is formed from hematite between 500 and 600 °C, and its appearance is consistent with a firing under moderately oxidizing conditions. Lead compounds result from the thermal decomposition of lead white originally used in the paintings.

Figure 6.8 depicts the voltammograms recorded for three samples from the more or less blackened blue-green regions of the Palomino frescoes. Here, the overlapping cathodic peaks of smalt and azurite at ca. −0.10 V precede a more or less sharp signal at ca. −0.55 V corresponding to the reduction of tenorite:

$$CuO_{solid} + 2H^{+}_{aq} + 2e^{-} \rightarrow Cu_{solid} + H_2O \quad (6.16)$$

The ratios smalt/azurite and azurite/tenorite yield information about the temperature reached during the thermal stress suffered by the frescoes. Thermochemical data (see additional literature) indicate that azurite was converted into tenorite up to 345 °C whereas smalt remained stable at least until 600 °C. Accordingly, on increasing temperature, the smalt/azurite ratio increases while the azurite/tenorite ratio decreases. Quantification was available from peak current/peak area data or by

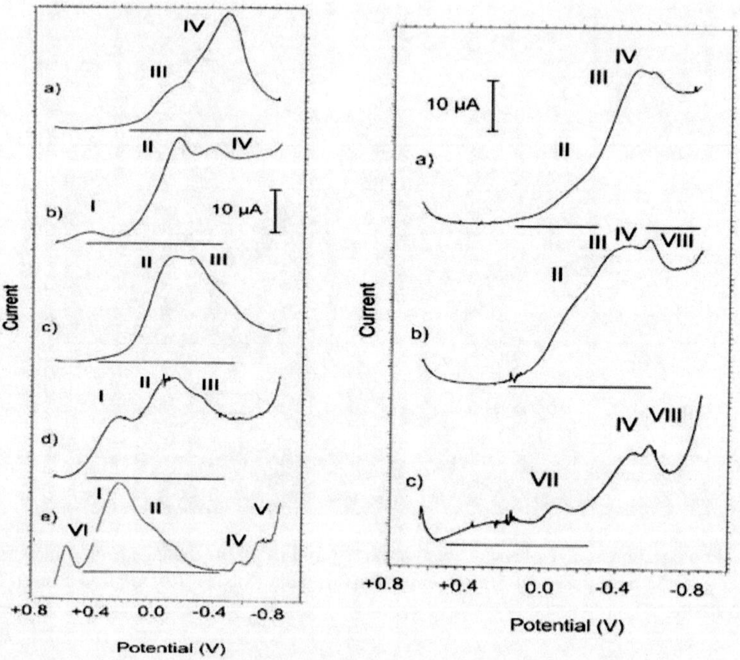

Fig. 6.7 Square wave voltammograms of paraffin-impregnated graphite electrodes modified with iron pigments and samples from the frescoes in the vault of the *Sant Joan del Mercat* church in Valencia (Spain). Left side: **a** Spanish hematite, **b** yellow French ochre, **c** Sienna raw, **d** Italian burst umber, and **e** greenish natural umber. Right side: Samples **a** TR4b, **b** TR5b, **c** TR2a from the frescoes. Electrolyte 0.10 M HCl. Potential scan initiated at 0.65 V in the negative direction; potential step increment 4 mV; square wave amplitude 25 mV; frequency 5 Hz. Adapted from ref. [10], with permission. Note that the axes are opposite to the IUPAC convention

means of the Tafel-type analysis of the voltammetric curves [11]. Figure 6.9 illustrates the variation of the azurite/tenorite peak area ratio with the azurite/tenorite molar ratio in synthetic mixtures and samples from Palomino's frescoes [11]. This figure contains the azurite/tenorite peak area ratio measured for a series of mixtures of known composition (solid figures). The data define a calibration curve where the independent variable is the abscissa (azurite/tenorite molar ratio) parameter. Then, measuring the peak area ratio of paint samples and adjusting the corresponding point to the trend line, we can obtain the composition of the sample. Interestingly, the data points for samples from the *Sant Joan del Mercat* church permitted to group the samples depending on their location in the vault. Those extracted from the central part (squares in Fig. 6.9) of the vault exhibited relatively low azurite/tenorite ratios, as expected for samples severely exposed to the gunfire. In contrast, the samples extracted from the lateral lunettes of the vault (triangles in Fig. 6.9) exhibited high azurite/tenorite ratios (triangles in Fig. 6.9), just as expected for samples submitted to a low thermal stress. These results can be used to draw a temperature map representative of the fire event.

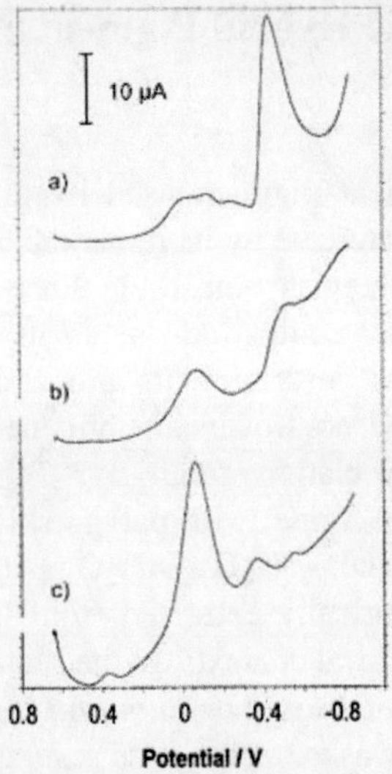

Fig. 6.8 Square wave voltammograms of paraffin-impregnated graphite electrodes modified with samples: **a** PVB9, **b** PV7, and **c** PV1 from the frescoes in the vault of the *Sant Joan del Mercat* church in Valencia (Spain) immersed into 0.50 M potassium phosphate buffer at pH 7.4. Potential scan initiated at 0.45 or 0.65 mV in the negative direction; potential step increment 4 mV; square wave amplitude 15 mV; frequency 2 Hz. Reproduced from ref. [11], with permission. Note that the axes are opposite to the IUPAC convention

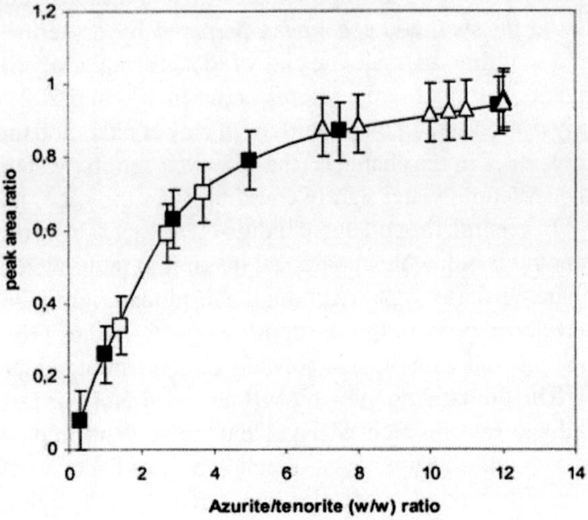

Fig. 6.9 Variation of the ratio between the peak areas for the azurite and tenorite signals in voltammograms such as in Fig. 6.8 with the azurite/tenorite molar ratio in synthetic mixtures of both components (black squares) and samples from the *Sant Joan del Mercat* church in Valencia (Spain) extracted from the central part (squares) and the lateral lunettes (triangles) of the vault. Reproduced from ref. [11], with permission

6.3 Organic–Inorganic Hybrid Pigments: The Maya Blue Problem

Maya blue (MB) is an ancient pigment used by the Mayas and other ancient Mesoamerican peoples characterized by its remarkable stability and peculiar color, varying from dark blue to greenish blue [12]. Since its discovery in 1931 [13], the pigment received increasing attention due to its peculiar properties and the absence of historical sources describing its preparation. In fact, the composition of the pigment remained controversial until the 1960s when it was recognized as resulting from the association of indigo (2-(3-oxo-2,3-dihydro-1H-indol-2-ylidene)-2,3-dihydro-1H-indol-3-one) with palygorskite, a channeled clay of ideal composition $Si_8(Mg_2Al_2)O_{20}(OH)_2(H_2O)_4 \cdot 4H_2O$ [14].

Indigo is a blue dye historically extracted from leaves of different plants in different geographical and cultural contexts. In the Mesoamerican area, the pigment was prepared from the leaves of the *añil* or *xiuquitlitl* (*Indigofera suffruticosa*) plant. In turn, the palygorskite clay was extracted from several sites in the region [15]. The ancient Mayas attributed MB a high symbolic value, applying it in ritual ceremonies and [12, 15].

MB is considered as ancestor of the currently called organic–inorganic hybrid materials. Its composition, structure, and optical properties, however, have become controversial in regard to three aspects: the location of the guest organic molecules on the palygorskite host, the type of dye–clay interaction, and the reasons for the hue variability of the pigment. For logical reasons, much of the research on MB was based on synthetic specimens prepared by dry crushing indigo with palygorskite and sepiolite, an analogue clay of ideal formula $Mg_4Si_6O_{15}(OH)_2 \cdot 6H_2O$, and subsequent heating at temperatures between 100 and 200 °C. This last process removes physisorbed water on the surface of clay crystals and much of the weakly bound water molecules in the channels (the so-called zeolitic water), but the clay retains much of the structural water tightly coordinated to the Mg^{2+} and Al^{3+} cations [16, 17].

In general, the content of indigo is ca. 1 wt% in its mixtures with palygorskite. This amount is below the theoretical maximum indigo loading to occupy all palygorskite channels (ca. 4 wt%). Although hybrid materials containing higher indigo loadings have been prepared in the context of MB studies [18, 19], the excess of indigo can be removed with organic solvents or concentrated HNO_3.

The dominating view on MB assumed more or less explicitly that only one dye, indigo, was present in MB and that the Mayans prepared a unique material by means of a unique –but unknown–recipe. The peculiar color of the MB would be the result of the shift of indigo absorption bands in the visible region by the effect of its interaction with the clay support [20]. There was, however, controversy on the location of the indigo molecules, either adsorbed onto the external surface of palygorskite crystals [21], entrapped into the palygorskite channels [16, 17, 19, 22–24], blocking the entrances at the edges of palygorskite channels [18, 25] and the grooves of channels in the external clay surface [26].

In this context, in 2006, we proposed an alternative view of MB assuming that the pigment contained a minor but significant fraction of dehydroindigo, an oxidized form of indigo [27]. In this model, the dehydroindigo/indigo ratio, in turn depending on the temperature of the thermal treatment used in pigment manufacturing, will be essential to determine the color of MB [27, 28]. The visible spectra of indigo dissolved in organic solvents display a maximum at ca. 600 nm whereas dehydroindigo solutions show absorbance maxima at ca. 450 nm. The superposition of the spectra of these components is close to the spectra of MB specimens, characterized by maxima at 425–500 and 640–660 nm, accounting for spectral shifts attributed to dye−clay interactions. Accordingly, indigo-rich specimens will be blue whereas the presence of increasing amounts of dehydroindigo would result in an increased greenish tint.

Figure 10a illustrates the square wave voltammetric response of solid indigo in contact with aqueous acetate buffer at pH 4.75. This technique permits the simultaneous recording of reductive and oxidative signals so that the two acute peaks appearing at −0.30 and 0.45 V versus Ag/AgCl can be described, respectively, in terms of the reduction of indigo to leucoindigo and the oxidation of indigo to dehydroindigo.

These are proton-assisted solid-state reduction and oxidation processes (see schematics in Fig. 6.11) showing a high degree of reversibility under our experimental conditions (see also Sect. 5.4). This permitted the determination of thermochemical parameters for the association of organic molecules to the clay framework [27].

The relevant point to emphasize was that the voltammograms of MB samples coming from different archaeological sites in Yucatán and Campeche showed, as can be seen in Fig. 6.6, peak current ratios differing from those recorded for pure indigo [27]. These differences were interpreted, using speciation criteria (see Sect. 4.4), as due to the presence of a significant fraction of dehydroindigo accompanying indigo.

Accordingly, the dehydroindigo model led to develop a more complex view of the MB. The application of electrochemical and spectroscopic techniques to archaeological samples indicated that the preparation of the MB evolved with time [28] and that this type of material was also used to prepare yellow pigments [29] and other materials [30].

Based on voltammograms such as in Fig. 6.6, the archaeological samples were divided into several electrochemical types distributed in the different archaeological sites as illustrated in Table 6.1. Chemometric analysis was based on the measurements of peak currents and peak potentials permitted to build the dendrograms in Fig. 6.12, corresponding to the hierarchical cluster analysis of voltammetric data [28].

The hierarchical cluster analysis of voltammetric data suggested that different characteristic types could be discriminated. This could be rationalized by assuming that different recipes were used to prepare MB. Crossing electrochemical data with the accepted chronology of the studies sites resulted in the proposal of an evolutionary scheme for the preparation of MB. This is illustrated in Fig. 6.13 where, similarly to the ramified schemes of phylogenetic analysis, the possible linkages between sites and preparative types are depicted [28].

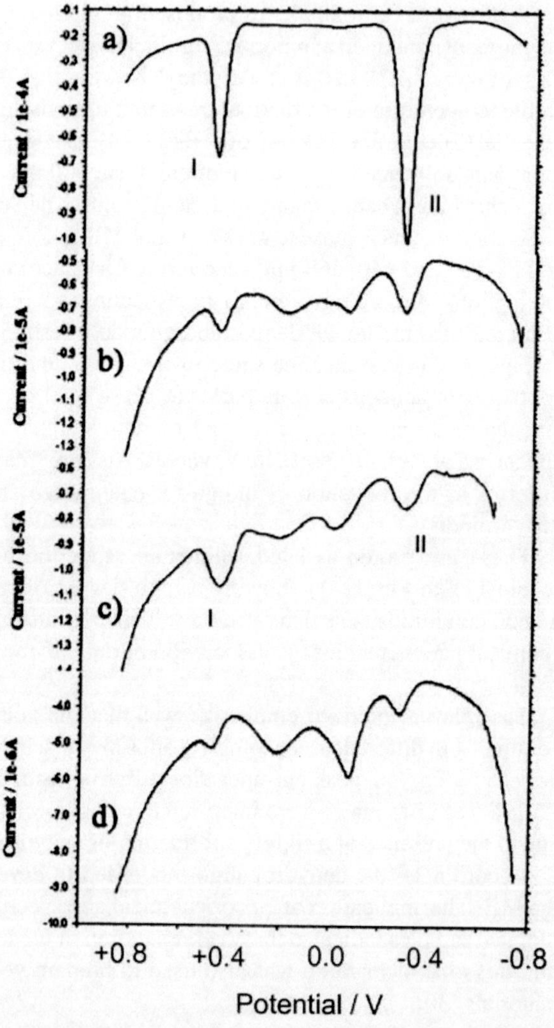

Fig. 6.10 Square wave voltammograms of paraffin-impregnated graphite electrodes modified with: **a** indigo microparticles, and MB samples from the sites of: **b** Calakmul, Substructure II-C, Late Preclassical period; **c** Dzibilnocac-2, Late Classical period; d) Mayapán-2, Early Postclassical period. Electrolyte 0.50 M sodium acetate buffer, pH 4.85. Potential scan initiated at − 0.75 V versus Ag/AgCl in the positive direction. Potential step increment 4 mV; square wave amplitude 25 mV; frequency 5 Hz. Reproduced from ref. [27], with permission. Note that the axes are opposite to the IUPAC convention

This approach, contrary to the monolithic, static, view of the Maya culture, provided a much more complex interpretation of the same, far from the traditional 'unicity paradigm' [31, 32]. This view was reinforced by the characterization of materials affine to MB in a context separate from the ritual, elitist one typically attributed to the MB [30]. Additionally, the synthetic schemes adopted by the Mesoamericans for the preparation of the pigment could be varied [29, 30, 32].

From the contemporary view of chemistry, the MB can be viewed as the protagonist of (at least) three 'revolutions' in chemical technology: the preparation of organic–inorganic hybrid materials, the thermal control of reactivity, and the modulation of chemical reactivity using solid supports. This last exploits the catalytic properties of the clays as well as their capability to act as structure-directing agents for

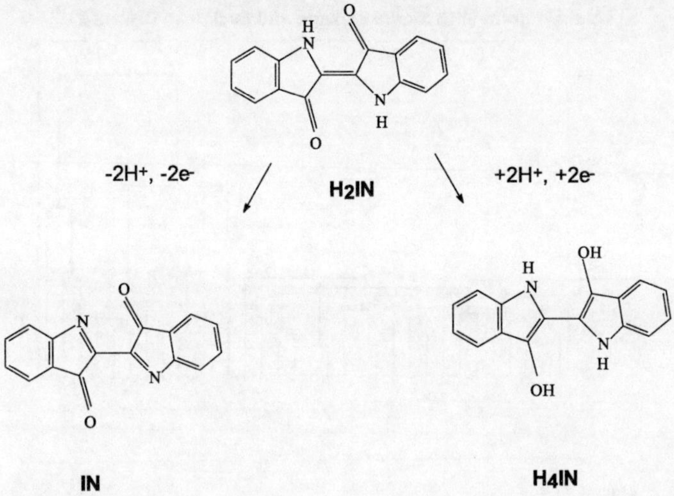

Fig. 6.11 Scheme for the proton-assisted solid-state reduction and oxidation processes experienced by microparticulate deposits of indigo in contact with aqueous electrolytes. Reproduced from ref. [27], with permission

Table 6.1 Distribution of the electrochemical types in the MB samples from different archaeological sites in Yucatán and Campeche. Reproduced from ref. [28], with permission

	Site	A2	A1	A4	A3	A5	B	C	D
Postclass	Mayapán					•	• •		•
Terminal class	Chichén Itzá						• •	• •	
Late classical	D'zula						•	• •	
	Ek Balam				•				
	Dzibilnocac					•			•
	Kulubá				•				
	Chacmultún				•				
	Mulchic	•				• ••			•
	El Tabasqueño	• •	•		• •	•			
	Acanceh		• •						
Early class	Calakmul A6		•			•			
Late preclass	Calakmul II-C			•					

synthesis [28]. In short, the Mesoamerican peoples created –obviously in a prescientific age–something like a 'Maya chemistry' rediscovered by modern science several centuries later [32, 33].

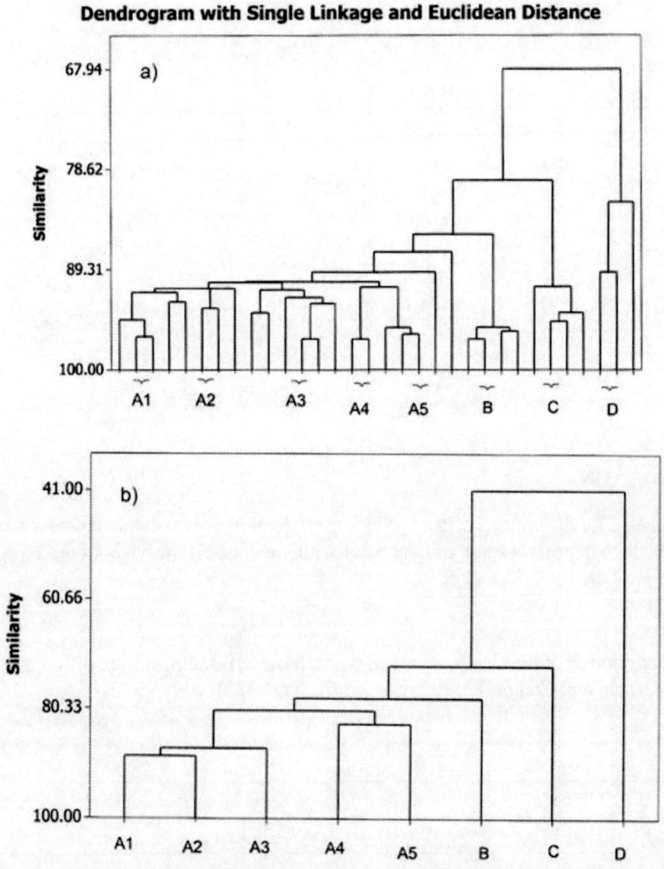

Fig. 6.12 Hierarchical Cluster Analysis with Euclidean distance of the autoscaled variables applied to voltammetric parameters recorded for **a** the entire set of archaeological samples and **b** the 'holotypes' characterizing the different electrochemical types. Reproduced from ref. [28], with permission

Electrochemical data on different indigo–clay systems also prompted a more complex view of MB as constituted by different organic components (indigo, dehydroindigo, indirubin, …) occupying different positions in the clay support, thus defining different topological redox isomers [34, 35].

The dehydroindigo model, despite subsequent reinforcements with chromatographic, mass spectroscopy, and other techniques [36, 37], was not unanimously accepted, alternating favorable [38–41] with neutral or opposite views [18, 19, 42–44]. In this context, the interpretation of the Raman spectra of MB was particularly relevant because it provides peculiar features differing from indigo in solution and in solid state and dehydroindigo [44–50].

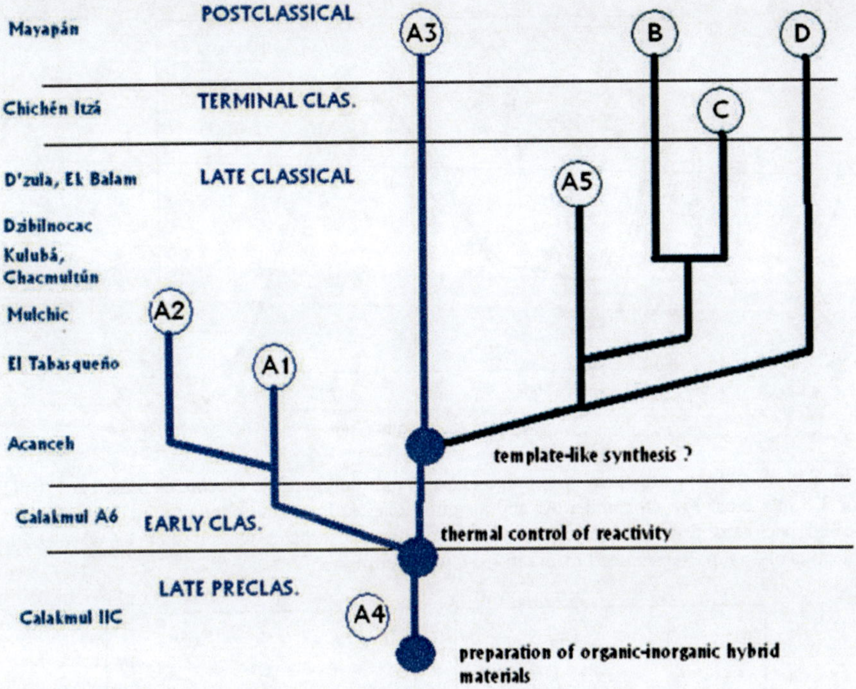

Fig. 6.13 Tentative chronological scheme for the evolution of MB in the Late Preclassical-Postclassical Maya periods based on the chemometric analysis of electrochemical data. Reproduced from ref. [28], with permission

The interpretation of data was complicated by the elusive nature of dehydroindigo, a highly reactive compound. In contradiction with the expectations, indigo plus palygorskite specimens prepared in the absence of oxygen also showed a blue-greenish hue, as illustrated in Fig. 6.14 [51]. Again, electrochemical data provided new insights into the MB problem. Figure 6.15 compares the cyclic voltammograms at glassy carbon electrode of water-containing (1% v:v) DMSO solutions of indigo, dehydroindigo ('fresh' and after exposure to aeration), and the DMSO extract of an indigo plus palygorskite specimen heated at 150 °C [51]. The voltammogram of indigo in anhydrous DMSO consists of the one-electron reduction to the corresponding radical anion [52], represented here by the essentially reversible couple C_1/A_1. In the presence of water traces, the voltammogram displays new features, which can be interpreted in terms of the proton-assisted oxidation of the indigo radical anion to indigo hemienol and subsequent superposition of different oxidation processes to dehydroindigo summarized in Fig. 6.16.

Accordingly, a new interpretation of MB was proposed based on the idea that indigo hemienol and dehydroindigo were successively formed during the heating of indigo plus palygorskite mixtures, i.e., that MB was formed via successive isomerization and redox tunning processes [51]. This interpretation was consistent with

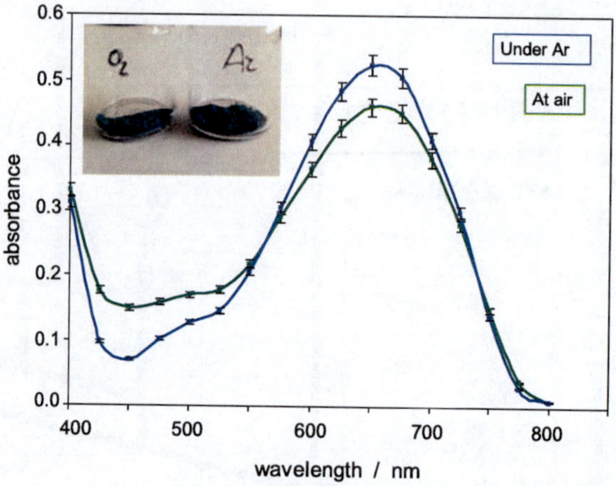

Fig. 6.14 Visible spectra of the two indigo (1 wt%) plus palygorskite specimens heated at 130 °C for 3 h in a closed reactor under Ar and air atmospheres. Inset: the photographic image of the heated specimens. Error bars correspond to three replicate absorbance measurements at wavelength intervals of 25 nm. Reproduced from ref. [51], with permission

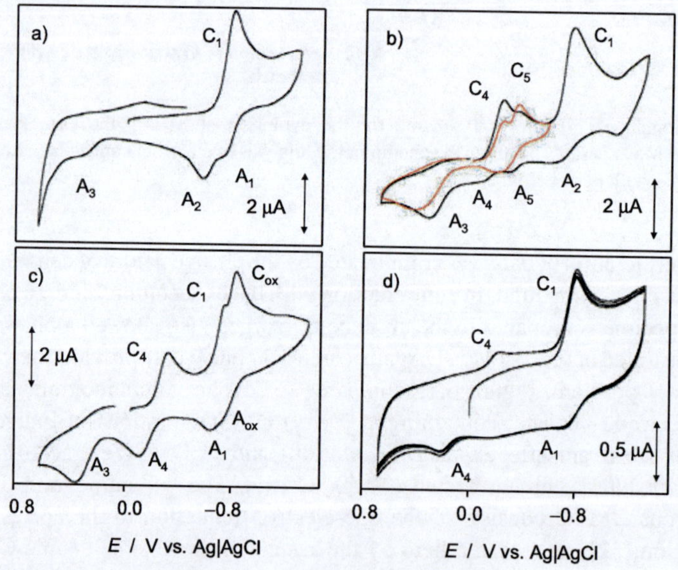

Fig. 6.15 Cyclic voltammograms at glassy carbon working electrode in 0.10 M Bu_4NPF_6/DMSO containing 1% v:v of water solutions. **a** 0.5 mM indigo; **b** freshly prepared 0.5 mM dehydroindigo; **c** the same solution after 5 min under Ar atmosphere; d) DMSO extract from an indigo (1%wt) plus palygorskite specimen treated at 150 °C. Potential scan initiated at 0.0 V versus Ag/AgCl in the negative direction; potential scan rate 50 mV s^{-1}. Reproduced from ref. [51], with permission. Note that the axes are opposite to the IUPAC convention

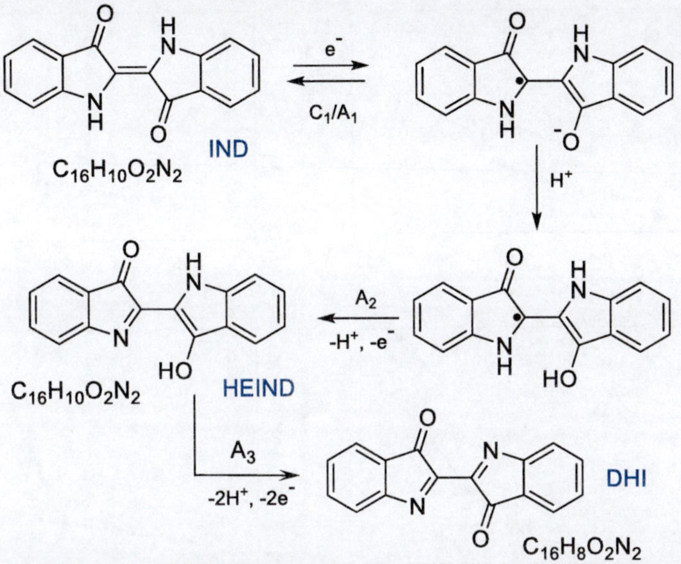

Fig. 6.16 Scheme to describe the indigo- and dehydroindigo-related electrochemical processes in Fig. 6.15. Reproduced from ref. [51], with permission

visible, infrared, and Raman spectroscopic data as well as with nuclear magnetic resonance (^{13}C NMR) data for indigo–sepiolite specimens (indigo–palygorskite specimens are NMR silent) [25, 53, 54]. This scenario was supported by voltammetric and spectral data denoting a two-step variation of different parameters with the heating temperature, illustrated in Fig. 6.17.

Interestingly, additional information can be obtained from electrocatalytic data on the oxygen reduction reaction process (ORR) at sample-modified electrodes in contact with air-saturated aqueous solutions. The voltammograms are dominated by the cathodic ORR process at potentials between −0.6 and −0.8 V versus Ag/AgCl (see Fig. 6.18). This is a multistep process by the one-electron reduction of O_2 to superoxide radical anion, which is followed by competing protonation, disproportionation, and subsequent electron transfer processes successively yielding H_2O_2 and H_2O. Intermediate reactive species such as hydroperoxide ($HO_2\cdot$) and hydroxyl ($HO\cdot$) radicals and hydroperoxide anion (HO_2^-) can also be formed depending on the experimental conditions. On comparing the voltammograms in Fig. 6.18, one can see that, apart from weak signals between −0.2 and −0.4 V, the main cathodic signal is more or less positively/negatively shifted and even appears as two superimposed signals. These features can be interpreted by assuming that (at least) two different electrocatalytic pathways operate. The first one involves the classical regeneration scheme involving the indigo/leucoindigo couple. The second pathway can be conceived as similar to that described by Compton et al. [55] for the catalysis of hematite on ORR in neutral media (see also Chap. 7). As hematite, the indigo species associated with the clay would act as a chemical catalyst accelerating the

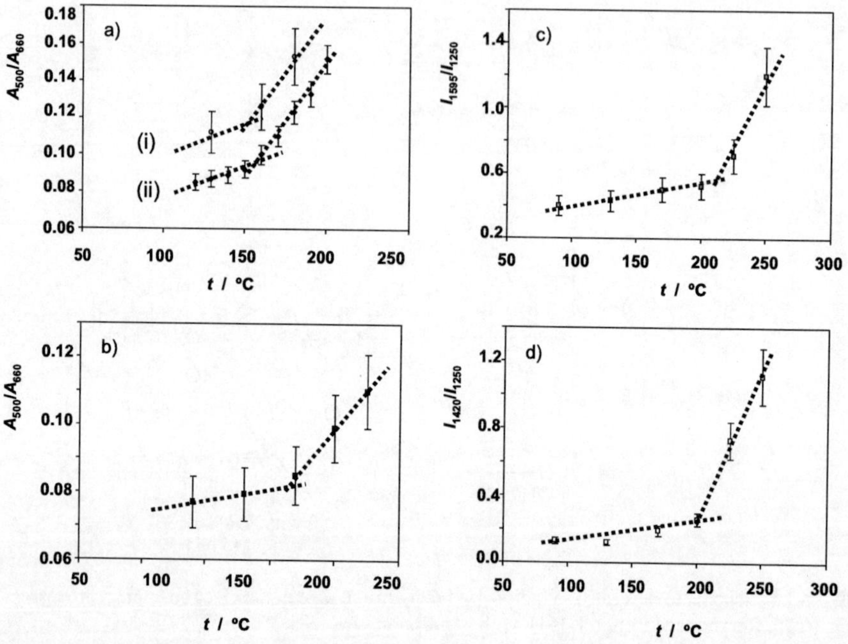

Fig. 6.17 **a** Variation of the ratio of absorbances at 500 and 660 nm in visible spectra of indigo plus palygorakite specimens from refs. [49] (i) and [51] (ii), and **b** ref. [44]. Variation of the intensity ratios of Raman bands at **c** 1595 and 1250 cm^{-1} and **d** 1420 and 1250 cm^{-1} in Raman spectra of similar replicants from ref. [44]. Error bars for our data correspond to dispersion from three replicate experiments. Error bars for literature data where uncertainty was not reported are depicted assuming relative uncertainties of 10%. Reproduced from ref. [51], with permission

disproportionation of H_2O_2 into H_2O and O_2. For our purposes, the relevant point to underline is that each one of the pathways was that prevailing when indigo or the indigo hemienol forms was the majority component of the hybrid material.

The resulting scenario can be reconciled with colorimetric data and tests of chemical attack for hybrid materials prepared by mixing indigo with different clays. Here, the size of the clay channels relative to that of the indigo molecule (4.8 Å) plays a crucial role [50]. The extreme case is represented by LTA zeolite. This has channels of 4.1 Å × 4.1 Å, too small to trap indigo molecules. The other extreme case is mordenite. This clay has 7.0 Å × 6.5 Å channels, and does not form stable hybrids (indigo is easily removed with organic solvents) because these are so wide to retain indigo molecules. In the intermediate cases of silicalite (channels of 5.3 Å × 5.6 Å), palygorskite (6.4 Å × 3.7 Å), and sepiolite (10.6 Å × 3.7 Å), there is the formation of stable hybrids (resistant to HNO_3 attack) [55]. Interestingly, silicalite hybrids retain the blue color after heating above 100 °C whereas palygorskite and silicalite ones become greenish [55]. This can be interpreted within the isomerization plus redox tunning model assuming that these reactions require a certain space to facilitate the re-accommodation of the new molecules. This can be made in the palygorskite and

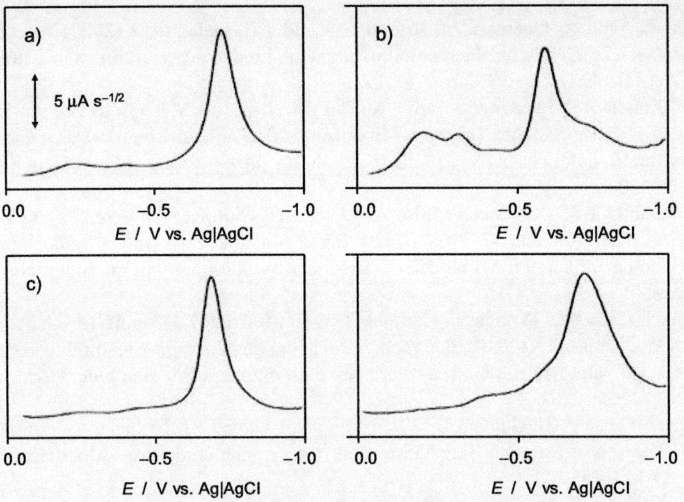

Fig. 6.18 Linear potential scan voltammograms, after semi-derivative convolution, of microparticulate deposits on glassy carbon electrode of **a** indigo, and **b**–**d** indigo (1%wt) plus palygorakite specimens prepared at **b** 25 and 180 °C under Ar and oxygen **d** atmospheres in contact with air-saturated 0.25 M HAc/NaAc aqueous buffer at pH 4.75. Potential scan initiated at 0.0 V in the negative direction; potential scan rate 50 mV s^{-1}. Reproduced from ref. [51], with permission. Note that the axes are opposite to the IUPAC convention

sepiolite, but the silicalite channels are so narrow to allow the required molecular rearranging, and indigo does not experience isomerization/oxidation reactions.

In this scenario, the MB can be seen as a polyfunctional hybrid material where different organic components are attached to the inorganic matrix occupying different locations. The case of MB illustrates the capabilities of electrochemistry to provide analytical information on organic–inorganic hybrid materials. In fact, solid-state electrochemical data can also be used to evaluate reaction kinetics [56] and acquire information on the in-depth distribution of the organic species [57].

References

1. Doménech-Carbó A, Doménech-Carbó MT, Castelló-Palacios A, Guerola-Blai V, Pérez E (2019) Electrochemical identification of painters/workshops: the case of Valencian Renaissance-Baroque painters (ca. 1550–ca. 1670). Electrochim Acta 297:685–695
2. De Viguerie L, Jaber M, Pasco H, Lalevée J, Morlet-Savay F, Ducouret G, Rigaud B, Pouget T, Sanchez C, Walter P (2017) A 19th century "ideal" oil paint medium: a complex hybrid organic-inorganic gel. Angew Chem Int Ed 56:1619–1623
3. Keune K, van Loon A, Boon JJ (2011) SEM Backscattered-Electron images of paint cross sections as information source for the presence of the lead white pigment and lead-related degradation and migration phenomena in oil paintings. Microsc Microanal 17:696–701

4. Mazzeo R, Prati S, Quaranta M, Joseph E, Kendix E, Galeotti M (2008) Attenuated total reflection micro FTIR characterisation of pigment–binder interaction in reconstructed paint films. Anal Bioanal Chem 392:65–76
5. Ortiz-Miranda AS, Doménech-Carbó A, Doménech-Carbó MT, Osete-Cortina L, Bolívar-Galiano F, Martín-Sánchez I, López-Miras MM (2016) Electrochemical characterization of biodeterioration of paint films containing cadmium yellow pigment. J Solid State Electrochem 20:3287–3302
6. Ortiz-Miranda AS, Doménech-Carbó A, Doménech-Carbó MT, Osete-Cortina L, Bolívar-Galiano F, Martín-Sánchez I (2017) Analyzing chemical changes in verdigris pictorial specimens upon bacteria and fungi biodeterioration using voltammetry of microparticles. Herit Sci 5: artic. 8
7. Doménech-Carbó A, Doménech-Carbó MT, Osete-Cortina L, Donnic M, Guasch-Ferré N, Gasol RM, Iglesias MA (2020) Electrochemical assessment of pigments-binding medium interactions in oil paint deterioration: a case study on indigo and Prussian blue. Herit Sci 8: artic. 71.
8. Grandjean F, Long GJ, Samain L (2012) The pivotal role of Mössbauer spectroscopy in the characterization of Prussian Blue and related iron cyanide complexes. Mössbauer Effect Ref Data J 35:143–153
9. Inga C, Ortíz E, Alvarez-Idaboy JR, Vivier-Bunge A (2012) Molecular description of indigo oxidation mechanisms initiated by OH and OOH radicals. J Phys Chem A 116:3643–3651
10. Doménech-Carbó A, Doménech-Carbó MT, Edwards HGM (2007) Identification of earth pigments by applying hierarchical cluster analysis to solid state voltammetry. Appl Sev Damaged Fres. Electroanal 19:1890–1900
11. Doménech-Carbó A, Doménech-Carbó MT, Edwards HGM (2008) Quantitation from Tafel analysis in solid-state voltammetry. application to the study of cobalt and copper pigments in severely damaged frescoes. Anal Chem 80:2704–2716
12. Reyes-Valerio C (1993) De Bonampak al Templo Mayor: el azul Maya en Mesoamérica. Siglo XXI, México
13. Merwin HE (1931) in Yucatan, Temple Warriors at Chitchen Itza. Morris HE, Charlot J, Morris AA (Eds). Carnegie Institution, Washington, p 356
14. Shepard AO (1962) Maya Blue: alternative hypotheses. Amer Antiq 27:565–566
15. Arnold DE, Branden JR, Williams PR, Feinman GM, Brown JP (2008) The first direct evidence for the production of Maya Blue: rediscovery of a technology. Antiquity 82:151–164
16. Chiari G, Giustetto R, Ricchiardi G (2003) Crystal structure refinement of palygorskite and Maya Blue from molecular modelling and powder synchrotron diffraction. Eur J Mineral 15:21–33
17. Fois E, Gamba A, Tilocca A (2003) On the unusual stability of Maya blue paint: molecular dynamics simulations. Micropor Mesopor Mat 57:263–272
18. Sánchez del Río M, Boccaleri E, Milanesio M, Croce G, van Beek W, Tsiantos C, Chyssikos GD, Gionis V, Kacandes GH, Suárez M, Garcia-Romero E (2009) A combined synchrotron powder diffraction and vibrational study of the thermal treatment of palygorskite–indigo to produce Maya blue. J Mater Sci 44:5524–5536
19. Tsiantos C, Tsampodimou M, Kacandes GH, Sánchez del Río M, Gionis V, Chryssikos GD (2012) Vibrational investigation of indigo–palygorskite association(s) in synthetic Maya blue. J Mater Sci 47:3415–3428
20. Reinen D, Köhl P, Müller C (2004) The nature of the colour centres in 'Maya Blue' the incorporation of organic pigment molecules into the Palygorskite lattice. Z Anorg Allg Chem 630:97–103
21. Van Olphen H (1967) Maya blue: a clay-organic pigment? Science 154:645–646
22. Kleber R, Masschelein-Kleiner R, Thissen J (1967) Etude et identification du "Bleu Maya." Stud Conservat 12:41–56
23. Giustetto R, Llabres i Xamena FX, Ricchiardi G, Bordiga S, Damin A, Gobetto R, Chierotti MR (2005) Maya Blue: A Computational and Spectroscopic Study. J Phys Chem B 109:19360–19368

24. Tilocca A, Fois E (2009) The color and stability of Maya Blue: TDDFT Calculations. J Phys Chem C 113:8683–8687
25. Hubbard B, Kuang W, Moser A, Facey GA, Detellier C (2003) Structural study of Maya Blue: textural, thermal and solidstate multinuclear magnetic resonance characterization of the palygorskite-indigo and sepiolite-indigo adducts. Clays Clay Miner 51:318–326
26. Chiari G, Giustetto R, Druzik J, Doehne E, Ricchiardi G (2008) Pre-columbian nanotechnology: reconciling the mysteries of the maya blue pigment. Appl Phys A Mater Sci Process 90:3–7
27. Doménech-Carbó A, Doménech-Carbó MT, Vázquez De Agredos-Pascual ML (2006) Dehydroindigo: a new piece into the Maya Blue puzzle from the voltammetry of microparticles approach. J Phys Chem B 110:6027–6039
28. Doménech-Carbó A, Doménech-Carbó MT, Vázquez De Agredos-Pascual ML (2007) Chemometric study of Maya Blue from the voltammetry of microparticles approach. Anal Chem 79:2812–2821
29. Doménech-Carbó A, Doménech-Carbó MT, Vidal-Lorenzo C, Vázquez De Agredos-Pascual ML (2011) From Maya Blue to 'Maya Yellow': a connection between ancient nanostructured materials from the voltammetry of microparticles. Angew Chem Int Ed 50:5741–5744
30. Doménech-Carbó A, Doménech-Carbó MT, Vidal-Lorenzo C, Vázquez De Agredos-Pascual ML (2012) Insights into the Maya Blue technology: Greenish Pellets from the ancient city of *La Blanca*. Angew Chem Int Ed 51:700–703
31. Doménech-Carbó A, Doménech-Carbó MT, Sánchez del Río M, Vázquez de Agredos-Pascual ML, Lima E (2009) Maya Blue as a nanostructured polyfunctional hybrid organic-inorganic material: the need to change paradigms. New J Chem 33:2371–2379
32. Doménech-Carbó A, Doménech-Carbó MT, Vidal-Lorenzo C, Vázquez De Agredos-Pascual ML, Osete-Cortina L, Valle-Algarra FM (2014) Discovery of indigoid-containing clay pellets from La Blanca: significance with regard to the preparation and use of Maya Blue. J Archaeol Sci 41:147–155
33. Doménech-Carbó A, Valle-Algarra FM, Doménech-Carbó MT, Osete-Cortina L, Domine M (2013) 'Maya chemistry' of organic–inorganic hybrid materials: isomerization, cyclicization and redox tuning of organic dyes attached to porous silicates. RSC Adv 3:20099–20105
34. Doménech-Carbó A, Doménech-Carbó MT, Sanchez del Rio M, Goberna S, Lima E (2009) Evidence of topological indigo/dehydroindigo isomers in Maya Blue-like complexes prepared from Palygorskite and Sepiolite. J Phys Chem C 113:12118–12131
35. Doménech-Carbó A, Doménech-Carbó MT, Sanchez del Rio M, Vázquez De Agredos-Pascual ML (2009) Comparative study of different indigo-clay Maya Blue-like systems using the voltammetry of microparticles approach. J Solid State Electrochem 13:869–878
36. Doménech-Carbó A, Doménech-Carbó MT, Valle-Algarra FM, Domine ME, Osete-Cortina L (2013) On the dehydroindigo contribution to Maya Blue. J Mater Sci 48:7171–7183
37. Doménech-Carbó A, Valle-Algarra FM, Doménech-Carbó MT, Domine ME, Osete-Cortina L, Gimeno-Adelantado JV (2013) Redox tuning and species distribution in Maya Blue-type materials: a reassessment. Appl Mater Sci Interf 5:8134–8145
38. Rondão R, de Melo JSS, Bonifacio VDB, Melo MJ (2010) Dehydroindigo, the forgotten indigo and its contribution to the color of Maya Blue. J Phys Chem A 114:1699−1708
39. Giustetto R, Wahyudi O (2011) Sorption of red dyes on palygorskite: synthesis and stability of red/purple Mayan nanocomposites. Micropor Mesopor Mater 142:221–235
40. Giustetto R, Wahyudi O, Corazzari I, Turci F (2011) Chemical stability and dehydration behavior of a sepiolite/indigo Maya Blue pigment. Appl Clay Sci 52:41–50
41. Sánchez-Ochoa F, Cocoletzi GH, Canto G (2017) Trapping and diffusion of organic dyes inside of palygorskite clay: the ancient Maya Blue pigment. Micropor Mesopor Mater 249:111–117
42. Dejoie C, Martinetto P, Dooryhée E, Strobel P, Blanca S, Bordat P, Brown R, Porcher F, Sanchez del Rio M, Anne M (2010) Indigo@Silicalite: a new organic-inorganic hybrid pigment. Appl Mater Interf 2:2308–2316
43. Ovarlez S, Giulieri F, Delamare F, Sbirrazzuoli N, Chaze A-M (2011) Indigo–sepiolite nanohybrids: temperature-dependent synthesis of two complexes and comparison with indigo–palygorskite systems. Micropor Mesopor Mater 142:371–380

44. Bernardino ND, Constantino VRL, De Faria DLA (2018) Probing the indigo molecule in Maya Blue simulants with resonance Raman spectroscopy. J Phys Chem C 122:11505–11515
45. Leona M, Casadio F, Bacci M, Picollo M (2004) Identification of the pre-Columbian pigment Maya Blue on works of art by Noninvasive UV-Vis and Raman spectroscopic techniques. J Am Inst Conserv 43:39–54
46. Vandenabeele P, Bodé S, Alonso A, Moens L (2005) Raman spectroscopic analysis of the Maya wall paintings in Ek'Balam, Mexico. Spectrochim Acta A 61:2349–2356
47. Sánchez del Río M, Picquart M, Haro-Poniatowski E, van Eslande E, Uc VH (2006) On the Raman spectrum of Maya blue. J Raman Spectrosc 37:1046–1053
48. Wiedemann HG, Brzezinka K-W, Witke K, Lamprecht I (2007) Thermal and Raman-spectroscopic analysis of Maya Blue carrying artefacts, especially fragment IV of the Codex Huamantla. Thermochim Acta 456:56–63
49. Grazia C, Buti D, Amat A, Rosi F, Romani A, Domenici D, Sgamellotti A, Miliani C (2020) Shades of blue: non-invasive spectroscopic investigations of Maya blue pigments. From laboratory mock-ups to Mesoamerican codices. Herit Sci 8: artic. 1
50. Doménech-Carbó A, Holmwood S, Di Turo F, Valle-Algarra FM, Edwards HGM, Doménech-Carbó MT (2019) Composition and color of Maya Blue: reexamination of literature data based on the dehydroindigo model. J Phys Chem C 123:770–782
51. Doménech-Carbó A, Costero A, Gil S, Montoya N, López-Lorente A, Sáez JA, Arroyo P, Doménech-Carbó MT (2021) Isomerization and Redox tuning: Reorganizing the Maya Blue puzzle from synthetic, spectral and electrochemical issues. J Phys Chem C 125:26188–26200
52. Bond AM, Marken F, Hill E, Compton RG, Hügel H (1997) The electrochemical reduction of indigo dissolved in organic solvents and as a solid mechanically attached to a basal plane pyrolytic graphite electrode immersed in aqueous electrolyte solution. J Chem Soc Perkin Trans 2:1735–1742
53. Giustetto R, Seenivasan K, Pellerej D, Ricchiardi G, Bordiga S (2012) Spectroscopic characterization and photo/thermal resistance of a hybrid palygorskite/methyl red Mayan pigment. Micropor Mesopor Mater 155:167–176
54. Raya J, Hirschinger J, Ovarlez S, Giulieri F, Chaze A-M, Delamare F (2010) Insertion of indigo molecules in the sepiolite structure as evidenced by ^{1}H–^{29}Si heteronuclear correlation spectroscopy. PhysChemChemPhys 12:14508–14514
55. Dejoie C, Dooryhée E, Martinetto P, Blanc S, Bordat P, Brown R, Porcher F, Sanchez del Rio M, Strobel P, Anne M, Van Eslande E, Walter P, Revisiting Maya Blue and designing hybrid pigments by Archaeomimetism. hal-00495128.v1,http://hal.archives-ouvertes.fr/
56. Doménech-Carbó A, Doménech-Carbó MT, Osete-Cortina L, Montoya N (2013) Application of solid-state electrochemistry techniques to polyfunctional organic-inorganic hybrid materials: the Maya Blue problem. Micropor Mesopor Mater 166:123–130
57. Doménech-Carbó A, Martini M, Valle-Algarra FM (2014) Determination of the depth profile distribution of guest species in microporous materials using the voltammetry of immobilized particles: application to lapachol attachment to palygorskite and kaolinite. Phys Chem Chem Phys 16:19024–19034

Additional Literature

58. Breitbach AM, Rocha JC, Gaylarde CC (2011) Influence of pigment on biodeterioration of acrylic paint films in Southern Brazil. J Coat Technol Res 8:619–628
59. Ciomartan DA, Clark RHJ, McDonald LJ, Odlyha MJ (1996) Chem Soc Dalton Trans 3639–3645
60. Cotte M, Checroun E, De Nolf W, Taniguchi Y, De Viguerie L, Burghammer M, Walter P, Rivard C, Salomé M, Janssens K, Susini J (2017) Lead soaps in paintings: friends or foes? Stud Conservat 62:2–23

61. Doménech-Carbó A, Doménech-Carbó MT, Vázquez De Agredos-Pascual ML (2007) Indigo/Dehydroindigo/Palygorskite complex in Maya Blue: an electrochemical approach. J Phys Chem C 111:4585–4595
62. Doménech-Carbó A, Doménech-Carbó MT, Vázquez de Agredos-Pascual ML (2009) Correlation between spectral, SEM/EDX and electrochemical properties of Maya Blue. A chemometric study. Archaeometry 51:1015–1034
63. Doménech-Carbó A, Doménech-Carbó MT, Edwards HGM (2011) On the interpretation of the Raman spectra of Maya Blue: a review on the literature data. J Raman Spectrosc 42:86–96
64. Doménech-Carbó MT, Casas-Catalán MJ, Doménech-Carbó A, Mateo-Castro R, Gimeno-Adelantado JV, Bosch-Reig F (2001) Analytical study of canvas painting collection from the Basilica de la Virgen de los Desamparados using SEM/EDX, FT-IR, GC and electrochemical techniques. Fresenius' J Anal. Chem 369:571–575
65. Doménech-Carbó MT, Edwards HGM, Hargreaves MD, Doménech-Carbó A (2012) An authentication case study: Antonio Palomino versus Vicente Guillo paintings in the vaulted ceiling of the Sant Joan del Mercat church (Valencia, Spain). J Raman Spectrosc 43:1250–1259
66. Doménech-Carbó MT, Osete-Cortina L, Doménech-Carbó A, Vázquez de Agredos –Pascual ML, Vdal-Lorenzo C (2014) Identification of indigoid compounds present in archaeological Maya blue by pyrolysis-silylation-gas chromatography-mass spectrometry. J Anal Appl Pyrol 105:355–362
67. Doménech-Carbó A, Doménech-Carbó MT, Osete-Cortina L, Valle-Algarra FM, Buti D (2014) Isomerization and redox tuning in 'Maya Yellow' hybrids from flavonoid dyes plus palygorskite and kaolinite clays. Micropor Mesopor Mater 194:135–145
68. Frost RL, Ding Z, Kloprogge JT, Martens WV (2002) Thermal stability of azurite and malachite in relation to the formation of mediaeval glass and glazes. Thermochim Acta 390:133–144
69. Fuentes ME, Peña B, Contreras C, Montero AL, Chianelli R, Alvarado M, Olivas R, Rodríguez LM, Camacho H, Montero-Cabrera LA (2008) Quantum mechanical model for Maya Blue. Int J Quantum Chem 108:1664–1673
70. Gómez-Romero P, Sánchez C (2005) Hybrid materials. functional properties. from Maya Blue to 21st century materials. New J Chem 29:57–58
71. Kalb L (1909) Über dehydro-indigo, ein neues oxidation-sprodukt des indigos. Chem Ber 42:3642
72. Kiseleva IA, Ogorodova LP, Melchakova LV, Bisengalieva MR, Becturganov NS (1992) Thermodynamic properties of copper carbonates—malachite $Cu_2(OH)_2CO_3$ and azurite $Cu_3(OH)_2(CO_3)_2$. Phys Chem Miner 19:322–333
73. Lee Y, Martín-Rey S, Osete-Cortina L, Martín-Sánchez I, Bolívar-Galiano F, Doménech-Carbó MT (2018) Evaluation of a gelatin-based adhesive for historic paintings that incorporates citronella oil as an eco-friendly biocide. J Adhes Sci Technol 32:2320–2349
74. Lima E, Guzmán A, Vera M, Rivera JL, Fraissard J (2012) Aged natural and synthetic Maya Blue-like pigments: what difference does it make? J Phys Chem C 116:4556–4563
75. Mansour SAA (1996) Thermoanalytical investigations of the decomposition course of copper oxysalts III. Copper(II) acetate monohydrate. J Thermal Anal 46:263–274
76. Mendoza-Flores A, Villalobos M, Pi-Puig T, Martínez-Villegas NV (2017) Revised aqueous solubility product constants and a simple laboratory synthesis of the Pb(II) hydroxycarbonates: Plumbonacrite and hydrocerussite. Geochim J 51:315–328
77. Noël J-M, Médard J, Combellas C, Kanoufi F (2016) Prussian Blue degradation during hydrogen peroxide reduction: a scanning electrochemical study on the role of the hydroxide ion and hydroxyl radical. ChemElectroChem 3:1178–1184
78. Noel JD, Wang Y, Giammar DE (2014) Effect of water chemistry on the dissolution rate of the lead corrosion product hydrocerussite. Water Res 54:237–246
79. Polette-Niewold LA, Manciu FS, Torres B, Alvarado M Jr, Chianelli RR (2007) Organic/inorganic complex pigments: ancient colors Maya Blue. J Inorg Biochem 101:1958–1973
80. Ruan HD, Frost RL, Kloprogge JT (2001) The behavior of hydroxyl units of synthetic goethite and its dehydroxylated product hematite. Spectrochim Acta A 57:2575–2586

81. Samain L, Gilbert B, Grandjean F, Long GJ, Strivay D (2013) Redox reactions in Prussian blue containing paint layers as a result of light exposure. J Anal Atom Spectrom 28:524–535
82. Seves AM, Sora S, Ciferr O (1996) The microbial colonization of oil paintings—a laboratory investigation. Int Biodeter Biodegr 37:215–224
83. Tomkinson J, Bacci M, Picollo M, Colognesi D (2009) The vibrational spectroscopy of indigo: a reassessment. Vibrat Spectrosc 50:268–276
84. Van der Snickt G, Dik J, Cotte M, Janssens K, Jaroszewicz J, De Wolf W, Groenewegen J, Van der Loeff L (2009) Characterization of a degraded cadmium yellow (CdS) pigment in an oil painting by means of synchrotron radiation based X-ray techniques. Anal Chem 81:2600–2610
85. Weerd J, Van Der Loon A, Boon JJ (2005) FTIR studies of the effects of pigments on the aging of oil. Stud Conservat 50:3–22
86. Witke K, Brzezinka K-W, Lamprecht I (2003) Is the indigo molecule perturbed in planarity by matrices? J Mol Struct 661–662:235–238
87. Yamazaki S, Sobolewski AL, Domcke W (2011) Molecular mechanisms of the photostability of indigo. Phys Chem Chem Phys 13:1618–1628
88. Zhang W, Hu S, Yin JJ, He W, Lu W, Ma W, Gu N, Zhang Y (2016) Prussian Blue Nanoparticles as Multienzyme Mimetics and Reactive Oxygen Species Scavenger. J Am Chem Soc 138:5860–5865

Chapter 7
Ceramic, Glass, and Glazed Materials I

7.1 Ceramic, Glass, and Glazed Heritage

Generically, ceramic heritage comprises a variety of manufactured productions having in common the use of silicate and aluminosilicate raw materials. For practical purposes, the ceramic heritage can be divided into ceramicware and bricks, glasses, and glazed materials. Pottery is of considerable importance in the archaeological context because of its variety and abundance. Predominantly based on style, pottery constitutes the primary source of information for dating in many archaeological contexts and offers valuable data on the technical, economical, and social ethnohistoric aspects associated with its production [1, 2].

Ceramic materials undergo more or less complex processes of degradation depending on their composition, manufacturing, and the environmental conditions. The ideal composition of glass ranges between 70–74 percent w/w silica, 16–22 percent alkali, or soda ash (sodium carbonate) or potash (potassium carbonate), and 5–10 percent flux lime (calcium oxide) [3, 4]. Glass and glazed materials are structurally characterized by the tetrahedral SiO_4 units.

Glazes are characterized by the presence of a Pb and/or Sn-based opacifier. Alkali glazes containing low amounts of PbO were produced in Iraq in the 8th century BCE [5], but the lead content was further progressively increased. The preparation of the glaze involves the formation of a frit, containing a lead and tin oxide mixture, the alkali, and the silica. The frit is applied as a suspension to a biscuit-fired body, and then a firing step is applied. The electroactive character of these species can be exploited for analytical and mechanistic purposes.

From the analytical point of view, the study of these materials is difficult by virtue of their complex chemical and mineralogical composition. Optical and electron microscopies are of obvious importance, and X-ray diffraction methods are accompanied by infrared and Raman spectroscopies and include isotope analysis [6]. Colorimetry [7, 8], neutron activation analysis [9], Mössbauer spectroscopy [10], among others, are also used [11, 12]. In general, multitechnique strategies have

A. Doménech-Carbó and M. T. Doménech-Carbó, *Electrochemistry for Cultural Heritage*, Monographs in Electrochemistry,
https://doi.org/10.1007/978-3-031-31945-7_7

to be adopted [13–15]. In this chapter, we will describe the application of electrochemical techniques, mainly VIMP, but also EIS, to acquire relevant archaeometric information on ceramic materials.

7.2 Detection and Characterization of Electroactive Species

The application of VIMP to ceramic heritage was initiated by the identification of Pb(II) and Pb(IV) centers in medieval glazes [16] and the characterization of different types of Mn(IV) centers in archaeological glass [17]. Promptly extended to pottery [18, 19], these studies were aimed at the identification and characterization of electroactive species in archaeological samples.

The darkening/browning of glass is attributed to the formation of iron and manganese compounds on the object's surface [3]. Here, a hydrated, gel layer of corrosion products is formed. The browning process involves the oxidation of Mn(II) to manganese dioxide from the corrosion gel layer, a process eventually promoted by microorganisms. VIMP allowed the characterization of different MnO_2 deposits in altered glazed samples from a set of archaeological submicrosamples of Roman and Middle Age sites in the vicinity of València and Saragossa (Spain) [17]. Representative data obtained on microparticulate deposits embedded into polymer films (Elvacite 2044) transferred onto the glassy carbon electrode in contact with acetic/acetate buffer are shown in Fig. 7.1. The voltammograms show overlapping cathodic signals distributed into two groups, one at peak potentials between 0.61 and 0.72 V versus saturated calomel electrode (SCE) and the other between 0.40 and − 0.10 V.

To interpret these voltammograms, it has to be taken into account that MnO_2 is a polymorph compound adopting different crystalline structures based on MnO_6 octahedra which share their edges and corners to create channels. Pyrolusite (β-MnO_2) is the most frequent mineral form, whereas γ-MnO_2 is characterized by a defective and more 'open' structure facilitating cation intercalation (K^+ in particular) to achieve electroneutrality. The electrochemistry of MnO_2 has been widely studied because of its use as electrode material in a variety of applications. In acidic media, the reduction of MnO_2 occurs in two consecutive steps yielding MnO(OH) solid and $Mn^{2+}{}_{aq}$, successively:

$$MnO_{2solid} + H^+_{aq} + e^- \rightarrow MnO(OH)_{solid} \quad (7.1)$$

$$MnO(OH)_{solid} + 3H^+_{aq} + e^- \rightarrow Mn^{2+}_{aq} + 2H_2O \quad (7.2)$$

These processes, however, can show considerable mechanistic complications, including the blocking effect exerted by the non-conducting MnO(OH) and the mediation of the disproportionation reaction [20–22]:

$$2MnO(OH)_{solid} + 2H^+_{aq} \rightarrow Mn^{2+}_{aq} + MnO_{2solid} + 2H_2O \quad (7.3)$$

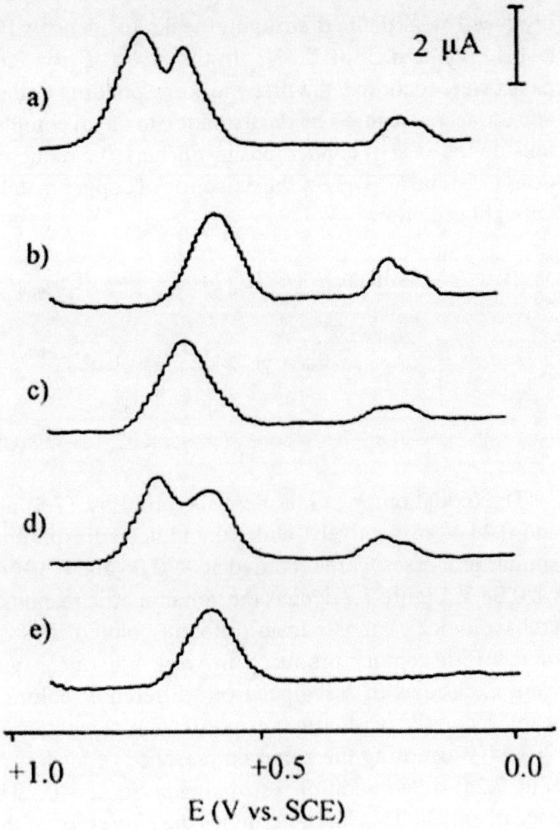

Fig. 7.1 Differential pulse voltammograms of sample-modified polymer film (Elvacite 2044) electrodes in contact with 0.50 M HAc/NaAc, pH 4.75. Archaeological samples from different Middle Age and Roman sites of **a** Paterna, **b** L'Almoina, **c** Paterna, **d** Llíria (all from Valencian Community, Spain), and **e** L'Almolda (Saragossa, Spain). Potential scan initiated at 0.95 V versus SCE in the negative direction; potential scan rate 20 mV s^{-1}; and pulse amplitude 80 mV. Adapted from ref. [17], with permission. Note that the axes are opposite to the IUPAC convention

Recently, the Gibbs energies for the electron an ion (proton) transfer of reaction (7.1) have been separately determined [23]. The results have shown that the Gibbs energy of proton transfer is strongly negative, that of the electron transfer slightly positive, so that the overall Gibbs energy is negative [23].

For our purposes, the relevant point to emphasize is that, as judged upon comparison with different MnO_2 blanks, pyrolusite displays a unique cathodic signal at ca. 0.70 V, whereas γ-MnO_2 forms exhibit a two-peak response in the same region of potentials. Accordingly, archaeological samples were divided into two basic types, and their voltammetric responses were correlated with the characteristics of the attack (acidity, rapidity of MnO_2 deposition) and the abundance in potassium of the glaze (determined by SEM/EDX analysis) [17].

VIMP was also used, mainly based on oxidative dissolution processes, to identify metal ions in archaeological glazes [24]. Here, the typical strategy of stripping analysis was adapted by applying first a negative potential prompting the reduction of the metal ions to metal, followed by a positive-going potential scan to record the electrochemical oxidation of the metal deposits. In contact with 0.10 M HCl, linear potential scan voltammograms at sample-modified graphite polyester composite electrodes produced well-defined stripping peaks to identify Pb (−0.44 V versus SCE), Sb (−0.11 V), and Sn (0.05 V). In the cases of iron, copper, and cobalt, two anodic peaks were recorded, the first one corresponding to the metal to metal ions oxidation and the second caused by the oxidation to metal complexes with strongly complexing ions. In the case of copper ions in silicates, the reduction is given by Eq. (7.4). Equations (7.5) and (7.6) show the oxidation of copper metal to Cu(I) and Cu(II) complexes with chloride ions:

$$\left\{Cu^{2+} - \text{silicate}\right\} + 2H^+_{aq} + 2e^- \rightarrow Cu_{solid} + \left\{2H^+ - \text{silicate}\right\} \quad (7.4)$$

$$Cu_{solid} + xCl^-_{aq} \rightarrow CuCl_x^{(x-1)-}{}_{aq} + e^- \quad (7.5)$$

$$CuCl_x^{(x-1)-}{}_{aq} + (y-x)Cl^-_{aq} \rightarrow CuCl_y^{(y-2)-}{}_{aq} + e^- \quad (7.6)$$

The oxidation processes described by Eqs. (7.5) and (7.6) take place at − 0.10 and 0.24 V, respectively, under the indicated experimental conditions. For iron the similar processes were recorded at − 0.06 and 0.30 V and for cobalt at − 0.29 and ca. 0.08 V. Figure 7.2 depicts the voltammetric responses recorded for three different archaeological samples from different colored tiles dated back to the second half of the 18th century produced in workshops near València (Spain). The samples were excised with a scalpel from differently colored regions of the tiles, which were decorated with vegetal motifs. All three samples exhibit a gross signal at − 0.44 V denoting the common presence of lead as an opacifier. The first sample (Fig. 7.2a) shows additional stripping peaks at − 0.29 and 0.10 V, denoting the presence of cobalt. This last determines the blue color of the sample. The second sample (Fig. 7.2b) shows a stripping signal at − 0.11 V, denoting the presence of antimony as a yellow pigmenting species. The third sample (Fig. 7.2c) shows peaks at − 0.29, − 0.11, and 0.10, thus revealing the presence of cobalt and antimony to produce the blue-greenish color of the leaves in the tile. The same methodology can be used for quantification purposes in glass and glazes [24].

Cathodic scan voltammograms can also be used to identify electroactive species in glass and glazes. An example of the application of this methodology combined with conventional SEM/EDX analysis was the study of *Moarraque* tiles from the Ali ibn Abi Talib Mosque ((مسجد علی ابن ابیطالب)) (Isfahan, Iran), dated back from the 1960s (see inset in Fig. 7.3). The study revealed that the analyzed pieces were made by combining the traditional technologies used by local workshops since the 14th century and modern techniques. Figure 7.3 shows the square wave voltammogram of a sample from a green tile attached to a paraffin-impregnated graphite electrode immersed in 0.25 M HAc/NaAc solution at pH 4.75. The voltammogram shows a series of overlapping cathodic peaks at ca. 0.40, 0.30, 0.15, − 0.10, − 0.55, and − 0.65 V. The signal at 0.40 V can be assigned to Pb(IV) centers. These are network-forming species (see next section) that accompany network-modified Pb(II) displaying an intense signal at − 0.60 V. The signal at 0.30 V can be attributed to the reduction of Fe(III) and the signal at − 0.10 V to that of Cu(II), the species

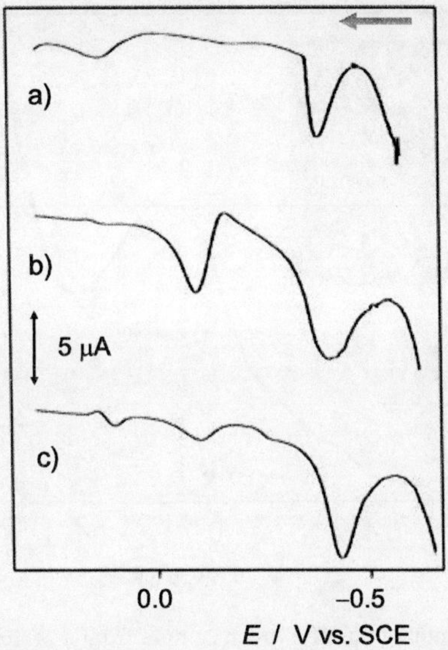

Fig. 7.2 Linear potential scan voltammograms of samples from 18th-century tiles produced in the Valencian region (Spain) attached to graphite—polyester composite electrodes in contact with 0.10 M HCl. The tiles were decorated with vegetal motifs. **a** Deep blue in the frame, inner petals, and bands; **b** yellow in the bands and petals; and **c** greenish-blue in the leaves. Potential scan initiated at − 0.70 V versus SCE in the positive direction; potential scan rate 10 mV s^{-1}. Reproduced from ref. [24], with permission. Note that the axes are opposite to the IUPAC convention

responsible for the green hue of the tile. The peaks at 0.15 and − 0.55 V correspond to the Sn(IV) to Sn(II) and Sn(II) to Sn metal reduction steps, consistently with the use of tin opacifiers in Islamic technology at least since the 9th–11th century [25].

7.3 Speciation

The previously mentioned examples illustrate the capability of the VIMP technique to identify mineral species and/or oxidation states [16, 17]. The speciation in terms of oxidation state was assessed for ceramic bodies attempting the quantification of the Fe(III)/Fe(II) ratio [18], of obvious interest for its implication with regard to the oxidizing/reducing conditions of firing. This aim, however, is made difficult by the complexity of the electrochemistry of ceramic pastes, discussed in the next sections.

In the case of lead glass and glazes, there are [PbO_4] unities in the silicon network accompanying Pb(II) centers with octahedral coordination [PbO_6], acting as network modifiers [5]. Tin-opacified glazes contain SnO_2 particles, but it is believed that there

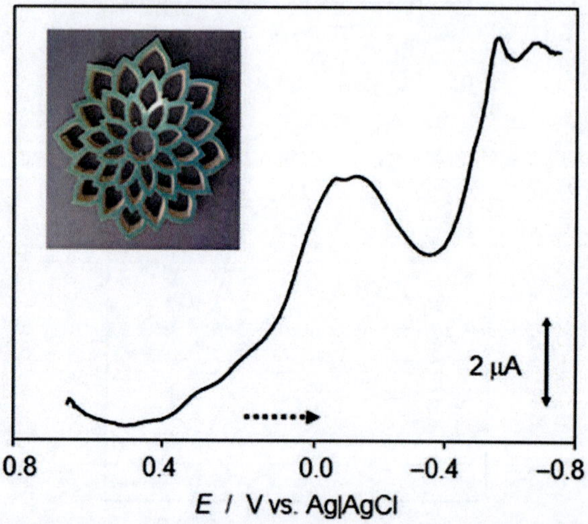

Fig. 7.3 Square wave voltammogram of a green sample from a *Moarraque* tile from the indoor decoration of the Ali ibn Abi Talib Mosque in Isfahan (Iran) attached to paraffin-impregnated graphite electrode immersed into 0.25 M HAc/NaAc aqueous buffer at pH 4.75. Potential scan initiated at 0.65 V in the negative direction; potential step increment 4 mV; square wave amplitude 25 mV; and frequency 5 Hz. Inset: photographic image of *Moarraque* tiles. The dotted arrow marks the direction of the potential scan. Note that the axes are opposite to the IUPAC convention

is at least some partial dissolution during the heating up and subsequent recrystallization during the final cooling down step [28]. Figure 7.4 permits the comparison of the negative-going square wave voltammograms of SnO_2, PbO_2, and PbO blanks attached to paraffin-impregnate graphite electrodes immersed into acetic/acetate buffer [29]. SnO_2 yields two main cathodic signals at − 0.14 and − 0.89 V versus Ag/AgCl, whereas PbO_2 displays peaks at 0.66 and − 0.54 V, eventually accompanied by minor signals at intermediate potentials. The voltammograms of PbO show a unique cathodic peak at − 0.54 V. This last process corresponds to the previously studied (see Chap. 3) PbO to Pb reduction. The electrochemistry of PbO_2 and SnO_2 (whose mineral forms are plattnerite and cassiterite, respectively) can be described in terms of a two-step reduction. In the case of PbO_2, this first step can be represented as:

$$PbO_{2solid} + 2xH^+_{aq} + 2xe^- \rightarrow PbO_{2-xsolid} + xH_2O \quad (7.7)$$

In principle, substoichiometric oxides are formed the final product being PbO:

$$PbO_{2-xsolid} + 2(1-x)H^+_{aq} + 2(1-x)e^- \rightarrow PbO_{solid} + (1-x)H_2O \quad (7.8)$$

The reduction of SnO_2 appears to involve some generation of intermediate $Sn^{2+}{}_{aq}$ species and can be described as [29]:

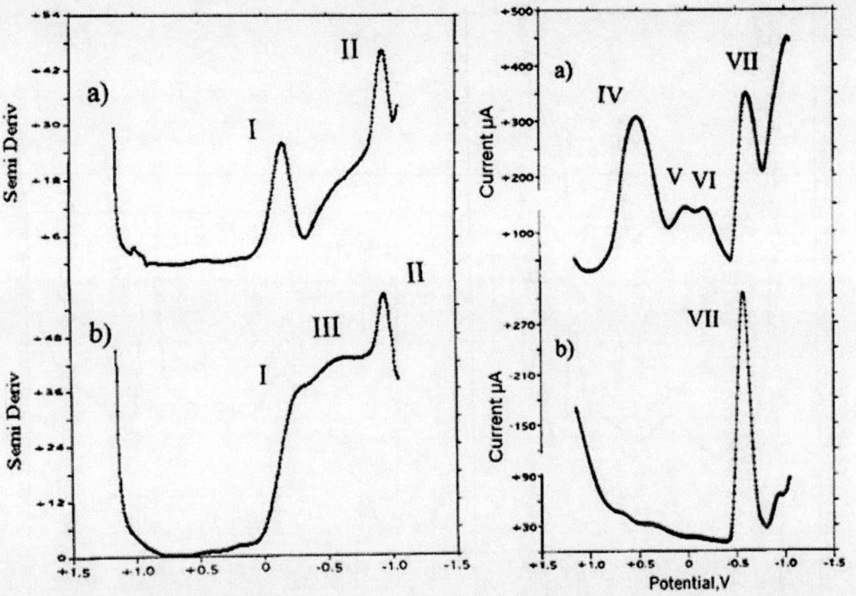

Fig. 7.4 Square wave voltammograms of paraffin-impregnated graphite electrodes modified with: left side: SnO_2; right side: **a** PbO_2, **b** PbO (litharge). Electrolyte: 0.50 M HAc/NaAc at pH 4.74. Potential scan initiated at 1.15 V versus Ag/AgCl in the negative direction; potential step increment 4 mV; square wave amplitude 25 mV; and frequencies of 5 Hz (left side, **a**), 50 Hz (right side, **b**), and 15 Hz (right side **a**, **b**). Adapted from ref. [29], with permission. Note that the axes are opposite to the IUPAC convention

$$SnO_{2solid} + 2(2-x)H^{+}_{aq} + 2e^{-} \rightarrow xSnO_{solid} + (1-x)Sn^{2+}_{aq} + (2-x)H_2O \quad (7.9)$$

$$SnO_{solid} + 2H^{+}_{aq} + 2e^{-} \rightarrow Sn_{solid} + H_2O \quad (7.10)$$

$$Sn^{2+}_{aq} + 2e^{-} \rightarrow Sn_{solid} \quad (7.11)$$

Glazed samples from the archaeological site of the “Testar” (Calle de Valencia 24, Manises, Spain) produced by medieval workshops displayed clearly different voltammetric responses for weathered and unweathered regions of the glazes. The former exhibited clear differences between samples although in general displayed the Sn – and Pb-localized signals previously described. The voltammograms of the weathered samples, however, were dominated by the initial SnO_2 reduction at – 0.14 V and the PbO to Pb reduction at – 0.54 V.

This voltammetry is consistent with SEM images revealing the important presence of cassiterite crystals in the corroded layers of the glaze (see Fig. 7.6). In order to achieve mechanistic information relative to the weathering process, the ratio between the peak currents measured for the SnO_2 reduction (labeled as I) and the PbO reduction (labeled as VII), $i_p(I)/i_p(VII)$, were measured in the weathered and unweathered

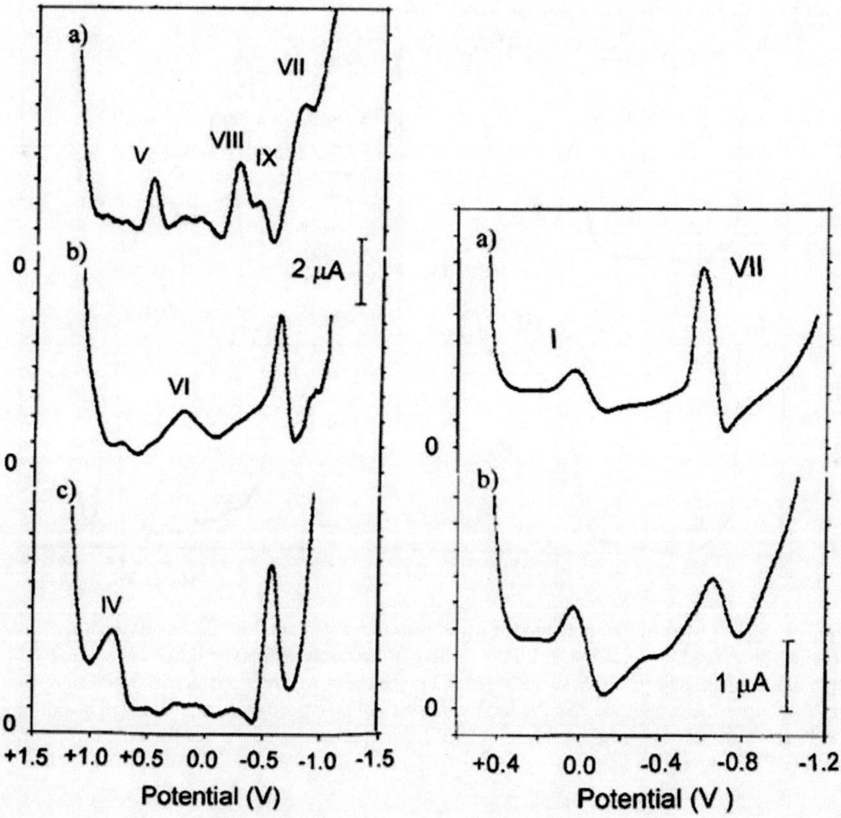

Fig. 7.5 Square wave voltammograms of paraffin-impregnated graphite electrodes modified with medieval glazed samples from the "Calle de Valencia 24" archaeological site (Manises, Spain). Left side: three unweathered samples; right side: two weathered samples. Electrolyte: 0.50 M HAc/NaAc at pH 4.74. Potential scan initiated at 1.15 V versus Ag/AgCl in the negative direction; potential step increment 4 mV; square wave amplitude 25 mV; and frequencies of 5 Hz (left side) and 15 Hz (right side). Adapted from ref. [29], with permission. Note that the axes are opposite to the IUPAC convention

regions of the samples that were correlated. Demonstrating the consistency of voltammetric measurements, the $(i_p(\mathrm{I})/i_p(\mathrm{VII}))_{unweath}$ ratio was linearly correlated with the Sn/Pb molar ratios determined from SEM/EDX data. This was not the case of the $(i_p(\mathrm{I})/i_p(\mathrm{VII}))_{weath}$, thus suggesting that different corrosion patterns occurred. Interestingly, as can be seen in Fig. 7.6, the plots of $(i_p(\mathrm{I})/i_p(\mathrm{VII}))_{unweath}$ versus $(i_p(\mathrm{I})/i_p(\mathrm{VII}))_{weath}$ grouped the studied samples (totaling 16) in three tendency lines [29].

This behavior can be interpreted on the basis of general ideas on glass weathering occurring in contact with aqueous solutions, moist air, or moist earth. In contact with acid and neutral solutions, the corrosion process starts with the replacement of alkali ions (and eventually alkaline earth ones) by hydrogen ions from hydrogen-bearing species. This can be conceived as an ion-exchange process where, simultaneously,

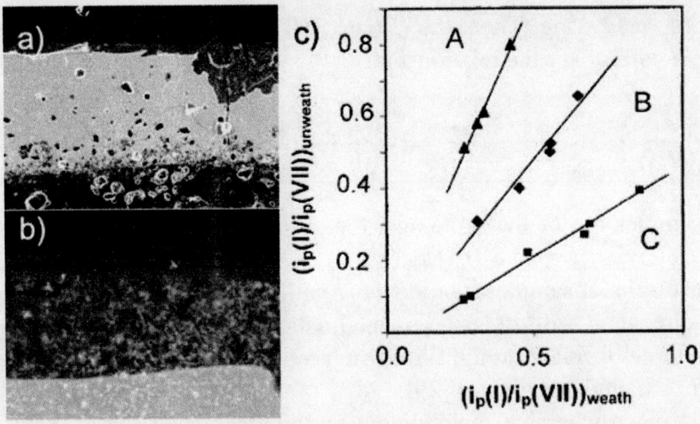

Fig. 7.6 **a** SEM secondary electron image of the glaze portion of fragment M.1 from the "Calle de Valencia 24" archaeological site (Manises, Spain). The layer of altered glaze is the dark gray region at the top of the photograph. **b** Magnified SEM image of the corroded region showing the dense array of cassiterite crystals. **c** Plots of the $i_p(I)/i_p(VII)$ ratio measured in the unweathered area versus the same ratio measured in the weathered area of the same sample. From voltammetric data such as in Fig. 7.5 (right side). Adapted from ref. [29], with permission

water molecules penetrate into the glaze. As a result, a gel layer is formed accompanied by a small volume expansion. In alkaline solutions, glass and glazes undergo a stronger attack by OH^- ions with faster destruction of the glass/glaze network by breakage of the Si–O-Si bonds [30, 31]. In the case of lead glass and glazes, the corrosion process involves the release of network-modifier Pb(II) centers while network-former Pb(IV) centers remain.

Accordingly and assuming that the peak currents are proportional to the concentrations of the different Sn and Pb forms in the weathered or unweathered regions, one can write:

$$\left(\frac{i_p(I)}{i_p(VII)}\right)_{unweath} = \frac{g_I c^o_{Sn}}{g_{VII} c^o_{Pb}} \tag{7.12}$$

where $c^o{}_{Sn}$ and $c^o{}_{Pb}$ represent the average concentrations of Sn and Pb in the unweathered glaze and g_I, g_{VII} are the electrochemical coefficients of the response of the previously described signals. If the weathering process involves the release of lead and the formation of a differently structured surface region enriched in cassiterite (see Fig. 7.6), one can expect that the concentrations of Sn and Pb in the corrosion layer depend on any kinetic law(s). Among a variety of possibilities facilitated by solid state kinetics models, one can write,

$$c_{Sn} = c^o_{Sn}\,(1 + k_{Sn}t^{\alpha}); \qquad c_{Pb} = c^o_{Pb}\left(1 - k_{Pb}t^{\beta}\right) \tag{7.13}$$

where c_{Sn}, c_{Pb} denote the concentrations of Sn and Pb in the corrosion layer after a corrosion time t and k_{Sn}, k_{Pb}, α, β, the kinetic parameters. The c^0 concentrations are those before weathering. Then, the $(i_p(\mathrm{I})/i_p(\mathrm{VII}))_{weath}$ determined in voltammetric experiments will satisfy the relationship:

$$\left(\frac{i_p(\mathrm{I})}{i_p(\mathrm{VII})}\right)_{weath} = \frac{g_{\mathrm{I}}c_{\mathrm{Sn}}}{g_{\mathrm{VII}}c_{\mathrm{Pb}}} = \left(\frac{1 + k_{\mathrm{Sn}}t^{\alpha}}{1 - k_{\mathrm{Pb}}t^{\beta}}\right)\left(\frac{i_p(\mathrm{I})}{i_p(\mathrm{VII})}\right)_{unweath} \tag{7.14}$$

If the samples are of the same age, the above equation predicts that the plots of $(i_p(\mathrm{I})/i_p(\mathrm{VII}))_{unweath}$ versus $(i_p(\mathrm{I})/i_p(\mathrm{VII}))_{weath}$ will provide a straight line passing through the origin for samples submitted to a common corrosive pattern. This prediction is in agreement with the experimental data in Fig. 7.6. The grouping of such data along three different straight lines suggests that three modes of deterioration operated for the studied samples [29].

The same electrochemical methodology and similar considerations can be applied to the study of the degradation of glass and glazes based on the record of iron- and manganese-related signals [32, 33]. Here, it is convenient to mention that there are two extreme types of glass [34]: soda glass, typically produced in Southern Europe and Near East, is made from crushed white pebbles and soda ash derived from burnt marine vegetation. The sodium content, as expressed in wt% of Na_2O, ranges between 10 and 20%. Potash glass is characteristic of interior Europe, being made from local sands and potash derived from wood ash and burnt inland vegetation. Here, the potassium content is scattered between 5 and 15 wt% K_2O. Three main preparative recipes were used in the past for glassmaking [35, 36], characterized by the *R* norm, defined as the alkali to alkaline earth ratio expressed in terms of the respective oxides. The Near East norm is characterized by *R* values around 1.5, while the Roman Provincial and Roman Capital norms correspond to larger *R* values (ca. 2.0–2.5 and ca. 3, respectively).

Iron(II) and manganese(II) often appear in glasses as a result of deliberate additions to decolorize it, although the former imparts a light greenish hue to the glass. The addition of MnO_2 was often made for oxidizing iron(II) to iron(III). This gives the glass a light yellow hue, but increasing the potash has been also reported as alternative procedure for reducing the Fe(II) content [28]. Glass corrosion results in the formation of brownish deposits of MnO_2 and/or Fe_2O_3 forms with variable degree of crystallinity and hydration. The complexity of this chemistry is increased in lead glasses, being reflected in voltammetric responses with high variability [32, 33], as can be seen in Fig. 7.7.

7.4 Characterization of Archaeological Glass Sites

In favorable cases, electrochemical data can be used to obtain structural information as well as to characterize individual archaeological sites and/or workshops. This is the case with lead silicate glasses. These appeared in Eastern Europe at the beginning of the 8th century BCE and can be divided into two groups, Eastern lead glasses, which originated from Eastern Europe and the Caucasian region, and lead-alkali glasses. The former group is characterized by high lead yields and low content of

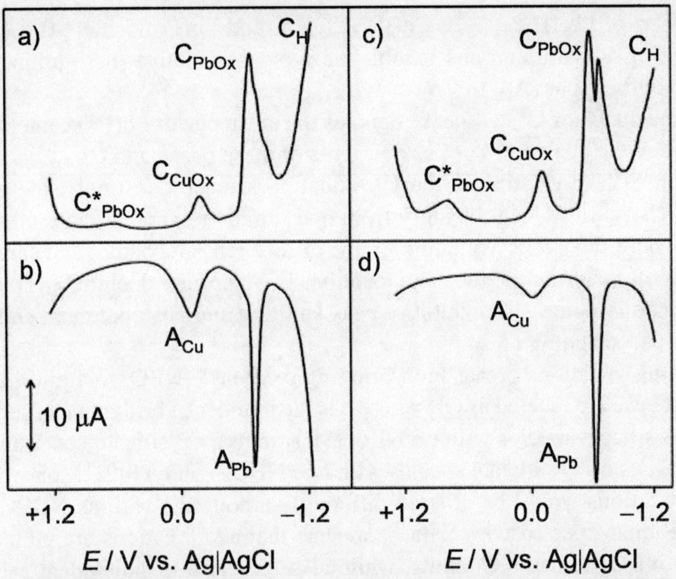

Fig. 7.7 Square wave voltammograms of samples **a**, **b** S01 and **c**, **d** S03 from the archaeological site of Puxmarina (Spain, dated back to the 9th–10th century), attached to graphite electrodes in contact with air-saturated 0.25 M HAc/NaAc aqueous buffer at pH 4.75. Potential scan initiated at **a**, **c** + 1.25 V in the negative direction; **b**, **d** − 1.25 V in the positive direction; potential step increment 4 mV; square wave amplitude 25 mV; and frequency 5 Hz. Reproduced from ref. [41], with permission. Note that the axes are opposite to the IUPAC convention

alkaline elements, whereas the latter were either soda, potash, or soda-potash-rich glasses [4].

The structural chemistry of lead glasses has become to some extent controversial with regard to the role played by the lead centers. It is generally accepted that some act as 6-coordinated network modifiers, while others act as network formers and have a lower coordination number (formally Pb^{4+} replaces Si^{4+}) [37]. Based on spectroscopic techniques and molecular dynamic calculations, it was suggested that the PbO_4 groups are the dominant structural units coexisting with PbO_3 units at low lead loadings [38, 39], but it has been also proposed that lead centers are linked between them by sharing an edge via two oxygen atoms to form a lead-based network constituted by Pb_2O_4 polyhedron units [37, 40]. For several authors [39], the network formation was driven by the interplay of SiO_4 tetrahedra and PbO_x polyhedra ($x = 3$–5), while the distribution of other (non-networking) units would be inhomogeneous.

In this context, the solid-state electrochemistry of a series of lead glass samples from different archaeological sites provided interesting structural and archaeometric information. Figure 7.7 shows the negative- and positive-going square wave voltammograms of two glass samples from the archaeological site of Puxmarina (Spain), dated back to the 9th–10th century (Caliphal age), in contact with acetate buffer [41]. The negative-going voltammograms yield cathodic signals associated with the reduction of Pb(II) (C_{PbOx}), Cu(II) (C_{CuOx}), and Pb(IV) centers (C^*_{PbOx}). The positive-going voltammograms exhibit the typical oxidative dissolution peaks of lead (A_{Pb}) and copper (A_{Cu}).

The appearance of C^*_{PbOx} peaks denotes the existence of Pb(IV) centers, and the ratio between the peak areas or peak currents of these peaks and C_{PbOx} ones mirrors the fraction of lead existing in the IV oxidation state. Interestingly, the profile of the signal C_{PbOx} varied significantly from one sample to another, as can be seen in Fig. 7.8, where the cyclic voltammograms of different archaeological samples (all from different historical periods and locations in Spain) are depicted [41]. Remarkably, the anodic region also exhibits a peak splitting and peak potential shifts in the signal for lead stripping (A_{Pb}).

Interestingly, cations create 'non-bridging oxygens' (NBOs) within the silicate structure. These are labeled as Q^n, where *n* is the number of bridging oxygens. Then, Q^4 denotes silicate species with no NBO (SiO_2 units), Q^3 silicate species with one NBO (Si_2O_5), and Q^2 silicate species with two NBOs (SiO_3) [42]. Monovalent and polyvalent cations would be ordered differently among the various NBOs. Mainly based on Raman spectroscopy data, it appears that alkali cations are preferentially bonded to NBO from the Q^3 units, while alkaline earth or polyvalent cations are preferentially bonded to NBO from the Q^2 units [43, 44]. Accordingly, the significant variability in the voltammetric pattern reflects the varied coordinative arrangements in the glasses.

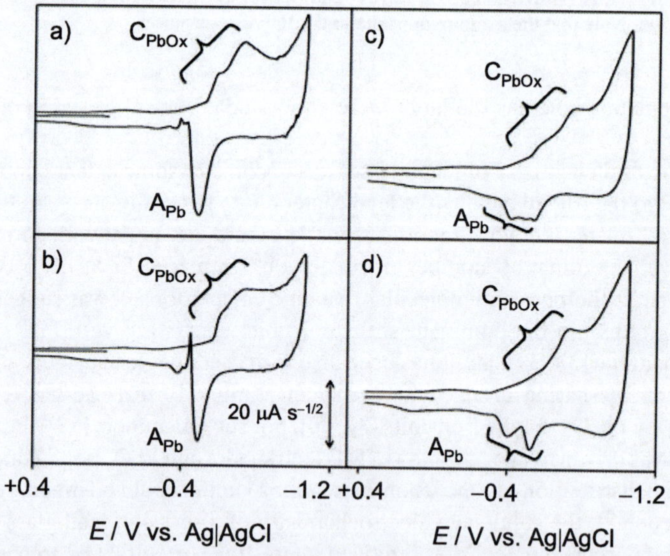

Fig. 7.8 Detail of cyclic voltammograms, after semi-derivative convolution, of graphite electrodes modified with samples **a** S05 (Burgo de Osma, 18th century); **b** S06 (Burgo de Osma, 18th); **c** S07 (Leon 13th–14th); and **d** S10 (La Granja de San Ildefonso, 20th) in contact with air-saturated 0.25 M HAc/NaAc aqueous buffer at pH 4.75. Potential scan rate 50 mV s^{-1}. Reproduced from ref. [41], with permission. Note that the axes are opposite to the IUPAC convention

In fact, coupled AFM – VIMP experiments permitted the distinction of different modes of lead metal generation via the C_{PbOx} processes. These can be seen in Fig. 7.9 where AFM amplitude error channel graphs recorded on deposits of two archaeological samples immobilized onto a graphite plate in contact with acetate buffer are depicted. The original deposits consisted of irregular glass grains distributed over the graphite terraces. After applying a reductive potential input, the glasses produced either irregular metallic lead grains in the vicinity of the parent glass grain, often accompanied by globular clusters, or fine grains eventually forming a dense array near the parent glass grains. The first type of deposits is similar to that formed in the topotactic reduction of litharge (see Fig. 2.5 in Sect. 2.2.3), while the other types of electrodeposition can be ascribed to a more or less important contribution of intermediate species in solution (see Sect. 3.9).

Conjointly considered, the voltammetric features can be correlated with Raman data to extract some structural information [41]. From the archaeometric point of view, the relevant aspect is that there is the possibility of electrochemical characterization of different archaeological sites. This is illustrated in Fig. 7.10, where a two-dimensional diagram representing the ratio between the current of the – 0.60 V

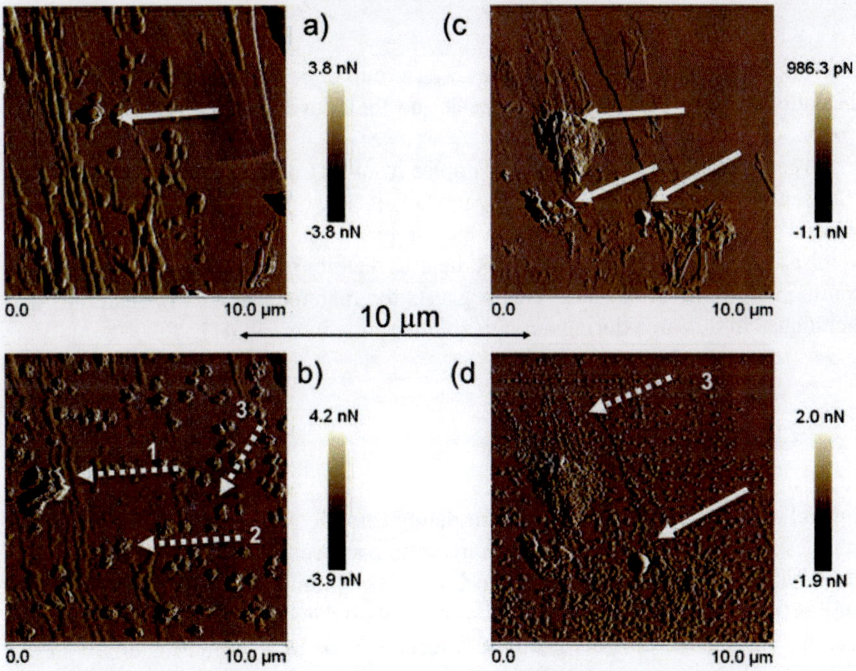

Fig. 7.9 AFM amplitude error channel graphs of deposits of samples **a** S01 (Puxmarina, 9th–10th century) and **b** S05 (Burgo de Osma, 18th century) on a graphite plate in contact with 0.25 M sodium acetate buffer, pH 4,75, **a**, **c** before, and **b**, **d** after application of a constant potential of – 0.70 V for 3 min. Continuous arrows: original glass grains; dotted arrows: metallic lead deposits of different types generated electrochemically. Reproduced from ref. [41], with permission

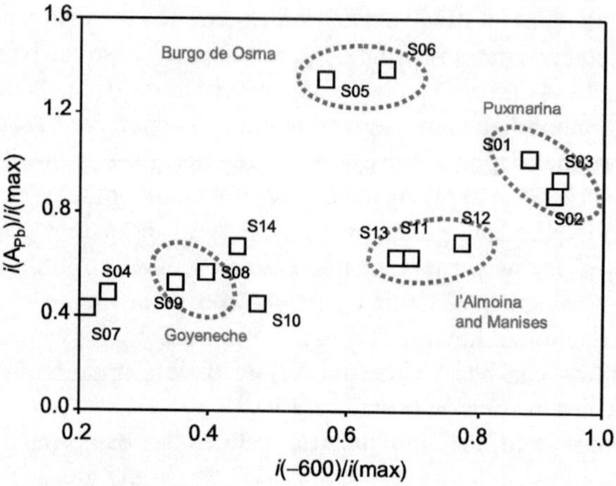

Fig. 7.10 Two-dimensional diagram from the voltammetric data such as in Fig. 7.8 for a set of lead glass samples from different archaeological sites in Spain. Plots of the ratio between the current at the − 600 mV peak and the maximum current in the C_{PbOx} group of peaks, *i*(−600)/*i*(max), versus the ratio between the main anodic peak and the above maximum current, *i*(A_{Pb})/*i*(max). Reproduced from ref. [41], with permission

peak and the maximum current in the C_{PbOx} group of peaks, *i*(−600)/*i*(max), versus the ratio between the main anodic peak and the above maximum current, *i*(A_{Pb})/*i*(max).

As can be seen in this figure, the samples from Puxmarina (Caliphal period, 9th–10th century), Burgo de Osma (18th), and Goyeneche (19th) fall in clearly separated regions of the diagram. Interestingly, the samples from L'Almoina (13th–15th) and Manises (16th–17th), all three located near to the city of Valencia (Spain), fall in a common area of the diagram. This suggests the maintenance of raw materials and techniques in this area during a relatively long time interval [41].

7.5 Glass Alteration and Dating

Several methods have been devised for dating ceramic materials, including thermoluminescence, fission tracks, obsidian hydration, and rehydroxylation methods [45]. These will be treated in more detail in Chap. 8, including the first method proposed for electrochemical dating [46]. This was based on the analysis of the voltammetric features associated with the catalytic effects exerted by ceramic materials on the oxygen evolution reaction (OER) in aqueous solutions. This method offers a particular advantage: the catalytic effects are easily determined in VIMP experiments at sample-modified graphite electrodes. The drawback is that there are several different electrocatalytically active species in the ceramic materials, thus making the method highly case-sensitive.

It has to be taken into account that, as previously noted, the alteration of glass in natural environments is a complex process depending on glass composition and heat treatment, temperature, time, pH and composition of the aqueous phase, exposure to microorganisms, sunlight, etc. [3]. The alteration of glass in contact with waters and soils involves the dissolution (congruent or incongruent) of the glass into the aqueous phase, glass hydration forming a hydrated porous silica network, and subsequent diffusion – hydration and dissolution, this last step being associated with preferential leaching and diffusion of sodium and potassium ions across the hydrated porous silica network [47–49].

Alteration of glasses in soils frequently results in the formation of zones rich in Mn, Ca, P, and Fe [50, 51]. This accumulation may be external in origin, since the above elements are common constituents of soil as a result of the breakdown of silicate minerals [52] so that, under anaerobic conditions, water-soluble Mn(II) and Fe(II) species predominate in waterlogged soils. Such species, however, are readily oxidized in aerobic environments to Mn(IV) and Fe(III), respectively. As previously noted, iron and manganese are frequently added to glass during its manufacturing. Fe(II) provides a green hue to the glass, whereas manganese dioxide is often added during glass production for oxidizing Fe(II) to Fe(III) then providing a purple tint from Mn(III) to offset the straw tint from the Fe(III) [28, 53]. The release of iron and manganese and the subsequent precipitation of black and brown insoluble compounds such as pyrolusite (MnO_2), manganite ($MnO(OH)$), and/or reddish-ochre iron oxides appears to be associated with discoloration and/or browning of archaeological glass. The exact redox state of iron in the melt depends on the redox potential in the furnace (strictly, the partial pressure of oxygen in the atmosphere [54]), but also on the total amount of iron present (so that, in principle, as the total iron increases, Fe(III) predominates over Fe(II)) and on the redox reactions between the various redox couples in the melt, the most important one being that between Mn and Fe. Formally:

$$Mn_2O_{3solid} + 2FeO_{solid} \rightarrow 2MnO_{solid} + Fe_2O_{3solid} \tag{7.15}$$

Under the conditions normally found in the glassmaking furnace, the equilibrium lies well on the right side, so much that Mn_2O_3 does not usually occur until all the iron present is in the Fe(III) form.

In summary, the chemical and mineralogical composition of the alteration layers of archaeological glass and, in particular, the relationship between the species in different oxidation states can provide information on the weathering process and, ultimately, on the age, of archaeological glass and glazed materials.

Because of its use of minimal amounts of sample, VIMP is of potential interest. The first approach to aging analysis can be made based on the record of the iron-centered voltammetric signals of glass samples. Figure 7.11 shows the square wave voltammetric response recorded for an archaeological sample of altered glass in contact with 0.10 M HCl electrolyte [55]. The voltammograms show a series of cathodic and anodic signals between the extreme rising currents associated with hydrogen evolution reaction (HER) and oxygen evolution reaction (OER). Apart from the gross peaks of ca. 0.5 V associated with oxygen functionalities in the graphite base electrode, the voltammograms show several peaks in the potential range between 0.4 and − 0.8 V versus Ag/AgCl attributable to different iron species.

The interpretation of these signals can be made following the guidelines applied in the characterization of different iron sites in ex-framework FeZSM-5 synthetic zeolites [56]. In principle, the reduction of framework Fe(III) and oxidation of framework Fe(II) species occurs at more positive potentials than the reduction of iron oxide forms produced as a result of weathering. These are respectively labeled as Fe(glass) and Fe(weath) in Fig. 7.11.

Formally, the cathodic processes involving iron species in the glass framework can be formally understood as reductive dissolutions:

$$\begin{aligned}\{Fe^{III}O_x\}_{glass} + 2xH^+_{aq} + (2x+z-3)e^- \\ \rightarrow (1-z)Fe^{3+}_{aq} + zFe^{2+}_{aq} + xH_2O\end{aligned} \tag{7.16}$$

and/or as proton-assisted reductive solid-state conversions:

$$\{Fe^{III}O_x\}_{glass} + zH^+_{aq} + ze^- \rightarrow \{Fe^{III}_{1-z}Fe^{II}_zO_{x-z}(OH)_z\}_{glass} \tag{7.17}$$

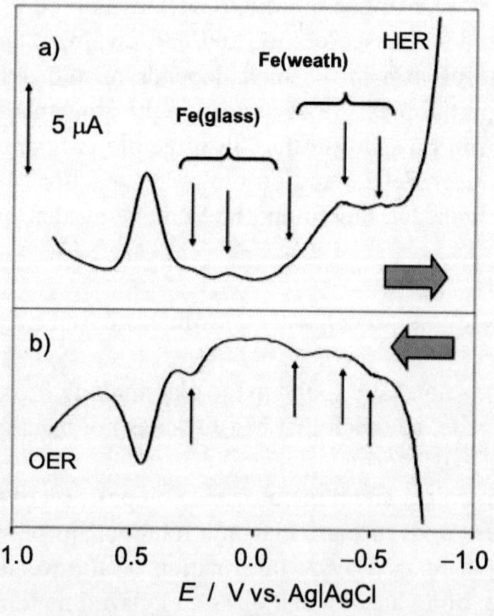

Fig. 7.11 Square wave voltammograms of graphite electrodes modified with a sample of the alteration layer of an irregular glass fragment of flask or jar recovered in the archaeological site of *Angel* Street in Manises, Spain, dated back to the 16th–17th century [55]. Electrolyte: 0.10 M HCl. Potential scan was initiated at **a** 0.85 V in the negative direction, and **b** − 0.85 V in the positive direction. The vertical arrows mark the iron-related voltammetric features distinguishing between those attributable to Fe(III) and Fe(II) centers in the glass and those attributable to Fe(III) oxides formed during the weathering process. The horizontal arrows mark the direction of the potential scan. Note that the axes are opposite to the IUPAC convention

whereas the oxidation of Fe(II) centers can either be assigned to oxidative dissolution:

$$\{Fe^{II}O_y\}_{glass} + 2yH^+_{aq} \rightarrow (1-w)Fe^{2+}_{aq} + wFe^{3+}_{aq} + yH_2O + we^- \quad (7.18)$$

and/or processes of the type,

$$\{Fe^{II}O_y\}_{glass} + wH_2O \rightarrow \{Fe^{III}_w Fe^{II}_{1-w} O_y(OH)_w\}_{glass} + wH^+_{aq} + we^- \quad (7.19)$$

without discarding processes assisted by alkali ions insertion/expulsion.

In turn, the reduction of 'external' Fe(III) oxide deposits formed during the weathering process can be formulated as follows:

$$Fe_2O_3 \cdot xH_2O_{solid} + 6H^+_{aq} + 2e^- \rightarrow 2Fe^{2+}_{aq} + (x+3)H_2O \quad (7.20)$$

The advance of the weathering process determines an increase of the intensity of the Fe(weath) signals relative to the Fe(glass) ones as well as an increase of the global Fe(III)/Fe(II) molar ratio.

Interestingly, voltammetric data can be correlated with SEM/EDX elemental analysis. An example is provided in Fig. 7.12, where data for a set of glass samples from the Roman archaeological site of *Sabaters* street in València (Spain) are shown [57]. The data correspond to a series of glass ingots and fragments from a workshop dated back to the late 3th to the 5th century. The set of samples contains local production and importation products fabricated in the three aforementioned norms: Near East, Roman Provincial, and Roman Capital, as judged by the *R* values (alkaline earth to alkaline ratio) determined from electron microscopy analysis, and are currently under study in the *Servei d'Investigació Arqueològica Municipal de València* (Spain).

Figure 7.12 shows the variation of the average *I*(ORR)/*I*(OER) ratio (ORR: oxygen reduction reaction, OER: oxygen evolution reaction) from three replicate measurements in cyclic voltammograms of sample-modified graphite electrodes in contact with air-saturated 0.10 M NaOH, with the percentage of SiO_2 determined by SEM/EDX quantitative analysis. One can see in this figure that the data points corresponding to archaeological glass samples fabricated according to the Near East norm define an apparently straight trend line clearly separated from that defined by samples fitting to the Roman Capital and Roman Provincial norms. Data in Fig. 7.12 can be interpreted as denoting that the activity of electrocatalytic sites in the glass depends on the composition and manufacturing of the material. As a general trend, the larger the SiO_2 content of the glass, the lower is the catalytic effect on ORR relative to OER, but this effect is different for objects prepared according to Near East norm and objects prepared following the Roman Capital and Roman Provincial norms. Although a much wider set of samples must be tested, these results suggest that, despite the fact that glass is not directly electroactive, electrocatalytic effects monitored by voltammetry, permit to discriminate between archaeological objects manufactured with different fabrication norms.

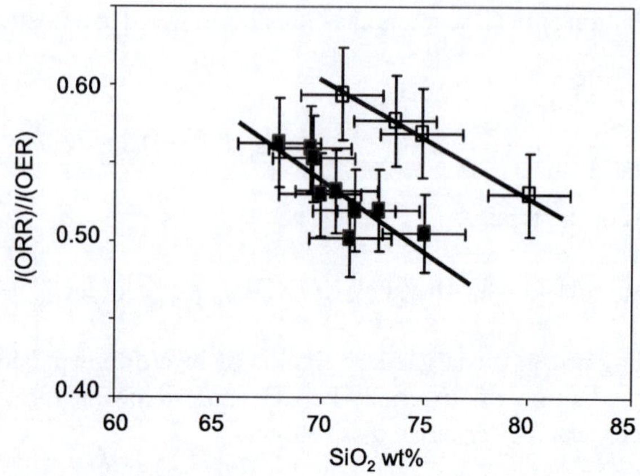

Fig. 7.12 Variation of the average *I*(ORR)/*I*(OER) ratio (ORR: oxygen reduction reaction, OER: oxygen evolution reaction), determined in cyclic voltammetric experiments of sample-modified graphite electrodes in contact with air-saturated 0.10 M NaOH at a potential scan rate of 50 mV s^{-1}, with the percentage of SiO_2 determined in SEM/EDX measurements in a set of archaeological glass samples from the Roman site of *Sabaters* street in València (Spain) dated back to the late 3th to the 5th century. Squares: Near East norm samples; solid squares: Roman Capital and Roman Provincial norms [57]

Interestingly, this grouping is consistent with that obtained from voltammetric data such as in Fig. 7.11, corresponding to square wave voltammograms in contact with 1.0 M HCl, using the sum of peak currents for the Fe(glass) and Fe(weath) signals, respectively, labeled as *I*(Fe(glass)) and *I*(Fe(weath)). The representation of *I*(Fe(glass)) versus *I*(Fe(weath)) is shown in Fig. 7.13, where one can see that, again, Near East norm samples can be separated from the Roman (Capital and Provincial) norms [57].

This result suggests that the 'corrosivity' of the glass, resulting in a more or less easy formation of iron oxide deposits, is in turn conditioned by the composition and manufacturing. Consequently, the *R* value, roughly representative of the more or less significant addition of flux lime, will influence the iron-related voltammetric response of glass samples.

The above result suggests that any proposal of a dating method should account for the composition and manufacturing protocols. Additionally, it has to be noted that the net amount of sample transferred onto graphite electrodes in VIMP experiments can also affect the peak current ratios as occurring in samples from metal corrosion products (see Chap. 9).

Despite the limitations imposed by all these factors, available data suggest that there is the possibility of proposing dating methodologies based on the increment of the relative height of 'external' iron oxide signals (as well as the catalytic effects of the different iron centers on ORR, OER, or other electrochemical processes) with age [58].

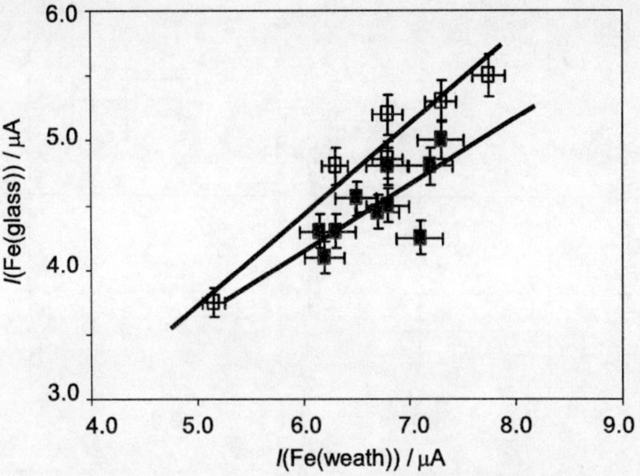

Fig. 7.13 Plots of *I*(Fe(glass)) versus *I*(Fe(weath)), determined in voltammograms in experimental conditions such as in Fig. 7.11, for a set of archaeological glass samples from the Roman site of *Sabaters* street in València (Spain) dated back between the late 3rd to the 5th century. Squares: Near East norm samples; solid squares: Roman Capital and Roman Provincial norms [57]

Figure 7.14 shows the data determined for a set of archaeological samples under study in the Spanish *Museu de Ceràmica de Manises* and the *Museu Nacional de Ceràmica de València*, corresponding to archaeological sites in the Comunitat Valenciana (Spain) covering ages between the 1st CE and 18th centuries [57]. Here, the ratio between the peak current for the reduction of iron oxides, *I*(Fe(weath)), and the sum of the peak currents for all iron-related voltammetric peaks, *I*(Fe(total)), is used as a 'regional' age marker. This means that, for samples of reasonably similar composition and manufacturing and submitted to reasonably uniform (in relation to the regional geological context and the time) aging conditions, the *I*(Fe(weath))/*I*(total) ratio increases monotonically with time.

The data in this figure are not normalized to the absolute iron composition in the glasses. This is a factor to be considered, because, assuming uniform conditions of degradation, the weathering rate relative to iron species should depend on the bulk concentration of such species in the glass.

One can see in this figure that the data points appear to define a trend band suggesting the possibility of developing strategies for age calibration after a refined analysis of composition and electrochemistry data. Eventually, electrochemical methods to determine the Fe(III)/Fe(II) ratio (see Chap. 3) can also be applied, and potential extensions to the geological context can be conceived.

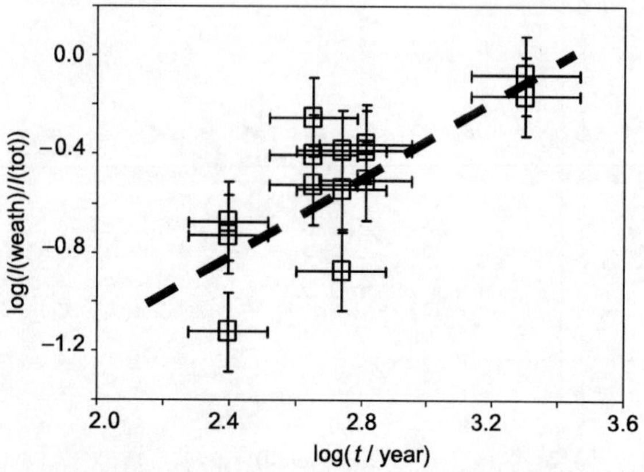

Fig. 7.14 Age variation of the *I*(Fe(weath))/*I*(total) ratio in double logarithmic scale measured in voltammograms such as in Fig. 7.11 for a set of archaeological glass samples corresponding to archaeological sites in the Comunitat Valenciana (Spain) covering the 1st CE to 18th centuries [57]. The dotted line represents the apparent linear trend of the represented data

References

1. Rice PM (1987) Pottery analysis: a sourcebook. The University of Chicago Press, Chicago
2. Eramo G, Mangone A (2019) Archaeometry of ceramic materials. Phys Sci Rev 4(11):331–355
3. Davison S, Newton RG (2003) Conservation and restoration of glass, 2nd edn. Butterworth-Heinemann, London
4. Janssens KHA (ed) (2013) Modern methods for analysing archaeological and historical glass. Wiley, New York
5. Mason RB, Tite MS (1997) The beginnings of tin-opacification of pottery glazes. Archaeometry 39(1):41–58
6. Artioli G (2010) Scientific methods and cultural heritage. Oxford University Press, Oxford
7. Mirti P, Davit P (2004) New developments in the study of ancient pottery by colour measurement. J Archaeol Sci 31:741–751
8. De Bonis A, Cultrone G, Grifa C, Langella A, Leone AP, Mercurio M, Morra V (2017) Different shades of red: the complexity of mineralogical and physico-chemical factors influencing the colour of ceramics. Ceram Int 43:8065–8074
9. Schwedt A, Mommsen H (2004) Clay paste mistures identified by Neutron Activation Analysis in pottery of a Roman workshop in Bonn, Germany. J Archaeol Sci 31:1251–1258
10. Shimada I, Häusler W, Hutzelmann W, Wagner U (2003) Early pottery making in Northern Coastal Peru. Part I: Mössbauer study of clays. Hyperfine Inter 150:73–89
11. Smit Z, Pelicon P, Holc M, Kos M (2002) PIXE/PIGE characterization of medieval glass. Nucl Inst Meth Phys Res B 189:344–349
12. Lühl L, Hesse B, Mantouvalou I, Wilke M, Mahlkow S, Aloupi-Siotis E, Kanngiesser B (2014) Confocal XANES and the attic black glaze: the three-stage firing process through modern reproduction. Anal Chem 86:6924–6930
13. Tite MS, Bimson M, Freestone IC (1982) An examination of the high gloss surfaces finishes on Greek Attic and Roman Samian wares. Archaeometry 24:117–126
14. Mangone A, Giannossa LC, Colafemmina G, Laviano R, Traini A (2009) Use of various spectroscopy techniques to investigate raw materials and define processes in the overpainting of Apulian red figured pottery (4th century BC) from southern Italy. Microchem J 92:97–102

15. Giannossa LC, Mininni RM, Laviano R, Mastrorocco F, Caggiani MC, Mangone A (2017) An archaeometric approach to gain knowledge on technology and provenance of Apulian red-figured pottery from Taranto. Archaeol Anthropol Sci 9:1125–1135
16. Doménech-Carbó A, Doménech-Carbó MT, Moya-Moreno M, Gimeno-Adelantado JV, Bosch-Reig F (2000) Voltammetric identification of lead (II) and (IV) in mediaeval glazes in abrasion-modified carbon paste and polymer film electrodes. Application to the study of alterations in archaelogical ceramic. Electroanalysis 12:120–127
17. Doménech-Carbó A, Doménech-Carbó MT, Osete-Cortina L (2001) Identification of manganese(IV) centers in archaeological glass using micro-sample coatings attached to polymer film electrodes. Electroanalysis 13:927–935
18. Doménech-Carbó A, Sánchez-Ramos S, Doménech-Carbó MT, Gimeno-Adelantado JV, Bosch-Reig F, Yusá-Marco DJ, Saurí-Peris MC (2002) Electrochemical determination of the Fe(III)/Fe(II) ratio in archaeological ceramic materials using carbon paste and composite electrodes. Electroanalysis 14:685–696
19. Sánchez-Ramos S, Bosch-Reig F, Gimeno-Adelantado JV, Yusá-Marco DJ, Doménech-Carbó A (2002) Application of XRF, XRD, thermal analysis and voltammetric techniques to the study of ancient ceramics. Anal Bioanal Chem 373:893–900
20. Ruetschi P, Giovaneli R (1988) Cation vacancies in MnO_2 and their influence of electrochemical reactivity. J Electrochem Soc 135:2663–2668
21. Bodoardo S, Brenet J, Maja M, Spinelli P (1994) Electrochemical behaviour of MnO_2 electrodes in sulphuric acid solutions. Electrochim Acta 39:1999–2004
22. Amarilla JM, Tedjar F, Poinsignon C (1994) Influence of KOH concentration on the γ-MnO_2 redox mechanism. Electrochim Acta 39:2321–2331
23. Malaie K, Scholz F, Schröder U, Wulff H, Kahlert H (2022) Determining the gibbs energy contributions of ion and electron transfer for proton insertion in ϵ-MnO_2. ChemPhysChem 23(24):e202200364
24. Doménech-Carbó A, Doménech-Carbó MT, Osete-Cortina L, Gimeno-Adelantado JV, Bosch-Reig F, Mateo-Castro R (2002) Electrochemical identification of metal ions in archaeological ceramic glazes by stripping voltammetry at graphite/polyester composite electrodes. Talanta 56:161–174
25. Doménech-Carbó MT, Doménech-Carbó A, Yusá-Marco DJ, Ahmadi H (2008) Characterization of Iranian Moarraque glazes by light microscopy, SEM-EDX and voltammetry of microparticles. J Cult Herit 9:e50–e54
26. Doménech-Carbó A, Doménech-Carbó MT, Osete-Cortina L, Gimeno-Adelantado JV, Sánchez-Ramos S, Bosch-Reig F (2003) Quantitation of metal ions in archaeological glass by abrasive stripping square-wave voltammetry using graphite/poliéster composite electrodes. Electroanalysis 15:1465–1475
27. Mason RB, Tite MS (1994) The beginnings of Islamic stonepaste technology. Archaeometry 36:77–91
28. Navarro JMF (1991) El vidrio. CSIC, Madrid
29. Doménech-Carbó A, Doménech-Carbó MT (2005) Electrochemical characterization of archaeological tin-opacified lead-alkali glazes and their corrosion processes. Electroanalysis 17:1959–1969
30. Scholze H (1977) Evidence of control of dissolution rates of glasses by H^+ mobility. J Am Ceram Soc 60:186
31. Silvestri A, Molin G, Salviulo G (2005) Archaeological glass alteration products in marine and land-based environments: morphological, chemical and microtextural characterization. J Non-Crystalline Solids 351:1338–1349
32. Doménech-Carbó MT, Doménech-Carbó A, Osete-Cortina L, Saurí-Peris MC (2006) A study of corrosion processes of archaeological glass from the Valencian region (Spain) and its consolidation treatment. Microchim Acta 154:23–142
33. Doménech-Carbó A, Doménech-Carbó MT, Osete-Cortina L (2008) Electrochemical study of iron oxide deposits in weathered archaeological glass. Int J Electrochem Sci 3:609–630
34. Brill RH (1992) Chemical analyzes of some glasses from Frattesina. J Glass Stud 34:11–22

35. Ščapova J (1990) Chemical technology and on spectral analysis of glass beads from Poland. In: Malinowski T (ed) Research on glass of the Lusatian and Pomeranian cultures in Poland. Academia Pedagogica, Słupsk, p 87
36. Cox GA, Ford BA (1993) The long-term corrosion of glass by ground-water. J Mat Sci 28:5637–5647
37. Takaishi T, Takahashi M, Jin J, Uchino T, Yoko T (2005) Structural study on $PbO–SiO_2$ glasses by X-ray and neutron diffraction and ^{29}Si MAS NMR measurements. J Am Ceram Soc 88:1591–1596
38. Rybicki J, Rybicka A, Witkowska A, Bergmanski G, Cicco AD, Minicucci M, Manzini G (2001) The structure of lead-silicate glasses: molecular dynamics and EXAFS studies. J Phys Condens Mater 13:9781–9797
39. Kohara S, Ohno H, Tanaka M, Usuki T, Morita H, Suzuya K, Akola J, Pusztai L (2010) Lead silicate glasses: binary network-former glasses with large amounts of free volume. Phys Rev B 82:134209
40. Mizuno M, Takahashi M, Takaishi T, Yoko T (2005) Leaching of lead and connectivity of plumbate networks in lead silicate glasses. J Am Ceram Soc 88:2908–2912
41. Doménech-Carbó A, Villegas MA, Agua F, Martínez-Ramírez S, Doménech-Carbó MT, Martínez B (2016) Electrochemical fingerprint of archaeological lead silicate glasses using the voltammetry of microparticles approach. J Am Ceram Soc 99:3915–3923
42. Matson DW, Sharma SK, Philpotts JA (1983) The structure of high-silica alkali-silicate glasses. A Raman spectroscopic investigation. J Non-Cryst Solids 58:323–352
43. Lee SK, Stebbins JJ (2003) Nature of cation mixing and ordering in na-ca silicate glasses and melts. J Phys Chem B 107:3141–3148
44. Robinet L, Bouquillon A, Hartwig J (2008) Correlations between raman parameters and elemental composition in lead and lead alkali silicate glasses. J Raman Spectrosc 39:618–626
45. Doménech-Carbó A (2015) Dating—an analytical task. ChemTexts 1:5
46. Scholz F, Schröder U, Meyer S, Brainina KhZ, Zakharchuk NF, Sobolev NV, Kozmenko OA (1995) The electrochemical response of radiation defects of non-conducting materials. An electrochemical access to age determinations. J Electroanal Chem 385:139–142
47. Bunker BC (1994) Molecular mechanisms for corrosion of silica and silicate glasses. J Non-Cryst Solids 179:300–308
48. Morgenstein ME, Wicket CL, Barkatt A (1996) Considerations of hydration-rind dating of glass artefacts: alteration morphologies and experimental evidence of hydrogeochemical soil-zone pore water control. J Archaeol Sci 26:1193–1210
49. Paul A (1990) Oxidation-reduction equilibrium in glass. J Non-Cryst Solids 123:254–362
50. Cox GA, Ford BA (1989) The corrosion of glass in the sea bed. J Mater Sci 24:3146–3153
51. Sterpenich J, Libourel G (2001) Using stained glass windows to understand the durability of toxic waste matrices. Chem Geol 174:181–193
52. Watkinson D, Weber L, Anheuser K (2005) Staining of archaeological glass from manganese-rich environments. Archaeometry 47:69–82
53. Schalm O, Proost K, De Vis K, Cagno S, Janssens K, Mees F, Jacobs P, Caen J (2010) Manganese staining of archaeological glass: the characterization of Mn-rich inclusions in leached layers and a hypothesis on its formation. Archaeometry 53:103–122
54. Ali AS, Khan I, Zhang B, Razum M, Pavic F, Santic A, Bingham PA, Nomura K, Kubuki S (2021) Structural, electrical and photocatalytic properties of iron-containing soda-lime aluminosilicate glass and glass-ceramics. J Non-Cryst Solids 553:120510
55. Doménech-Carbó A; Doménech-Carbó MT. Unpublished results
56. Doménech-Carbó A, Pérez-Ramírez J, Ribera A, Mul G, Kapteijn F, Arends IWCE (2002) Electrochemical characterization of iron sites in ex-framework FeZSM-5. J Electroanal Chem 519:72–84
57. Doménech-Carbó A; Doménech-Carbó MT, Martínez B. Unpublished results
58. Doménech-Carbó A (2012) Chapter 7: Electrochemical techniques. In: Edwards HGM, Vandenabeele P (eds) Analytical archaeometry, selected topics. The Royal Society of Chemistry, pp 210–241

Additional Literature

59. Al-Saad Z (2002) Chemical composition and manufacturing technology of a collection of various types of Islamic glazes excavated from Jordan. J Archaeol Sci 29:803–810
60. Andaloro E, Belfiore CM, De Francesco AM, Jacobsen JK, Mittica GP (2011) A preliminary archaeometric study of pottery remains from the archaeological site of Timpone della Motta, in the Sibaritide area (Calabria—southern Italy). Appl Clay Sci 53:445–453
61. Bouhjar F, Bessaïs B, Marí B (2018) Ultrathin-layer α-Fe_2O_3 deposited under hematite for solar water splitting. J Solid State Electrochem 22:2347–2356
62. Burgio L, Clark RJH, Firth S (2001) Raman spectroscopy as a means for the identification of plattnerite (PbO_2), of lead pigments and of their degradation products. Analyst 126:222–227
63. Costagliola P, Baldi G, Cipriani C, Pecchioni E, Buccianti A (2000) A. Mineralogical and chemical characterisation of the Medicean glass mosaic tesserae and mortars of the Grotta del Buontalenti, Giardino di Boboli, Florence, Italy. J Cult Herit 1:287–299
64. De Guzmán RN, Shen Y-F, Shaw BR, Suib SL, O'Young C-L (1993) Role of cyclic voltammetry in characterizing solids: natural and synthetic manganese oxide octahedral molecular sieves. Chem Mater 5:1395–1400
65. De Vito C, Medeghini L, Garruto S, Coletti F, De Luca I, Mignardi S (2018) Medieval glazed ceramic from Caesar's Forum (Rome, Italy): production technology. Ceram Int 44:5055–5062
66. Doménech-Carbó MT, Aura-Castro E, Lopez-Ballester E, Peris-Martínez, V, Gimeno-Adelantado JV, Bosch-Reig F (1998) Study of alterations and characterization of manufacturing techniques in mediaeval glazed ceramic from Manises (Valencia). In: Proceedings of interim meeting ICOM-CC (ICOM), Helsinki, pp 97–105
67. Doménech-Carbó A, Doménech-Carbó MT, Osete-Cortina L (2001) Electrochemical identification of iron centres in archaeological glass using microsamples attached to graphite-polyester composite electrodes. In: Proceedings of the 1st international conference "Hyalos-Vitrum-Glass", Rhodes, pp 345–347
68. Doménech-Carbó A, Pérez-Ramírez J, Kapteijn F, Ribera A, Moulijn JA (2002) Characterization of iron centers in ex-framework FeZSM-5 catalyst by electrochemical methods. Catal Lett 78:302–312
69. Doménech-Carbó A, Torres FJ, de Sola ER, Alarcón J (2006) Electrochemical detection of high oxidation states of chromium(IV and V) in chromium-doped cassiterite and tin-sphene ceramic pigmenting systems. Eur J Inorg Chem 2006(3):638–648
70. Doménech-Carbó MT, Mai-Cerovaz C, Doménech-Carbó A (2022) Application of focused ion beam-field emission scanning electron microscopy-X-ray microanalysis in the study of the surface alterations of archaeological tin-glazed ceramics. Ceram Int 48:14067–14075
71. Freestone IC, Wolf S, Thirlwall M (2009) Isotopic composition of glass from the Levant and the south-eastern Mediterranean region. In: Degryse P, Henderson J, Hodgins G (eds) Isotopes in vitreous materials. Leuven University Pres, Leuven, pp 31–52
72. Genga A, Siciliano M, Tepore A, Mangone A, Triani A, Laganara C (2008) An archaeo-metric approach about the study of medieval glass from Siponto (Foggia, Italy). Microchem J 90:56–62
73. Helebrant A, Matouek J (1995) Prediction of glass corrosion under static and dynamic conditions. Glastech Ber Glass Sci Technol 68:207
74. Mirti P, Gulmini M, Davit P (2002) Colorants and opacifiers in seventh and eighth century glass investigated by spectroscopic techniques. Anal Bioanal Chem 372:221–229
75. Molera J, Vendrell-Saz M (2001) Chemical and textural characterization of tin glazes in Islamic ceramics from eastern Spain. J Archaeol Sci 28:331–340
76. Peter LM (2013) Energetics and kinetics of light-driven oxygen evolution at semiconductor electrodes: the example of hematite. J Solid State Electrochem 17:315–326
77. Robinet L, Coupry C, Eremin K, Hall C (2006) The use of Raman spectrometry to predict the stability of historic glasses. J Raman Spectrosc 37:789–797

78. Sánchez-Ramos S, Bosch-Reig F, Gimeno-Adelantado JV, Yusá-Marco DJ, Doménech-Carbó A (2002) Study and dating of meadiaeval ceramic tiles by analytical determination of enamels with atomic absorption spectroscopy, X-ray fluorescence and electron probe microanalysis. Spectrochim Acta B 57:689–700
79. Wilke M, Partzsch GM, Bernhardt R, Lattard D (2005) Determination of the iron oxidation state in basaltic glasses using XANES at the K-edge. Cheml Geol 220:143–161

Chapter 8
Ceramics, Glasses, and Glazed Materials II

8.1 Pottery, an Overview

Pottery is constituted by ceramic materials possessing not only high compositional complexity and mineralogical variability, but also significant variations in their microstructure. This can be defined, following Rice [1], as the "internal arrangement of crystalline and amorphous materials, voids (pores), and the boundaries between them in a polycrystalline and (usually) polyphasic medium". The mineralogical composition usually comprises quartz, sedimentary carbonates, feldspars (plagioclase and K-feldspar), clays, and nodules of iron and manganese compounds. Among them, hematite (Fe_2O_3) is by far the most abundant iron compound in ceramic bodies, but other iron minerals (magnetite, Fe_3O_4, goethite FeO(OH), augite, $Ca_{0.61}Mg_{0.76}Fe_{0.49}(SiO_3)$, fayalite, Fe_2SiO_4; chrysolite, $(Mg,Fe)_2SiO_4$) can be also found. The proportion of iron (and manganese) minerals in ceramic bodies is, however, relatively low, typically reaching values below 10 wt% (expressed as FeO).

8.2 Electrochemistry of Pottery

In principle, only iron and manganese compounds will display a reasonably sensitive voltammetric response by applying VIMP techniques. Iron-rich minerals are used in the manufacture of bricks and tiles whereas iron-poor minerals are used for manufacturing porcelains, glasses, and glazes. The proportions of iron(III) and iron(II) raw materials modulate significantly the properties of the resulting ceramic material [2].

Figure 8.1 shows the square wave voltammograms recorded at graphite electrodes modified with two commercial hematite sources and two samples of red-figured Apulian ceramic fragments recovered from the archaeological sites of Altamura and Taranto (Eastern Italy) in contact with air-saturated 0.10 M HCl aqueous solution [3].

A. Doménech-Carbó and M. T. Doménech-Carbó, *Electrochemistry for Cultural Heritage*, Monographs in Electrochemistry,
https://doi.org/10.1007/978-3-031-31945-7_8

This type of ceramicware was produced within the 5th and 4th centuries BCE and has received considerable attention due to its high quality and sophisticated decoration which includes a black gloss mainly composed of hercynite, an iron spinel ($FeAl_2O_4$) embedded in vitreous or partly crystalline matrices (see Fig. 8.2) [4–7].

The voltammograms show high currents at the potential limits due to the OER (ca. 1.0 V versus Ag/AgCl) and HER (ca. − 1.0 V) processes. Upon scanning the potential in the negative direction, voltammetric peaks at ca. 0.80 (C_{HE}), 0.40 (C_g), and − 0.65 V (C_{ORR}) appear, the last being preceded by more or less marked peaks between 0.0 and − 0.45 V (C_H). The peak at − 0.65 V corresponds to the reduction of dissolved oxygen and is clearly enhanced in hematite-modified electrodes relative to unmodified ones. The signal Cg can be assigned to the oxidation/reduction of oxygen functionalities in the graphite surface [8]. The signal C_H can be assigned to the reductive dissolution of iron oxide forms, including, of course, hematite (see Sect. 2.2.2) whereas the signal C_{HE} at ca. 0.80 V has a well-defined counterpart in

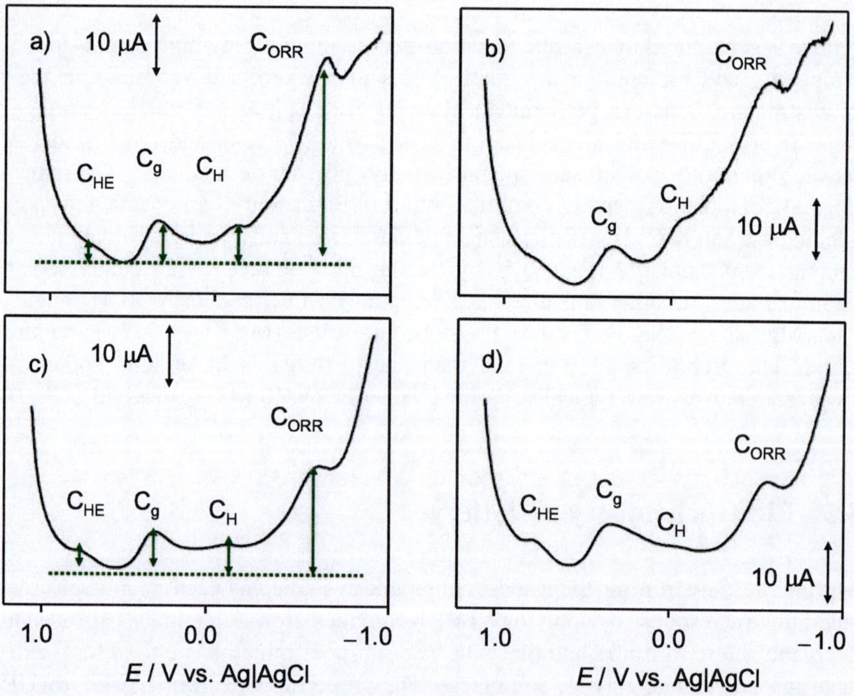

Fig. 8.1 Square wave voltammograms at graphite electrodes modified with **a** anhydrous hematite (Probus reagent), **b** hematite pigment (Kremer 48600) and red-figured Apulian ceramic samples, **c** A5 (Altamura site, Italy), and **d** 212411 (Taranto, Italy) in contact with air-saturated 0.10 M HCl aqueous solution. Potential scan initiated at 1.25 V in the negative direction; potential step increment 4 mV; square wave amplitude 25 mV; frequency 5 Hz. Dotted lines represent the baselines used for peak current measurements. This baseline has been used only for a formal evaluation. It is not the true baseline of the electrode, which is not accessible by measurements (see Sect. 4.3.3). Reproduced from ref. [3], with permission. Note that the axes are opposite to the IUPAC convention

Fig. 8.2 Photographic image of an Apulian Red-Figure Krater (inv. number FG48524). Tomb ONC28 from Arpi (FG, Italy). Reproduced from ref. [3], with permission

positive-going voltammograms and can be associated with redox processes involving Fe(IV) and/or Fe(V) oxidation states [9–12].

The last process is superimposed on the cathodic wave corresponding to the ORR process. This follows a multistep mechanism that is catalytically enhanced by hematite. According to Compton et al. [13, 14] in neutral media, hematite acts as a chemical catalyst accelerating the disproportionation of H_2O_2 into H_2O and O_2. This catalytic pathway is shown in Fig. 8.3a [3]. The initial step of the ORR process is the one-electron reduction of O_2 to superoxide radical anion further yielding H_2O_2 through a cascade mechanism. One of the intermediate steps is the disproportionation of H_2O_2 into H_2O and O_2. On catalyzing this process, hematite promotes the regeneration of O_2 accelerating the overall ORR process. This catalytic effect appears to be significantly affected by the degree of hydroxylation of the hematite surface [11, 12] and is increased by the generation of Fe(II) centers [15]. This is in agreement with the recognized ability of Fe(II) spinels to catalyze the ORR process [16]. Then, a second catalytic pathway could involve the electrochemical generation of Fe(II) active sites superimposed on the reductive dissolution of hematite. This process takes place at potentials near to those where the ORR process occurs and can be represented as [3]:

$$Fe_2O_{3solid} + xH^+_{aq} + xe^- \rightarrow \left\{Fe^{III}_{2-x}Fe^{II}_xO_{3-x}(OH)_x\right\}_{surf} \qquad (8.1)$$

These active sites can react with O_2 generating superoxide radical anion ($O_2\cdot^-$), thus accelerating the first step of the electrochemical ORR process:

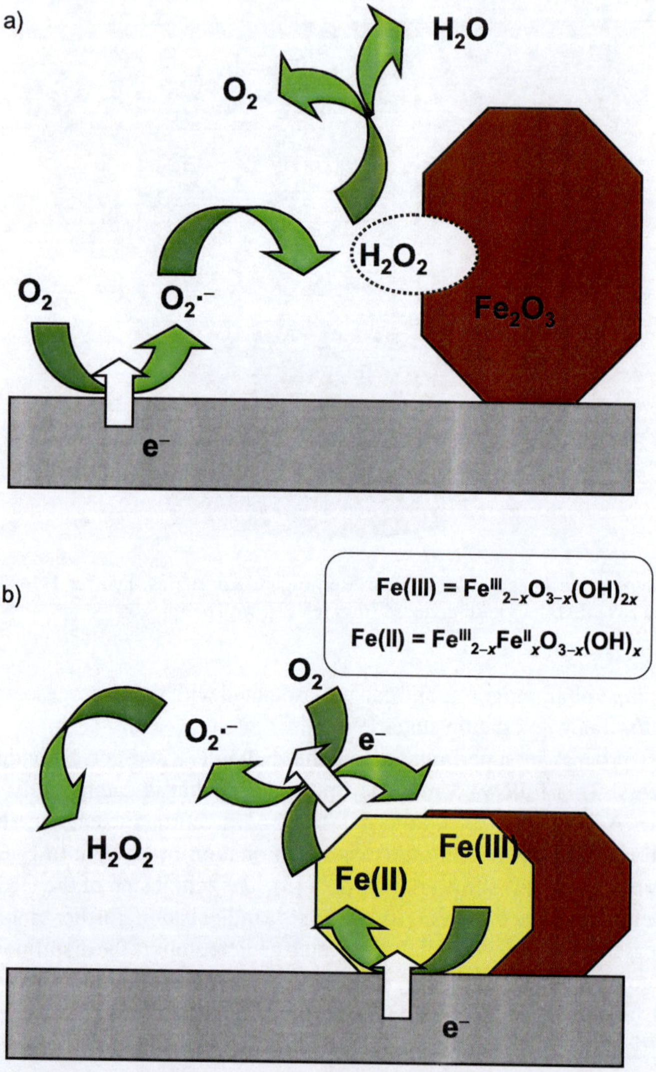

Fig. 8.3 **a** Scheme of the chemical catalysis of ORR by hematite. Here, O_2 is reduced to $O_2\cdot^-$, further protonated to H_2O_2 whose disproportionation into O_2 and H_2O is catalyzed by hematite **b** Tentative scheme for the electrocatalytic ORR process mediated by hydroxylated hematite forms. This pathway involves the electrochemical reduction of Fe(III) hematite surface sites to Fe(II) ones and their reduction of O_2 to $O_2\cdot^-$. This accelerates the first step of the ORR process while the Fe(II) catalyst is electrochemically regenerated. The intermediate steps yielding H_2O_2 from $O_2\cdot^-$ are omitted for simplicity. Reproduced from ref. [3], with permission

$$\begin{aligned}\{Fe^{III}_{2-x}Fe^{II}_{x}O_{3-x}(OH)_{x}\}_{surf} &+ xH_2O + x\,O_2 \\ &\rightarrow \{Fe^{III}_{2}O_{3-x}(OH)_{2x}\}_{surf} + xH^{+}_{aq} + x\,O_2{\cdot}^{-}_{aq}\end{aligned} \tag{8.2}$$

Conceivably, the catalytic Fe(II) centers are electrochemically regenerated within the same potential range:

$$\begin{aligned}\{Fe^{III}_{2}O_{3-x}(OH)_{2x}\}_{surf} &+ xH^{+}_{aq} + xe^{-} \\ &\rightarrow \{Fe^{III}_{2-x}Fe^{II}_{x}O_{3-x}(OH)_{x}\}_{surf} + xH_2O\end{aligned} \tag{8.3}$$

completing the electrocatalytic cycle for the ORR. This second pathway is schematically depicted in Fig. 8.3b [3].

The above scheme suggests an interpretation of the C_{HE} process in terms of the oxidation/reduction of hydroxylated iron species or even hematite. Tentatively, these processes can be represented as [3]:

$$Fe^{III}_{2}O_{3solid} + xH_2O \rightarrow \{Fe^{IV}_{x}Fe^{III}_{2-x}O_3(OH)_{x}\}_{surf} + xH^{+}_{aq} + xe^{-} \tag{8.4}$$

$$\{Fe^{III}_{2}O_{3-x}(OH)_{2x}\}_{surf} \rightarrow \{Fe^{IV}_{x}Fe^{III}_{2-x}O_3(OH)_{x}\}_{surf} + xH^{+}_{aq} + xe^{-} \tag{8.5}$$

On the basis of these considerations, one can assume that the signal C_{ORR} (including the occluded signal C_H) is mainly caused by the content of anhydrous, crystalline hematite in the ceramic material, whereas the signal C_{HE} is due to the content of hydrated iron oxide forms. Although other components of the ceramic bodies can exert electrocatalytic activity, the SECM study of microparticulate deposits of hematite and ceramic was consistent with these assumptions [3, 17]. This can be seen in Figs. 8.4 and 8.5. The former illustrates the SECM color plot and the topographic profiles of a microparticulate deposit of crystalline hematite fixed onto a carbon paste bed in contact with air-saturated 0.10 M HCl. The tip potential was fixed at − 0.65 V so that the reduction of dissolved oxygen acts as the redox probe. When no potential was applied to the hematite deposit, the grains appear as positive feedback features (Fig. 8.4a, b) in agreement with the catalytic effect exerted by hematite on the ORR process. Upon applying to the substrate (i.e., to the hematite particles) a potential of − 0.70 V, the SECM pattern becomes moderately changed, being limited to small narrowing of the hematite grains and the disappearance of several small crystals.

Similar SECM patterns were obtained for ceramic samples. However, when successive cathodic and anodic potentials were applied to the substrate, significant changes were recorded. These can be seen in Fig. 8.5 for a microparticulate deposit of an Apulian red-figured sample from the archaeological site of Arpi. As before, when no potential was applied to the substrate, the mineral particles appear as positive feedback features (Fig. 8.5a, b), several of which vanish when a potential of − 0.70 V was applied to the substrate (Fig. 8.5c). These features, as in the case of pure hematite in Fig. 8.4, can be attributed to the easier reductive dissolution of hydrated, less-crystalline hematite forms (see Sect. 2.2.2). When a potential of 1.0 V

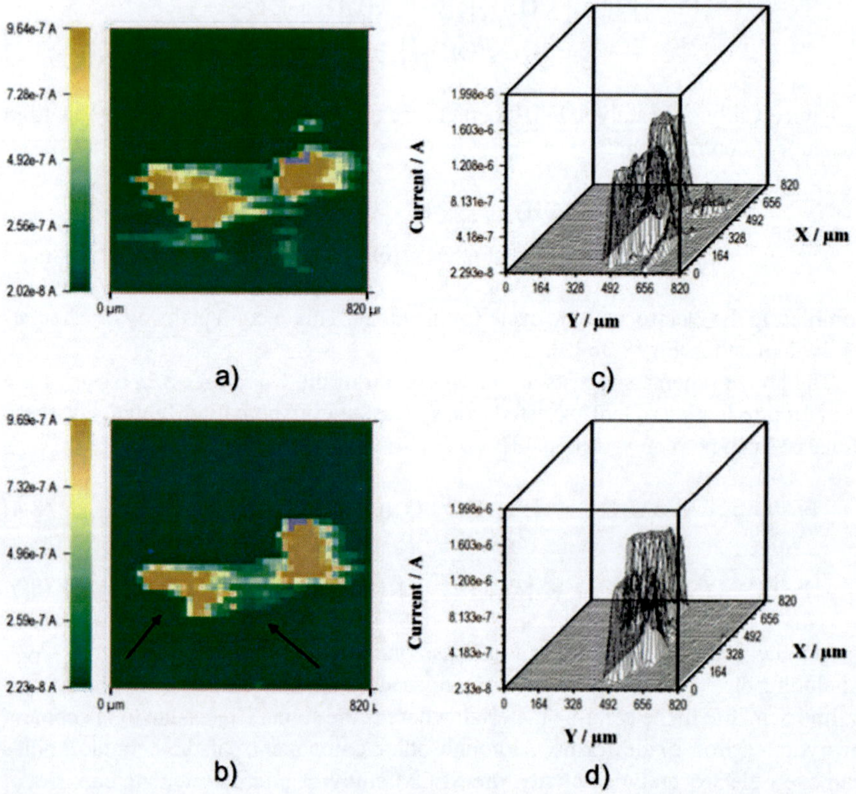

Fig. 8.4 **a, b** SECM color plots and **c, d** topographic profiles of a microparticulate deposit of crystalline hematite fixed onto a carbon paste bed in contact with air-saturated 0.10 M HCl. The tip potential was fixed at − 0.65 V; substrate potential **a, c** 0.0 V; **b, d** − 0.70 V. Reproduced from ref. [3], with permission

is subsequently applied to the same substrate (Fig. 8.5d), several of the above-peaked features remain while any others disappear. This potential (in the region of the C_{HE} signal) is positive enough to oxidize the more or less hydroxylated Fe(III) surface species, in agreement with the oxidative pathway described by Eqs. (8.4) and (8.5).

8.3 Characterization of Archaeological Sites

The above results indicate that hematite (but probably several other minority iron compounds also) displays a rich electrochemistry, including significant catalytic effects on the ORR. This electrochemistry will be dependent on the degree of hydroxylation, crystallinity, and presence of Fe(II) species, but also on factors such as the substitution of Fe^{3+} by Al^{3+} (see again Sect. 2.2.2). Additionally, the presence of

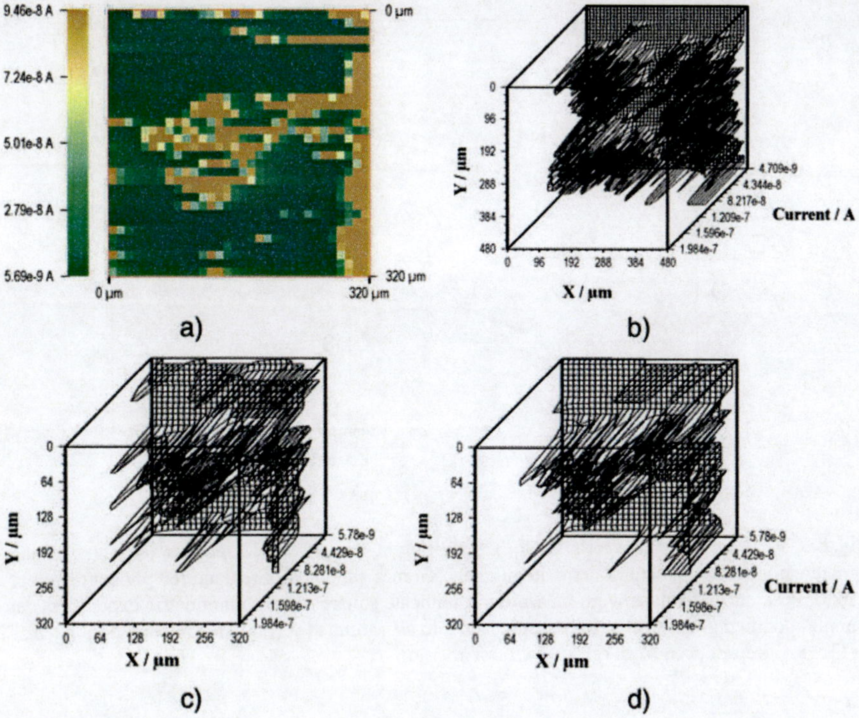

Fig. 8.5 **a** SECM color plot and **b–d** topographic profiles displayed by a microparticulate deposit of Apulian red-figured pottery (I1-01 from the archaeological site of Arpi, Italy) fixed onto a carbon paste bed in contact with air-saturated 0.10 M HCl. $E_T = -0.65$ V; E_S was successively fixed from 0.0 V **a**, **b** to $-$ 0.70 V **c** and 1.0 V **d**. Reproduced from ref. [3], with permission

other mineral components with electrocatalytic activity and the porosity, compactness, etc., and properties of the ceramic body can also modify the electrochemical response. Accordingly, one can expect that the voltammetric response recorded under fixed electrochemical conditions is sensitive to minute changes in the properties of the ceramic body. As a result, an electrochemical characterization of archaeological handcrafts based on the VIMP methodology can be made [3, 17–22].

The sensitivity of the voltammetric response to these factors can be seen in Fig. 8.6, where the peak currents measured for the C_{ORR} ($I(C_{ORR})$) versus those measured for the CHE signal ($I(C_{HE})$) are depicted for three different commercials in conditions such as in Fig. 8.1. In this figure, the peak currents measured in replicate experiments at freshly modified graphite electrodes are depicted, clearly suggesting that a linear relationship exists between $I(C_{ORR})$ and $I(C_{HE})$.

These variations (that are in principle linear) can be interpreted by considering the modeling of VIMP processes described in Chap. 3. Since the hematite particles form a discontinuous deposit onto the graphite surface, the redox process—including

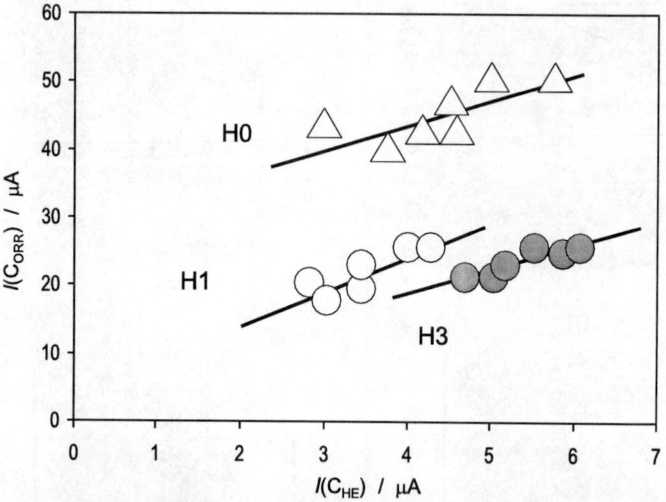

Fig. 8.6 Plots of $I(C_{ORR})$ versus $I(C_{HE})$ for different hematites: H0, Probus reagent; H1 and hematite pigments H1, natural hematite pigment Kremer 48600; H3 Ercolano red pigment Kremer 41600. Peak current values were measured in replicate square wave voltammetric experiments at sample-modified graphite electrodes immersed into air-saturated 0.10 M HCl in conditions such as in Fig. 8.1. Reproduced from ref. [3], with permission

electrocatalytic processes—initiates at the particle/base electrode/electrolyte three-phase junction. In the case of electrochemical processes A such as described by Eqs. (8.1) to (8.5), the peak current (I_A) can be taken as proportional to the total perimeter of the three-phase junction of the particles and, ultimately, to the surface area S covered by the same, i.e.:

$$I_A = g_A p S^{\alpha} \tag{8.6}$$

where g_A is an electrochemical coefficient of response, and p and α are geometrical factors relating the perimeter of the particles with their surface in contact with the base graphite electrode. Notice then, for instance, if cuboid particles of size a are organized over a graphite plate, the particle perimeter is $4a$ and the covered area a^2 so that the perimeter is given by $S^{1/2}/4$; i.e., $p = ¼$, $\alpha = ½$. Note that in the case of ceramic samples, the particles of the electroactive mineral species (typically, hematite crystals) are embedded into an electrochemically silent substrate that covers a surface S. This means that the coefficient g_A will be also representative of the fraction of electroactive species in the ceramic body.

In the case of HER, OER, and ORR, the measured currents will in principle result from the sum of the contributions of the process occurring at the regions of the graphite surface exposed to the electrode, and the catalytic process initiated at the hematite grains/graphite/electrolyte boundary. Assuming that the total surface area of the graphite electrode is S_o, the former will be proportional to the non-covered graphite surface, i.e., $S_o - S$. The catalytic contribution will be proportional to the perimeter of the three-phase junction and, ultimately, to the covered surface S. The peak current (I_B) for the second electrochemical process can be expressed as:

$$I_B = g_B(S_o - S) + h_B p S^{\alpha} \tag{8.7}$$

where g_B and h_B represent the electrochemical coefficients of the response of the B-process at the graphite surface and the catalytic process, respectively.

If we perform replicate VIMP experiments, one can assume that only S will vary. In the more favorable situation, $\alpha \approx 1$ and S can easily be eliminated between Eqs. (8.6) and (8.7), to give [3]:

$$I_B = I_A\left(\frac{ph_B - g_B}{ph_A - g_A}\right) + S_o\left(g_B - g_A\frac{ph_B - g_B}{ph_A - g_A}\right) \tag{8.8}$$

This equation predicts a linear dependence of I_B on I_A where the slope and the intercept will be characteristics of each mineral or ceramic material. This is obviously a rough approximation. Note that, due to the abrasive character of the electrode conditioning, the graphite surface undergoes more or less intense scratching. This results not only in the variation of S_o between different experiments but on the number of exposed graphite sites (edges, planes) which have different electroactivity.

If the signals corresponding to two different electroactive species (R and T) displaying separate voltammetric peaks are considered, the peak currents will be,

$$I_R = h_R p_R S^{\alpha} \; ; \quad I_T = h_T p_T S^{\beta} \tag{8.9}$$

where all symbols have the meaning equivalent to that described for Eq. (8.7). In this case, the peak currents of the R- and T-related voltammetric processes will satisfy the relationship:

$$I_T = \left(\frac{h_T p_T}{h_R^{\beta/\alpha} p_R^{\beta/\alpha}}\right) I_R^{\beta/\alpha} \tag{8.10}$$

thus predicting a potential variation of I_T on I_R.

Voltammetric data for archaeological samples from different geographical areas and techniques have been studied using this methodology combining different electrolytes (HCl, H_2SO_4) and substrate electrodes (paraffin-impregnated graphite, gold). These include Apulian red-figured pottery (5th–4th BCE) [17, 18], pottery from the Roman sites of Nerva's Forum (second half of 9th–early 11th), Caesar's Forum (second half of 9th-early 11th) and Magna Mater Temple (3rd century) [19], Inka pottery from Qotakalli archeological site in the Cusco (Perú) between the initial and colonial periods [20], Phoenician red slip ware from different sites of the Mediterranean area dated back between the 8th to the 6th century BCE [21], and pottery from the site of Sagunt (Spain), mainly of Roman age [22].

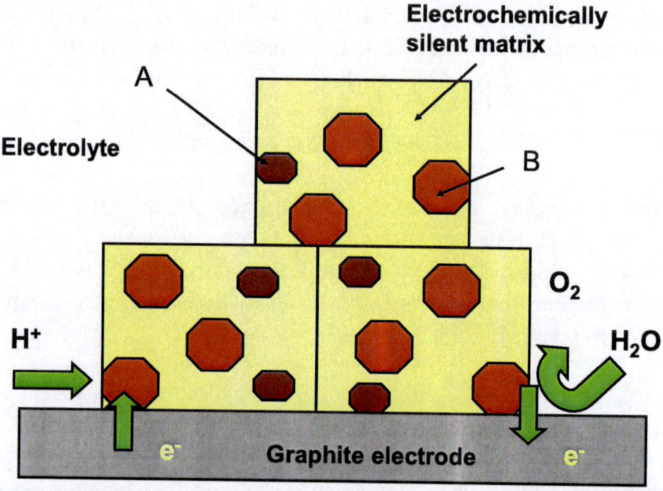

Fig. 8.7 Scheme for the description of the electrochemistry of pottery samples. Grains of two different iron minerals (A, B) are embedded into a non-electroactive matrix. The crystals in contact with the base electrode and the electrolyte display signals for the reduction of Fe(III) minerals (or oxidation of Fe(II) ones). These processes involve coupled proton/electron transfers. Catalytic effects on OER are also schematized. Reproduced from ref. [21], with permission

Figure 8.7 shows a scheme for the description of the electrochemistry of electroactive minerals embedded into an electrochemically silent ceramic matrix in pottery samples used to develop Eqs. (8.6) to (8.10) [21]. The figure schematizes the proton-assisted reduction of Fe(III) minerals and electrocatalytic effects on OER. Of course, oxidation of Fe(II) minerals and electrocatalytic effects on HER, ORR, and eventually, other electrochemical processes are also possible.

Figure 8.8 illustrates the discriminating capability of the VIMP methodology by comparing the experimental $I(C_{ORR})$ versus $I(C_{HE})$ data recorded in square wave voltammograms in conditions such as in Fig. 8.1 for Apulian red-figured pottery from the archaeological site of Taranto. Here, the data points corresponding to samples extracted from the black gloss are grouped in a clearly different trend line than those defined by the samples extracted from the ceramic body of the archaeological specimens. One can see that the $I(C_{ORR})$ values for the black gloss are up to those for the ceramic body. This feature can be interpreted as evidencing the larger catalytic effect exerted by the components of the black gloss (mainly, hercynite, a Fe(II) compound) relative to the components of the ceramic body (in principle, dominated by hematite).

Interestingly, the described methodology can discriminate between different provenances or manufacturing centers. An example is shown in Fig. 8.9, where the data from square wave voltammograms (again in conditions such as in Fig. 8.1) of the black gloss of Apulian red-figured pottery from different archaeological sites is depicted. One can see in Fig. 8.9a that the data points of Taranto are clearly separated from those of fragments recovered at the ONC28 tomb in Arpi, which in turn are

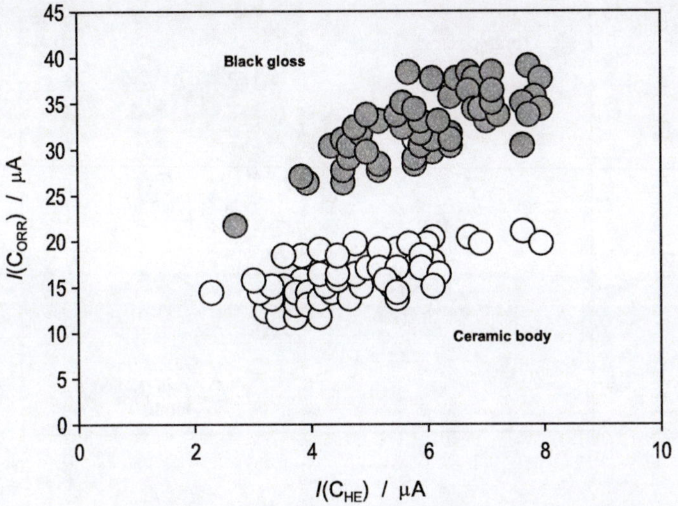

Fig. 8.8 Plots of $I(C_{ORR})$ versus $I(C_{HE})$ recorded in square wave voltammograms in conditions such as in Fig. 7.117.11 for Apulian red-figured pottery from the archaeological site of Taranto corresponding to samples extracted from the black gloss of and the ceramic body of ceramicware fragments. Adapted from ref. [17], with permission

separated from those of extracted from the Niobidi's tomb in the same archaeological site. In all cases, experimental data can reasonably be fitted to linear dependences such as described by Eq. (8.8).

Figure 8.9b refers to an aspect under debate [23–26]: the extent of the influence of the Attic production (where this red-figured pottery was originated) on the production of Apulian workshops. This figure shows the voltammetric data for two Apulian sites, Monte Sannace and Altamura, with one Attic site, Pella. Clearly, the data points for the site of Monte Sannace diverge from those of the Altamura site. These are close but separated from those of the site of Pella and the samples from Niobidi's tomb in Arpi. These features suggest that the Apulian workshops produced local variations of the original Attic manufacturing recipes more or less close to these (vide infra).

Figure 8.10 illustrates the voltammetry of hematite and archaeological samples of Phoenician red slip ceramics from the site of Motya (Sicily, Italy) in contact with 0.10 M H_2SO_4. Here, signals labeled as C2–C4 and A2–A3 can be assigned to the reduction/oxidation of iron compounds. These samples are representative of the typical Phoenician ceramic production, the red slip ware, characterized by shiny deep red coating on its surface. These ceramics can be considered as a marker of the expansion of the Phoenicians along the trade routes all over the Mediterranean since the 14th century BCE but concentrated between the 8th and 6th BCE [27].

Figure 8.11 condensates the data recorded for a series of 80 red slip ware objects unearthed in archeological sites of different countries of the Mediterranean world. This figure depicts the variation of the peak current for signal A2 (i(A2)) versus the peak current for the signal A1 (i(A1)) recorded in square wave voltammograms of

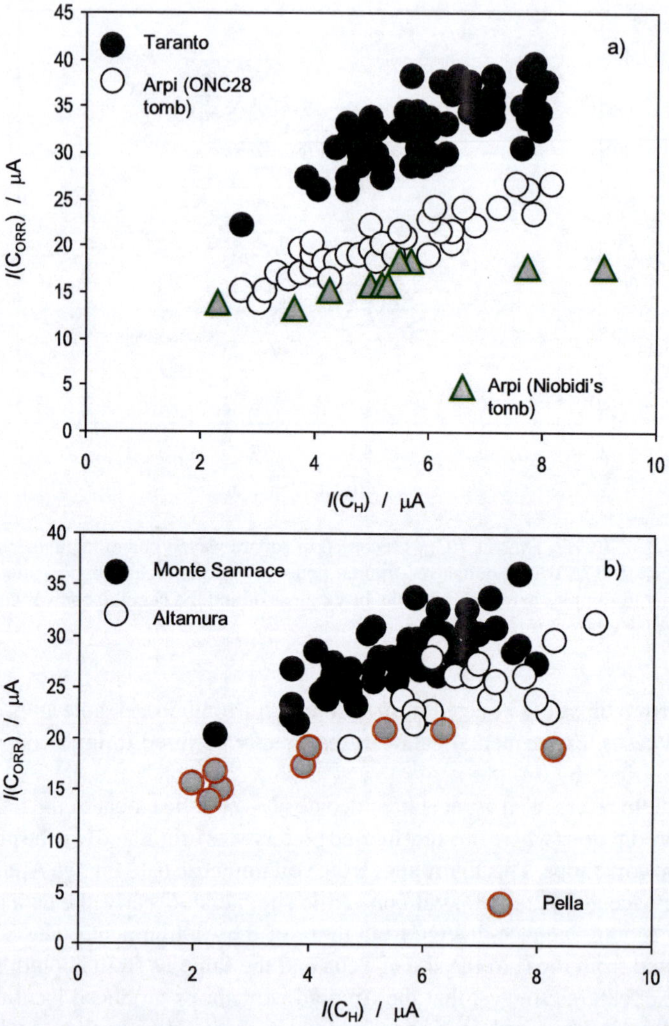

Fig. 8.9 Plots of $I(\mathrm{C_{ORR}})$ versus $I(\mathrm{C_H})$ recorded in square wave voltammograms in conditions such as in Fig. 8.1 for the black gloss of Apulian red-figured pottery samples from the archaeological sites of **a** Taranto and Arpi; **b** Monte Sannace, Altamura and Pella, all in Italy. From ref. [17], with permission

Phoenician red slip wares in conditions such as in Fig. 8.10c,d [21]. Remarkably, all data points fall within the same band of the diagram suggesting that this pair of signals does not permit discrimination between archaeological sites. In contrast, if the peak current for the signals C3 and C1 are balanced (i(C1) and *i*(C3), respectively), one obtains satisfactory discrimination between sites. This can be seen in Fig. 8.12 in which the i(C3) versus i(C1) representation is presented for a set of plates from the site of Motya (6th–8th BCE) and two different sites in Cadiz (Spain), the Necropolis

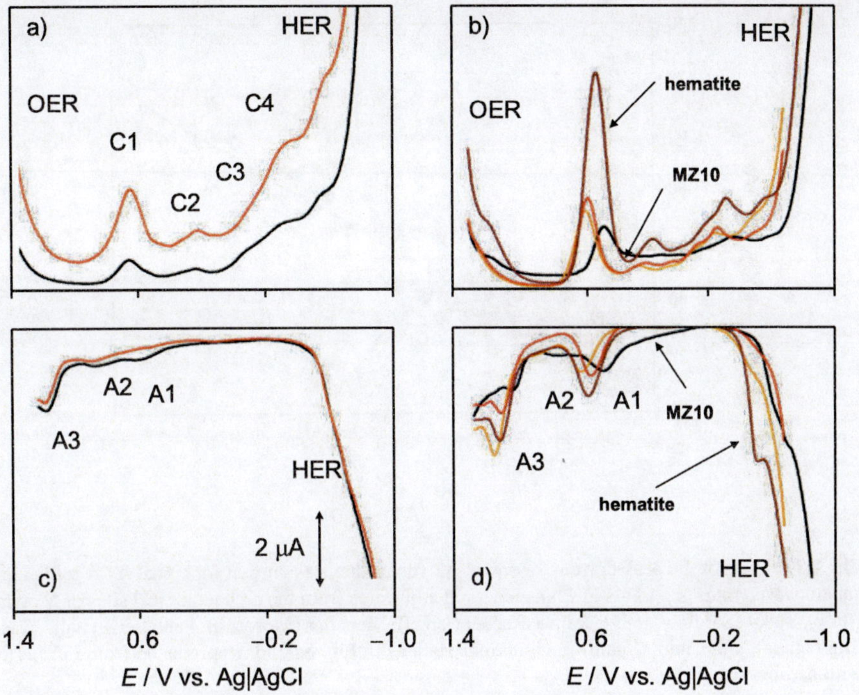

Fig. 8.10 Square wave voltammograms of sample **a, b** MD.07.2219/56 (two replicate measurements) from the archaeological site of Motya (Sicily, Italy), **c, d** hematite, sample MC.11.2491/51 from the same site and, two replicate measurements on a 50%wt mixture of hematite and that sample attached to graphite electrode in contact with air-saturated 0.1 M H_2SO_4. Potential scan initiated at **a** 1.25 V in the negative direction; **b** − 0.85 V in the positive direction; potential step increment 4 mV; square wave amplitude 25 mV; frequency of 5 Hz. Reproduced from ref. [21], with permission. Note that the axes are opposite to the IUPAC convention

dated back to the 6th BCE century and the Cánovas del Castillo site located at the center of the modern city, occupied since the 8th century BC.

One can see that the data points of the samples from Motya tend to assemble along a curved trend line separated from the line on which the data points of samples from Cadiz are. The curved paths can be fitted to potential functions as predicted by Eq. (8.10). Interestingly, the data of the two Cadiz sites, although on the same line, are in different regions of the same. Since these samples correspond to red slip productions separated ca. 200 years, it seems reasonable to hypothesize that the manufacturing technique was based on similar raw materials but with some variation in the conditions of firing.

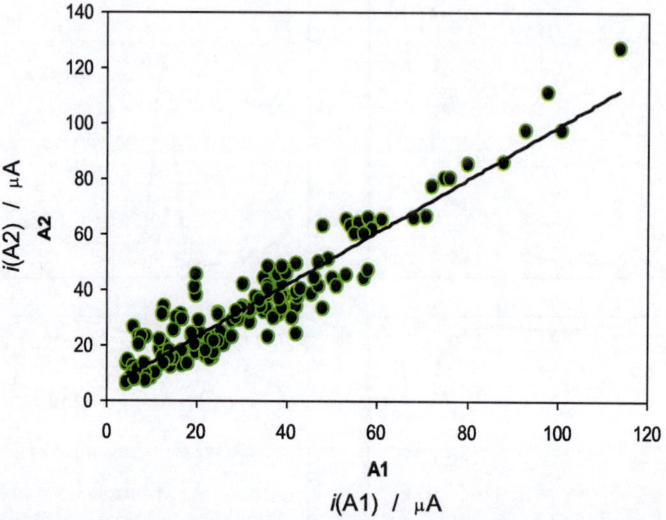

Fig. 8.11 Plots of the peak current for signal A2 versus the peak current for signal A1 recorded in square wave voltammograms of Phoenician red slip wares from the archaeological sites of Motya (Italy), Sulky (Sardinia, Italy), Ramat-Rachel (Israel), Mogador (Morocco), Pantelleria (Italy), Tas Silg (Malta), and Cadiz (Spain) in conditions such as in Fig. 8.9c, d. Reproduced from ref. [21], with permission

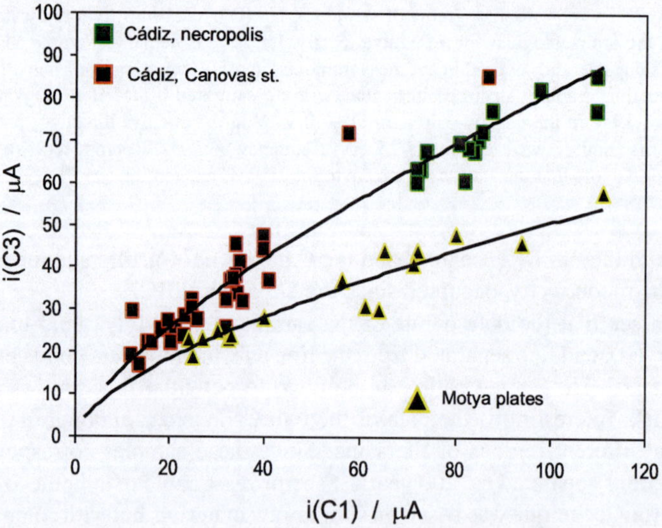

Fig. 8.12 Plots of the peak current for signal C3 versus the peak current for the signal C1 recorded in square wave voltammograms of a series of Phoenician red slip wares from the archaeological sites of Motya (Italy) and Cadiz (Spain, necropolis and Cánovas del Castillo street sites) in conditions such as in Fig. 8.10a, b. Reproduced from ref. [21], with permission

8.4 Information on Manufacturing Techniques

This matter deals with essential aspects of ceramic ware production: the origin and processing of raw materials and the firing conditions (type of furnace, firing temperature, oxidizing/reducing conditions of firing, single-step or multiple-step firing) [26–29]. Correlating electrochemical data with those provided by microscopy, diffraction, and spectroscopic techniques can provide information about these technological aspects. For instance, estimates of the Fe(III)/Fe(II) ratio in ceramic bodies, which focused on the early VIMP studies [30, 31], yield information on the oxidizing/reducing conditions of firing. The complexity of the composition of ceramic samples as well as the influence of their physical, 'textural' properties (granulometry of the particles, porosity, compactness, …) requires, however, balancing the information from different analytical techniques.

Within this frame, it is worth noting that the significance of electrochemical data can be expanded, apart from using multivariate chemometric techniques, by means of an appropriate selection of the voltammetric parameters. An example is given in Fig. 8.13, where the peak current of the C_{ORR} process ($I(C_{ORR})$) is plotted versus the peak current of the Cg signal ($I(C_g)$) from experimental data for Apulian red-figured pottery in experimental conditions such as in Fig. 8.1. This representation, when applied to the entire set of samples from the archaeological sites of Altamura (Bari), Arpi (Foggia), Conversano (Bari), Monte Sannace (Gioia del Colle, Bari), Egnatia (Fasano, Brindisi), and Taranto, permits to discriminate between those made by local techniques in Apulian workshops and those made using the Attic mode of fabrication (coming from the Attic site of Pella and different Apulian sites) [17].

Voltammetric data permit to hypothesize about the manufacturing conditions. In the case of the Apulian red-figured pottery previously mentioned, the averaged value of the $I(C_{ORR})/I(C_{HE})$ data for each site can be seen as representing the compromise between the accessibility of the metal centers to the oxidation/reduction processes and the degree of surface hydroxylation. Then, in the case of the black gloss, the $I(C_{ORR})/I(C_{HE})$ will increase by increasing the Fe(II)/Fe(III) ratio (roughly, the Fe(II) spinel to hematite ratio). In the case of the ceramic body, an increase of the $I(C_{ORR})/I(C_{HE})$ will denote, roughly, an increase in the hematite to hydroxylated hematite forms ratio. This will correspond to an increase in the firing temperature. Figure 8.14 depicts a two-dimensional diagram grouping the Apulian samples according to these parameters [17].

8.5 Impedance Analysis

As previously discussed (see Sect. 2.3), impedance measurements, in particular, electrochemical impedance spectroscopy (EIS) can be used as an analytical technique to acquire information of archaeometric interest. Although the use of EIS is essentially focused on the study of metallic objects [32, 33], this technique can also be used in

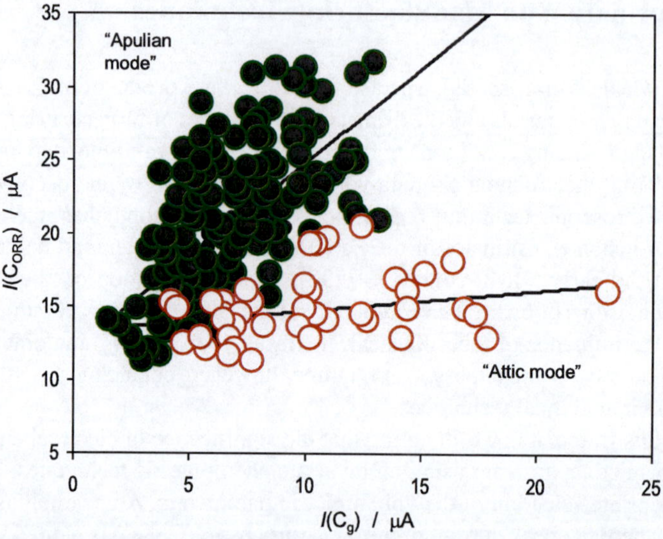

Fig. 8.13 $I(C_{ORR})$ versus $I(C_g)$ representation for samples of Apulian red-figured pottery recorded in experimental conditions such as in Fig. 8.1. Reproduced from ref. [17], with permission

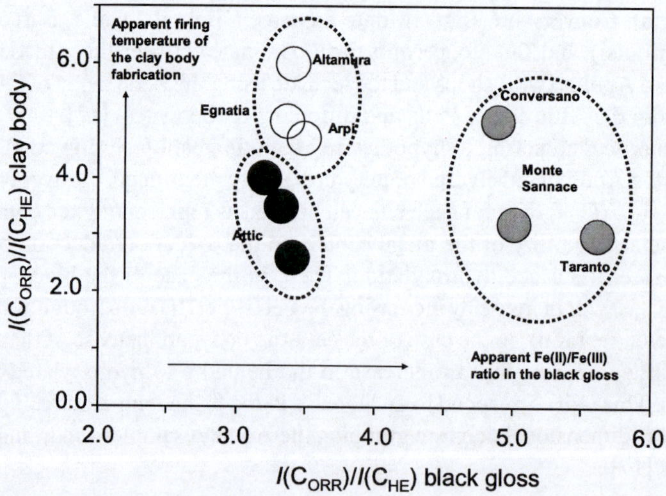

Fig. 8.14 Two-dimensional diagram grouping the Apulian red-figured pottery samples according to averaged $I(C_{ORR})/I(C_H)$ data for the black gloss and the clay body studied in ref. [17]. Adapted from ref. [17], with permission

the study of ceramic materials [18–20]. This is based on the idea that EIS, although sensitive to the chemical/mineralogical composition of solid materials [34], is mainly responsive to their textural (compaction, hardness, crystallinity, porosity) properties. Accordingly, EIS can be viewed as complementary to VIMP that, although influenced by these textural properties, is mainly responsive to the composition of the sample.

The importance of these textural properties is illustrated in Fig. 8.15, where electron microscopy images of cross-sections of different samples of Apulian red-figured pottery are shown. The two first images, corresponding to samples from the archaeological site of Taranto, show the absence **a** and presence **b** of an engobe layer. The other two images illustrate the difference in the thickness of the surface black gloss, thick **c** and thin **d**, respectively. Additionally, one can see that the presence of more or less abundant inclusions of quartz, feldspars and carbonate grains, and the porosity of the ceramic body varies significantly from one to another site.

The main difficulty in applying EIS to ceramic materials, however, derives from the intrinsic difficulty of modeling the impedance spectra. Effectively, in the studied systems, the powdered ceramic material is distributed over the surface of a paraffin-impregnated graphite electrode forming a discontinuous distribution of particles. This implies that charge transfer can occur through different paths involving the bare and sample-covered regions of the base electrode exposed to the electrolyte solution [35–37].

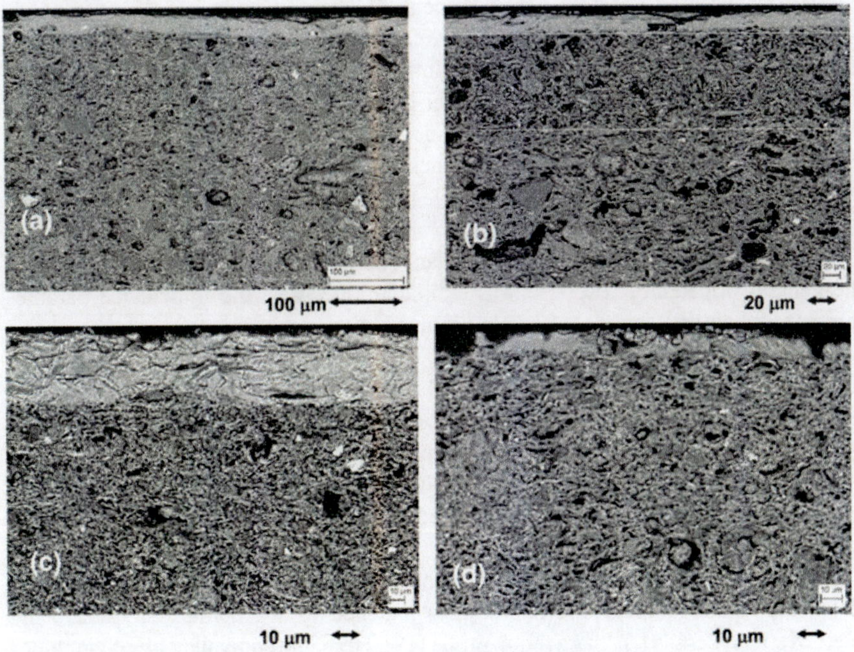

Fig. 8.15 SEM-BSE photomicrographs of cross-sections of Apulian red-figured pottery from **a**, **b** Taranto, **c** Monte Sannace, and **d** Arpi. Reproduced from ref. [18], with permission

Figure 8.16 shows the EIS analysis applied to a series of pottery samples from Caesar Forum (second half of 9th–early 11th CE), the Magna Mater Temple, and Nerva Forum, dating back to the 3rd century and the second half of 9th–early 11th CE. Here, air-saturated 0.10 M H_2SO_4 was used as the electrolyte and the ORR process was used as a redox probe by means of the application of a bias potential of $-$ 0.40 V versus Ag/AgCl. The interval of frequencies was limited to the 10^5 to 10^{-1} Hz range because of the large dispersion recorded at lower frequencies. The Nyquist plot of the negative imaginary component of total impedance, $-Z_{imag}$, versus the real component of the same, Z_{real}, consisted of a capacitive semicircle at high frequencies followed by a depressed capacitive loop at middle and low frequencies, clearly denoting the presence of (at least) two relaxation processes. Consistently, the Bode plots of the (minus) phase angle ($-\varphi$) versus the logarithm of frequency (log$f$) displayed two relative maxima at high and intermediate frequencies. It is pertinent to underline that electrode modification was carried out by means of an abrasive transfer of the ceramic particles to the graphite electrode surface, common with VIMP experiments. This is an aspect to be accounted for in interpreting experimental data (vide infra).

Modeling experimental impedance spectra required relatively complicated equivalent circuits such as depicted as an inset in Fig. 8.16. This circuit was composed of the usual solution resistance (R_s), double-layer capacitance (C_{dl}), and charge transfer resistance (R_{ct}) elements, adding a parallel combination of a capacitor (C_{cer})—eventually replaced by a constant phase element, Q_{cer}— and resistance (R_{cer}) in series with R_{ct}. The R_{ct} plus C_{dl} pair can be associated with charge transfer through the electrolyte/uncovered graphite electrode interface. The R_{cer}, C_{cer} association can be ascribed to charge transfer through the electrolyte/ceramic particle/base electrode three-phase junction. The equivalent circuit was completed with a Warburg element in series with the charge transfer resistance. This element represents the diffusive constraints associated with the interparticle regions interposed between the grains of ceramic material.

As can be seen by comparing the impedance spectra of different samples, although the general impedance response was similar in all cases, the impedance parameters differed among samples. Sample grouping was performed using two empirical parameters: the logarithm of the absolute value of total impedance at the limit of low frequencies (0.10 Hz), logZ_{low}, and the logarithm of the frequency at which a maximum (negative) phase angle was recorded (marked in Fig. 8.16c), logf_{max}, a strategy that parallels those used to discriminate porous oxides of various compositions [36]. The resulting sample grouping is depicted in Fig. 8.17. Comparison with previous mineralogical data [28] suggested that logf_{max} was mainly influenced by the porosity of the ceramic body.

Figure 8.18 shows the impedance data for a sample of Apulian red-figured pottery (from the archaeological site of Taranto) attached to a graphite electrode using $[Fe(CN)_6]^{3-}$/$[Fe(CN)_6]^{4-}$ as a redox probe [19]. Here, the equivalent circuit includes resistance/capacitance associations for charge transfer through the electrolyte/base electrode interface (R_{ct}/C_{dl} plus Warburg impedance) and through the ceramic particles (R_p/Q_p). The model includes an additional resistance/capacitance association

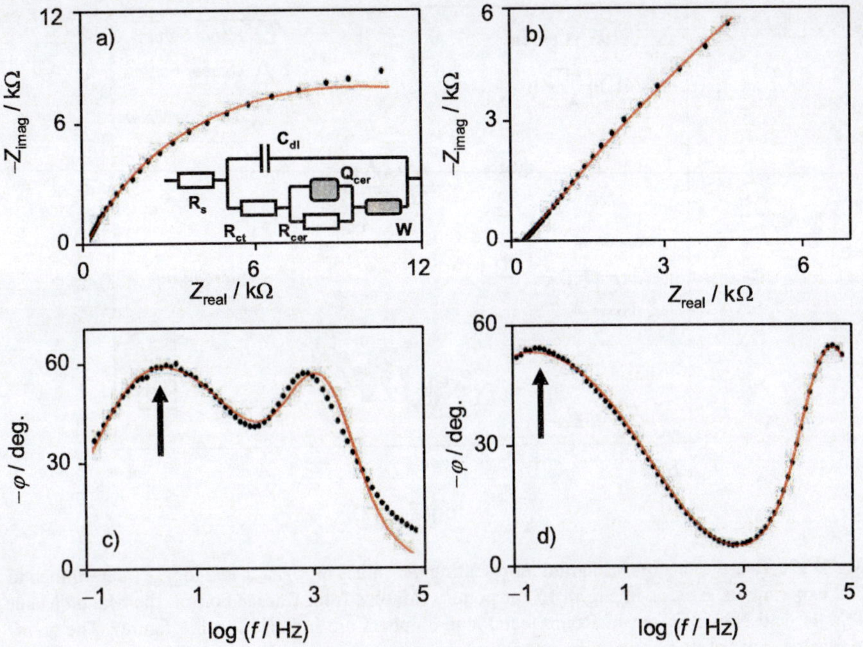

Fig. 8.16 **a, b** Nyquist and **c, d** Bode plots of EIS data for microparticulate deposits of samples **a, c** TMM86 8 (Magna Mater temple) and **b, d** US1425 (Nerva's Forum) attached to graphite electrode in contact with air-saturated 0.1 M H_2SO_4. Bias potential − 0.40 V versus Ag/AgCl. Experimental data points (circles) are superimposed on the continuous lines corresponding to the fit of experimental data to the equivalent circuit depicted as an inset. Arrows mark the maximum (minus) phase angle at low/intermediate frequencies used for grouping purposes. Reproduced from ref. [19], with permission

(R_1/Q_1). The resistance R_1 can be conceived as the charge transfer resistance associated with electron transfer at the electrode/ceramic particle/electrolyte solution three-phase boundary and C_1 represents the capacitive effects associated with charge separation effects in this region. These two parameters were just those displaying systematic variations between the different archaeological sites as can be seen in Fig. 8.19.

Experimental data can satisfactorily be fitted to linear variations of lnR_1 on lnC_1 each one characterizing a given archaeological site. These variations can tentatively be interpreted as resulting from the modification of the graphite/ceramic particles interface due to the more or less intense scratching produced during the abrasive electrode conditioning. This is illustrated in Fig. 8.20. Obviously, ceramic samples containing gross inclusions of hard minerals such as quartz will produce significant scratching and the concomitant modification of the impedance parameters.

Accordingly, the interfacial resistance R_1 can be modeled as caused by a set of cylindrical plates of averaged height L and radius r at the graphite surface. The resistance R of a plate can be expressed as:

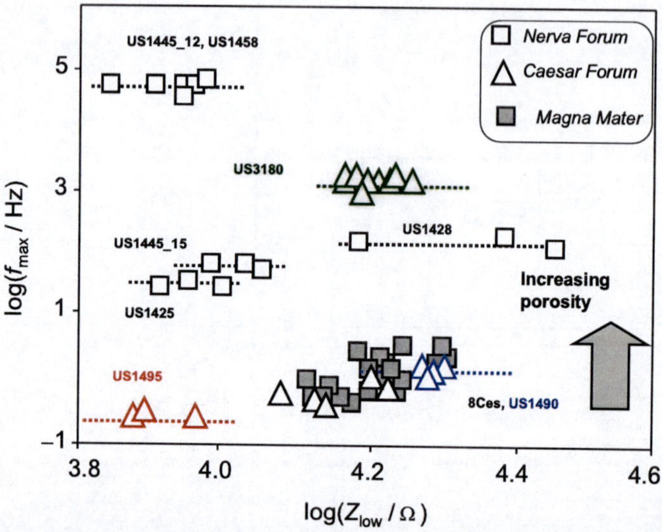

Fig. 8.17 Two-dimensional diagram representing the values of $\log Z_{low}$ and $\log f_{max}$ determined in EIS experiments such as in Fig. 8.15 for pottery samples from Caesar Forum, the Magna Mater Temple, and Nerva Forum in Rome (catalogue numbers are indicated in the figure). The arrow marks the order of porosity determined from mineralogical analysis in ref. [28]. Reproduced from ref. [19], with permission

$$R = \rho \frac{L}{S} \tag{8.11}$$

where ρ is the resistivity and $S\ (= \pi r^2)$ is the surface area of the plate. In the simplest case, the charge separation at the interface can be represented by a capacitor whose capacity C can be expressed as:

$$C = \varepsilon \frac{S}{\delta} \tag{8.12}$$

where ε is the permittivity of the medium interposed between charges and δ is the averaged separation between the surfaces. Eliminating S between Eqs. (8.11) and (8.12) leads to,

$$R = \left(\frac{\rho \varepsilon L}{\delta}\right) \frac{1}{C} \tag{8.13}$$

This equation predicts a linear dependence of R with C^{-1} corresponding to a linear variation of $\ln R_1$ on $\ln C_1$ of slope − 1. This simple approach can be refined considering the rough nature of the generated surfaces. This factor does not significantly influence R_1 but should affect the capacity as described for porous electrodes [36, 38]. This separation from the ideal capacitive behavior is taken into account by constant

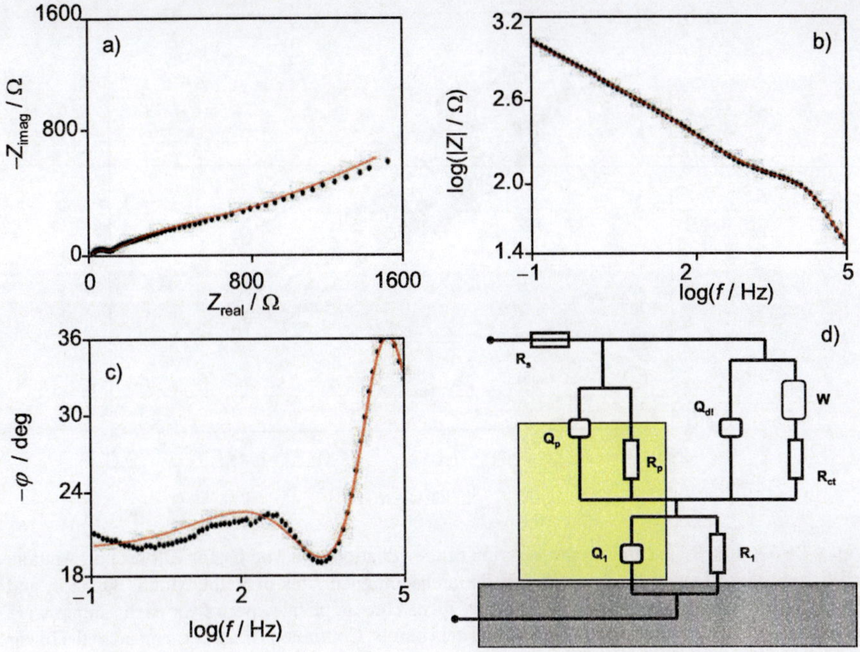

Fig. 8.18 Impedance spectrum of Apulian red-figured pottery sample 227130 from the archaeological site of Taranto attached to graphite electrode in contact with 5 mM $K_3[Fe(CN)_6]/K_4[Fe(CN)_6]$ plus 0.10 M KCl solution. Bias potential 0.20 V versus Ag/AgCl. **a** Nyquist plot; **b** Bode plot of log(modulus of total impedance) versus log(frequency); **c** Bode plot of negative phase angle versus log(frequency); **d** equivalent circuit. Circles: experimental data points; continuous lines: theoretical spectrum using this equivalent circuit. Reproduced from ref. [18], with permission

phase elements (CPEs) whose impedance Z varies with the angular frequency ω as:

$$Z = Q(j\omega)^{-\alpha} \tag{8.14}$$

where Q represents the inverse of the differential capacitance of the interface existing in the ideal case ($\alpha = 1$). This parameter can be related to the effective capacitance of the element (C_{eff}) with an associated resistance R_{eff} by means of the relationship [39]:

$$R_{\text{eff}} = Q^{-1/(1-\alpha)} C_{\text{eff}}^{\alpha/(1-\alpha)} \tag{8.15}$$

This expression corresponds to a potential dependence of R_{eff} on C_{eff} which, in principle, can be seen as a generalization of Eq. (8.13) for a non-ideal capacitive behavior. Experimental data in Fig. 8.18 are in agreement with this formalism identifying R_1 and C_1 with R_{eff} and C_{eff}, respectively.

For our purposes, the relevant aspect to emphasize is that the parameters Q and α in Eq. (8.15) are characterizing the properties of the graphite/ceramic particle

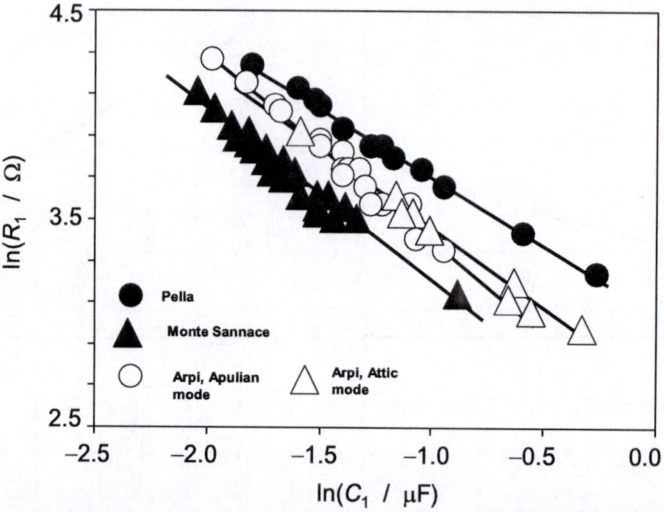

Fig. 8.19 Double logarithmic representation of the variation of R_1 on C_1 for EIS data on Apulian red-figured pottery samples recovered in the archaeological sites of Pella, Monte Sannace, and Arpi, Experimental conditions such as in Fig. 8.18. Three replicate spectra for each sample were performed, all being represented as individual data points. Continuous lines correspond to the linear fit of experimental data. Adapted from ref. [18], with permission

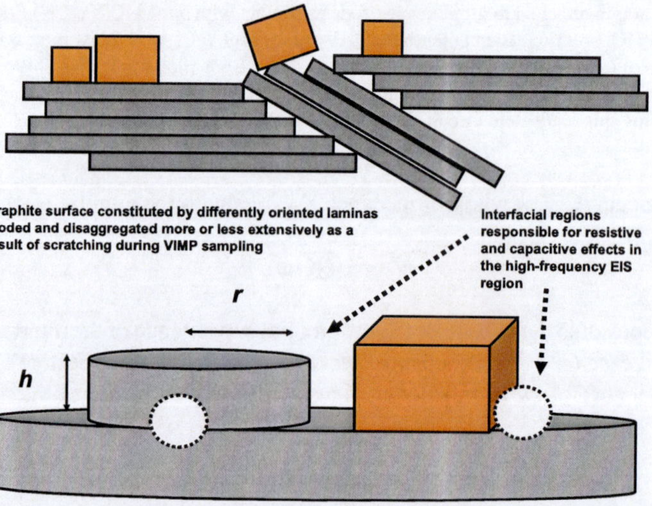

Fig. 8.20 Scheme for the interpretation of the impedance parameters associated with the base electrode/ceramic particles interface. Reproduced from ref. [19], with permission

interface and in turn are depending on the previously mentioned textural properties (hardness, porosity, inclusions, etc.) of the clay body. In summary, these parameters can specify the different ceramic productions.

8.6 Dating

Dating ceramic materials is possible via different methods. Apart from the thermoluminescence and fission tracks 'physical' methods [40], two 'chemical' methods are available. The obsidian hydration method [41] is based on the measurement of the thickness of hydrated layers formed in this volcanic material, formed by the rapid cooling of silica-rich lava upon microscope examination or other techniques [42]. The process occurs under diffusion control so that the increase in thickness of the hydration layer is proportional to the square root of the age (in the following, $t^{1/2}$ law). The rehydroxylation method [43] is based on the fact that the firing process carried out during pottery manufacturing releases chemically combined water in the form of OH hydroxyls from the clay. After cooling, the ceramic material experiences a continuous rehydroxylation process following a $t^{1/4}$ kinetic law. The age can be estimated by combining mass measurements on a ceramic specimen and the same after being dehydroxylated by heating at ca. 500°C and completely rehydroxylated under controlled conditions. Obviously, both methods are sensitive to the conditions of aging (temperature, humidity). Although they are submitted to refinements and corrections, they constitute interesting approaches for dating archaeological ceramics [44, 45].

Just the first dating method based on electrochemistry was proposed for dating ceramics using the VIMP approach [46]. The method was based on the detection of the electrocatalytic effects exerted by ceramic materials on the oxygen evolution reaction (OER) in aqueous alkaline media. The inspiring idea was that the damage caused to ceramic materials by different agents (cosmic radiation, …) will result in the creation of defect sites in the material able to act as catalytic centers.

Apart from the excursion on glass dating based on VIMP monitoring of different iron species [47], treated in the precedent chapter, no other attempts to exploit electrochemical capabilities have been reported.

There is a possibility, however, to combine the philosophy of the rehydroxylation method with that of electrochemical dating. The obvious difficulty is the considerable variety of compositions of ceramic materials and, consequently, the variety of possible electrocatalysts. Here, the focus will be made on hematite electrochemistry based on the sensitivity of its 'direct' voltammetric response and its electrocatalytic effects on ORR to minute changes in the degree of hydration, crystallinity, etc. (see Fig. 8.5) [3].

Firing experiments performed on hematite-containing minerals revealed that the voltammetric response in contact with air-saturated 0.10 M HCl was relatively insensitive to heating until ca. 850°C but varied drastically up to 900°C. As can be seen

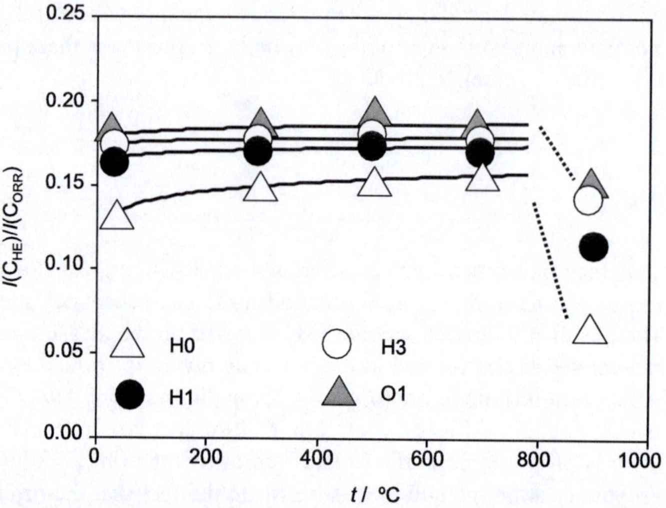

Fig. 8.21 Plots of the $I(C_{HE})/I(C_{ORR})$ ratio versus heating temperature for different hematite-containing materials after thermal treatment for 24 h. H0, Probus reagent; H1 and hematite pigments H1, natural hematite pigment Kremer 48,600; H3 Ercolano red pigment Kremer 41600; O1: French ochre pigment Kremer 40030. Peak current values were measured in replicate square wave voltammetric experiments at sample-modified graphite electrodes immersed into air-saturated 0.10 M HCl in conditions such as in Fig. 8.1. Reproduced from ref. [3], with permission

in Fig. 8.21 [3], the ratio $I(C_{HE})/I(C_{ORR})$ decreases abruptly for all materials above this temperature.

The above results can be interpreted on the basis of kinetic experiments submitting hematite to heating at different temperatures under a uniform rate. As shown in Fig. 8.22, a common $I(C_{HE})/I(C_{ORR})$ versus heating time curve was initially recorded at 500 and 900°C. Under the used experimental conditions, heating at 500°C results in a monotonically rising variation of the peak current ratio. However, in agreement with data in Fig. 8.21, after 1 h of heating, the process at 900°C is accompanied by a drastic decrease of the $I(C_{HE})/I(C_{ORR})$ ratio. These features can be understood by assuming that at temperatures around 500°C, hematite experiences a 'smooth' dehydroxylation (removing resulting in an increase of the $I(C_{HE})/I(C_{ORR})$ ratio with time. Heating above 900°C produces an initial loss of water until complete removal and a second, drastic, dehydroxylation takes place. Now, the dehydroxylated product —presumably non-hydroxylated hematite with high degree of crystallinity— increases notably the electrocatalytic ability toward the ORR process so that the $I(C_{HE})/I(C_{ORR})$ ratio decreases with time. This is in agreement with theoretical DFT studies concluding that hematite surfaces can possess variable degrees of hydroxylation involving chemisorbed water molecules, the hydrogenation of undercoordinated subsurface oxygen atoms, and the formation of hydroxyl groups singly and doubly coordinated to iron centers [49]. The catalytic features summarized in Fig. 8.22 could be attributed to the temperature-dependent sequential loss of these differently hydroxylated species. Tentatively, the catalytic behavior observed upon heating at 500°C results from the loss of singly coordinated hydroxyl units. Upon heating at 900°C, this process is faster and, when singly bound hydroxyl units are exhausted, the loss of doubly bound hydroxyls starts. This last process results in the exposure of 'anhydrous' hematite surfaces to the electrolyte solution and the concomitant drastic change in catalytic activity shown in Fig. 8.22. Additionally, FESEM images of hematite specimens (see Fig. 8.23) showed that the original pigment forms aggregates of irregular crystals which become spheroidal after heating at 900°C [3]. This suggests a process of interparticle sintering consistent with infrared studies of hydroxyl units in hematite and goethite [50, 51] which could also affect the electrocatalytic response of the mineral.

These experiments suggest the possibility of developing dating methodologies based on the idea that during the firing process of manufacturing, hematite becomes lightly (firing temperatures below ca. 900°C) or intensively (firing temperatures above 900°C) dehydroxylated. The mineral will be progressively rehydroxylated with time, the rate of this process being dependent on the hematite content in the ceramic body. Accordingly, a possible evaluation of the age would result from the determination, via electrochemical assays such as in Fig. 8.22, of the rehydroxylation rate. This process, however, actually depends on a variety of factors (composition of the ceramic paste, firing temperature, oxidizing/reducing conditions of firing, time of firing, ...) so a series of refinements are obviously needed. Preliminary data reveal significant differences in the heating curves of 'red' pottery samples of different ages

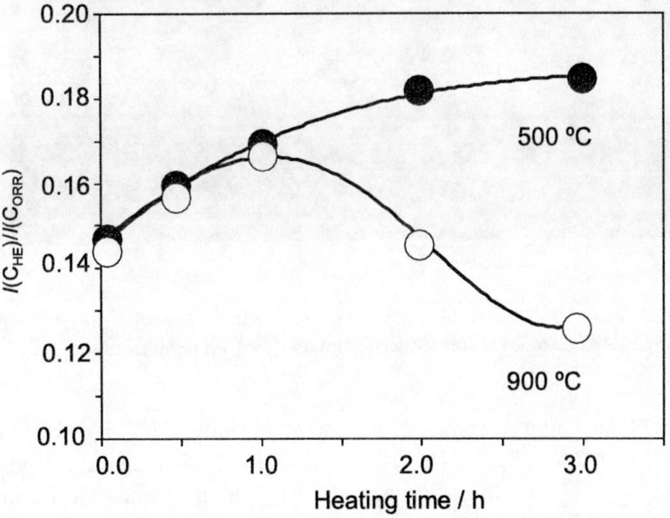

Fig. 8.22 Variation of the $I(C_{HE})/I(C_{ORR})$ ratio with the heating time in voltammograms of hematite-modified graphite electrodes in contact with air-saturated 0.10 M HCl (experimental conditions such as in Fig. 8.1) for specimens uniformly heated in a furnace at 500 and 900°C for several times [48]

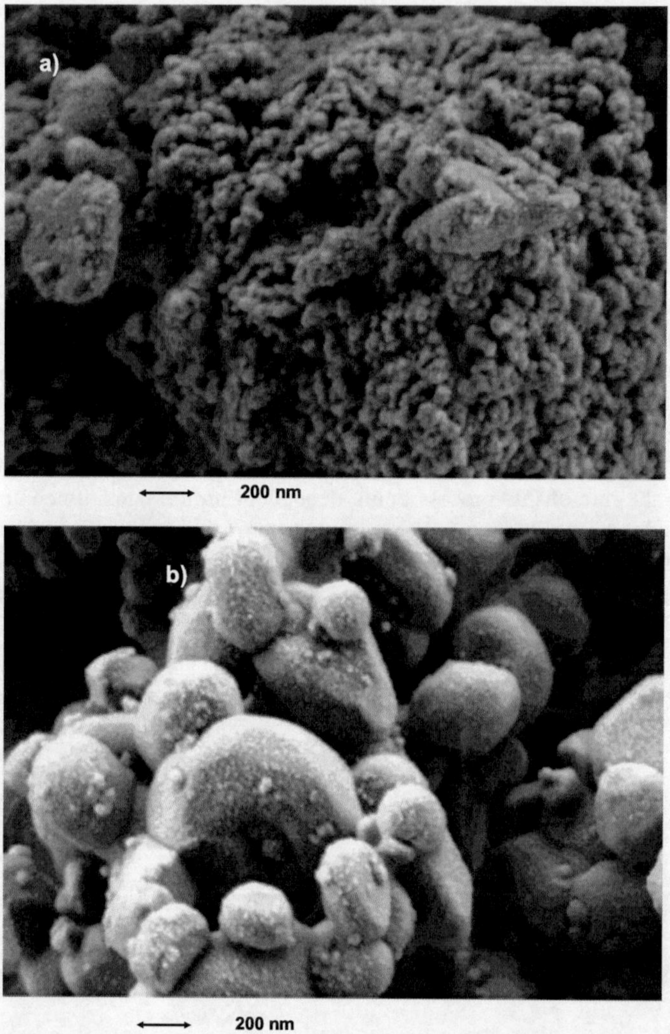

Fig. 8.23 FESEM images of carbon-coated samples of natural hematite pigment **a** before and **b** after heating at 900°C for 24 h. Reproduced from ref. [3], with permission

(see Fig. 8.24 for pottery samples from Altamura, Italy, 5th–4th century BCE, La Bastida de les Alcusses, Spain, 4th century BCE, and Xàtiva, Spain, 8th–9th century CE, squares). One can see that the heating curves exhibit a profile more complicated than that recorded for hematite, thus denoting the complexity of the ceramic system. Current research is aimed to interpret such data and evaluate their use for dating purposes.

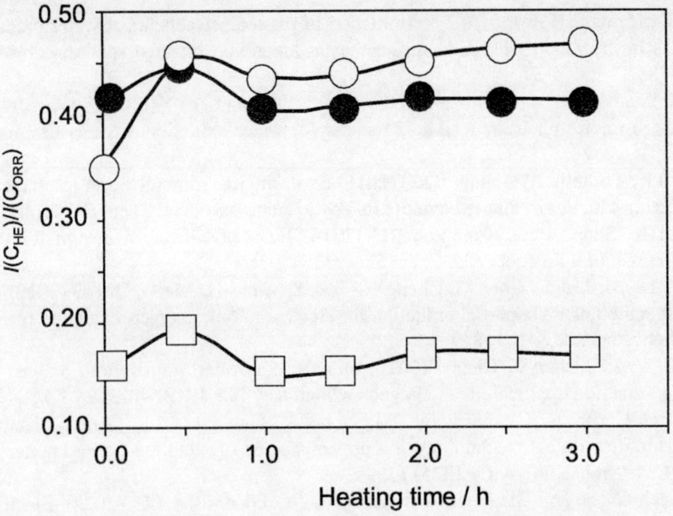

Fig. 8.24 Variation of the $I(C_{HE})/I(C_{ORR})$ ratio with the heating time in voltammograms of sample-modified graphite electrodes in contact with air-saturated 0.10 M HCl (averaged values for five replicate measurements in experimental conditions such as in Fig. 8.1) for 'red' pottery samples from Altamura, Italy (5th–4th century BCE, solid circles), La Bastida de les Alcusses, Spain (4th century BCE, circles), and Xàtiva, Spain (8th–9th century CE, squares), uniformly heated in a furnace at 900°C for several times [48]

References

1. Rice PM (1987) Pottery analysis: a sourcebook. The University Chicago press, Chicago
2. Karamanov A, Pelino M (2001) Crystallization phenomena in iron-rich glasses. J Non-Cryst Solids 281:139–151
3. Doménech-Carbó A, Giannuzzi M, Mangone A, Giannossa LC, Di Turo F, Cofini E, Doménech-Carbó MT (2022) Hematite, an electrocatalytic marker for the study of archaeological clay bodies: a VIMP and SECM study. ChemElectroChem 9:59–68
4. Tite MS, Bimson M, Freestone IC (1982) An examination of the high gloss surfaces finishes on Greek Attic and Roman Samian wares. Archaeometry 24:117–126
5. Maniatis Y, Aloupi E, Stalios AD (1993) New evidence for the nature of the Attic black gloss. Archaeometry 35:23–34
6. Mirti P, Perardi A, Gulmini M (2006) A scientific investigation of the provenance and technology of a black-figure amphora attributed to the Priam group. Archaeometry 48:31–43
7. Thorn J, Glascock M (2010) New evidence for Apulian red-figure production centres. Archaeometry 52:777–795
8. Noked M, Soffer A, Aurbach D (2011) The electrochemistry of activated carbonaceous materials: past, present, and future. J Solid State Electrochem 15:1563–1578
9. Le Formal F, Grätzel M, Sivula K (2010) Controlling photoactivity in ultrathin hematite films for solar water-splitting. Adv Funct Mater 20:1099–1107
10. Cummings CY, Marken F, Peter LM, Upul Wijayantha KG, Tahir AAJ (2011) New insights into water splitting at mesoporous α-Fe_2O_3 films: a study by modulated transmittance and impedance spectroscopies. J Am Chem Soc 134:1228–1234

11. Zandi O, Hamann TW (2016) Determination of photoelectrochemical water oxidation intermediates on haematite electrode surfaces using operando infrared spectroscopy. Nat Chem 8:778–783
12. Tang C, Sun B, Li M, Zhang J, Fan X, Gao F, Tong Y, Dong L, Li Y (2019) Surface hydroxylated hematite promotes photoinduced hole transfer for water oxidation. J Mater Chem A 7:8050–8054
13. Shimizu K, Tschulik K, Compton RG (2016) Exploring the mineral-water interface: reduction and reaction kinetics of single hematite (α-Fe_2O_3) nanoparticles. Chem Sci 7:1408–1414
14. Shimizu K, Sepunaru L, Compton RG (2016) Innovative catalyst design for the oxygen reduction reaction for fuel cells. Chem Sci 7:3364–3369
15. Wan H, Lv M, Liu X, Chen G, Zhang N, Cao Y, Wang H, Ma R, Qiu G (2019) Activating hematite nanoplates via partial reduction for electrocatalytic oxygen reduction reaction. ACS Sust Chem Eng 7:11841–11849
16. Zhao Q, Yan Z, Chen C, Chen J (2017) Spinels: controlled preparation, oxygen reduction/evolution reaction application, and beyond. Chem Rev 117:10121–10211
17. Doménech-Carbó A, Giannuzzi M, Mangone A, Giannossa LC (2022) Electrochemical methods to discriminate technology and provenance of Apulian red-figured pottery. I: VIMP and SECM. Archaeometry 64:1325–1339
18. Doménech-Carbó A, Giannuzzi M, Mangone A, Giannossa LC (2022) Electrochemical methods to discriminate technology and provenance of Apulian red-figured pottery. II: EIS. Archaeometry 64:1124–1137
19. Di Turo F, Montoya N, Piquero-Cilla J, De Vito C, Coletti F, De Luca I, Doménech-Carbó A (2018) Electrochemical discrimination of manufacturing types of pottery from Magna Mater Temple and Fora of Nerva and Caesar (Rome, Italy). Appl Clay Sci 162:305–310
20. Doménech-Carbó A, Latorre-Riveros L, Huanasponcco W, Quispe D, Doménech-Carbó MT, Cabrera CR, Gutiérrez MC, Carmona K, Pérez A (2019) Solid-state electrochemical analysis of Inka pottery from Qotakalli archaeological site in the Cusco (Perú) area. J Solid State Electrochem 23:1541–1552
21. Fabrizi L, Nigro L, Cappella F, Spagnoli F, Guirguis M, de Villedry AMN, Doménech-Carbó MT, De Vito C, Doménech-Carbó A (2020) Discrimination and provenances of Phoenician Red Slip Ware using both the solid state electrochemistry and petrographic analyzes. Electroanalysis 32:258–270
22. Ramaciotti M, Gallello G, Navarro D, Doménech-Carbó A, Roldán C, Hernández E, Garrigues S, Pastor A (2020) An innovative multi-analytical approach based on spectroscopic and electrochemical techniques to study a complex Roman amphorae collection. Appl Clay Sci 198:105857
23. Mirti P, Gulmini M, Perardi A, Davit P, Elia D (2004) Technology of production of red figure pottery from Attic and southern Italian workshop. Anal Bioanal Chem 380:712–718
24. Mangone A, Giannossa LC, Colafemmina G, Laviano R, Traini A (2009) Use of various spectroscopy techniques to investigate raw materials and define processes in the overpainting of Apulian red figured pottery (4th century BC) from southern Italy. Microchem J 92:97–102
25. Mangone A, Caggiani MC, Giannossa LC, Eramo G, Redavid V, Laviano R (2013) Diversified production of red figured pottery in Apulia (Southern Italy) in the late period. J Cult Herit 14(1):82–88
26. Robinson E (2014) The early phases of Apulian red figure. In: Schierup S, Sabetai V (eds) The regional production of red-figure pottery: Greece, Magna Graecia and Etruria. Aarhus University Press, Aarhus, pp 217–233
27. Bikai PM (1978) The pottery of Tyre. Aris and Phillips, Warminster, p 75
28. De Vito C, Medeghini L, Garruto S, Coletti F, De Luca I, Mignardi S (2018) Medieval glazed ceramic from Caesar's Forum (Rome, Italy): production technology. Ceram Int 44:5055–5062
29. Walton M, Trentelman K, Cummings M, Poretti G, Maish J, Saunders D (2013) Material evidence for multiple firings of ancient Athenian red-figure pottery. J Am Ceram Soc 96:2031–2035

30. Doménech-Carbó A, Sánchez-Ramos S, Doménech-Carbó MT, Gimeno-Adelantado JV, Bosch-Reig F, Yusá-Marco DJ, Saurí-Peris MC (2002) Electrochemical determination of the Fe(III)/Fe(II) ratio in archaeological ceramic materials using carbon paste and composite electrodes. Electroanalysis 14:685–696
31. Sánchez-Ramos S, Bosch-Reig F, Gimeno-Adelantado JV, Yusá-Marco DJ, Doménech-Carbó A (2002) Application of XRF, XRD, thermal analysis and voltammetric techniques to the study of ancient ceramics. Anal Bioanal Chem 373:893–900
32. Cano E, Lafuente D, Bastidas DM (2010) Use of EIS for the evaluation of the protective properties of coatings for metallic cultural heritage: a review. J Solid State Electrochem 14:381–391
33. Ramírez-Barat B, Cano E (2018) In situ electrochemical impedance spectroscopy measurements and their interpretation for the diagnostic of metallic cultural heritage: a review. Chem Electro Chem 5:2698–2716
34. Redondo-Marugán J, Piquero-Cilla J, Doménech-Carbó MT, Ramírez-Barat B, Al-Sekheneh W, Capelo S, Doménech-Carbó A (2017) Characterizing archaeological bronze corrosion products intersecting electrochemical impedance measurements with voltammetry of immobilized particles. Electrochim Acta 246:269–279
35. Retter U, Widmann A, Siegler K, Kahlert H (2003) On the impedance of potassium nickel(II) hexacyanoferrate(II) composite electrodes—the generalization of the Randles model referring to inhomogeneous electrode materials. J Electroanal Chem 546:87–96
36. Chen G, Waraksa CC, Cho H, Macdonald DD, Mallouk TE (2003) EIS Studies of porous oxygen electrodes with discrete particles. I. Impedance of oxide catalyst supports. J Electrochem Soc 150:E423–E428
37. Bredar ARC, Chown AL, Burton AR, Farnum BH (2020) Electrochemical impedance spectroscopy of metal oxide electrodes for energy applications. ACS Appl Energ Mater 3:66–98
38. Go J–Y, Pyun S–I (2007) A review of anomalous diffusion phenomena at fractal interface for diffusion-controlled and non-diffusion-controlled transfer processes. J Solid State Electrochem 11:323–334
39. Hirschhorn B, Orazem ME, Tribollet B, Vivier V, Frateur I, Musiani M (2010) Determination of effective capacitance and film thickness from constant-phase-element parameters. Electrochim Acta 55:6218–6227
40. Doménech-Carbó A (2015) Dating—an analytical task. ChemTexts 1:5
41. Friedman I, Smith RL (1960) A new dating method using obsidian: part I, the development of the method. Amer Antiq 25:476–522
42. Liritzis I, Diakostamatiou M (2002) Towards a new method of obsidian hydration dating with secondary ion mass spectrometry via a surface saturation layer approach. Archaeometry 2:3–20
43. Savage S, Wilson MA, Carter MA, Hoff WD, Hall C, McKay B (2008) Moisture expansion and mass gain in fired clay ceramics: a two-stage $(time)^{1/4}$ Process. J Phys D Appl Phys 41:055402
44. Liritzis I, Laskaris N (2011) Fifty years of obsidian hydration dating in archaeology. J Non-Cryst Solids 357:2011–2023
45. Hamilton A, Hall C (2012) A review of rehydroxylation in fired-clay ceramics. J Am Ceram Soc 95:2673–2678
46. Scholz F, Schröder U, Meyer S, Brainina KHZ, Zakharchuk NF, Sobolev NV, Kozmenko OA (1995) The electrochemical response of radiation defects of non-conducting materials. An electrochemical access to age determinations. J Electroanal Chem 385:139–142
47. Doménech-Carbó A (2012) Chapter 7: Electrochemical techniques. In: Edwards HGM, Vandenabeele P (eds) Analytical archaeometry, selected topics. The Royal Society of Chemistry, pp 210–241
48. Doménech-Carbó A, Giannuzzi M, Mangone A, Giannossa LC, Mai-Cerovaz C, Doménech-Carbó MT. Unpublished results
49. Guo H, Barnard AS (2011) Surface structure and environment-dependent hydroxylation of the nonpolar hematite (100) from density functional theory modeling. J Phys Chem C 115:23023–23029

50. Rendón JL, Serna CJ (1981) IR spectra of powder hematite: effects of particle shize and shape. Clay Miner 16:375–381
51. Ruan HD, Frost RL, Kloprogge JT (2001) The behavior of hydroxyl units of synthetic goethite and its dehydroxylated product hematite. Spectrochim Acta A 57:2575–2586

Additional Literature

52. Andaloro E, Belfiore CM, De Francesco AM, Jacobsen JK, Mittica GP (2011) A preliminary archaeometric study of pottery remains from the archaeological site of Timpone della Motta, in the Sibaritide area (Calabria—southern Italy). Appl Clay Sci 53:445–453
53. Bowen PK, Drelich J, Scarlett TJ (2013) Modeling rehydration/rehydroxylation mass-gain curves from davenport ceramics. J Am Ceram Soc 96:885–891
54. Chatfield M (2010) Tracing firing technology through clay properties in Cuzco, Perú. J Archaeol Sci 37:727–736
55. Coletti F (2012) La ceramica invetriata di età tardoantica a Roma: nuovi dati da recenti scavi stratigrafici. Rei Cretariae Romanae Fautorum Acta 42:181–191
56. De Benedetto GE, Laviano R, Sabbatini L, Zambonin PG (2002) Infrared spectroscopy in the mineralogical characterization of ancient pottery. J Cult Herit 3:177–186
57. De Luca I (2006) Ritrovamenti dei secoli IX-X dai Fori di Cesare e di Nerva. In: Meneghini R, Santangeli Valenzani R (eds) Roma. Lo scavo dei Fori Imperiali 1995–2000. Contesti ceramici, Ecolefracais de Rome, Roma, pp 92–108
58. De Vito C, Medeghini L, Mignardi S, Orlandi D, Nigro L, Spagnoli F, Lottici PP, Bersani D (2014) Technological fingerprints of black-gloss ware from Motya (Western Sicily, Italy). Appl Clay Sci 88–89:202–213
59. Fysh SA, Fredericks PM (1983) Fourier transform infrared studies of aluminous goethites and hematites. Clays Clay Miner 31:377–382
60. Gallet Y, Le Goff M (2015) Rehydration and rehydroxylation in ancient ceramics: new constraints from mass gain analyzes versus annealing temperaturas. J Am Ceram Soc 98:2738–2744
61. Guirguis M (2005) Storia degli studi e degli scavi a Sulky e a Monte Sirai. Riv Studi Fenici 33:13–31
62. Klahr B, Hamann T (2014) Water oxidation on hematite photoelectrodes: insight into the nature of surface states through in situ spectroelectrochemistry. J Phys Chem C 118:10393–10399
63. Maggetti M, Galetti G (1981) Archäometrische Untersuchungen und spätlateinezeitlicher Keramik von Base-Gasfabrik und Sissach-Brühl. Archäologisches Korrespondenzblatt 11:321–328
64. Mangueira GM, Teixeira S, Silva FA, Franco RWA (2016) Fe^{3+} in pottery: distinction of the use for cooking and production parameters. Appl Clay Sci 129:88–91
65. Medeghini L, Lottici PP, De Vito C, Mignardi S, Bersani D (2014) Micro-Raman spectroscopy and ancient ceramics: applications and problems. J Raman Spectrosc 45:1244–1250
66. Nigro L (2013) Before the Greeks: the earlist phoenicien settlement in Motya—recent discoveries by Rome "La Sapienza" expedition. Vicino Oriente XVII:39–74
67. Niveau-de-Villedary y Mariñas AM (2018) Gadir revisited. A proposal for reconstruction of the Archaic Phoenician foundation. Vicino Oriente 32:91–109
68. Sillar B (2000) Dung by preference: the choice of fuel as an example of how Andean pottery production is embedded within wider technical, social and economical practices. Archaeometry 42:43–60
69. Stevenson C, Novak SW (2011) Obsidian hydration dating by infrared spectroscopy: method and calibration. J Archaeol Sci 38:1716–1726
70. Unali A (2017) Sulky—Sant'Antioco. In: Guirguis M (a cura di): La Sardegna fenicia e punica. Sassari, pp 128–142

Chapter 9
Organic Materials

9.1 Introduction

There is a variety of organic materials playing a role in cultural heritage, from structural and decorative components of buildings, through furniture, textiles, pigments, bindings, and varnishes, to paper, etc. We can include within this extensive group tar and food residuals and charcoal found on archaeological sites. The majority of the materials of interest are of vegetal origin; for this reason, the electrochemistry of organic materials associated with cultural heritage productions is closely related to so-called vegetal electrochemistry [1], but also an 'animal electrochemistry' could be developed for leather, animal food residuals, etc.

The electrochemical study of organic pigments and binding media was treated in Sects. 5.4, 5.5, and 6.2 while hybrid organic–inorganic materials were studied in Sect. 6.3. In this chapter, we focus our attention on ligneous materials. Although there are till now only a few works devoted to analyzing the rich and complex electrochemistry of these systems in the archaeometric domain [2–4], this can be considered a promising research field that should be exploited in the future.

9.2 Vegetal Electrochemistry

The electrochemistry of plants has a wide variety of aspects, from electrophysiology to membrane potentials and ion transport studies. The first approaches to an electroanalytical view of plants were aimed at the voltammetric characterization of taxonomic groups and the establishment of phylogenetic relationships [5, 6]. These were based on the methodology of VIMP. The experimental protocol consisted of recording the voltammetric response of microparticulate deposits of products resulting from the evaporation of ethanolic (or other solvents) extracts of plants. These extracts were prepared by maceration of leaves, seeds, roots, etc., with the organic solvent in an

A. Doménech-Carbó and M. T. Doménech-Carbó, *Electrochemistry for Cultural Heritage*, Monographs in Electrochemistry,
https://doi.org/10.1007/978-3-031-31945-7_9

agate mortar. Then, a drop of the extract was transferred onto the surface of an inert electrode (typically, glassy carbon) and the solvent was allowed to evaporate in air. As a result, a thin film of solid microparticles of vegetal components scarcely soluble in water was formed onto the base electrode. The final step consisted of recording of the voltammetric response of that deposit in contact with an aqueous electrolyte (typically acetate and phosphate buffers).

Figure 9.1 shows an example of the obtained voltammetric response for ethanolic extracts of leaves of a) *Euonymus japonicus* (family Celastraceae), b) *Lobularia maritima*, c) *Koelreuteria bipinnata*, and d) *Begonia acerifolia*, four plants of the Celastraceae, Brassicaceae, Sapindaceae, and Begoniaceae families, respectively [5]. One can see that the voltammetric profiles were clearly different allowing for characterization of taxons using bivariant and multivariant chemometric methods. Depending on the vegetal groups, this electrochemotaxonomy can arrive at the level of species or even varieties whereas cluster analysis permits the construction of electrophylogenetic trees comparable with those determined from the genetic analysis [5, 6].

The recorded voltammetric responses are mostly due to the oxidation of different vegetal components. These include metabolites and structural components of plants; among others, flavonoids, flavonols, flavones, indigoids, antocyannins, and lignins. Here, the voltammetric analysis is based on the responses associated with lignin

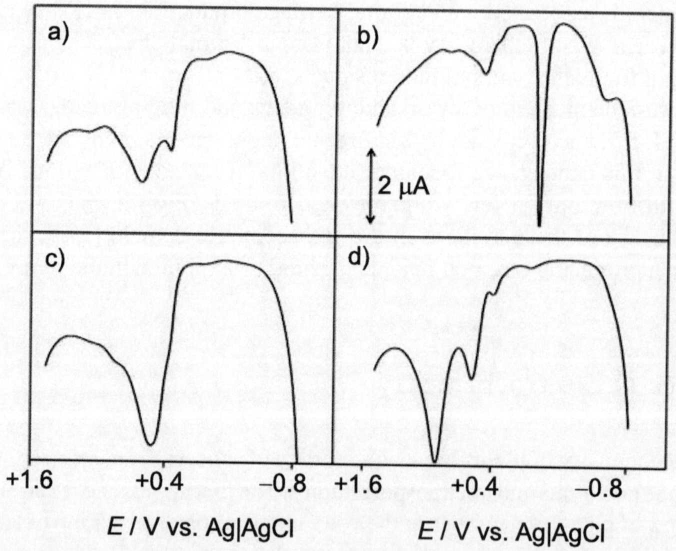

Fig. 9.1 Square wave voltammograms of microparticulate deposits on glassy carbon electrode of ethanolic extracts of leaves of **a** *Euonymus japonicus*, **b** *Lobularia maritima*, **c** *Koelreuteria bipinnata*, and **d** *Begonia acerifolia* plants. Electrolyte: 0.25 M aqueous acetate buffer, pH 4.75. Potential scan initiated at − 1.05 V versus Ag/AgCl in the positive direction; potential step increment 4 mV; square wave amplitude 25 mV; frequency 5 Hz. Reproduced from ref. [5], with permission. Note that the axes are opposite to the IUPAC convention

components. Lignins are three-dimensional aromatic polymer/oligomer products whose biosynthesis results from an enzyme-mediated dehydrogenative polymerization of three phenylpropanoid monomers: coumaryl, coniferyl, and sinapyl alcohol [7]. Lignins possess an amorphous hydrophobic structure that gives the polymer a mechanical strength and can undergo reactions of substitution, oxidation, and hydrolysis often resulting in depolymerization leading to the formation of aromatic monomers. Lignins are electroactive, yielding well-defined voltammetric responses [7, 9, 10]. Figure 9.2a shows schematically the electrochemical oxidation of syringic acid, one of the structural subunits of lignins. Flavonoid-type compounds have a widely studied electrochemistry in both solution and solid state ([11–14], see also Chap. 5). Figure 9.2b illustrates the oxidation of the 3', 4'-dihydroxybenzoic moiety of baicalein, a representative component of the group [2]. Typically, *o*-catechol units are oxidized to the corresponding *o*-quinones, but there is a possibility of competing oxidative pathways.

As can be seen in Fig. 9.1, the oxidation processes take place, typically, in the − 0.5 to 1.0 V versus Ag/AgCl region of potentials and exhibit a variable degree of reversibility. Naturally, pristine compounds with quinonic units can yield cathodic signals. Since the majority of the electroactive components are partially soluble in water, proton-assisted redox processes can involve 'pure' solid-to-solid transformations and reductive/oxidative dissolutions.

An interesting aspect of the above 'plant-VIMP' is that there is a possibility of expanding its analytical capabilities by exploiting the electrochemical generation

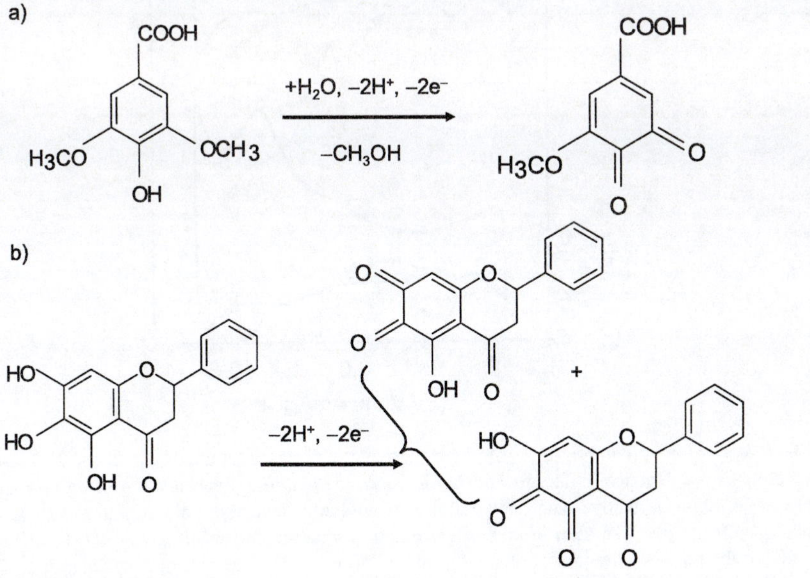

Fig. 9.2 Scheme for the electrochemical oxidation of **a** syringic acid and **b** baicalein. Reproduced from ref. [2], with permission

of reactive oxygen species (ROS) [15, 16]. This possibility has been explored by different research groups [17–19] and is based on the evidence that during the electrochemical reduction of oxygen dissolved in aqueous electrolytes, superoxide radical anion, hydroperoxide radical, and hydroxyl radical can be generated as intermediate species [20].

Figure 9.3 shows the cyclic voltammograms of ethanolic extracts of leaves of seedlings from two species of the Inulae family (*Dittrichia viscosa* and *Limbarda crithmoides*) in contact with air-saturated acetate buffer at pH 4.75 [16]. In these voltammograms, the potential scan initiated at 0.0 V versus Ag/AgCl in the positive direction so that in the initial scan a series of anodic peaks (A_A) were recorded. These can be attributed to the oxidation of pristine lignins, flavonoids, etc., components of leaves. Some of such peaks are coupled with cathodic counterparts (C_A) recorded in the subsequent cathodic scan.

These signals precede the prominent cathodic wave at ca. − 0.6 V corresponding to the reduction of dissolved oxygen (C_{ox}). In the next (2nd) anodic scan, several A_A anodic peaks diminish or even vanish and additional peaks (A_B) appear. Since

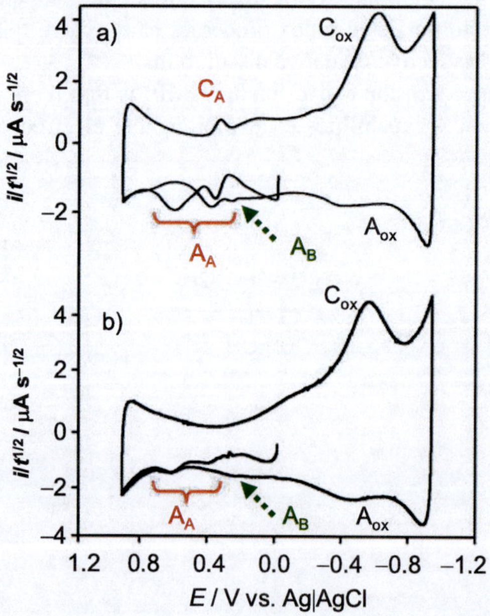

Fig. 9.3 Cyclic voltammograms of microparticulate films of ethanolic extracts of leaves of seedlings of **a** *Dittrichia viscosa* and **b** *Limbarda crithmoides* deposited on the glassy carbon electrode. Semi-derivative data convolution was applied. Electrolyte: air-saturated 0.25 M HAc/NaAc buffer at pH 4.75 after 7 days of germination. Potential scan initiated at 0.0 V versus Ag/AgCl in the positive direction; potential scan rate 50 mV s^{-1}. The green arrows mark the signals appearing after electrochemical ROS generation. Reproduced from ref. [16], with permission. Note that the axes are opposite to the IUPAC convention

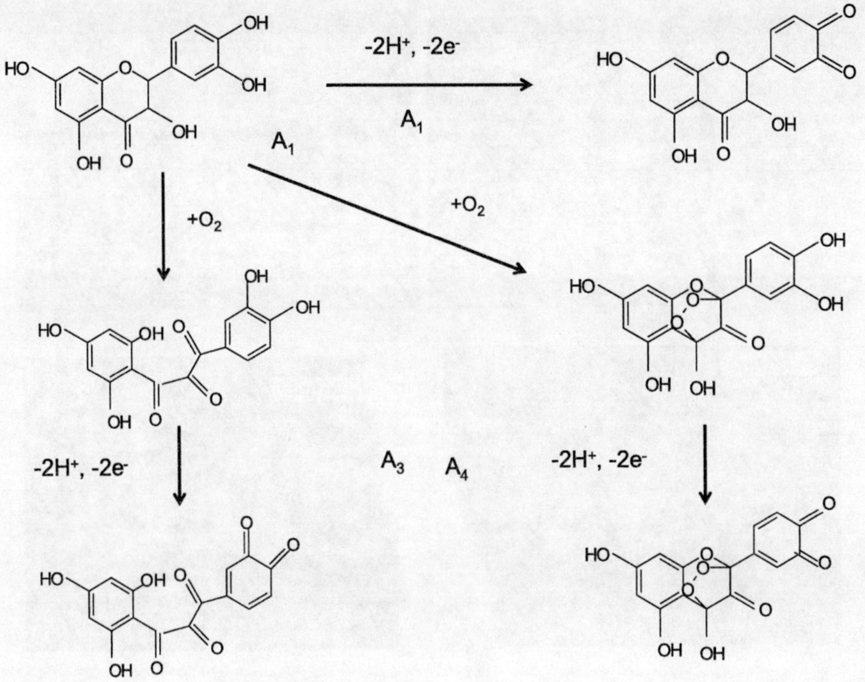

Fig. 9.4 Scheme for the electrochemical oxidation of quercetin and one of its products of reaction with ROS. Reproduced from ref. [15], with permission

in the Cox process a series of reactive ROS have been generated electrochemically, these new anodic signals can be attributed to the oxidation of products of the reaction between ROS and the original plant components. The reaction of ROS with flavonoids can yield oxygen-bridged species [21] which in turn, as depicted in Fig. 9.4, can undergo electrochemical oxidation [15]. For analytical purposes, the aspect to underline is that electrochemical ROS generation can produce new voltammetric signals able to be used for taxonomical characterization and reactivity studies [1].

9.3 Wooden Objects

A representative image of wood samples studied with the above methodology is shown in Fig. 9.5, corresponding to three altarpieces fabricated in the late 15th century and the beginning of the 17th century in different locations of the Comunitat Valenciana (Spain). The extraction/deposition protocol was applied to ca. 0.5 mm needles extracted with the help of a scalpel from the inner parts of the piece either being exposed (see Fig. 9.5a) or covered by ground layers (see Fig. 9.5b). In the first

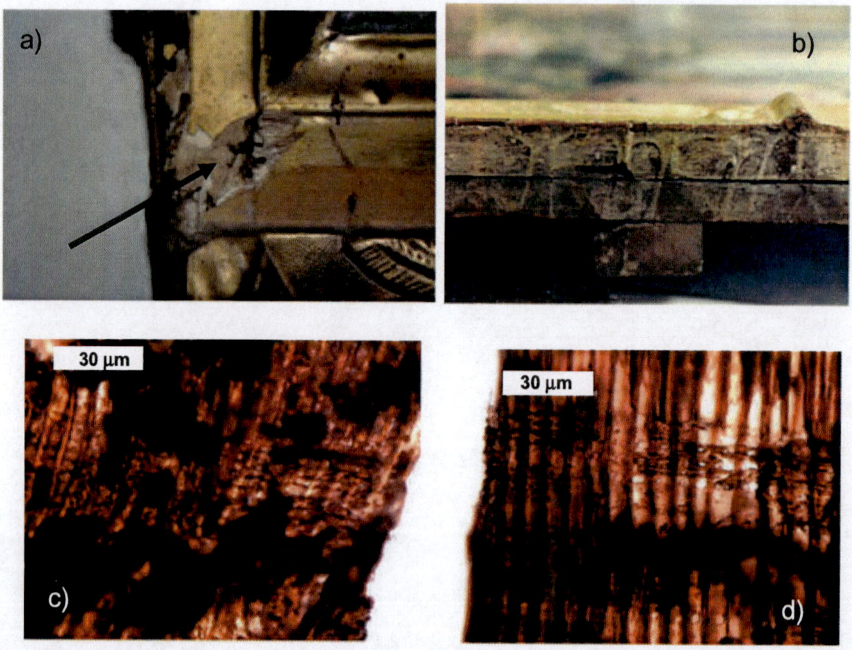

Fig. 9.5 Photographs of regions of the altarpieces: **a** *Taules de Sant Vicent, Sant Jaume* church, Algemesí, Spain, workshop of Francisco Ribalta, 1603–1605) and **b** *Retaule de les Ànimes, Sant Martí* church, Sogorbe, Spain, workshop of Francisco Pérez, ca. 1625). **c**, **d** Optical microscopy images of the radial section of the sample from the first altarpiece **c** and a sample from the *Retaule de la Verge de Gràcia, Sant Miquel Arcángel* church (Enguera, Spain, Paolo de San Leocadio, 1482–1484) **d**. The arrow marks the region of the first altarpiece where the studied sample was taken. Reproduced from ref. [2], with permission

case, sampling was allowed in the fractured region whereas, in the second, almost all accessible wood surface area was covered by ground layers of the golden panel covering the wooden structure.

The study was aimed at identifying the vegetal species of wood. This analytical task is routinely performed by anatomical observation. This defines the anthracological methods based on the study of anatomical details in radial and cross-sections of wood samples [22, 23]. This methodology requires highly specialized technicians, and the taxonomical level (family, genus, species) often depends on the quality of the sample (conservation, size). The anatomical similarities between different taxonomic groups can limit the identification to the genus or family or even larger levels (for instance, limiting the discrimination to distinguish between Conifers and Angiosperms). In this context, the term taxon is frequently used to design an identification unit regardless the 'true' taxonomical level of identification [24, 25].

In the case of wood samples in Fig. 9.5, optical microscopy images under diascopic illumination with parallel polarized light of radial sections show a set of large pinoid pits occupying practically the entire cross-field. The analysis of anatomical details permitted the identification of the *Pinus sylvestris* species in samples in Fig. 9.5c, d [2].

Figure 9.6 compares the square wave voltammetric responses recorded for microparticulate deposits resulting from the ethanolic extracts of four samples extracted from different regions of a wooden cabinet dated back to the 15th century using 1.0 M H_2SO_4 as the electrolyte. Three replicate measurements on freshly modified electrodes are superimposed for each sample. The wood material was identified as pertaining to *Pinus* species and the voltammetric records exhibited reasonable repeatability for each sample. All samples exhibited the same series of voltammetric peaks but there were clear differences between the relative intensity of such peaks in samples extracted from different locations of the same furniture. The samples extracted from external, air-exposed regions displayed ill-defined profiles as compared to the inner samples, thus suggesting that the level of exposition to the environment plays a crucial role in the voltammetric response.

This aspect is of crucial importance because it indicates that the comparison between different samples must take into account not only the taxonomic group but also the details of the manufacturing process (use of protective varnishes, for instance) and the exposure to the environment.

The identification strategy can be improved by combining the data from extracts using different organic solvents. Figure 9.7 shows the cyclic voltammetry of microparticulate films from ethanolic and acetone extracts of *Quercus robur* wood in contact with air-saturated aqueous acetate buffer. As in Fig. 9.3, the voltammograms show a series of anodic peaks in the 0.0 to 1.0 V region of potentials (A_{pp}) accompanying the prominent ORR wave (C_{ox}). Again, the generation of ROS during this last process determines changes in the A_{pp} region in the second anodic scan.

Under these circumstances, and despite the aforementioned difficulties, judicious sampling can provide reasonable electrochemical identification of wood samples. This can be seen in Fig. 9.8 where the linear potential scan voltammograms of ethanolic extracts of *Quercus robur* and *Quercus pyrenaica* from the Repository of the Universitat Politècnica de València (Spain) are compared with wood samples from a study desk, anonymous, 17th–18th century (Oaxaca, Mexico) and a coffered ceiling of the *Igreja do Salvador*, anonymous, 14th century (Coimbra, Portugal). A comparison of these voltammograms suggests that the archaeological samples can be respectively applied to each one of the *Quercus* species [2].

A refined identification can be obtained using bivariant and multivariant chemometric methods using a matrix of data constituted by the currents measured at intervals of 4 mV, $I(E)$, in voltammograms such as in Fig. 9.8. Each voltammogram provides a set of n currents which can be normalized relative to the maximum current measured in each voltammetric record, i(max). As a result, a set of n normalized currents, $I(E)$, ($=i(E)/i(\text{max})$) is obtained for each electrolyte solution and/or each extracting solvent. If m solvents (or m electrolytes) are used, a matrix of data of dimensions $m \times n$ is obtained and chemometric methods can be applied to it as

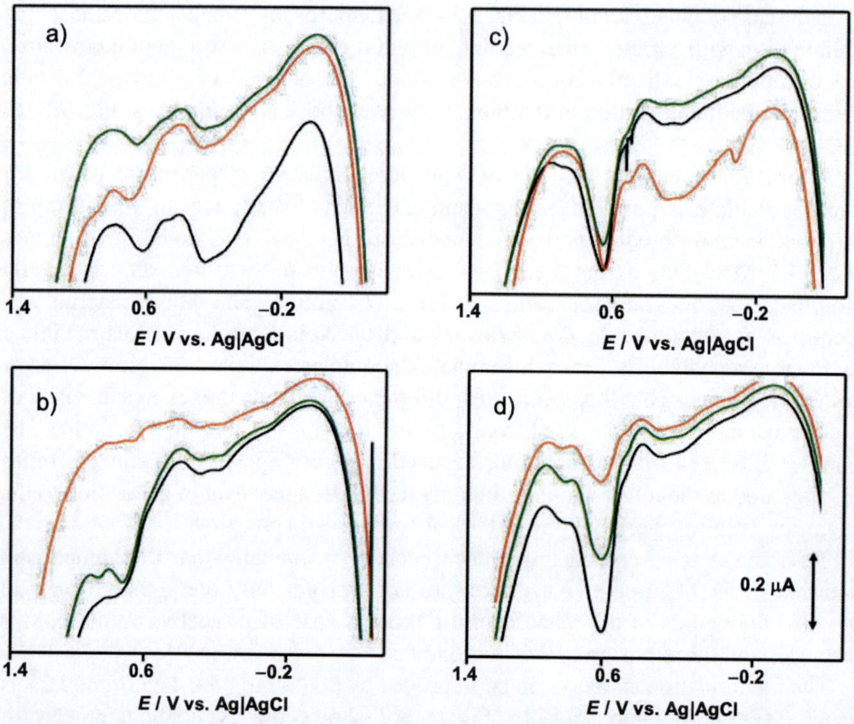

Fig. 9.6 Square wave voltammograms of microparticulate deposits on glassy carbon electrode resulting from the ethanolic extracts of samples from different **a**, **b** inner and **c**, **d** external regions of a cabinet (*bargueño*) dated back to the 15th century in contact with air-saturated 1.0 M H_2SO_4. Three replicate measurements are superimposed for each sample. Potential scan initiated at − 0.75 V versus Ag/AgCl in the positive direction; potential step increment 4 mV; square wave amplitude 25 mV; frequency 5 Hz. Note that the axes are opposite to the IUPAC convention

well as matching strategies to compare samples and reference woods [2]. A simple, visual strategy can be used by comparing the two-dimensional diagrams in which the $I(E)$ values in two different solvents (or two different electrolytes) are plotted [5, 26]. Figure 9.9 depicts several of these representations using the normalized currents obtained for microparticulate deposits of ethanol, acetone, and chloroform extracts (I(EtOH, I(acetone), I(chloroform), respectively) of *Quercus robur* and *Olea europea*.

As is usual in VIMP, the ratios between selected pairs of peak currents can also be used for grouping wood samples. Figure 9.10 shows how the ratios between the currents at 0.80 V (I(800)) and 0.45 V (I(450) and the maximum current in linear potential scan voltammograms in conditions such as in Fig. 9.8 permits to discriminate between different taxonomic groups of woods.

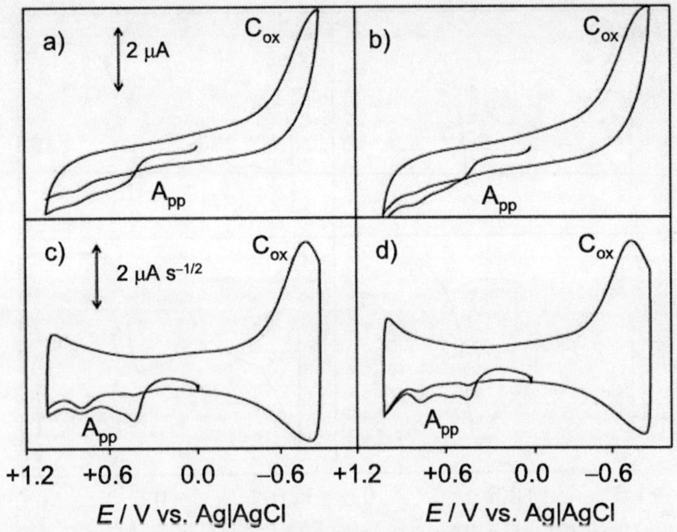

Fig. 9.7 Cyclic voltammograms of microparticulate deposits from extracts of *Quercus robur* wood on glassy carbon electrode in contact with air-saturated 0.25 M aqueous acetate buffer, pH 4.75. Voltammograms before (**a**, **b**) and after (**c**, **d**) application of semi-derivative data convolution of **a**, **c** ethanolic and **b**, **d** acetone extracts. Potential scan initiated at 0.0 V in the positive direction; potential scan rate 50 mV s^{-1}. Reproduced from ref. [2], with permission. Note that the axes are opposite to the IUPAC convention

9.4 Paper

Paper is an important material in cultural heritage. It supports manuscripts, artworks photographs, as well as decorative elements [27]. In particular, the high historic value of ancient manuscripts [28] has prompted the interest in studying their degradation [29], deacidification [30], and cleaning [31]. These studies are classically focused on spectroscopic techniques [32], but electrochemical techniques are of obvious potential interest as suggested by the study on papers used in conservation and restoration [4].

Ancient papers were prepared from vegetal materials such as mulberry, rice, bamboo and hemp, and jute, mostly from used rags. At the beginning of the 19th century, the paper was prepared at a large scale from wood pulp, but this increased the problems related to paper quality and degradation. The ancient papers were constituted exclusively or almost exclusively by cellulose, making them more resistant over time than modern papers. This is due to the fact that non-cellulosic substances (lignin, hemicellulose, resins, gums) are much more easily attacked by chemical reagents than cellulose [33]. Additionally, modern papers are submitted to bleaching, thus increasing the presence of other non-cellulosic substances. Moreover, kaolinite or gypsum can be added to the paper surface to improve whiteness, smoothing, and printability.

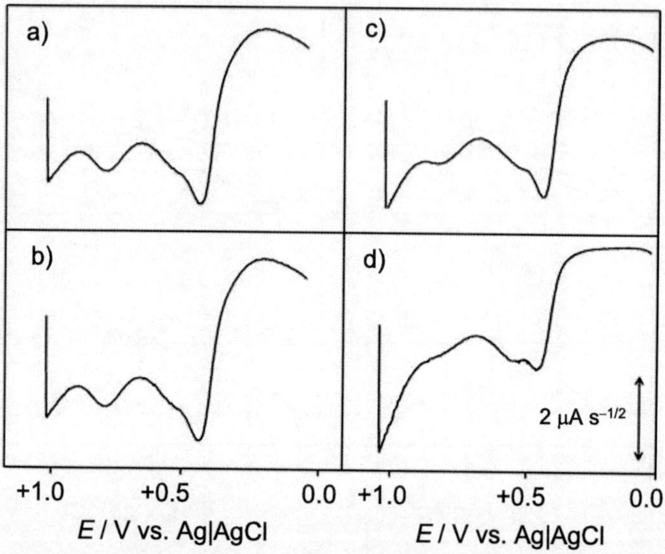

Fig. 9.8 Linear potential scan voltammograms of microparticulate films of ethanolic extracts of: **a** *Quercus robur*, **b** *Quercus pyrenaica*; and samples from **c** a study desk, anonymous, 17th–18th century (Oaxaca, Mexico), and **d** a coffered ceiling of the *Igreja do Salvador*, anonymous, 14th century (Coimbra, Portugal). Semi-derivative convolution was applied to voltammetric data. Electrolyte: air-saturated 0.25 M aqueous acetate buffer, pH 4.75. Potential scan initiated at 0.0 V in the positive direction; potential scan rate 50 mV s^{-1}. Reproduced from ref. [2], with permission. Note that the axes are opposite to the IUPAC convention

Among the wide variety of papers currently used, the so-called Japanese paper is of interest in the field of conservation and restoration where it is used in the restoration of graphic documents, the performance of laminations and consolidations, and the protection of pictorial work and photographs. In archaeology, it is used as an intermediate protective layer and/or in chemical treatments or consolidation. Figure 9.11 shows a sequence of photographic images illustrating the successive steps in the reintegration of paper in a watercolor drawing [4].

Commercial reintegration papers come from China, Japan, and Korea where the traditional production of this type of paper has different local traditions [34]. The electrochemical discrimination between different commercial types was carried out following a protocol similar to that used for wooden artifacts. A sample of 1 cm^2 of paper was macerated with 0.5 mL of acetone in an agate mortar adding, and a fraction of the resulting suspension was allowed to evaporate at the air over the surface of a glassy carbon electrode. As in the case of woods, repeatability and reproducibility require that the base electrode was cleaned and conditioned accurately. 1.0 M H_2SO_4 was used as the electrolyte. Figure 9.12 shows the voltammetric profiles recorded between 0.0 and 1.45 V versus Ag/AgCl in cyclic voltammograms of four different commercial paper specimens [4]. The voltammograms consist of a series of anodic peaks at essentially the same potentials but differing in their relative intensities.

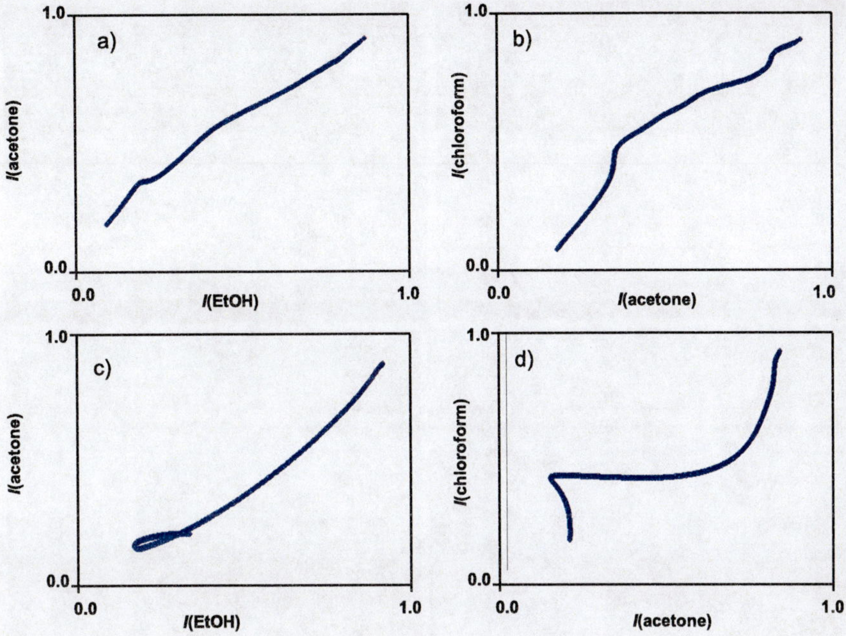

Fig. 9.9 Two-dimensional diagrams representing the values of the normalized currents in linear potential scan voltammograms extracts of: **a**, **b** *Quercus robur* and **c**, **d** *Olea europea* in conditions such as in Fig. 9.8 combining data (see text) of extracts in ethanol, acetone, and chloroform

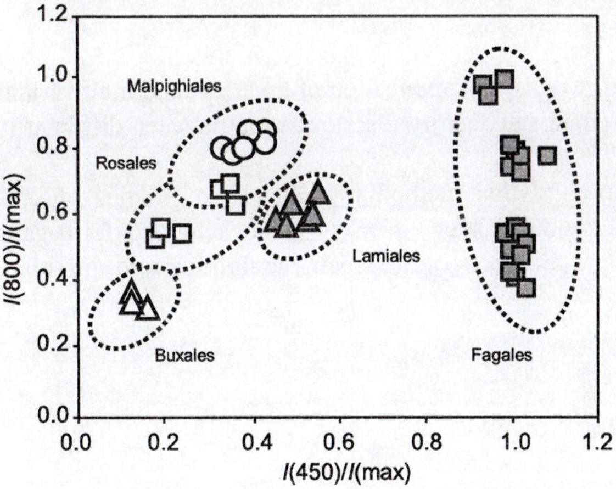

Fig. 9.10 Representation of $I(800)/i(\max)$ versus $I(450)/i(\max)$ determined in linear potential scan voltammograms, after semi-derivative convolution, of ethanolic extracts of wood samples in conditions such as in Fig. 9.8. Data of three replicate experiments for each species are represented. Reproduced from ref. [2], with permission

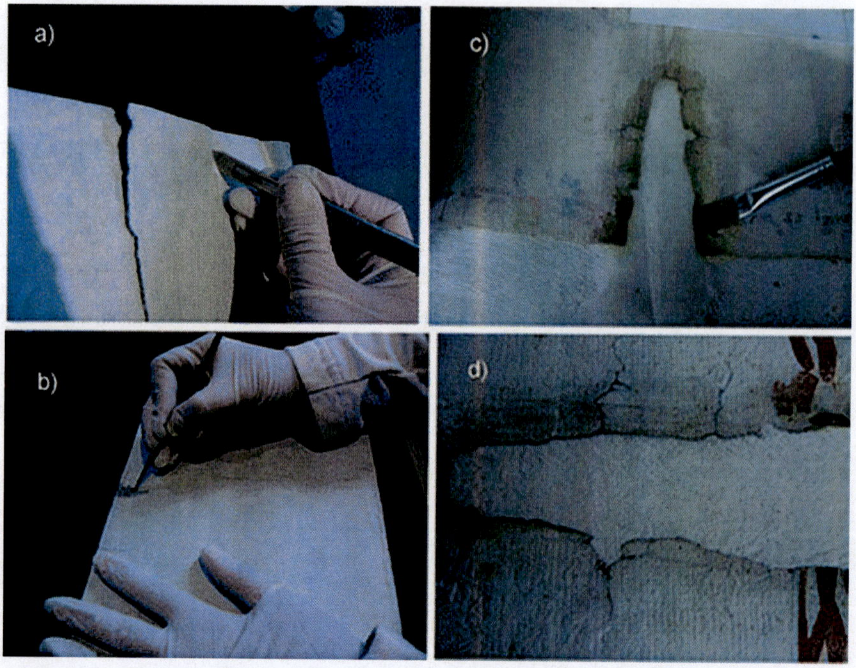

Fig. 9.11 Photographic sequence illustrating the process of elaboration of graft for the reintegration of paper in a watercolor drawing. **a** Determine the graft form, **b** defragment of Japanese paper, **c** adhesion of the fragment, and **d** final result of the reintegration process. Reproduced from ref. [4], with permission

As in the case of woods, the application of bivariant and multivariant chemometric techniques permitted satisfactory discrimination between different types of papers [4].

The application of this methodology to other varieties of paper as well as directly related products such as inks (gallic acid and ferro-gallic derivatives, typical components are electroactive) seems straightforward and subject of the future research.

9.5 Charcoal

Charcoal can be defined as the product resulting from the carbonization of vegetal matter at relatively high temperatures under quasi-anaerobic conditions [35]. Its composition significantly depends on the starting material and the conditions of carbonization that, in principle, must be up to 300°C. Charcoal (eventually from animal origin) is frequent in archaeological sites and constitutes an important source of archaeometric information [24], in particular, with regard to paleoenvironmental

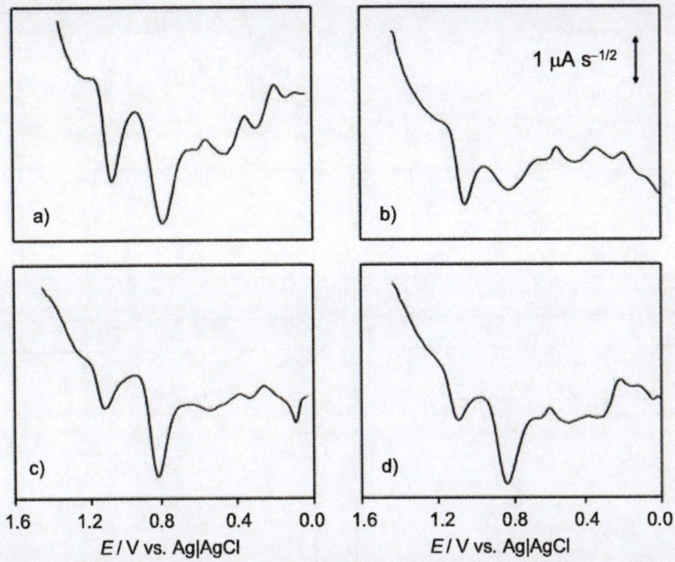

Fig. 9.12 Semi-derivative convolution of the forward scan of a cyclic voltammogram of microparticulate deposits from acetone extracts of different Japanese paper specimens transferred onto glassy carbon electrode in contact with 1.0 M H_2SO_4. **a** Pulp type, **b** kozo type, **c** Manila type, and **d** mitsumata type. Semi-derivative convolution was applied to voltammetric data. Potential scan initiated at 0.0 V in the positive direction; potential scan rate 50 mV s^{-1}. Only the region of potentials between 0.0 and 1.45 V is shown. Reproduced from ref. [4], with permission. Note that the axes are opposite to the IUPAC convention

[36] and paleotechnological data [37]. Additionally, the study of charcoal facilitates a better selection of samples for radiocarbon (^{14}C) dating.

One of the main aspects involved in the scientific examination of charcoal is the elucidation of the taxonomic affiliation of the original vegetal matter. This is, of course, the subject of anthracological analysis [23] complemented with other techniques, in particular infrared spectroscopy [35]. It is important to recognize that, although temperature, including maximum temperature, is an important parameter in charcoal chemistry, the fire intensity, which is the time-averaged energy flux in W m^{-1}, [38] and the concept of charring intensity [39] are parameters to be accounted for.

Figure 9.13 shows the attenuated total reflectance – Fourier transform infrared (ATR-FTIR) spectra of a sample of wood, identified as *Pinus halepensis* species, extracted from a Valencian altarpiece fabricated in 1728, and a series of charcoal samples also anthracologically identified as the proceeding from the same species.

The bands between 3600 and 2900 cm^{-1} are mainly related to OH, CH_3, CH_2, and CH groups and their intensity decreases on increasing the temperature of the pyrolysis due to the loss of water impregnating the cell wall of wood, the oxidation or hydrolysis of OH groups of hemicelluloses and the alteration of the crystallinity of cellulose. The band at 1600 cm^{-1} can be assigned to C = O bonds of carboxyl

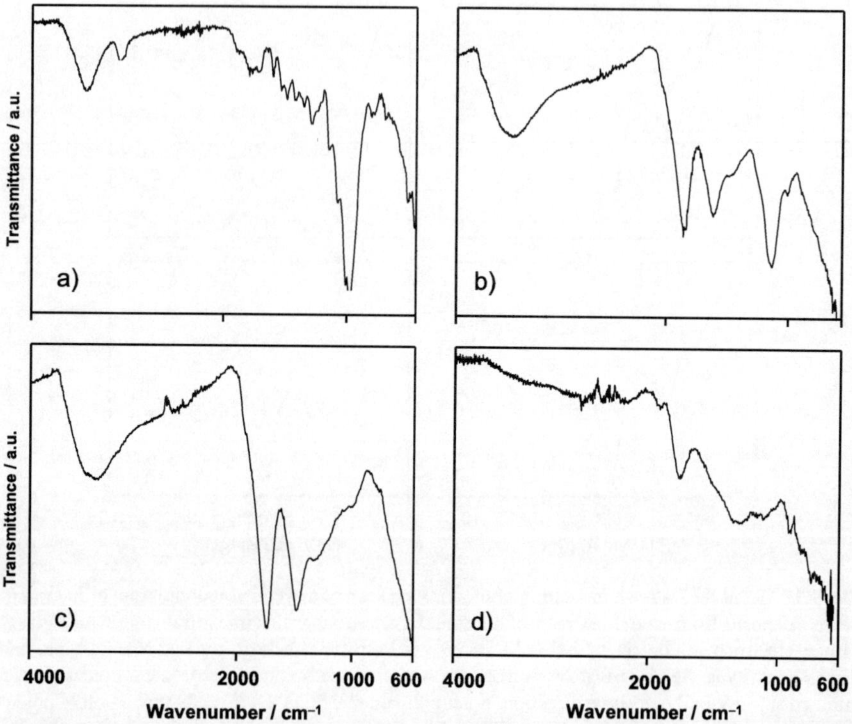

Fig. 9.13 ATR-FTIR spectra of **a** wood sample from the altarpiece in *the Sant Josep* church in Torrent, Spain, 1728; **b–d** charcoal samples from the repository of the Department of Archaeology of the University of Valencia (Spain). **b** Neolithic site of Costamar-Torre de la Sal, Cabanes, Spain ca. 4500 BCE; **c** Iberian site of La Bastida de les Alcusses, Moixent, Spain, ca. 2450 BCE; **d** site of Castell, Villena, Spain, ca. 1600 CE. All specimens were identified as *Pinus halepensis*

acids and the vibration of ketones of lignin and hemicelluloses and to the vibration of $C = C$ groups of aromatic rings, usually related to the presence of lignin. This also results from the formation of aromatic compounds originated by the elimination of hydrogen and oxygen of aliphatic compounds during the pyrolysis. This band is much more intense for charcoal than for wood. The carbonyl band at 1740 cm^{-1} is of low intensity in charcoal, a feature that can be due to the cleavage of acetyl/acetoxy groups in xylan.

The bands between 1460 and 1420 cm^{-1} can be attributed to monomer methoxyl groups of lignin whereas the bands in the 1375 to 1325 cm^{-1} region can be associated with $C - H$ vibration of polysaccharides, hemicelluloses, and lignin components of wood. The bands at 1240 and 1110 cm^{-1} can be assigned to oxygenated groups of hemicelluloses. These bands, as well as all bands at wavenumbers lower than 1000 cm^{-1} are diminished or even eliminated in charcoal relative to wood with the increase of pyrolysis temperature, thus indicating the degradation of wood components.

The infrared spectra in Fig. 9.13 denote the high complexity of the charcoal composition. The differences among the spectra in Fig. 9.13b–d can be attributed to the different temperatures (or better, charring intensity) of the carbonization process.

These differences should be also detectable in the voltammetric response as revealed from available data [40]. In turn, in cases where the carbonization conditions were quite similar, the discrimination of species is difficult from ATR-FTIR spectra, as illustrated in Fig. 9.14 for two charcoal samples identified as *Pinus brutia* and *Pinus pinea* species, respectively [40].

The application of the voltammetric strategies used for wood and paper to charcoal produces, however, promising results, as can be seen in Fig. 9.15. Here, the voltammetric response of microparticulate deposits on the glassy carbon electrode of the ethanolic extracts of charcoal samples in Fig. 9.14 in contact with 0.10 M H_2SO_4 aqueous solution is depicted. As in the case of wood and paper, well-defined anodic signals are recorded in the potential range between 0.0 and 1.4 V versus Ag/AgCl. The anodic peaks for both species appear at the same potential but the relative peak intensities vary from one to another sample. Although studying a wide set of samples is needed to substantiate the influence of the charring intensity and the

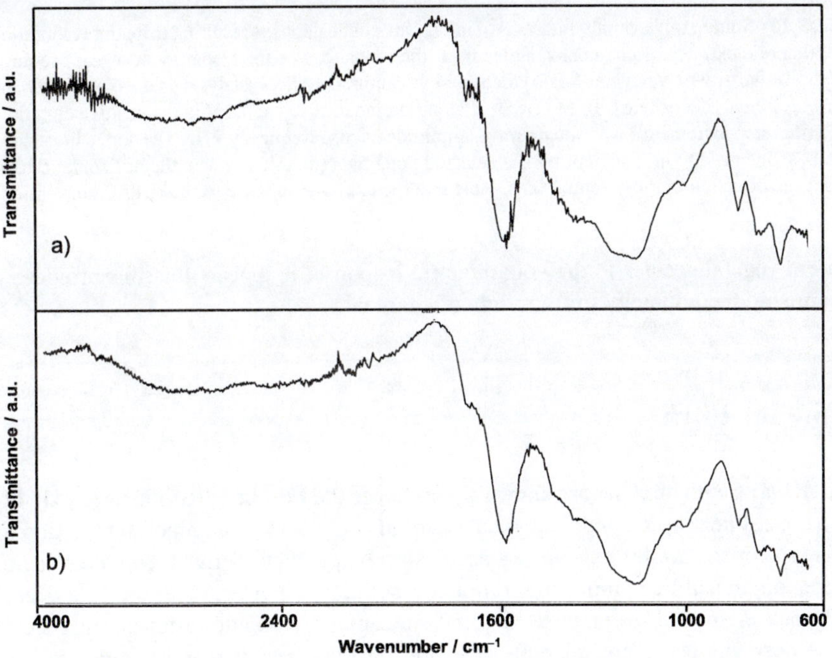

Fig. 9.14 ATR-FTIR spectra of charcoal samples from the Repository of the Department of Archaeology of the University of Valencia (Spain) identified via anthracological analysis as **a** *Pinus brutia* and **b** *Pinus pinea*

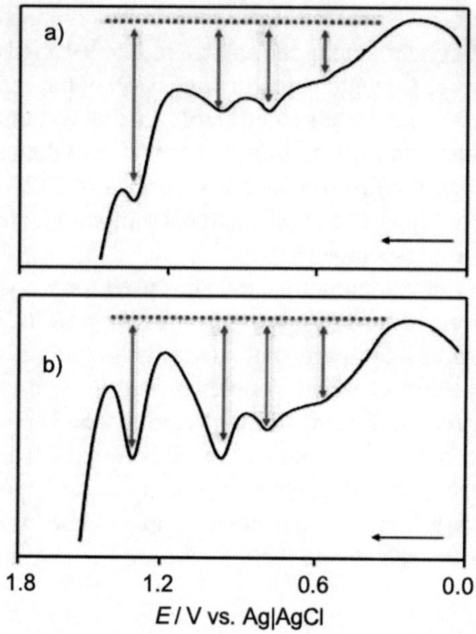

Fig. 9.15 Square wave voltammograms of microparticulate deposits resulting from the evaporation of ethanolic extracts of charcoal samples from the repository of the Department of Archaeology of the University of Valencia (Spain) identified via anthracological analysis as **a** *Pinus brutia* and **b** *Pinus pinea.* Electrolyte 0.10 M H_2SO_4. Potential scan initiated at 0.0 V in the positive direction; potential step increment 4 mV; square wave amplitude 25 mV; frequency 5 Hz. The dotted line marks the baseline used to measure currents at selected potentials (double arrows); the horizontal arrows mark the direction of the potential scan. Note that the axes are opposite to the IUPAC convention

parent vegetal species in the voltammetric response, it appears that this provides a promising methodology for the study of charcoal.

9.6 Tar Pitch

Birch barks with birch tar are known at least since the Middle Paleolithic age [41, 42] and constitute relevant archaeological materials [43]. Today, an important fraction of this type of material is extensively used as asphalt cement derived from petroleum. Here, organosulfur components (aliphatic sulfides, alkyl aryl sulfides, alkyl aryl sulfones, elemental sulfur, disulfides and mercaptans, aromatic sulfones, and others) have been subject to considerable attention for their role as marker compounds to identify the source of a spilled crude oil, with obvious environmental and forensic applications [44].

There is a variety of organosulfur compounds in technological and archaeological asphalt-type materials, but their aging involves the reaction of thiophenes and

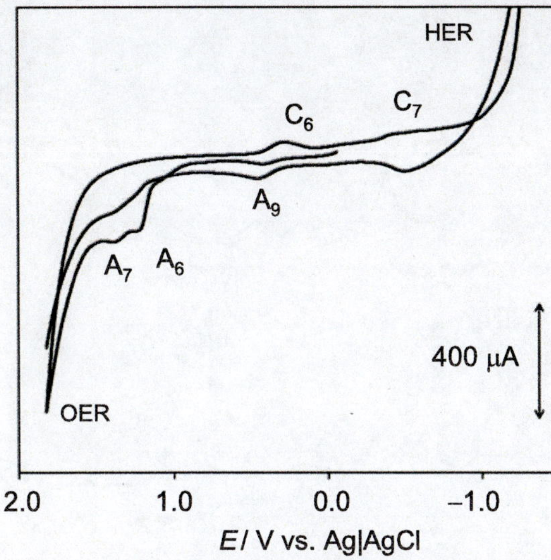

Fig. 9.16 Cyclic voltammogram of a graphite electrode modified with 3–phenyl thiophene in contact with 0.10 M H_2SO_4 aqueous solution. Potential scan initiated at 0.0 V in the positive direction; potential scan rate 50 mV s^{-1}. Reproduced from ref. [48], with permission. Note that the axes are opposite to the IUPAC convention

sulfides with oxygen leading to sulfoxides and sulfones [45, 46]. Due to the multi-component nature of asphalts, chromatographic methods have been widely applied for analytical purposes. Additionally, many electroanalytical studies in the solution phase have been described for detecting organosulfur compounds in petroleum and petroleum derivatives [47]. Asphalts and tar pitch can also be studied using solid-state electrochemistry by means of VIMP techniques [48].

Figure 9.16 shows the cyclic voltammogram recorded for a deposit of 3–phenyl thiophene on a graphite electrode in contact with 0.10 M H_2SO_4 solution [48]. The voltammogram shows a series of peaks positioned between the extreme HER and OER processes. The main anodic processes can be attributed to the successive oxidations to the corresponding sulfoxide and sulfone (labeled in the figure as A_6 and A_7), whereas the C_6/A_9 couple can correspond to a solution-phase process with the formation of a radical cation [48].

The study of asphalt aging combining VIMP and thermal stress and photochemical stress normalized aging protocols revealed the existence of a variety of possible processes, schematized for thiophenes in Fig. 9.17 [49]. For our purposes, it is particularly relevant to emphasize the electrochemical formation of polythiophenes mimicking that promoted by ultraviolet light and the possible involvement of ROS in aging processes.

In this context, VIMP was used to characterize organosulfur components in tar pitch ovens found at the Spanish archaeological sites of *Cueva de Nerja* (Paleolithic age, ca. 35000 cal BCE), *l'Illeta dels Banyets* (Alacant, Spain, 4th century BCE) and Gestalgar (12th–13th CE). Figure 9.18 shows the photographic images of the Roman oven at *l'Illeta dels Banyets* and the Islamic oven found at Gestalgar, Spain. The SEM images of the bituminous remains in the same figure show a set of more or less irregular amorphous pellets often accompanied by crystalline mineral components.

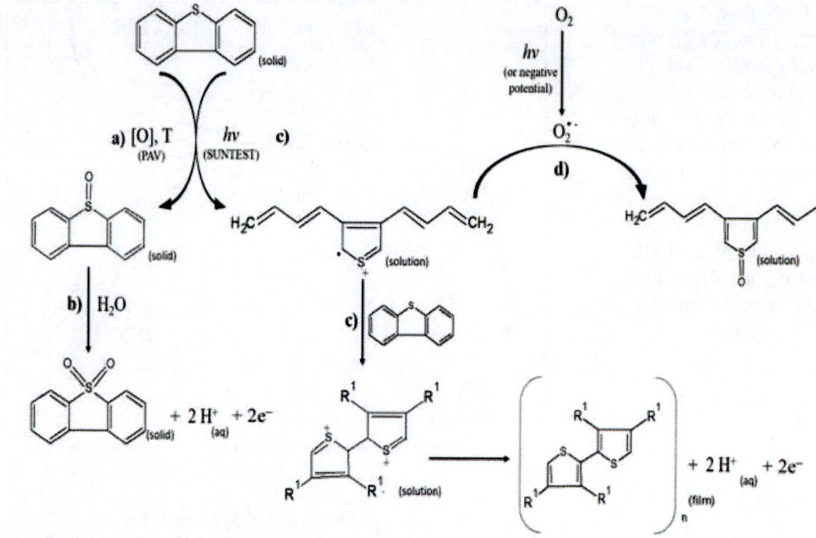

Fig. 9.17 Tentative scheme of chemical and photochemical processes involving thiophenes occurring during asphalt aging based on reported electrochemical data. Reproduced from ref. [49], with permission

For voltammetric analysis, the working electrode was prepared by macerating ca. 30 μg of bituminous sample in contact with 1 mL of toluene and dipping the solution onto the graphite electrode surface.

Here, repetitive cyclic voltammetry was used to test electropolymerization processes. When this type of process occurs, successive layers of the generated polymer are deposited onto the electrode surface and, if the polymer is electroactive, the corresponding voltammetric signals increase in height during successive potential cycles.

This behavior can be seen in Fig. 9.1a for a sample of un-aged asphalt in contact with 0.10 M H_2SO_4. In the first positive-going potential scan, two apparently irreversible anodic waves at ca. 1.0 and 1.2 V appear. These correspond to the initial oxidation of organosulfur compounds presumably to radical cations that rapidly polymerize. The generated polymer displays an essentially reversible reduction/oxidation process at ca. 0.5 V that is enhanced in successive potential scans. Comparison with reference compounds [48, 49] permits the attribution of the observed electrochemistry to the formation of polythiophenes.

The relevant aspect for archaeometric purposes is that in un-aged samples (Fig. 9.19a) there is no signal at 0.5 V in the initial anodic scan. In contrast, artificially aged and 'old' samples produce a more or less intense signal in this scan (see

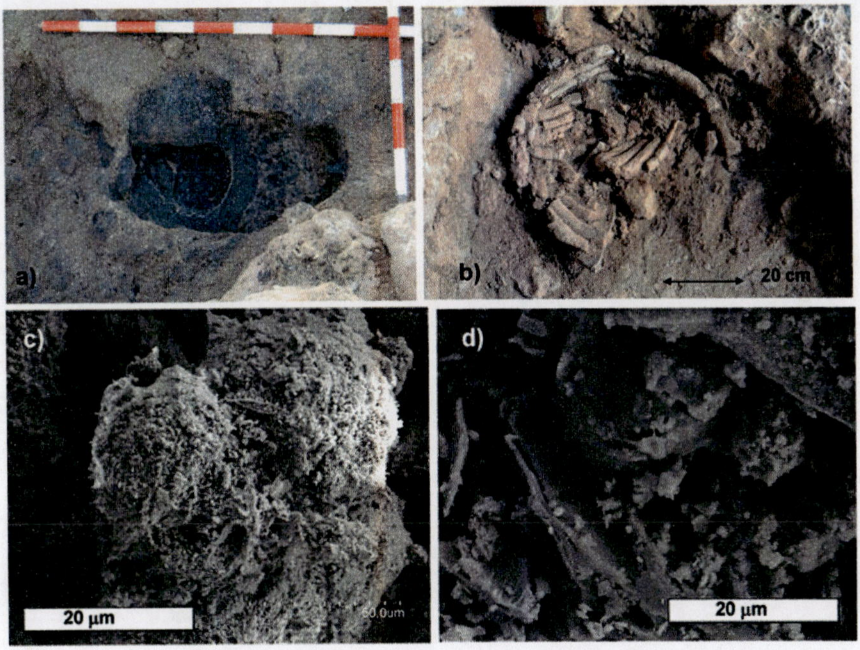

Fig. 9.18 Photographic images of **a** Roman oven at *l'Illeta dels Banyets* (Alacant, Spain) and **b** Islamic oven of tannur type, Gestalgar, Spain. SEM images of samples **c** from *l'Illeta dels Banyets* and **d** *Cueva de Nerja* (Málaga, Spain). Reproduced from ref. [3], with permission

Fig. 9.18b–d). Accordingly, the more or less intense appearance of the polythiophene signal in the initial scan of the voltammograms can be considered an age marker.

9.7 Dating

Several age markers can be considered in order to apply electrochemistry for dating organic materials. In the case of wooden objects, vanillin, a product of the degradation of lignins, has been proposed and can be used as a chronological marker. The essential idea is that the rate of loss of vanillin from lignin decreases with time following a first-order rate law [50]. Given the electroactive character of vanillin [51], an estimate of the vanillin/lignin ratio from voltammetric data is potentially usable for dating wood of art objects such as sculptures, panel paintings, and furniture. Preliminary data on wood samples using the extraction methodology described in Sect. 9.3 support the interest in this approach currently submitted to research [52].

In the case of tar pitch, the ratio between the area or peak current for the A_{PT} signal in the first and second scans, $I(A_{PT}(1))/I(A_{PT}(2))$, can be taken as a quantitative age marker. As discussed in the preceding section, this ratio will increase with the age [3]. This idea is consistent with reported methods of dispersion and emulsion

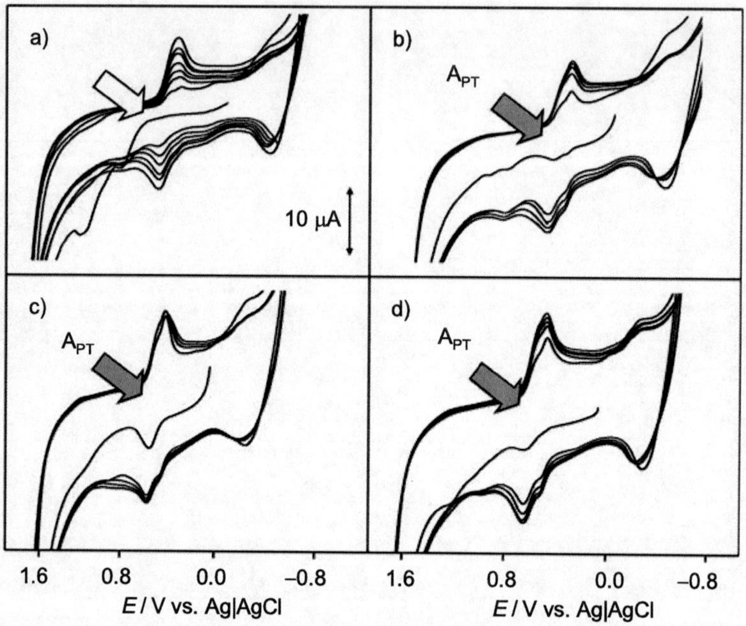

Fig. 9.19 Repetitive cyclic voltammetry of microparticulate deposits of samples: **a** un-aged asphalt cement supplied by the Petrobras Research and Development Center (CENPES), Rio de Janeiro, Brazil; **b** Islamic Owen from Gestalgar site; **c** *Cueva de Nerja* site; and **d** *Illeta dels Banyets* site, on graphite electrodes. Electrolyte, 0.10 M H_2SO_4 aqueous solution. Potential scan initiated at 0.0 V in the positive direction; sweep rate 50 mV s^{-1}. Adapted from ref. [3], with permission. Note that the axes are opposite to the IUPAC convention

chemical polymerization of thiophene in organic/aqueous interfaces denoting that polymerization is catalyzed by oxidizing agents [53], reactive oxygen species, in particular [54]. It should be noted that, since polymerization reactions can in principle occur in the surface of tar droplets, the depth reached in the sampling will be more or less relevant depending on the nature of the tar deposit (asphalts, microparticulate carbonaceous materials, …). Few available data on carbonaceous remains support the suitability of this dating method illustrated in Fig. 9.20 [3].

For charcoal dating, a method is under study which is based on the exploitation of the electrocatalytic effects of carbonaceous materials on different electrochemical processes. Here, one of the problems to solve is to elucidate the influence of the carbonization conditions (summarized as charring intensity [39]).

In principle, the ORR can be used as an electrochemical probe. This process is sensitive to the composition of the carbon material and, among other factors, to the pretreatment applied to it generating N and/or O functionalities [55]. Wood pyrolization eliminates volatile components and increases the proportion of conjugated carbon atoms based on sp^2 electronic configurations and can produce porous carbons with increasing conductivity and specific surface area [56]. These factors, as well as the presence of heteroatoms, increase the catalytic activity of the ORR process.

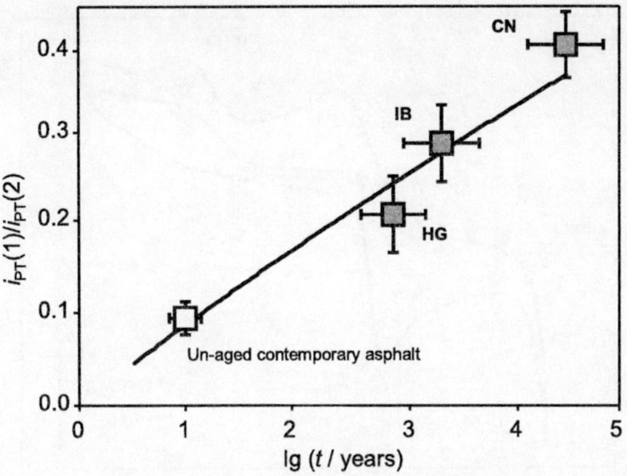

Fig. 9.20 Calibration graph for dating tar pitch-type materials using the $I(A_{PT}(1))/I(A_{PT}(2))$ ratio versus the logarithm of age (log$t$). Data points for contemporary asphalts and carbonaceous remains from a Paleolithic oven of the *Cueva de Nerja* (Málaga, Spain) dated back to 3530 ± 360 BCE (CN), a Islamic oven (HG) of *tannur* type found in Gestalgar (12th–13th, Valencia, Spain), and a Punic oven (IB) for preparing tar pitch from the archaeological site of *l'Illeta dels Banyets* (4th century BC, Alicante, Spain). Adapted from ref. [3], with permission

In this regard, the pyrolysis process yielding charcoal mimics, to some extent, the thermal activation of carbon electrodes. This generates three basic types of functionalities [55]: (i) oxygenated groups, mainly of carbonyl type, (ii) carboxyl groups, and (iii) quinone/hydroquinone units. The former can be eliminated by degassing above 800°C and their electrochemical activity is low. The carboxyl groups, more electroactive, can be degasified above 400°C whereas the quinone/hydroquinone groups are those more electrochemically active, being responsible for background signals displayed by carbonaceous electrodes in contact with aqueous electrolytes [57, 58].

The essential idea for age calibration purposes is that, after pyrolysis, the charcoal buried in humid soil will experience a continuous process of chemical modification as a result of the contact with water. These can result in generation of oxygen functionalities modifying its electrocatalytic response. Of course, there are a variety of possible chemical processes depending on the composition of the pristine charcoal and the burial conditions (temperature, aeration, ...). Additionally, it has to be taken into account that the electrochemical activity of graphite appears to be larger at the edge planes than at the basal planes, although this activity seems to be strongly dependent on the considered redox pair [57, 59]. This means that the progressive generation of oxygen functionalities during charcoal aging may result either in an increase of catalytic ability due to the formation of new catalytically active centers, and the decrease of such ability by the decrease in the number of graphitic active sites. This is a complex situation to be elucidated for dating purposes.

Current studies are underway using the ORR, OER, and HER processes as a redox probe including the $Fe(CN)_6^{3-}/Fe(CN)_6^{4-}$ pair as an internal standard. To increase repeatability, thin films of charcoal plus paraffin oil or nujol oil pastes deposited onto

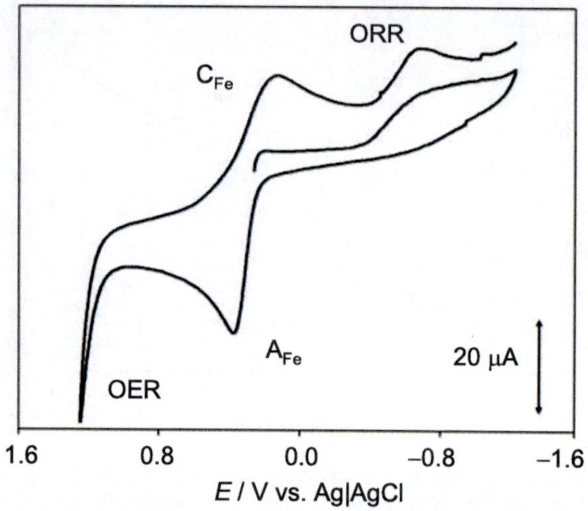

Fig. 9.21 Cyclic voltammogram recorded for a thin film of charcoal (from the archaeological site of Callosa de Segura (Spain), dated back to the Bronze age (2100–2000 CE) plus paraffin oil paste deposited onto glassy carbon electrode. The remains were identified as corresponding to *Pinus halepensis* species. Electrolyte: air-saturated 5.0 mM $K_4Fe(CN)_6$ plus 0.50 M H_2SO_4 aqueous solution. Potential scan initiated at 0.0 in the negative direction; potential scan rate 50 mV s^{-1} [60]. Note that the axes are opposite to the IUPAC convention

glassy carbon electrode were used as working electrodes immersed into solutions with different supporting electrolytes. Figure 9.21 shows the cyclic voltammetric response obtained for a charcoal sample from the Bronze age in air-saturated 5.0 mM $K_4Fe(CN)_6$ plus 0.50 M H_2SO_4 aqueous solution [60]. The voltammogram exhibits the well-defined, essentially reversible $Fe(CN)_6^{3-}/Fe(CN)_6^{4-}$ couple (cathodic and anodic signals C_{Fe}, A_{Fe}, respectively) accompanying the ORR wave and the rising current for OER.

Remembering that under ordinary VIMP conditions there is no possibility of controlling the net amount of solid sample transferred onto the base electrode (see Chap. 4), only peak current ratios can satisfactorily be handled. In the case of the voltammetry of charcoal such as in Fig. 9.21, the ratio between the peak currents for the ORR and C_{Fe} or A_{Fe} signals was taken as an age marker at the expense of the account for the influence of the conditions in which the charcoal sample was produced and aged. Since it is difficult to reproduce the amount of sample deposited onto the base glassy carbon electrode, a fraction of which is directly exposed to the electrolyte solution, we can perform considerations similar to those described in Sect. 8.3.

Now, let us assume that under fixed electrochemical conditions (electrolyte, potential scan rate, etc.), the peak current measured for a given process X at sample-coated glassy carbon electrode will be, assuming that the coating covers a region of effective area S:

$$I(\mathrm{X}) = g_{\mathrm{GCE}}(\mathrm{X})(S^{\circ} - S) + g_{\mathrm{char}}(\mathrm{X})S \tag{9.1}$$

where S° is the area of the base electrode and $g_{\mathrm{GCE}}(\mathrm{X})$ and g_{char},(X), are the coefficients of electrochemical response for the process X at the glassy carbon and the charcoal paste. The second is the parameter of interest because it will reflect the changes in the composition, age, etc., of the sample. If now we record the peak current for a second electrochemical process, Y, we can write:

$$I(\mathrm{Y}) = g_{\mathrm{GCE}}(\mathrm{Y})(S^{\circ} - S) + g_{\mathrm{char}}(\mathrm{Y})S \tag{9.2}$$

Eliminating S between Eqs. (9.1) and (9.2) yields:

$$\begin{aligned} I(Y) = {} & g_{\mathrm{GCE}}(\mathrm{Y})S^{\circ} \\ & - \frac{(g_{\mathrm{char}}(Y) - g_{\mathrm{GCE}}(Y))g_{\mathrm{GCE}}(X)S^{\circ}}{g_{\mathrm{char}}(X) - g_{\mathrm{GCE}}(X)} \\ & + I(X)\frac{g_{\mathrm{char}}(Y) - g_{\mathrm{GCE}}(Y)}{g_{\mathrm{char}}(X) - g_{\mathrm{GCE}}(X)} \end{aligned} \tag{9.3}$$

This equation predicts a linear variation of I(Y) on I(X) whose slope and ordinate at the origin will be functions of $g_{\mathrm{char}}(\mathrm{X})$ and g_{char},(Y), the parameters presumed to be age-dependent. Figure 9.22 shows the variation of the peak currents for the ORR (I(ORR)) and A_{Fe} ($I(A_{Fe})$) processes recorded in voltammograms such as in Fig. 9.21 for *Pinus halepensis* charcoal of the archaeological sites of *Cova des Moro* (Manacor, Spain, 13th CE) and Callosa de Segura (Spain, 2100–2000 CE) [60]. One can see that, in agreement with the previous theoretical treatment, a linear variation of I(ORR) on $I(A_{Fe})$ is obtained. Available data for charcoal samples identified as remains of *Pinus halepensis* suggest that this methodology, combined with ATR-FTIR in order to evaluate the charring intensity of the pyrolytic process, can be a potentially useful approach for dating purposes [60].

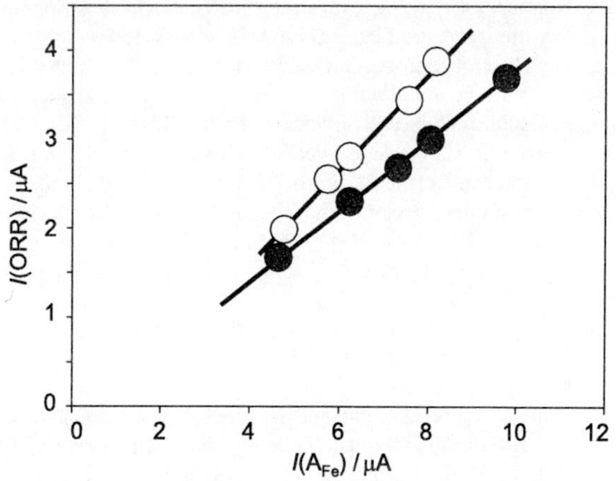

Fig. 9.22 Plots of I(ORR) versus n $I(A_{Fe})$ recorded in replicate cyclic voltammetric experiments such as in Fig. 9.21 for charcoal samples of *Pinus halepensis* from archaeological sites of solid circles: *Cova des Moro* (Manacor, Spain, 13th CE) and circles: Callosa de Segura (Spain, 2100–2000 CE)

References

1. Doménech-Carbó A (2021) Electrochemistry of plants: basic theoretical research and applications in plant science. J Solid State Electrochem 25:2747–2757
2. Doménech-Carbó A, Doménech-Carbó MT, Ferragud-Adam X, Ortiz-Miranda AS, Montoya N, Pasíes-Oviedo T, Peiró-Ronda MA, Vives-Ferrándiz J, Carrión Y (2017) Identification of vegetal species in wooden objects using in situ microextraction-assisted voltammetry of microparticles. Anal Methods 9:2041–2048
3. Doménech-Carbó A, Da Silveira GD, Medina-Alcaide MA, Martinez-Carmona A, López-Serrano D, Pasíes-Oviedo T, Algarra-Pardo VM, De Carvalho LM, Montoya N (2018) Polythiophenes as markers of asphalt and archaeological tar pitch aging. Characterization using solid state electrochemistry. Electrochem Commun 87:18–21
4. Di Turo F, Mai C, Haba A, Doménech-Carbó A (2019) Discrimination of papers used in C&R by the means of the voltammetry of immobilized microparticles technique. Anal Methods 11:4431–4439
5. Doménech-Carbó A, Ibars AM, Prieto-Mossi J, Estrelles E, Scholz F, Cebrián-Torrejón G, Martini M (2015) Electrochemistry-based chemotaxonomy in plants using the voltammetry of microparticles methodology. New J Chem 39:7421–7428
6. Doménech-Carbó A, Ibars AM, Prieto-Mossi J, Estrelles E, Doménech-Carbó MT, Ortiz-Miranda AS, Lee Y (2017) Access to phylogeny from voltammetric fingerprints of seeds: the *Asparagus* case. Electroanalysis 29:643–650
7. Milczarek G (2009) Preparation, characterization and electrocatalytic properties of an iodinellignin-modified gold electrode. Electrochim Acta 54:3199–3205
8. Deepa AK, Dhepe PL (2014) Solid acid catalyzed depolymerization of lignin into value added aromatic monomers. RSC Adv 4:12625–12629
9. Cottyn B, Rivard M, Majira A, Beauhaire J, Allais F, Martens T, Baumberger S, Ducrot PH (2015) Comparative electrochemical study on monolignols and Horticulture relevant for the comprehension of the lignification. Phytochem Lett 13:280–285
10. Admassie S, Nilsson TY, Inganäs O (2014) Charge storage properties of biopolymer electrodes with (sub)tropical lignins. Phys Chem Chem Phys 16:24681–24684
11. Kilmartin PA, Hsu CF (2003) Characterisation of polyphenols in green, oolong, and black teas, and in coffee, using cyclic voltammetry. Food Chem 82:501–512
12. Timbola AK, de Souza CD, Giacomelli C, Spinelli A (2006) Electrochemical oxidation of quercetin in hydro-alcoholic solution. J Braz Chem Soc 17:139–148
13. Ramesova S, Sokolova R, Tarabek J, Degano I (2013) The oxidation of luteolin, the natural flavonoid dye. Electrochim Acta 110:646–654
14. Komorsky-Lovric S, Novak I (2013) Abrasive stripping voltammetry of myricetin and dihydromyricetin. Electrochim Acta 98:153–156
15. Doménech-Carbó A, Gavara R, Hernández P, Domínguez I (2015) Contact probe voltammetry for *in situ* monitoring of the reactivity of phenolic tomato (*Solanum lycopersicum* L.) compounds with ROS. Talanta 144:1207–1215
16. Doménech-Carbó A, Cervelló P, González JM, Soriano P, Estrelles E, Montoya N (2019) Electrochemical monitoring of ROS influence on seedlings and germination response to salinity stress of three species of the tribe Inuleae. RSC Adv 9:17856–17867
17. Marsellis B, Garcia-Gomez J, Michaud PA, Rodrigo MA, Comninellis C (2003) Electrogeneration of hydroxyl radicals on boron-doped diamond electrodes. J Electrochem Soc 150:D79–D83
18. Enache TA, Chiorcea-Paquim AM, Fatibello-Filho O, Oliveira-Brett AM (2009) Hydroxyl radicals electrochemically generated in situ on a boron doped diamond electrode. Electrochem Commun 11:1342–1345
19. Kapałka A, Foti G, Comninellis C (2009) The importance of electrode material in environmental electrochemistry: formation and reactivity of free hydroxyl radicals on boron-doped diamond electrodes. Electrochim Acta 54:2018–2023

20. Xu J, Huang W, McCreery RL (1996) Isotope and surface preparation effects on alkaline dioxygen reduction at carbon electrodes. J Electroanal Chem 410:235–242
21. Krishnamachari V, Levine LH, Paré PW (2002) Flavonoid oxidation by the radical generator AIBN: a unified mechanism for quercetin radical scavenging. J Agric Food Chem 50:4357–4363
22. Schweingruber FH (1990) Anatomy of European woods. Hupt, Stuttgart
23. Nigra BT, Faull KF, Barnard H (2015) Analytical chemistry in archaeological research. Anal Chem 87:3–18
24. Vernet J-L, Ogereau P, Figueiral I, Yanes CM, Uzquiano P (2001) Guide d'identification des charbons de bois préhistoriques du sud-ouest de l'Europe. CNRS, Paris
25. Carrión Y, Rosser P (2010) Revealing Iberian woodcraft: conserved wooden artefacts from south-east Spain. Antiquity 84:747–764
26. Martini M, de Carvalho LM, Blasco-Blasco A, Doménech-Carbó A (2015) Screening and authentication of herbal formulations based on microextraction-assisted voltammetry of microparticles. Anal Methods 7:5740–5747
27. Madariaga JM (2015) Analytical chemistry in the field of cultural heritage. Anal Methods 7:4848–4876
28. Chiriu D, Ricci PC, Cappellini G, Carbonaro CM (2018) Ancient and modern paper: study on ageing and degradation process by means of portable NIR μ-Raman spectroscopy. Microchem J 138:26–34
29. Bronzato M, Calvini P, Federici C, Dupont A-L, Meneghetti M, Di Marco V, Biondi B, Zoleo A (2015) Degradation by-products of ancient paper leaves from wash wàters. Anal Methods 7:8197–8205
30. Poggi G, Giorgi R, Mirabile A, Xing H, Baglioni P (2017) A stabilizer-free non-polar dispersion for the deacidification of contemporary art on paper. J Cult Herit 26:44–52
31. Casoli A, Isca C, De Iasio S, Botti L, Iannuccelli S, Residori L, Ruggiero D, Sotgiu S (2014) Analytical evaluation, by GC/MS, of gelatine removal from ancient papers induced by wet cleaning: a comparison between immersion treatment and application of rigid Gellan gum gel. Microchem J 117:61–67
32. Pottier F, Kwimang S, Michelin A, Andraud C, Goubard F, Lavédrine B (2017) Independent macroscopic chemical mappings of cultural heritage materials with reflectance imaging spectroscopy: case study of a 16th century Aztec manuscript. Anal Methods 9:5997–6008
33. Campanella MPSL, Casoli A, Colombini MP, Bettolo RM, Matteini M, Migneco LM, Montenero A, Nodari L, Piccioli C, Zappalà MP, Portalone G, Russo U (2007) Chimica per l'arte. Zanichelli, Milan
34. Yum H (2011) History and function of dispersion aids used in traditional East Asian papermaking. J Am Inst Conserv 34:202–208
35. Labbé N, Harper D, Rials TJ (2006) Chemical structure of wood charcoal by infrared spectroscopy and multivariate analysis. J Agric Food Chem 54:3492–3497
36. Ascough PL, Bird MI, Scott AC, Collinson ME, Cohen-Ofri I, Snape CE, Le Manquais K (2010) Charcoal reflectance measurements: implications for structural characterization and assessment of diagenetic alteration. J Archaeol Sci 37:1590–1599
37. Gosling WD, Cornelissen HL, McMichael CNH (2019) Reconstructing past fire temperatures from ancient charcoal material. Palaeogeogr Palaeoclimatol Palaeoecol 520:128–137
38. Keeley JE (2009) Fire intensity, fire severity and burn severity: a brief review and suggested usage. Int J Wild Fire 18:116–126
39. Pyle LA, Hockaday WC, Boutton T, Zygourakis K, Kinney TJ, Masiello CA (2015) Chemical and isotopic thresholds in charring: implications for the interpretation of charcoal mass and isotopic data. Environ Sci Technol 49:14057–14064
40. Doménech-Carbó A, Carrión Y, Olivares-Altalaguerri S, Ramírez-Rodríguez M. Unpublished results
41. Grünberg JM (2002) Middle Palaeolithic birch-bark pitch. Antiquity 76:15–16
42. Mazza PPA, Martini F, Sala B, Magi M, Colombini MP, Giachi G, Landucci F, Lemorini C, Modugno F, Ribechini E (2006) A new Palaeolithic discovery: Tar-hafted stone tools in a European Mid-Pleistocene bone-bearing bed. J Archaeol Sci 33:1310–1318

43. Pollard AM, Heron C (2008) Archaeological chemistry, 2nd edn. RSC Publishing, Cambridge
44. Andersson JT, Hegazi AH, Roberz B (2006) Polycyclic aromatic sulfur heterocycles as information carriers in environmental studies. Anal Bioanal Chem 386:891–905
45. Green JB, Yu SKT, Pearson CD, Reynolds JW (1993) Analysis of sulfur compound types in asphalt. Energy Fuels 7:119–126
46. de Carvalho LM, Nascimento PC, Bohrer D (2015) Distribution of total sulfur in acidic, basic, and neutral fractions on Brazilian asphalt cements and its relationship to the aging process. Energ Fuels 29:1431–1437
47. Da Silveira GD, Hoinacki CK, Goularte RB, Nascimento PC, Bohrer D, Cravo M, Leite LFM, de Carvalho LM (2017) A cleanup method using solid phase extraction for the determination of organosulfur compounds in petroleum asphalt cements. Fuel 202:206–215
48. Da Silveira GD, de Carvalho LM, Montoya N, Doménech-Carbó A (2017) Solid state electrochemical behavior of organosulfur compounds. J Electroanal Chem 806:180–190
49. Da Silveira GD, de Carvalho LM, Montoya N, Doménech-Carbó A (2018) Evaluation of aging processes of petroleum asphalt cements by solid state electrochemical monitoring. Electrochim Acta 270:461–470
50. Rogers RN (2005) Studies on the radiocarbon simple from the shroud of Turin. Thermochim Acta 425:189–194
51. Deng P, Xu Z, Zeng R, Ding C (2015) Electrochemical behavior and voltammetric determination of vanillin based on an acetylene black paste electrode modified with graphene–polyvinylpyrrolidone composite film. Food Chem 180:156–163
52. Doménech-Carbó A, Doménech-Carbó MT, Ferragud-Adam X, Ramírez-Rodríguez M. Unpublished results
53. Yang R, Wang S, Zhao K, Li Y, Li C, Xia Y, Liu Y (2017) Comparison of oxidation polymerization methods of thiophene in aqueous medium and its mechanism. Polym Sci B 59:16–27
54. Nowaczyk J, Kadac K, Olewnik-Kruszkowska E (2015) Emulsion polymerization of thiophene—the new way of conducting polymers shyntesis. Avd Sci Technol Res J 9:118–122
55. Noked M, Soffer A, Aurbach D (2011) The electrochemistry of activated carbonaceous materials: past, present, and future. J Solid State Electrochem 15:1563–1578
56. Barrière F, Downard AJ (2008) Covalent modification of graphitic carbon substrates by non-electrochemical methods. J Solid State Electrochem 12:1231–1244
57. Chen P, McCreery RL (1996) Control of electron transfer kinetics at glassy carbon electrodes by specific surface modification. Anal Chem 68:3958–3965
58. Barisci JN, Wallace GG, Baughman RH (2000) Electrochemical studies of single-wall carbon nanotubes in aqueous solutions. J Electroanal Chem 488:92–98
59. Davies TJ, Hyde ME, Compton RG (2005) Nanotrench arrays reveal insight into graphite electrochemistry. Angew Chem Int Ed 44:5121–5126
60. Doménech-Carbó A, Doménech-Carbó MT, Carrión Y, Ramírez-Rodríguez M, Olivares-Alatalaguerri S. Unpublished results

Additional Literature

61. Hernández VA, Hermes M, Milchev A, Scholz F (2009) The overall adhesion-spreading process of liposomes on a mercury electrode is controlled by a mixed diffusion and reaction kinetics mechanism. J Solid State Electrochem 13:639–649
62. Bezerra RMR, Neves ACO, Pimenta AS, Lima KMG (2015) Estimation of Brazilian charcoal properties using attenuated total reflectance-Fourier transform infrared (ATR-FTIR) spectrometry coupled with multivariate analysis. Anal Methods 7:5695–5701
63. Braunger ML, Barros A, Ferreira M, Olivati CA (2015) Electrical and electrochemical measurements in nanostructured films of polythiophene derivatives. Electrochim Acta 165:1–6

64. Cohen-Ofri I, Weiner L, Boaretto E, Mintz G, Weiner S (2006) Modern and fossil charcoal: aspects of structure and diagenesis. J Archaeol Sci 33:428–439
65. de Carvalho LM, Hilgemann M, Scholz F, Kahlert H, da Rosa MB, Wuster M, Lindequist U, do Nascimento PC, Bohrer D (2010) Electrochemical assay to quantify the hydroxyl radical scavenging activity of medicinal plant extracts. Electroanalysis 22(4):406−412
66. Davrieux F, Rousset PLA (2010) Discrimination of native wood charcoal by infrared spectroscopy. Quim Nova 33:1093–1097
67. Doménech-Carbó A, Cebrián_Torrejón G, Lopes-Souto A, de-Moraes MM, Jorge-Kato K, Fechine-Tavares J, Barbosa-Filho JM (2015) Electrochemical ecology: VIMP monitoring of plant defense against external stressors. RSC Adv 5(75):61006−61011
68. Doménech-Carbó A, Dias D, Donnici M (2021) In vivo electrochemical monitoring of signaling transduction of plant defense against stress in leaves of *Aloe vera* L. Electroanalysis 33:1024–1032
69. Domínguez I, Doménech-Carbó A (2015) Screening and authentication of tea varieties based on microextraction-assisted voltammetry of microparticles. Sensors Actuator B 210:491–499
70. Doménech-Carbó A, Montoya N, Soriano P, Estrelles A (2018) An electrochemical analysis suggests role of gynodioecy in adaptation to stress in *Cortaderia selloana.* Curr Plant Biol 16:9–14
71. Domínguez-Delmás M, Bossema FG, van der Mark B, Kostenko A, Coban SB, van Daalen S, van Duin P, Batenburg KJ (2021) Dating and provenancing the *Woman with lantern* sculpture—a contribution towards attribution of Netherlandish art. J Cult Herit 50:179–187
72. Fu L, Zheng Y, Zhang P, Zhu J, Zhang H, Zhang L, Su W (2018) Embedding leaf tissue in graphene ink to improve signals in electrochemistry-based chemotaxonomy. Electrochem Commun 92:39–42
73. Ghica ME, Oliveira-Brett AM (2005) Electrochemical oxidation of rutin. Electroanalysis 17:313–318
74. Han B, Daheur G, Sablier M (2016) Py-GCxGC/MS in cultural heritage studies: an illustration through analytical characterization of traditional East Asian handmade papers. J Anal Appl Pyrolysis 122:458–467
75. Kuang F, Zhang D, Li Y, Wan Y, Hou B (2009) Electrochemical impedance spectroscopy analysis for oxygen reduction reaction in 3.5% NaCl solution. J Solid State Electrochem 13:385–390
76. Le Bourvellec C, Hauchard D, Darchen A, Burgot J-L, Abasq M-L (2008) Validation of a new method using the reactivity of electrogenerated superoxide radical in the antioxidant capacity determination of flavonoids. Talanta 75:1098–1110
77. Ma R, Lin G, Zhou Y, Liu Q, Zhang T, Shan G, Yang M, Wang J (2019) A review of oxygen reduction mechanisms for metal-free carbon-based electrocatalysts.npj Comput Mat 5:78
78. Ortiz-Miranda AS, König P, Kahlert H, Scholz F, Osete-Cortina L, Doménech-Carbó MT, Doménech-Carbó A (2016) Voltammetric analysis of *Pinus* needles with physiological, phylogenetic and forensic applications. Anal Bioanal Chem 408:4943–4952
79. Ouyang C, Wang X (2020) Recent progress in pyrolyzed carbon materials as electrocatalysts for the oxygen reduction reaction. Inorg Chem Front 7:28–36
80. Piech R, Bás B, Rubiak WW, Paczosa-Bator B (2012) Fast cathodic stripping volammetric determination of elemental sulfur in petroleum fuels using renewable mercury film silver based electrode. Fuel 97:876–878
81. Romagnoli M, D'Alessandro L, Seccaroni C, Moioli P, Poggi D, Grabner M (2016) The Pietà di Ragusa panel: a science-based contribution to its dating by dendrochronology, wood anatomy and pigment analysis. J Cult Herit 21:767–774
82. Scholz F, González GLL, de Carvalho LM, Hilgemann M, Brainina KhZ, Kahlert H, Jack RS, Minh DT (2007) An electrochemical system to detect free radicals and radical scavengers in solution. Angew Chem Int Ed 46:8079–8082
83. Serafim DM, Stradiotto NR (2008) Determination of sulfur compounds in gasoline using mercury film electrode by square wave voltammetry. Fuel 87:1007–1013

Chapter 10
Metallic Heritage: Electrochemistry of Corrosion Products

10.1 Introduction

The production of metal artifacts marks one of the crucial steps in the evolution of human civilization. Except the 'noble' metals appearing in nature in their native, elemental state (gold mainly, often silver, and rarely copper and mercury), all metals have to be synthesized from their compounds existing in the earth's crust: oxides, sulfides, carbonates, sulfates, … In contact with the environment, metals tend in general to oxidize to different metal compounds.

Then, metallurgy is a science/technology struggling against thermodynamics. This means that metallic heritage is submitted to continuous degradative interaction with the environment, ultimately resulting in the mineralization of the metal, i.e., its more or less complete conversion into oxidized compounds. With the awareness of this 'universal' problem associated with the use of metals, a variety of strategies of protection have been developed. These, of course, have to be adapted to the wide variety of metals and their applications. Here, we will focus the attention on applications in the study and preservation of cultural heritage goods.

The study of metallic heritage is complicated, however, by the same complexity as the processes of metal corrosion. Although in cases such as atmospheric contact at high temperature, where 'chemical' corrosion occurs, in general, under ordinary conditions of contact with the atmospheric environment, metal corrosion can be seen as an electrochemical process [1]. This will be influenced by the chemical properties of the metal and its corrosion products, the microstructure of the metal—in turn resulting from the thermomechanical conditioning applied during the manufacturing process—and the characteristics of the environment which define the so-called corrosion history of the object under study. There are a variety of corrosion modalities acting at different micro- and mesostructural levels (intergranular corrosion, pitting corrosion, crevice corrosion, …). The interaction between the metal and the environment not only results in chemical, electrochemical, or even biological corrosion processes but also subsequent pedological processes (recrystallization, segregation,

A. Doménech-Carbó and M. T. Doménech-Carbó, *Electrochemistry for Cultural Heritage*, Monographs in Electrochemistry,
https://doi.org/10.1007/978-3-031-31945-7_10

desegregation, cementation, monolithizations) giving rise to a more or less complex patina of corrosion products.

10.2 Corrosion of Metal Objects

Restricting our description to ordinary electrochemical corrosion, this process involves the disposal of oxygen and an electrolyte. There is also the possibility of anaerobic corrosion where other chemical species replace oxygen as the oxidant, processes occurring in seawater in principle promoted by the action of sulfate-reducing bacteria [2].

In the case of uniform corrosion of a metal M covered by a thin humidity layer acting as the electrolyte, there is an anodic reaction initially yielding metal ions, the metal dissolution reaction (MDR)

$$M_{solid} \rightarrow M^{n+}_{aq} + ne^{-} \tag{10.1}$$

coupled to the cathodic reaction of oxygen reduction (ORR):

$$O_{2aq} + 2H_2O + 4e^{-} \rightarrow 4OH^{-}_{aq} \tag{10.2}$$

In principle, anodic and cathodic sites are spatially separated on the metal surface and randomly distributed over atomic distances. In this situation, the system is characterized by a mixed potential and a current density, macroscopically uniform throughout the metal surface [3]. The MDR and ORR processes are essentially irreversible [4], but the former is generally less hindered (exhibiting both a lower activation overpotential and a larger reactant yield), the access of dissolved oxygen to the metal surface acting as a rate-determining step.

Among a wide variety of models, one of the simplest approaches can be formulated as a one-dimensional process in which the metal surface is covered by a thin, uniform electrolyte layer with high saline concentration and constant thickness where only the processes described by Eqs. (10.1) and (10.2) operate and oxygen diffusion is under a pseudo-steady state [5]. Then, the Butler – Vomer kinetic equation applied to the MDR process can be written as:

$$I\,(M/M^{n+}) = I^{\circ}\,(M/M^{n+})\exp\left[\frac{nF(1-\alpha(M/M^{n+})}{RT}(E_{corr} - E^{\circ}\,(M/M^{n+}))\right] \tag{10.3}$$

Here, $I^{o}(M/M^{n+})$ represents the exchange current of the MDR process, E_{corr} the corrosion potential, $E^{o}(M/M^{n+})$ the standard potential of the M/M^{n+} couple, and $\alpha(M/M^{n+})$ the electron transfer coefficient of the same. For the ORR process, one can write [5, 6],

$$I\,(O_2/OH^-) = I^{\circ}\,(O_2/OH^-)$$
$$\exp\left[\frac{-4F\alpha(O_2/OH^-)}{RT}\left(E_{corr} - E^{\circ}\,(O_2/OH^-)\right)\right]\left(\frac{c_{x=0}}{c_{sat}}\right) \quad (10.4)$$

where the exchange current, electron transfer coefficient, and standard potential of the O_2/OH^- couple are introduced. The oxygen concentration at the metal/electrolyte interface, $c_{x=0}$, and the saturation concentration of oxygen in the electrolyte solution, c_{sat}, are in turn depending on the temperature and salinity.

The description of the overall corrosion process, however, is conditioned by a variety of factors. These include the diffusive regime of oxygen through the electrolyte layer, transport of ionic species, precipitation of insoluble metal salts, as well as the kinetics of the ORR, which can involve a direct four-electron path or a stepped two-electron, two-electron path with intermediate formation of H_2O_2. Additionally, the corrosion process advances via the formation of an oxide layer which can be described in terms of the permeation of oxygen and ionic species through a porous layer [7]. In this case, a contact potential difference appears between the metal and the adjacent oxide layer, which is assumed to be semiconducting. This can be defined as the difference between the work functions of the two surfaces when there is no current passing across the interface [7].

In general, the metal forms a primary corrosion patina of metal oxide. Depending on the characteristics of the metal oxide, it acts as a protective barrier (Al, Pb) or is permeated by oxygen and ions facilitating the advance of the corrosion process (Fe as an extreme case). For the purposes of this book, the relevant point to emphasize is that the corrosion of metals is a complex multistep process determining the formation of a stratified multiphase system. There is an enormous variety of corrosion patterns depending on the metal or alloy and the environmental conditions. In this section, we limit the description of the essentials proposed by the two most widely studied cases, iron and copper/bronze.

In the case of iron, it has been proposed that the process starts with MDR oxidation,

$$Fe_{solid} \rightarrow Fe^{2+}_{aq} + 2e^- \quad (10.5)$$

followed by a series of competing reactions [8]:

$$Fe^{2+}_{aq} + 2H_2O \rightarrow Fe(OH)_{2solid} + 2H^+_{aq} \quad (10.6)$$

$$4Fe^{2+}_{aq} + O_{2aq} + 6H_2O \rightarrow 4FeO(OH)_{solid} + 8H^+_{aq} \quad (10.7)$$

$$Fe^{2+}_{aq} + xH_2O \rightarrow Fe(OH)_x^{(2-x)+}{}_{aq} + xH^+_{aq} \quad (10.8)$$

forming goethite, α-FeO(OH), and/or lepidocrocite, γ-FeO(OH), or more or less undefined iron oxide hydrates. The process is catalytically accelerated by chloride ions forming akaganéite (chloride-containing β-FeO(OH) forms, presumably via complexation of Fe(II) ions:

$$Fe^{2+}_{aq} + xCl^{-}_{aq} \rightarrow FeCl_{x}^{(2-x)+}{}_{aq} \quad (10.9)$$

$$4FeCl_{x}^{(2-x)+}{}_{aq} + O_{2aq} + 6H_2O \rightarrow 4\{FeO_{1-x}(OH)_{1+x}Cl_x\}_{solid} + 4(2-x)H^{+}_{aq} \quad (10.10)$$

This is important because it acts as a chloride reservoir (the representative formula is $FeO_{0.833}(OH)_{1.167}Cl_{0.167}$) to promote corrosion [9, 10]. A contrary role is exerted by magnetite (Fe_3O_4) which acts as a protective layer that is presumed to be formed by the reaction,

$$Fe^{2+}_{aq} + 2FeO(OH)_{solid} \rightarrow Fe_3O_{4solid} + 2H^{+}_{aq} \quad (10.11)$$

The detailed mechanism, however, is not yet well understood and is quite sensitive to the local conditions of corrosion. Depending on the conditions, a variety of mineral compounds, including Fe(II) and Fe(III) chlorides and hydroxides, siderite ($FeCO_3$), vivianite ($Fe_3(PO_4)_2{\cdot}8H_2O$), among others, can be formed.

The structure of the corrosion layers of iron objects under burial conditions has been modeled [11, 12] in terms of a dense product layer (DPL) covering the metal core and in turn being covered by a layer of transformed medium (TM), as depicted in Fig. 10.1. The DPL zone consists of ferric oxyhydroxides, mainly goethite and lepidocrocite, and can contain slag inclusions from the metal as well as foreign mineral inclusions like quartz. The TM constitutes a transition zone between the DPL and the soil containing metal corrosion products and soil materials (calcite, clays, quartz, ...). The metal core can be covered by a layer of conductive oxides (CO, mainly magnetite) and oxide marblings (magnetite and/or maghemite). Importantly, the DPL/TM interface corresponds, approximately, to the original surface of the object. In cases of moderate mineralization, the shape of the object is maintained and its morphological examination is possible. Iron corrosion gives rise to porous corrosion layers characterized by the formation of 'rust' and a significant alteration of the original metallic surface ultimately leading to the complete mineralization of the artifact.

The corrosion structures of copper and bronze range between two extreme types: noble patinas or even and coarse or burgeoning ones [13, 14]. The former retains the shape of the object and looks like shiny and lustrous. Here, the patina possesses a bilayer structure composed of an outer layer of about 5–50 μm thick and an internal layer in contact with the alloy. The inner layer is composed of cuprite. The outer layer also presents low copper contents but high tin contents.

'Coarse' patinas are characterized by a thickness between 200 μm and several mm. They exhibit a three-layer structure, with a primary patina of cuprite, covered by

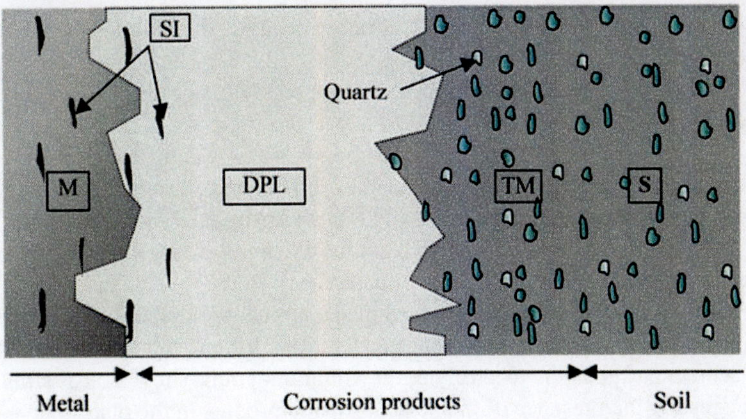

Fig. 10.1 Scheme of the structure of iron corrosion layers existing on iron artifacts buried in the soil. Reproduced from Ref. [11], with permission

a secondary patina which contains other corrosion products depending on the composition of the environment: mainly copper hydroxycarbonates (minerals of the malachite family) in calcareous soils, copper hydroxychlorides in coastal environments (atacamite, $Cu_2(OH)_3Cl$), and copper hydroxysulfates (brochantite, $Cu_4(OH)_6SO_4$) in several environments [13], in all cases accompanied by tenorite. A ternary patina is often formed containing pedological products resulting from the interaction with the surrounding soil. In cases of high corrosion (often termed 'bronze disease'), there is the formation of corrosion deposits with the destruction of the original metallic surface with the formation of crusts, pits, crevices, lamellar plates, etc.

The corrosion of alloys can offer additional complexities. Thus, tin in bronze artifacts, and zinc in brass, are in general selectively dissolved into the outer medium [15, 16]. These processes, called destannification and dezincation, respectively, both belonging to the family of dealloying processes, do not always occur [15, 17]. Another particular case of interest is that of lead in leaded bronze alloys. Here, lead forms globular features more or less regularly distributed in the bronze matrix [15, 18, 19]. These are accompanied by unalloyed copper inclusions of [20], attributed either to a pseudomorphical replacement process associated with destannification or copper redeposition after lead dissolution [21].

10.3 Identification of Metals and Corrosion Products

As described in Chap. 4 for pigments, the VIMP methodology can be applied to identify metals and alloys in works of art and archaeological objects [22, 23], with several works being focused on the identification of corrosion products in copper and bronze objects [24–27], iron [28], silver [29], and lead [30]. 'One-touch' sampling strategies based on 'graphite pencil' electrodes permitted to minimize the amount of sample needed for electrochemical measurements [22, 30, 31], whereas the application of Tafel-type representations applies for the resolution of mixtures of corrosion products [32, 33].

An example of the application of this methodology for identifying corrosion products in archaeological objects is provided in Fig. 10.2. Here, the square wave voltammograms of a blank of lead(II) acetate and a sample extracted by means of the 'one-touch' sampling from an Iberian plate from the *Pico de los Ajos* (Yátova, Valencian Community, Spain) archaeological site, dated back to the fourth century BCE, are shown. This is one of the epigraphic lead plates typical of the Iberian culture developed in the Eastern Iberian Peninsula between the 4th and the 2nd century BCE. Using acetate buffer as the electrolyte, the voltammograms show a clear similarity, thus suggesting the presence of this lead corrosion product in the plate.

Additionally, this example illustrates one of the problems encountered in the conservation of metallic heritage. The plate was exhibited for some time in wooden expositors where vapors of acetic acid produced by the wood act as corrosion agents [34, 35].

The case of lead corrosion products is particularly favorable for electrochemical identification. Figure 10.3 shows two other examples of identification of corrosion products, litharge, and anglesite, respectively, in Iberian lead plates.

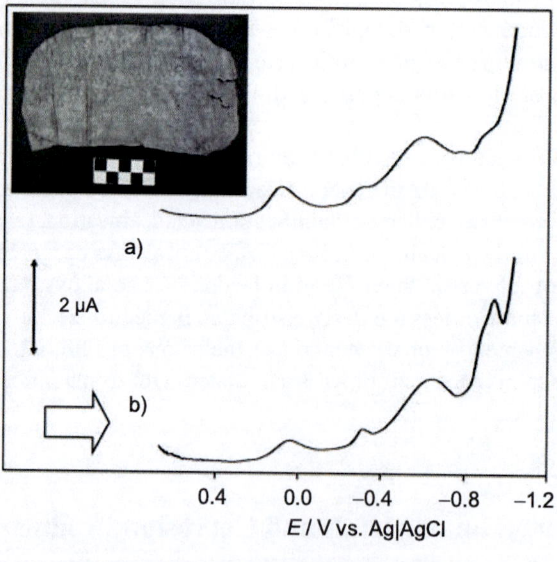

Fig. 10.2 Square wave voltammograms of: **a** $Pb(AcO)_2$ and **b** sample from *Pico de los Ajos* Iberian plate, on graphite electrodes immersed into 0.50 M HAc/NaAc aqueous buffer solution at pH 4.85. Potential scan initiated at 0.45 V in the negative direction; potential step increment 4 mV; square wave amplitude 25 mV; and frequency 5 Hz. Inset: the photographic image of the lead plate. Sample by courtesy of the Museu de Prehistòria de València (Spain). Adapted from Ref. [30], with permission. Note that the axes are opposite to the IUPAC convention

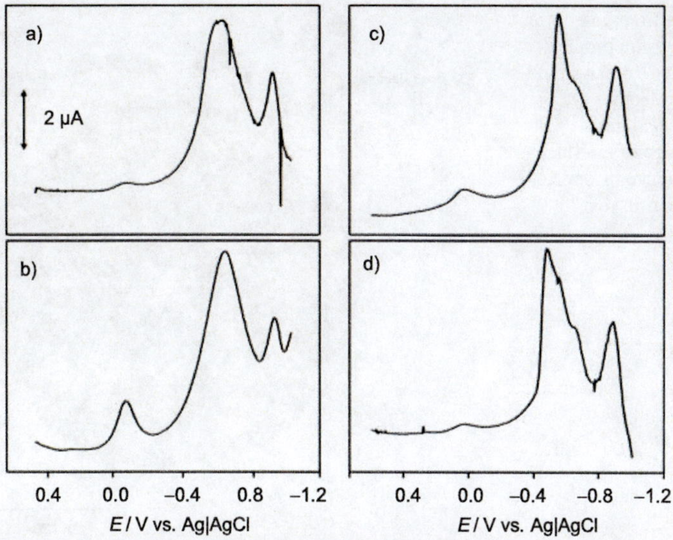

Fig. 10.3 Square wave voltammograms of: **a** PbO (litharge), **b** patinated fragment separated from a plate of the *La Punta d'Orleyl* Iberian site (Borriana, Spain), **c** anglesite, and **d** concretion from the above plate separated during its cleaning treatment. Conditions such as in Fig. 10.2. Sample by courtesy of the Museu Municipal de Borriana (Castelló, Spain). Adapted from Ref. [30], with permission. Note that the axes are opposite to the IUPAC convention

On the contrary, the identification of iron corrosion products is not straightforward because of the variety of products and the sensitivity of the voltammetric response to the degree of hydration, crystallinity, etc. Despite these problems, mapping of corrosion products on archaeological artifacts is possible using VIMP. This is illustrated in Fig. 10.4a for a shield boss recovered in the *Torre la Sal* archaeological site (Oropesa, Spain), corresponding to the Iberian-Roman period (fourth century BCE). Magnetite, siderite, ochres, hematite, and jarosite ($KFe_3(OH)_6(SO_4)_2$) were identified and distributed in different locations on the rough surface [28]. Figure 10.4b shows the X-ray image of the shield showing the high degree of mineralization. This image illustrates a recurrent problem in iron archaeology, the determination of the existence and, if any, the location of a metallic nucleus. We will treat again this matter in the next chapter.

Artificial patinas, mainly on bronze and copper, have been fabricated since antiquity for protective and esthetic purposes [36, 37]. Identification using VIMP has been recently reported [38].

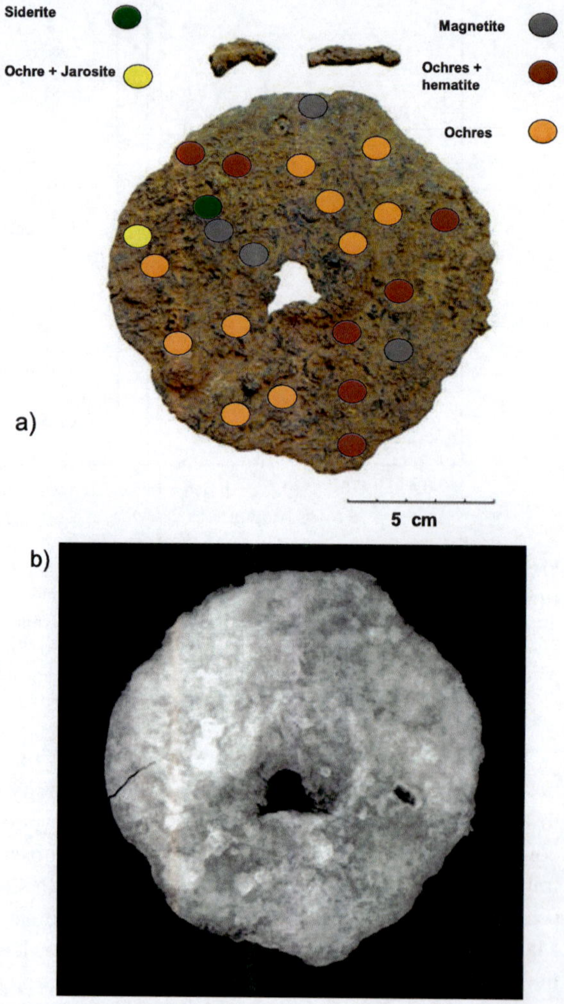

Fig. 10.4 **a** Mapping of the surface corrosion products on an Iberian-Roman shield boss derived from electrochemical data. **b** X-ray image of the shield boss. Reproduced from Ref. [28], with permission

10.4 In-Depth Electrochemistry

The VIMP sampling on metallic artifacts consists of pressing a graphite bar (diameters typically between 0.5 and 2 mm) over (preferentially) a flat region of the metal patina. As deduced from the charge passed under the voltammetric peaks, this procedure permits to extract a few nanograms of sample from the corrosion layers, usually consisting of a set of laminas on micrometric size. A typical SEM image was presented in Fig. 4.4, corresponding to the surface of a commercial graphite bar after sampling of an artificially prepared red sulfate-type patina over bronze [38].

These laminas are released from the external region of the metal patina. This offers a more or less stratified section as shown by electron microscopy images recorded in trenches created by fast ion bombardment coupled to field emission scanning electron microscopy (FIB-FESEM). Figure 10.5 shows the secondary electron image of an Iberian bronze coin from the archaeological site of Cástulo (Jaen, Spain) dated back to the 2nd–first century BCE. The equipment generates a focused beam of Ga ions that excavates a trench on the coin surface showing its stratified section. By comparing the image with that in Fig. 10.6, corresponding to an Iberian coin from the site of Iltirta (Lleida, Spain), one can see that there are differences in the faceted aspect of the trench. These coins differ significantly in their Sn content (low in the Cástulo coin, high in the Iltirta coin) in turn reflected in their voltammetric response [39] .

Figure 10.7 depicts a (highly) idealized representation of the sampling procedure resulting in the release of a few laminas from the external patina [38]. The relevant point to emphasize is that, depending on the characteristics of the patina and the exerted pressure, the sampling will extract more or less number of more or less thick laminas, and that these laminas possess a stratified composition and

Fig. 10.5 FESEM-FIB of an Iberian coin (cat. n. 25654, Museu de Prehistòria de València, Spain) from Cástulo (Jaen, Spain). The image shows a ca. 10 µm length and ca. 15 µm depth generated by FIB in a flan region of the coin. Adapted from Ref. [39], with permission

Fig. 10.6 FESEM-FIB of an Iberian coin (cat. n. 41868, Museu de Prehistòria de València, Spain) from Iltirta (Lleida, Spain). The image shows a ca. 10 µm length and ca. 15 µm depth generated by FIB in a flan region of the coin. Adapted from Ref. [39], with permission

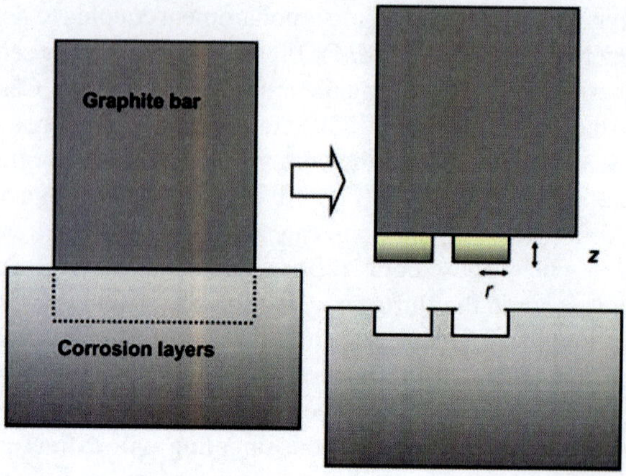

Fig. 10.7 Idealized scheme of the sampling procedure by means of a graphite bar on the surface of a metallic object with the release of few laminas from the external patina. Adapted from Ref. [38], with permission

textural (compactness, crystallinity, porosity, …) properties. Notice that, since the electrochemical processes initiate at the plate/graphite/electrolyte solution three-phase boundary, the initial potential scan will provide the response of the more external layers of the patina. In successive scans (vide infra), the response of deeper regions of the patina will be recorded.

These ideas are consistent with experimental data in Fig. 10.8, where a series of square wave voltammograms after sampling at different pressures on two artificial bronze patinas, nitrate—(a) and chloride-based (b), is depicted. The nitrate patina is mainly composed by gerhardtite, $Cu_2(NO_3)(OH)_2$, whose reduction can be described as a proton-assisted, two-electron reduction process [38] :

$$Cu_2(NO_3)_2(OH)_{2\,solid} + 2H^+_{(aq)} + 4e^- \rightarrow 2Cu_{solid} + 2NO_3^-{}_{aq} + 2H_2O \tag{10.12}$$

As can be seen in Fig. 10.8, on increasing the pressure during the sampling process, the intensity of the cathodic peak progressively increases showing peak splitting [38]. This can be attributed to the in-depth variation of the properties of the gerhardtite patina and/or the appearance of the signal corresponding to a primary patina of cuprite, presumably formed.

This scheme confirmed the results obtained by means of the experimental arrangement illustrated in Fig. 10.9 [40]. This permits the layer-by-layer analysis of the patina of metal objects immersed in an electrolyte bath by means of the application of successive potential scans. The essential idea is that, as shown in Fig. 10.10, the application of cathodic potentials produces a partial dissolution of the external corrosion layer and the progressive penetration of the graphite bar into the patina. As a result, the voltammograms are representative of progressively deeper regions of the corrosion layer.

The arrangement in Fig. 10.9 is, however, not appropriate for the study of archaeological objects. As will be seen in Chap. 11, there is the possibility of working with one drop of electrolyte or replacing the liquid electrolyte with jellified electrolytes. Here,

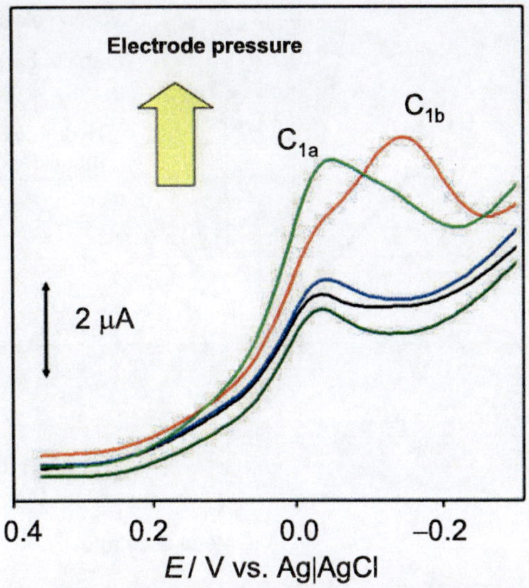

Fig. 10.8 Square wave voltammograms of graphite electrodes in contact with 0.25 M HAc/NaAc buffer at pH 4.75 after sampling with different pressures onto an artificial nitrate patina prepared on bronze. Potential scan initiated at 0.65 V in the negative direction; potential step increment 4 mV; square wave amplitude 25 mV; and frequency 5 Hz. Adapted from Ref. [38], with permission. Note that the axes are opposite to the IUPAC convention

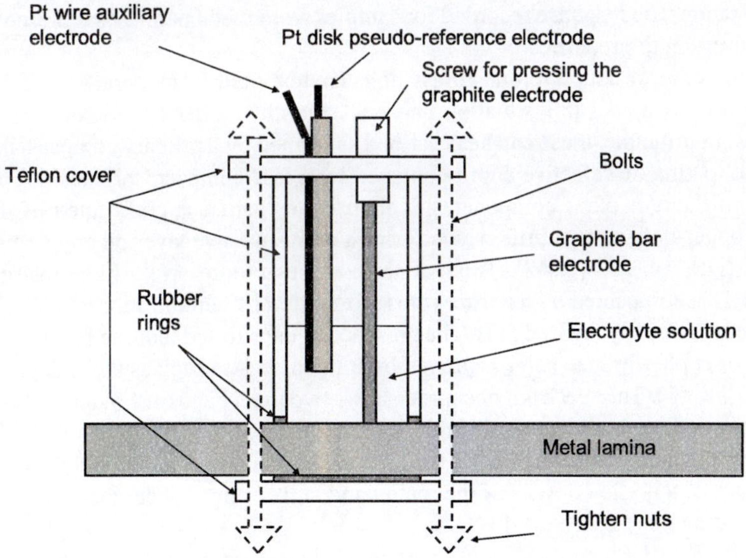

Fig. 10.9 Scheme for an electrochemical cell used for layer-by-layer studies on metal laminas covered by a corrosion layer. The graphite bar is put in contact with the corrosion patina covering the metallic lamina so that the electroactive region is concentrated in the graphite/corrosion layer/electrolyte solution junction. Reproduced from Ref. [40], with permission

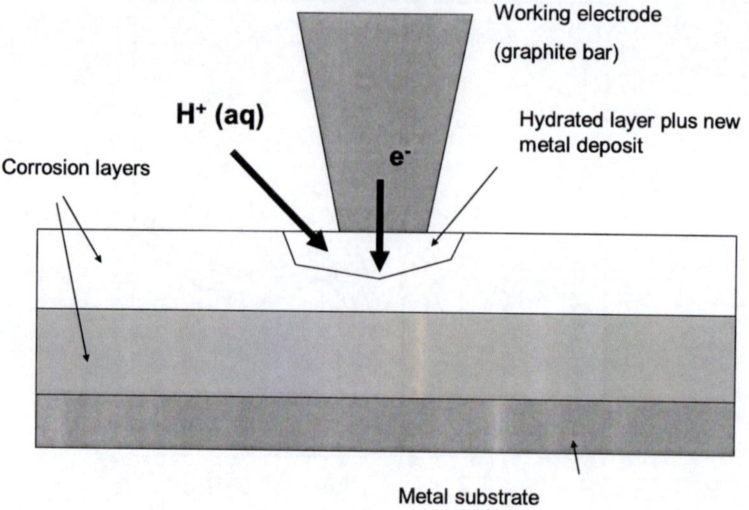

Fig. 10.10 Schematic representation of the layer-by-layer voltammetry using the experimental arrangement in Fig. 10.9. In successive potential scans, the corrosion layers are partially dissolved and the graphite electrode penetrates more deeply into the patina. Adapted from Ref. [40], with permission

the relevant aspect is that voltammetric data are coincident in supporting the idea that the voltammetric response recorded for samples from metal patinas is representative of their in-depth properties.

In the case of copper/bronze/brass objects submitted to moderate corrosion, it is possible to model this situation using a 'smooth' model of the corresponding patinas. In principle, these can be described as formed by a primary compact 'impermeable' patina of defective cuprite covered by a less compact, 'permeable' patina of cuprite and/or other components. This primary patina is constituted of defective (copper-deficient) cuprite which forms a native passive layer on copper/bronze objects [13, 14, 25, 41, 42]. This patina increases progressively its porosity, and cuprite is accompanied by tenorite, whose formation by aerobic cuprite oxidation is thermodynamically favored [43]. The presence of cuprite and tenorite in the external layers of copper/bronze coins is proved by Raman spectra such as those depicted in Fig. 10.11 [43]. The spectrum of cuprite shows two main bands at 114 and 220 cm^{-1}, accompanied by shoulders at 145 and 186 cm^{-1} and by broad bands at 420, 525, and 625 cm^{-1} which can be attributed to Cu_2O amorphous or finely divided [42]. The spectrum of tenorite consists of the main band at 297 cm^{-1}, accompanied by weak signals at 346 and 631 cm^{-1} [44].

Figure 10.12 shows the typical voltammetric pattern recorded in contact with air-saturated acetate buffer for moderately corroded copper and bronze coins, here corresponding to a coin minted in the Honan Province during the last Chinese emperors Kuang Hsu (光绪) and Hsüan T'ung (宣統) [45]. The voltammogram

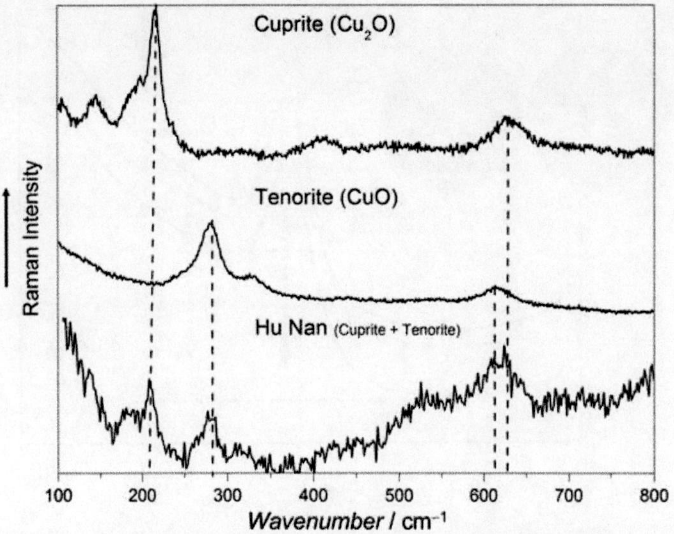

Fig. 10.11 Raman spectra of **a** cuprite, **b** tenorite, and **c** a Chinese ten cash coin emitte by the Peiyang Province in the Kuang Hsü Chinese emperor's era. Reproduced from Ref. [43], with permission

shows a cathodic peak at ca. −0.10 V versus Ag/AgCl (C_1) accompanied by shoulders ca. −0.30 V (C_2) and −0.70 V (C_{ox}). These signals can be attributed to the reduction of cuprite and tenorite, existing in the patina plates adhered to the graphite electrode and the ORR process, respectively. In replicate experiments on samples from the same coin and samples from different coins not only did the absolute values of the intensities of these signals ($i(C_1)$, $i(C_2)$) varied but also their relative values. Interestingly, comparing data points for coins of different mints permitted to define mint-characteristic tendency curves in graphs such as depicted in Fig. 10.13 corresponding to Chinese ten cash coins emitted around 1900 [45] and a series of Spanish *maravedís* produced in 1661 [46].

Qualitatively, data in Fig. 10.13 suggest that, on increasing the net amount of sample transferred onto the graphite electrode (and hence, increasing the absolute intensity of the voltammetric peaks), the tenorite/cuprite ratio decreases. This is the result expected if tenorite was formed at the expense of cuprite so that the tenorite/cuprite ratio is larger in the external regions of the corrosion layer than in the internal ones. Accordingly, one can assume that the compactness and crystallinity decrease from inner to external layers, whereas the proportion of tenorite increases. This situation is depicted in Fig. 10.14.

These features can be interpreted by adopting an idealized model assuming that the graphite bar extracts a set of N laminas from the corrosion layers of the metal object and that the measured currents for the different processes can be taken as proportional to the net amount of the corresponding electroactive component of the patina. Accordingly, the peak current of the process C_1, $i(C_1)$, will be representative of

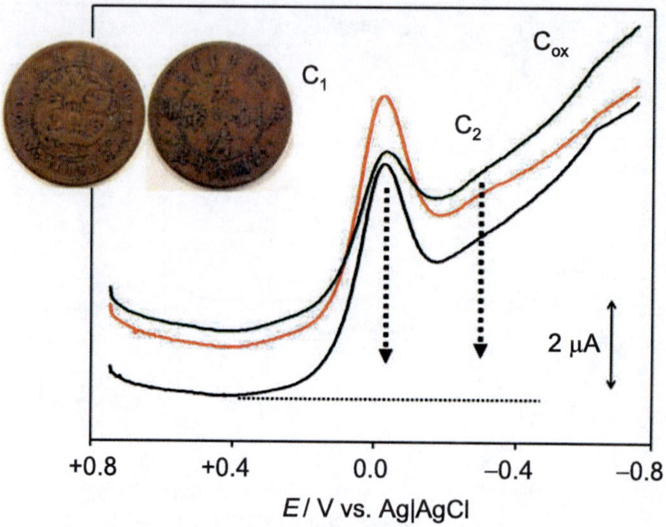

Fig. 10.12 Square wave voltammograms (three replicate independent measurements at freshly sample-modified electrodes) of samples from a coin emitted by the Honan Province during the last Chinese emperor's era were attached to a graphite electrode. Electrolyte: air-saturated 0.25 M HAc/NaAc aqueous buffer, pH 4.50. Potential scan initiated at 0.75 V in the negative direction; potential step increment 4 mV; square wave amplitude 25 mV; and frequency 5 Hz. Horizontal dotted lines represent the baselines for peak current measurements. Inset: the photographic image of the obverse (left) and reverse (right) coin. Adapted from Ref. [45], with permission. Note that the axes are opposite to the IUPAC convention

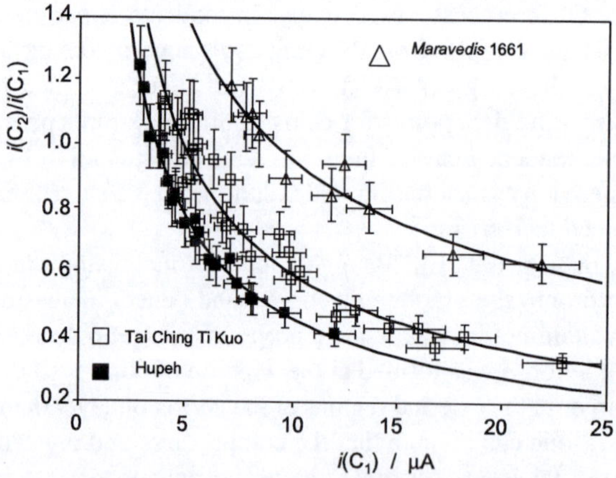

Fig. 10.13 Plots of the $i(C_2)/i(C_1)$ versus $i(C_1)$ for Tai Ching Ti Kuo and Hupeh coins and Spanish *maravedis* minted in Madrid in 1661. Data recorded in voltammograms in conditions such as in Fig. 10.12; three independent measurements for each coin are represented. Continuous lines correspond to the fit of each series of data to a potential law. Error bars correspond to estimated uncertainties of ± 10%. Reproduced from Ref. [45], with permission

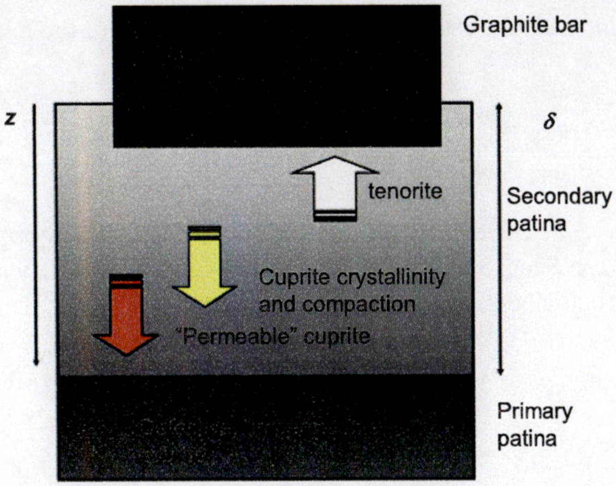

Fig. 10.14 Scheme of the in-depth distribution of electroactive species in the corrosion layers of copper/bronze objects undergoing moderate atmospheric corrosion

the net amount (mole number) of cuprite, m_{cup}, transferred to the graphite electrode,

$$i(C_1) = g_{cup} m_{cup} \tag{10.13}$$

whereas the current measured for the C_2 shoulder, $i(C_2)$ (baseline for current measurements shown in Fig. 10.12), will be representative of the contribution of tenorite, m_{ten}. Then:

$$i(C_2) = g_{ten} m_{ten} \tag{10.14}$$

In the above equations, g_{cup}, g_{ten}, represent the respective electrochemical coefficients of response. As in other previously studied systems, these coefficients depend on the electroactive material, the electrolyte, and the experimental conditions of the voltammetric experiment (scan rate, etc.).

Now, it is assumed that the sampling process results in the adherence to the graphite electrode of a set of N circular laminas of radius r and a certain thickness clearly lower than r. In a given voltammetric experiment, the measured cathodic currents correspond to the abstraction of cuprite and tenorite since the external corrosion layer ($z = 0$) until a certain depth, z. Ideally, the system can be described in terms of elemental cylinders of volume dV, thickness dz, and base area S, so that d$V = S$dz.

Assuming that tenorite is formed in the secondary patina at the expense of the originally formed cuprite, the number of tenorite grains per volume unit, n_{ten}, will increase from zero in the primary/secondary patina boundary, located at $z = \delta$, to a value n_{ten}^{sup} at the external surface of the secondary patina ($z = 0$). In turn, the number of copper centers in cuprite grains per volume unit, n_{cup}, will decrease from its value in the primary patina, n_{cup}^{o}, to the value reached at the external surface, n_{cup}^{sup}. Taking a normalized depth $x = 1 - z/\delta$, the depth variation of the amounts of cuprite and tenorite in each plate can be expressed by means of $F(x)$ functions representative of the variation of the density of copper centers with depth. Then, the values of m_{cup},

m_{ten}, can be estimated as the absolute values of the integrals:

$$m_{cup} = \int_{x=1}^{x} NS\delta[n^{o}_{cup} - n^{sup}_{ten}F(x)]dx; \; m_{ten} = \int_{x=1}^{x} NS\delta n^{sup}_{ten}F(x)dx \quad (10.15)$$

The currents associated with the reduction of copper oxides will be proportional (see Sect. 3.9) to their respective net amounts. Then, introducing the respective coefficients of electrochemical response, g_{cup}, g_{ten}, the currents $i(C_1)$, $i(C_2)$ will be:

$$i(C_1) = \int_{x=1}^{x} g_{cup}NS\delta[n^{o}_{cup} - n^{sup}_{ten}F(x)]dx; \; i(C_2) = \int_{x=1}^{x} g_{ten}NS\delta n^{sup}_{ten}F(x)dx \quad (10.16)$$

Then, the $i(C_2)/i(C_1)$ ratio is independent of the $N\delta S$ product:

$$\frac{i(C_2)}{i(C_1)} = \frac{g_{ten}\int_{x=1}^{x} n^{suo}_{ten}F(x)dx}{g_{cup}\int_{x=1}^{x}[n^{o}_{cup} - n^{sup}_{ten}F(x)]dx} \quad (10.17)$$

One possible function is $F(x) = x^{\alpha}$. In this case, Eq. (10.17) becomes:

$$\frac{i(C_2)}{i(C_1)} = \frac{g_{ten}n^{sup}_{ten}(1 - x^{1+\alpha})}{g_{cup}\left[n^{o}_{cup}(1 - x)(1 + \alpha) - n^{sup}_{ten}(1 - x^{1+\alpha})\right]} \quad (10.18)$$

If the cuprite to tenorite conversion is low, as expected for moderately corroded artifacts, x can be approximated by a function of $i(C_1)$ and ultimately express the $i(C_2)/i(C_1)$ ratio as a function of $i(C_1)$. This generalized model can be extended to situations where other voltammetric signals associated with silver and lead appear [39, 45, 46].

10.5 Archaeometric Issues

Apart from dating, monographically treated in the last chapter of this book, there are a series of archaeometric targets where electrochemical techniques can offer relevant information. These include authentication, characterization of manufacturing techniques, and tracing provenances.

Authentication of metal artifacts is a difficult task requiring the analysis of the composition of the base metal, its metallographic structure, and, eventually, isotope analysis. This last yields insights into metal provenance issue source tracing by comparing isotopic signatures of environmental samples, metallurgical residues, and potentially related ore deposits. Lead isotopes are the most widely used for this purpose due to their stability and minimal metal segregation. There are, however, several factors influencing isotope analysis, namely heterogeneity in the ore composition, preparation of metal artifacts by mixing different ores, using additives during the preparation process, metal recycling, isotopic fractionation during metal production, and paleopollution [47, 48]. Of course, the use of non-destructive techniques is desirable for authentication based on the analysis of the metal patina of interest [47]. Studies on lead artifacts have been reported [49, 50]. Here, authentication criteria are based on the record of the voltammetric signatures characterizing Pb(IV) and the analysis of the extent of the formation of 'porous' lead oxide over the primary patina of compact litharge.

VIMP data can provide information on subtle changes in the manufacturing of metal objects. In the case of a series of Chinese ten cash coins emitted by different provinces in the last emperor's era, four electrochemical types were discriminated based on the representations in Fig. 10.13 and their double logarithmic equivalents in Fig. 10.15.

Remarkably, these electrochemical types were consistent with the geographical distribution in northern, central, and southern areas of China, illustrated in Fig. 10.16. The differences between different mints can be attributed to minor changes in the composition and variations in the metallography of the coins produced by the different provincial mints. Following the scheme in Fig. 10.14, the different thermomechanical protocols of coin production must result in differences in the compaction, crystallinity, porosity, etc. of the secondary cuprite patina and the tenorite/cuprite ratio, giving rise to the observed differences in the voltammetric pattern.

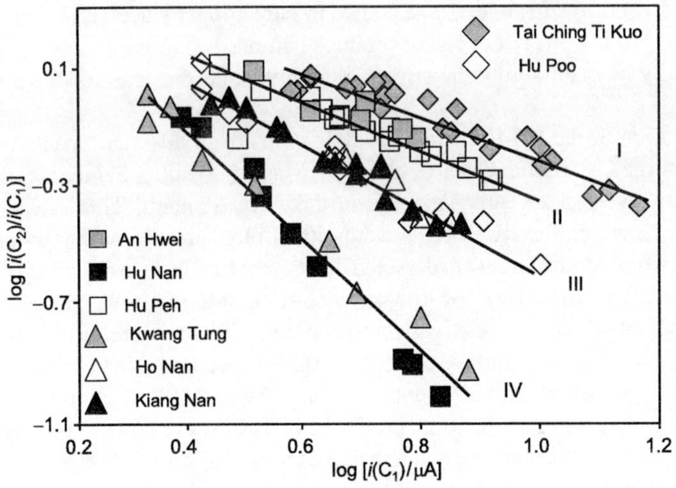

Fig. 10.15 Variation of log[i(C_2)/i(C_1)] on log[i(C_1)] for ten cash copper coins minted in different Chinese provinces during the last emperor's period. From voltammetric data such as in Fig. 10.12. Adapted from Ref. [45], with permission

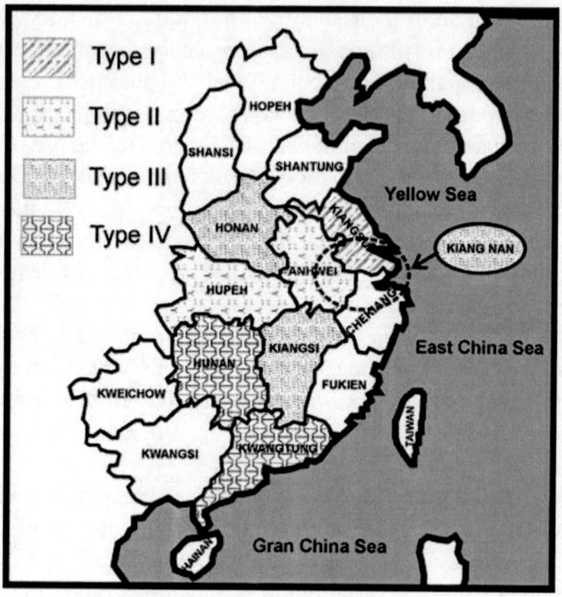

Fig. 10.16 Geographical distribution of the different electrochemical types assigned to the coinage emitted by different Chinese provinces during the last emperor's era. Adapted from Ref. [45], with permission

Another interesting case study is that of Spanish *maravedís* produced between 1661 and 1664. In 1660, the Spanish king Felipe IV ordered to coin a new series of fleece attempting to promote a deep monetary modification and face the existing economical crisis. This consisted of the introduction of silver in the most widely used currency, the *maravedís*, which was coined in pure copper since 1602. The currency was initially cast coins but began to produce hammered coins as early as 1661, and the difficulty of maintaining the silver value motivated the cease of the production in 1664.

Application of the VIMP methodology to coins from different Spanish mints (A Coruña, Burgos, Córdoba, Cuenca, Granada, Madrid, Trujillo, Segovia, Sevilla, and Valladolid) revealed an early decrease in the silver content. This can be seen in Fig. 10.17 in which the $i(C_2)/i(C_1)$ versus $i(C_1)$ plots corresponding to *maravedís* fabricated in the Madrid mint are depicted. Voltammetric data permit group coins into three groups. The first series, constituted by coins mint in 1661, was characterized by the presence of silver with relatively high content. This was detected by the appearance of an oxidative dissolution peak at ca. 0.25 V versus Ag/AgCl accompanying the stripping of copper at 0.0 V (and confirmed by SEM/EDX analysis) [46]. The second group corresponds to coins produced in 1661 and 1662 containing a lower silver content. The third group, coinage dated from 1662 to 1664, was characterized by the absence of silver.

It is interesting to note that, in this case, cathodic signals associated with AgCl reduction are masked by the cuprite reduction signal at −0.10 V. Data in Fig. 10.17 denote that the tenorite/cuprite ratio increases on lowering the silver content and revealed significant differences between different mints. Voltammetric data proved

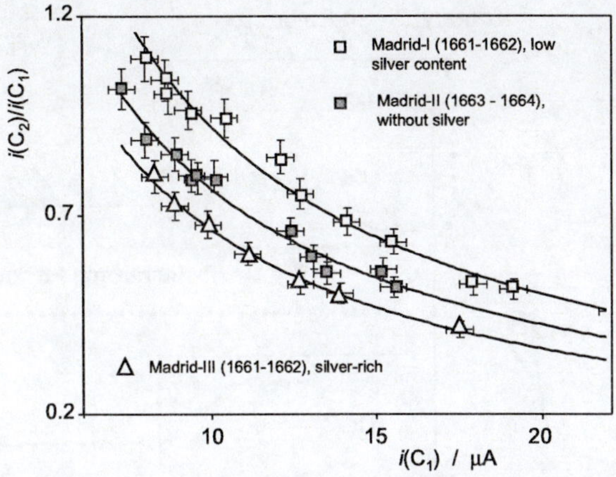

Fig. 10.17 Plots of $i(C_2)/i(C_1)$ versus $i(C_1)$ for *maravedís* produced in the Madrid mint in 1661 (circles) in 1661 and 1662 (triangles) and 1662–1664 (squares) from voltammograms in conditions equivalent to those in Fig. 10.13. Sampling was performed on two or three spots for each coin. Continuous lines correspond to the fit of such subseries of data to a potential law. Error bars corresponded to deviations of 5% in the values of the represented quantities. Adapted from Ref. [46], with permission

the existence of a hiatus in the production of currency as early as 1661. This can be attributed to the lowering in the silver content and/or change from hammer to mill fabrication [46].

A third peculiar case of currency analysis was reported for Iberian coins of the sites of Iltirta, Cástulo, and Obulco (Spain) [39]. The study of these productions dated back to the 2nd–first century BCE is difficult because they were very irregular and the abundance of forgeries produced in other locations in the same historical period. Interestingly, currency produced in Iltirta (near the Ebro valley at the northeast of the Iberian Peninsula) consisted mainly of ternary Cu–Sn–Pb, in contrast with those of Caástulo and Obulco. These cities were located in the Guadalquivir Valley (South of the Iberian Peninsula), far from the sources of cassiterite in the northwest of the Iberian Peninsula [48], then producing Sn-free coins [51]. This was clearly reflected in the voltammetric patterns balancing the signals for the reduction of copper and lead corrosion products. Again, VIMP data permitted a reasonable grouping of samples and suggested, tentatively, a possible evolutionary scheme in the production of such sites characterized by the progressive increase in lead content (Fig. 10.18) [39]. This can be clearly seen in the square wave voltammograms depicted in this figure, recorded under conditions similar to those used in Fig. 10.12. This scheme could provide the basis for a chronological ordering of the coins, but it is also conceivable that the differences in the lead loading were the result of local fluctuations in the disposal of raw materials and/or the incidence of forgeries, logically made with lower copper contents.

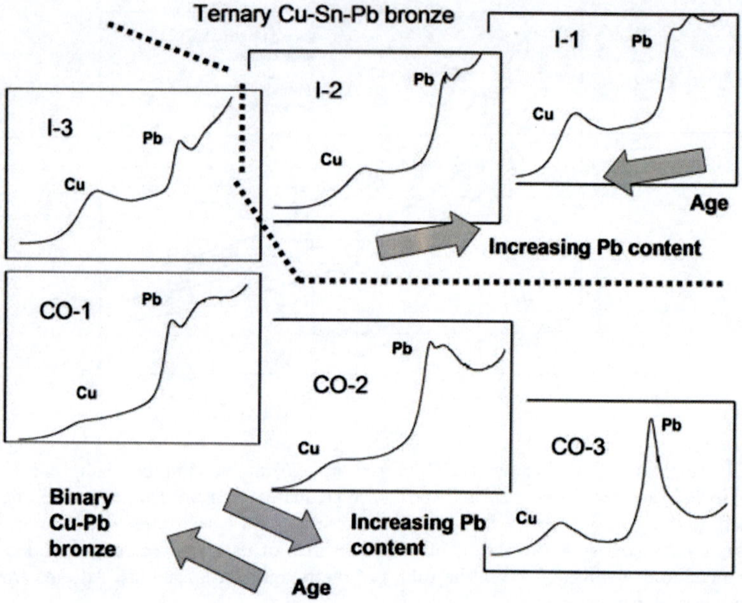

Fig. 10.18 Tentative scheme for the possible evolution of minting in the Iberic cities of Iltirta, Cástulo, and Obulco is derived from voltammetric data in conditions similar to those in Fig. 10.12. I: Iltirta; CO: Cástulo or Obulco; Cu and Pb mark the cathodic signals corresponding to the reduction of copper and lead corrosion products, respectively. Adapted from Ref. [39], with permission

The essential guidelines of this methodology have been also applied to the analysis of the first Polish coins [52], *Antoninianus* silver coins [53], Roman brass coins [54], and coffin portraits from the National Museum in Krakow [55]. A study on archaeological artifacts permitted the correlation of VIMP data with lead isotope analysis for determining the provenance of archaeological bronzes [56].

10.6 Impedance Analysis

Electrochemical impedance spectroscopy (EIS) is extensively used in studies on metal corrosion, being generally applied to metal probes. The application of EIS to archaeological objects and works of art, as well as to representative fragments, will be discussed in the next chapter. This requires that the objects have an accessible conducting region suitable for establishing electrical contact. This is not possible in highly mineralized artifacts. However, there is the possibility of application of impedance techniques combining EIS with VIMP sampling to acquire archaeometric information from samples extracted from the corrosion layers.

As previously studied, the VIMP methodology can be applied to the identification of corrosion products even when several electroactive compounds are distributed in stratified lamellae. There are cases, however, in which the high corrosion level and/or the complexity of the corrosion pattern limit the analytical opportunity in the grouping of a series of archaeological objects. In these cases, EIS measurements on microparticulate deposits of minerals and samples extracted from the corrosion layers of the artifacts can be of interest [57–59]. An example of this methodology was the study of a series of bronze artifacts from the Roman archaeological site Gadara (Jordan), which dated back to the fourth century CE [59].

Apart from the general demands for 'true' impedance spectroscopy described in Chap. 2, the VIMP-EIS intersection offers two main problems: the variable amount of sample deposited onto the base electrode in replicate experiments and the complexity of the modeling of the necessary equivalent circuit. Effectively, the systems under study constitute discontinuous, inhomogeneous deposits onto the graphite electrode allowing for different charge transfer pathways.

There is an additional matter to be considered in the analysis of corrosion products. In usual EIS experiments, the open-circuit potential or the equilibrium potential of a redox probe in the solution (typically 1–5 mM equal concentrations of $K_3Fe(CN)_6$ and $K_4Fe(CN)_6$ in the presence of a more concentrated supporting electrolyte) is applied as a bias potential on which the sinusoidal excitation signal (typically 5–10 mV of amplitude) is superimposed. To study microparticulate deposits of corrosion products, we can apply different bias potentials in order to promote Faradaic processes. In the case of blanks of cuprite, tenorite, malachite, and samples from Gadara, 0.10 M $HClO_4$ was used as the electrolyte, the bias potential being −0.25 V versus Ag/AgCl. Under these conditions, impedance spectra such as in Fig. 10.19 were obtained.

Although under these experimental conditions, the impedance measurements cannot be strictly considered as impedance spectra (see Sect. 2.3), the operation proceeds under essentially steady-state conditions, and the EIS formalism can reasonably be applied. As can be seen in Fig. 10.19, the Nyquist plots of the (minus) imaginary impedance component versus the real impedance component can be described as two consecutive capacitive loops.

Based on EIS studies on metal hexacyanoferrates [60] and porous oxides [61, 62], the spectra can be modeled in terms of a solution resistance (R_s) in series to two parallel branches. The first one contains a parallel association of the charge transfer resistance (R_{ct}) at the graphite/electrolyte interface and the double-layer capacitance (C_{dl}). The second contains a parallel association of a constant phase element (Q_p) and a resistance (R_p) and a series resistance (R_i). The R_p/Q_p elements can be assigned to the presence of insulating and/or semiconducting particles partially covering the graphite electrode. In turn, the resistance R_i can be associated with the occurrence of a Faradaic process as a result of the application of the indicated bias potential. This resistance can represent that of the ionomeric layer produced (see Sect. 2.2) during the reduction of the copper compound, as well as the charge transfer resistance associated with the charge transfer occurring at the particle/base electrode/electrolyte three-phase boundary. This equivalent circuit is drawn in Fig. 10.20a.

Several samples, however, required the use of a second equivalent circuit. This includes a parallel combination of new resistance (R_{mct}) and capacitor (C_{mdl})

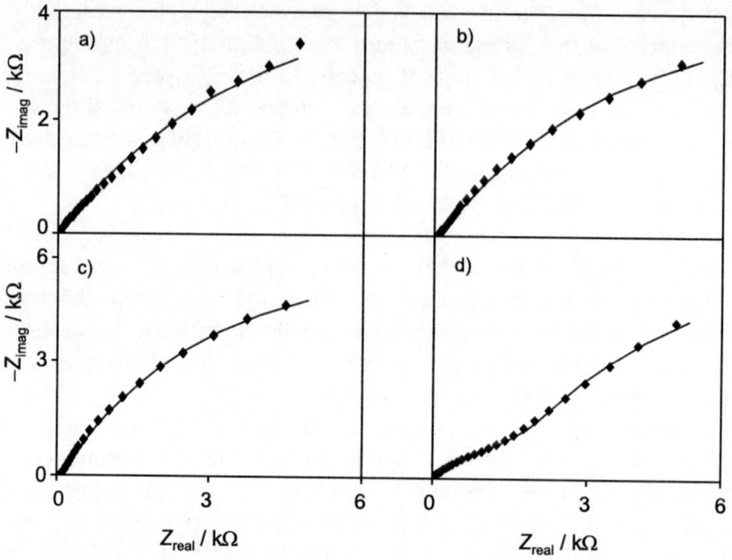

Fig. 10.19 Nyquist representation of experimental impedance data (circles) recorded for **a** cuprite and samples **b** S2, **c** W2, and **d** P11 from the corrosion layers of bronze artifacts of the Roman archaeological site of Gadara (Jordan). The continuous lines depict the theoretical spectra after fitting experimental data to the equivalent circuits in Fig. 10.20a (**a**–**c**) and Fig. 10.20b (spectrum **d**). Reproduced from Ref. [59], with permission

elements. These can be interpreted as representative of the charging effects at the metal/electrolyte interface assuming that metal deposits can also be formed via the release of metal ions from the ionomeric layer (see section). This provides an opportunity for a new charge transfer route as schematized in Fig. 10.20b.

To compare the impedance parameters determined in replicate experiments, one can assume that R_p can be associated with the ohmic resistance (Ω) of the microparticulate deposit of corrosion product or reference compound. This resistance can be expressed as,

$$R_p = \frac{\delta_p}{\vartheta_p A \sigma_p} \tag{10.19}$$

where δ_p (m), σ_p (Ω^{-1} m^{-1}) are, respectively, the average thickness of the particle deposit and the conductivity of the material. ϑ_p represents the fraction of the covered surface of the graphite electrode, whose area is A (m^2). On the first examination, it is possible to assume that the microparticulate deposit of corrosion products forms less than a monolayer of crystals over the graphite surface. Increasing the number of sample results in the addition of new layers to the deposit and increases the surface coverage. However, only the first line of crystals should contribute effectively to the Faradic response (see Sect. 2.2) and, hence, to the EIS response so that δ_p can approximately be taken as a constant.

Following the same line of reasoning, the resistance associated with the ionomeric layer, R_{ion}, can be approached by the expression,

$$R_{ion} = \frac{\delta_{ion}}{\vartheta_{ion} A \sigma_{ion}} \tag{10.20}$$

where δ_{ion} denotes the thickness of the ionomeric layer, σ_{ion} its conductivity, and ϑ_{ion} the fraction of surface area which can be associated with the ionomeric layer. Eliminating A between the above equations, one obtains:

$$R_p = \frac{\delta_p \vartheta_{ion} \sigma_{ion}}{\delta_{ion} \vartheta_p \sigma_p} R_{ion} \tag{10.21}$$

This equation is independent of the amount of sample transferred onto the base electrode surface and predicts a linear variation of R_p on R_{ion} assuming that all other parameters remain essentially constant in replicate experiments. In principle, R_{ion} can be identified as the resistance R_i in the equivalent circuit in Fig. 10.20a, but it is conceivable to add an additional term to R_i. This term can be associated with the charge transfer resistance for charge transport through the metal deposit, R_{ctp}, so that $R_i = R_{ion} + R_{ctp}$. Then, Eq. (10.21) can be rewritten as:

$$R_p = \frac{\delta_p \vartheta_{ion} \sigma_{ion}}{\delta_{ion} \vartheta_p \sigma_p} (R_i - R_{ctp}) \tag{10.22}$$

In turn, the capacitance C_p (F) associated with the CPE labeled as Q_p can be expressed as:

$$C_p = \varepsilon_p \frac{A \vartheta_p}{\delta_p} \tag{10.23}$$

Here, ε_p is the dielectric permittivity (F m^{-1}) of the particles. Combining Eqs. (10.19) and (10.23), one obtains:

$$C_p = \frac{\varepsilon_p}{\sigma_p} \frac{1}{R_p} \tag{10.24}$$

This equation predicts a linear relationship between C_p and $R_p{}^{-1}$ from replicate measurements on microparticulate deposits of corrosion products and reference materials. Experimental data reasonably agreed with predictions from Eqs. (10.22) and (10.24). Figure 10.21 shows the variation of R_p on R_i, using the resistance values calculated from the fitting of experimental EIS data to the equivalent circuit in Fig. 10.20a, for different copper minerals and samples of bronze artifacts of the Roman archaeological site of Gadara (Jordan) [59]. The impedance spectra were

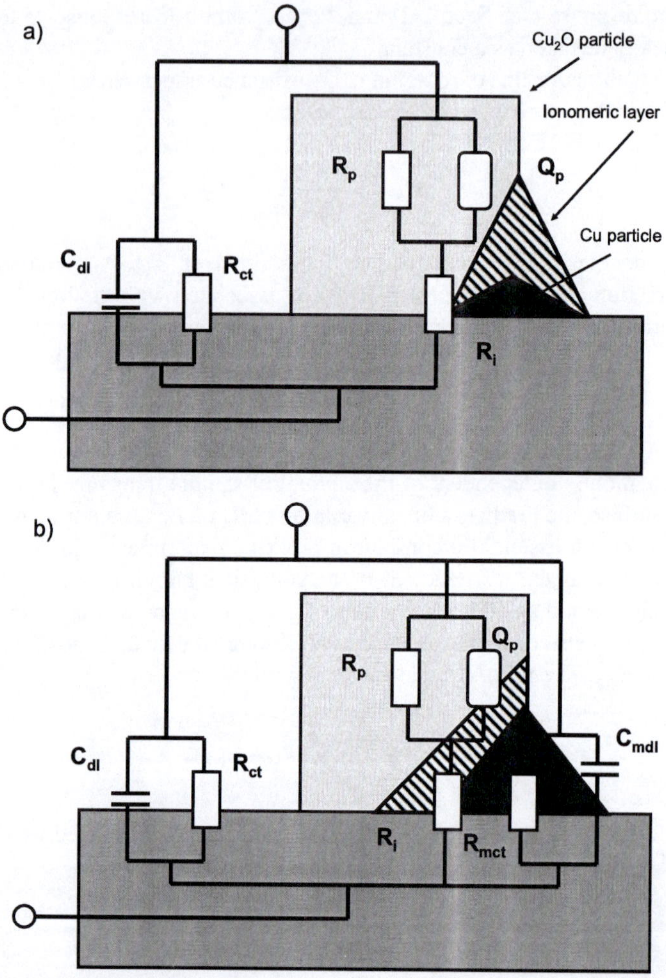

Fig. 10.20 Equivalent circuits were used to model the impedance spectra of reference compounds and samples from the corrosion layers of bronze artifacts of the Roman archaeological site of Gadara (Jordan). Reproduced from Ref. [59], with permission

recorded under the conditions of Fig. 10.19 for microparticulate deposits of samples and reference compounds having different amounts of solid. Experimental data agree well with the theoretical expectations from Eq. (10.22) assuming that each component has characteristic values of the conductivity, thickness, and surface coverage parameters and that R_{ctp} is different for each corrosion product but essentially independent on its amount, in agreement with coupled AF-VIMP data [59]. Data in Fig. 10.21 permit the identification of tenorite as the main component of the indicated samples.

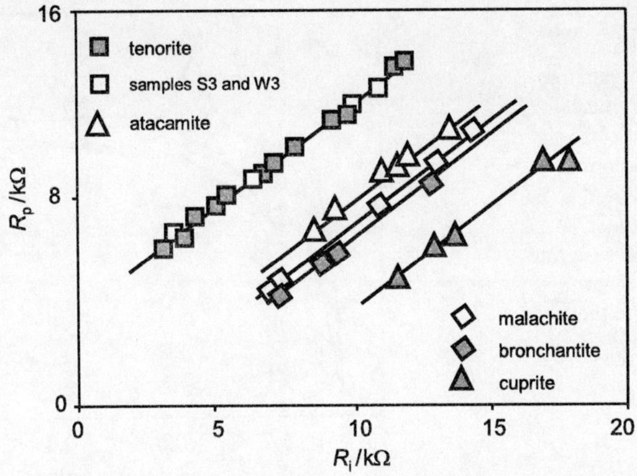

Fig. 10.21 Plots of R_p versus R_i values calculated from curve fitting of experimental EIS data in conditions such as in Fig. 10.19 to the equivalent circuit depicted in Fig. 10.20a for a series of copper corrosion products and samples from the corrosion layers of bronze artifacts of the Roman archaeological site of Gadara (Jordan). Reproduced from Ref. [59], with permission

10.7 Gold Electrochemistry

The 'noble' character of gold makes it reluctant to undergo corrosion processes equivalent to those described for copper, iron, lead, silver, etc. Its thermochemical stability in an aerobic environment, combined with its relative rarity—although with relative accessibility—in the natural context prompted its extensive use in almost all civilizations.

Despite its chemical inertness, gold displays rich electrochemistry that can be exploited in the field of cultural heritage. Figure 10.22 shows a typical voltammogram of a commercial polycrystalline gold electrode in 0.10 M H_2SO_4. The initial anodic scan shows a well-defined peak at ca. 1.2 V (A_1) preceding the OER rising current. In the subsequent cathodic scan, a symmetric peak appears at ca. 0.8 V (C_1) preceding the rising ORR current at ca. −0.3 V. Interestingly, in the second anodic scan, additional oxidation peaks appear between 0.10 and 1.0 V (A_2).

This voltammetry, although widely studied, is not completely understood. The more extended view attributes the anodic process A_1 to the electrochemical oxidation of metallic gold forming a monolayer of various gold oxides, and this process can be represented, for simplicity, as [63]:

$$Au_{solid} + 3H_2O \rightarrow Au(OH)_{3surf} + 3H^+_{aq} + 3e^- \quad (10.25)$$

The cathodic process C_1 involves the reduction of these oxidized gold species to gold metal. This process is clearly separated from the previous anodic process A_1, a feature that has received several interpretations. One possible explanation is that the electrochemically generated monolayer of hydrated gold oxide forms can be

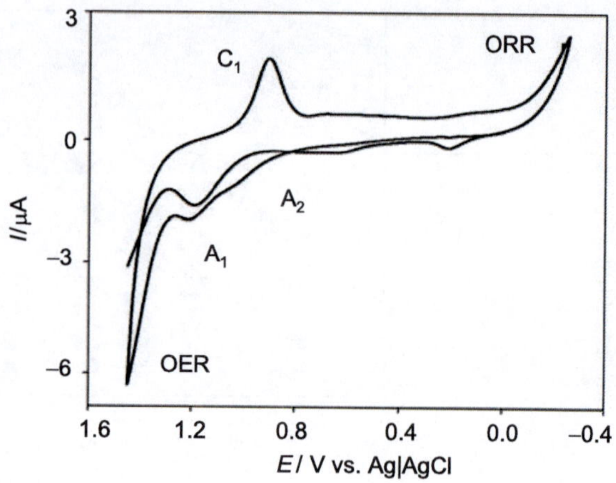

Fig. 10.22 Cyclic voltammogram of a commercial polycrystalline gold electrode in contact with air-saturated 0.10 M H_2SO_4 aqueous solution. Potential scan initiated at 0.0 V in the positive direction; potential scan rate 50 mV s^{-1}. Note that the axes are opposite to the IUPAC convention

dehydrated to form Au_2O_3 [64, 65]:

$$Au(OH)_{3surf} \rightarrow (1/2)Au_2O_{3surf} + (3/2)H_2O \tag{10.26}$$

experiencing successive electrochemical reactions at more positive potentials:

$$Au(OH)_{3surf} + Au_{solid} + H_2O \rightarrow 2AuO(OH)_{surf} + 3H^+_{aq} + 3e^- \tag{10.27}$$

$$5AuO(OH)_{surf} \rightarrow 2Au_2O_{3surf} + Au_{solid} + 2O_{2aq} + 5H^+_{aq} + 5e^- \tag{10.28}$$

The result is the generation of Au_2O_3 which can be reduced to gold via the process C_1.

$$Au_2O_{3surf} + 6H^+_{aq} + 6e^- \rightarrow 2Au_{solid} + 3H_2O \tag{10.29}$$

Additionally, note that there is a catalytic effect [Eq. (10.28)] on the OER process. Depending on the experimental conditions, the signal A_1 can show peak splitting. This can be interpreted as resulting from the exposure of different crystalline gold planes [66].

An aspect of particular interest is the appearance of new anodic signals in the second (and successive) scans. These signals (A_2) can be attributed to the oxidation of active surface sites (Au*) generated during electrochemical runs [66–68]. These active sites can be seen as coordinatively unsaturated sites whose electrochemical oxidation can be represented as:

$$Au^* \xrightarrow{-3e^-} Au^{3+} \xrightarrow{+(n+3)H_2O} Au^{3+} \cdot (n+3)H_2O \xrightarrow{-3H^+} Au(OH)_3 \cdot nH_2O \tag{10.30}$$

For electroanalytical purposes, the sensitivity can be increased using HCl electrolytes. Here, gold oxidation also involves more or less extensive oxidative dissolution to form Au(III) chloride complexes. This process appears to proceed through a multistep pathway involving chloride adsorption on gold resulting in the formation of Au(I) and Au(III) complexes in the solution phase [69]. For simplicity, the overall oxidation process can be described as:

$$Au_{solid} + xCl^{-}_{aq} \rightarrow AuCl_{x}{}^{(x-3)-}{}_{aq} + 3e^{-} \tag{10.31}$$

In HCl media, well-defined voltammetric responses were recorded at graphite electrodes after VIMP sampling on gold artifacts [70–74]. Examples of several archaeological samples are depicted in Fig. 10.23 [74]. Interestingly, these voltammograms exhibit the main oxidation process (A_1) accompanied by the anodic signals corresponding to the oxidation of active sites (A_2) and a stripping signal near 0.0 V (A_{AgCu}). This signal can in principle be assigned to the oxidation of silver and copper (usual minority components of gold samples) to the corresponding monovalent chlorides (M = Ag, Cu):

$$M(Au)_{solid} + Cl^{-}_{aq} \rightarrow MCl_{solid} + e^{-} \tag{10.32}$$

The relative intensity of the A_{AgCu} and A_1 peaks can be taken as representative of the (Ag plus Cu) to Au molar ratio in the metal, whereas the ratio between the peak currents for signals A_2 and A_1 is representative of the metallographic properties of the artifact and the surface aging.

Gold aging will be treated in detail in Chap. 12 in relation to dating. The essential idea is that there are at least three operating mechanisms: adsorption of oxygen functionalities, generation of surface defect sites, and migration and release of salts of copper and silver, ubiquitous components of gold [73, 74]. Here, one can remark that there is possible an in-depth analysis of voltammetric signals and that balancing the relative values of the peak currents, including those of OER, permits the characterization of different productions. Studies on gildings in Altarpieces [70], gold embroidery threads from archaeological textiles [71], and natural gold samples [72] have been reported.

Figure 10.24 presents a two-dimensional diagram in which the production of different archaeological gold sites falls in separated regions [74]. Here, the peak currents measured in voltammograms such as in Fig. 10.23 were combined. These were those recorded for the signal A_1 in the first ($i_{Au1}(1)$) and second ($i_{Au1}(2)$) anodic scans, the signal A2 in the second scan ($i_{Au2}(2)$), and the A_{AgCu} peak in the second scan ($i_{AgCu}(2)$). In Fig. 10.24, the $i_{AgCu}(2)/i_{Au1}(1)$ ratio, roughly representative of the silver plus copper content relative to gold, is represented versus the $[i_{Au1}(2) + i_{Au2}(2)]/i_{Au1}(1)$ ratio, roughly representative of the degree of alteration of the gold surface. One can see that samples from the Mapungubwe Hill grave area (dated back to 1200–1290 CE) and a ring of unknown age (*Ring im Grafenwald*, Rheinische Landesmuseum Trier) possess low silver plus copper contents but with different degrees of surface alteration. Higher silver plus copper contents

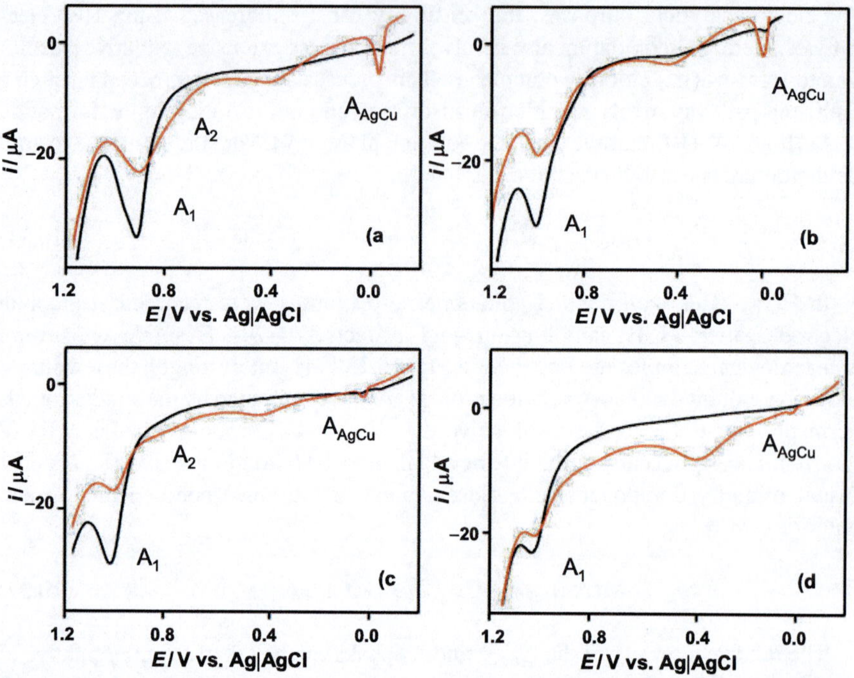

Fig. 10.23 First (black line) and second (red line) anodic linear potential scan voltammograms of microparticulate deposits of samples from different archaeological sites attached to graphite electrodes immersed into 0.10 M HCl. **a** Lamina dated back to 1300−1400 BCE, Mannheim museum, cat. n. MA184428); **b** Philippine piloncito (900−1300 CE); **c** Ingot from Santa Llúcia, Alcalà de Xivert, Museu de Belles Arts de Castelló, Spain catt. n. SL118-1 g (dated back to 575 BCE); and **d** *Ring im Grafenwald*, Rheinische Landesmuseum Trier, unknown age. Potential scan rate 50 mV s^{-1}. Adapted from Ref. [74], with permission. Note that the axes are opposite to the IUPAC convention

with different degrees of surface alteration were detected in samples from ingots recovered at the site of Santa Llúcia (Alcalà de Xivert, Spain), a lamina dated back to 1300−1400 BCE (Mannheim museum, MA03), and a ring fabricated in 1988 (Z01). On the whole, these data illustrate the capability of the VIMP methodology to discriminate production techniques/archaeological sites using a few nanograms of the sample.

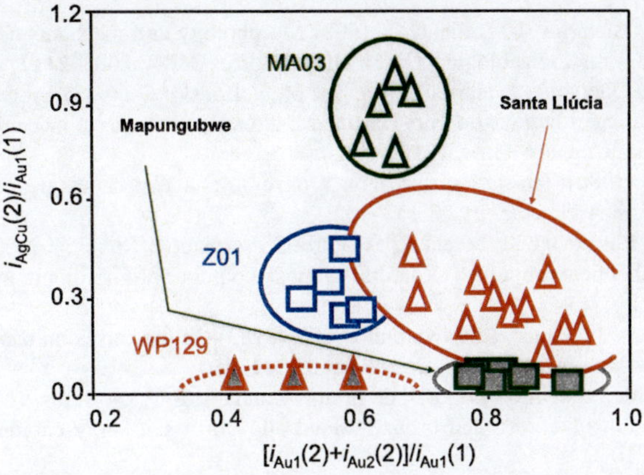

Fig. 10.24 Two-dimensional diagram used to discriminate the production of different archaeological gold sites combining the peak currents recorded in the first and second scan in voltammograms such as in Fig. 10.23. Data for the sites of Santa Llúcia (Alcalà de Xivert, Spain), a lamina dated back to 1300–1400 BCE (Mannheim museum, MA03), gold foils from the Mapungubwe Hill grave area (1200–1290 CE), a ring fabricated in 1988 (Z01), and a ring of unknown age (*Ring im Grafenwald*, Rheinische Landesmuseum Trier, WP129). Adapted from Ref. [74], with permission

References

1. Dillmann P, Wantkinson D, Angelini E, Adriaens A (2013) Corrosion and conservation of cultural heritage metallic artefacts. European Federation of Corrosion, Woodhead, Cambrige
2. Hamilton WA (1994) Biocorrosion: the action of sulphate reducing bacteria. Biochemistry of microbial degradation. Kluwer Academic Publishers, Dordrecht, pp 555–570
3. Perez N (2004) Electrochemistry and corrosion science, 1st edn. Kluwer Academic Publishers, Boston
4. Damjanovic A (1969) In: Bockris JOM, Conway BE (eds) Modern aspects of electrochemistry, vol 5. Plenum Press, New York
5. Venkatraman MS, Cole IS, Emmanuel B (2011) Model for corrosion of metals covered with thin electrolyte layers: pseudo-steady state diffusion of oxygen. Electrochim Acta 56:7171–7179
6. Oldham KB, Myland JC (1994) Fundamentals of electrochemical science. Academic Press, San Diego, p 175
7. Venkatraman MS, Cole IS, Emmanuel B (2011) Corrosion under a porous layer: a porous electrode model and its implications for self-repair. Electrochim Acta 56:8192–8203
8. Selwyn LS, Sirois PJ, Argyropoulos V (1999) The corrosion of excavated archaeological iron with details on weeping and akaganéite. Stud Conservat 44:217–232
9. Refait P, Génin JMR (1997) The mechanisms of oxidation of ferrous hydroxychloride β-$Fe_2(OH)_3Cl$ in aqueous solution: the formation of akaganeite versus goethite. Corros Sci 39:539–553
10. Ståhl K, Nielsen K, Jiang J, Lebech B, Hanson JC, Norby P, van Lanschot J (2003) On the akaganéite crystal structure, phase transformations and possible role in post-excavational corrosion of iron artifacts. Corros Sci 45:2563–2575
11. Neff D, Dillmann P, Bellot-Gurlet L, Béranger G (2005) Corrosion of iron archaeological artefacts in soil: characterisation of the corrosion system. Corros Sci 47:515–535
12. Neff D, Dillmann P, Decostes M, Beranger G (2006) Corrosion of iron archaeological artefacts in soil: estimation of the average corrosion rates involving analytical techniques and thermodynamic calculations. Corros Sci 48:2947–2970

13. Robbiola L, Blengino J-M, Fiaud C (1998) Morphology and mechanisms of formation of natural patinas on archaeological Cu-Sn alloys. Corros Sci 40:2083–2111
14. Robbiola L, Queixalos I, Hurtel LP, Pernot M, Volfovsky C (1988) Etude de la corrosion de bronzes archéologiques du Fort-Harrouard: altération externe et mécanisme d'altération stratifiée. Stud Conserv 33:205–215
15. Meeks ND (1986) Tin-rich surfaces on bronze—some experimental and archaeological considerations. Archaeometry 28:133–162
16. Chiavari C, Rahmouni K, Takenouti H, Joiret S, Vermaut P, Robbiola (2007) Composition and electrochemical properties of natural patinas of outdoor bronze monuments. Electrochim Acta 52:7760–7769
17. Soto L, Franey JP, Graedel TE, Kammlott GW (1983) On the corrosion resistance of certain ancient chinese bronze artifacts. Corros Sci 23:241–250
18. Cura D'Ars de Figueiredo Junior J, de Freitas Cunha Lins V, De Bellis VM (2007) Surface characterization of a corroded bronze-leaded alloy in a salt spray cabinet. App Surf Sci 253:7104–7107
19. Ingo GM, Plescia P, Engelini E, Riccucci C, De Caro T (2006) Bronze roman mirrors: the secret of brightness. Appl Phys A 83:611–615
20. Wang Q, Merkel JF (2001) Studies on the redeposition of copper in Jin bronzes from Tianma-Qucun, Shanxi, China. Stud Conservat 46:242–250
21. Quaranta M, Catelli E, Prati S, Sciutto G, Mazzeo R (2014) Chinese archaeological artefacts: microestructure and corrosion behaviour of high-leaded bronzes. J Cult Herit 15:283–291
22. Costa V, Urban F (2005) Lead and it alloys: metallurgy, deterioration and conservation. Rev Conserv Int Inst Conserv 6:48–62
23. Costa V, Leyssens K, Adriaens A, Richard N, Scholz F (2010) Electrochemistry reveals archaeological materials. J Solid State Electrochem 14:449–451
24. Souissi N, Bousselmi L, Khosrof S, Triki E (2004) Voltammetric behaviour of an archeaological bronze alloy in aqueous chloride media. Mater Corros 55:284–292
25. Serghini-Idrissi M, Bernard MC, Harrif FZ, Joiret S, Rahmouni K, Srhiri A, Takenouti H, Vivier V, Ziani M (2005) Electrochemical and spectroscopic characterizations of patinas formed on an archaeological bronze coin. Electrochim Acta 50:4699–4709
26. Doménech-Carbó A, Doménech-Carbó MT, Martínez-Lázaro I (2008) Electrochemical identification of bronze corrosion products in archaeological artefacts. A Case Study Microchim Acta 162:351–359
27. Satovic D, Martinez S, Bobrowski A (2010) Electrochemical identification of corrosion products on historical and archaeological bronzes using the voltammetry of micro-particles attached to a carbon paste electrode. Talanta 81:1760–1765
28. Doménech-Carbó A, Lastras M, Rodríguez F, Osete-Cortina L (2013) Mapping of corrosion products of highly altered archaeological iron using voltammetry of microparticles. Microchem J 106:41–50
29. Capelo S, Homem PM, Cavalheiro J, Fonseca ITE (2013) Linear sweep voltammetry: a cheap and powerful technique for the identification of the silver tarnish layer constituents. J Solid State Electrochem 17:223–234
30. Doménech-Carbó A, Doménech-Carbó MT, Peiró-Ronda MA (2011) 'One-touch' voltammetry of microparticles for the identification of corrosion products in archaeological lead. Electroanalysis 23:1391–1400
31. Blum D, Leyffer W, Holze R (1996) Pencil-leads as new electrodes for abrasive stripping voltammetry. Electroanalysis 8:296–297
32. Doménech-Carbó A, Doménech-Carbó MT, Pasíes-Oviedo T, Bouzas-Bello MC (2011) Application of modified tafel analysis to the identification of corrosion products on archaeological metals using voltammetry of microparticles. Electroanalysis 23:2803–2812
33. Doménech-Carbó A, Doménech-Carbó MT, Redondo-Marugán J, Osete-Cortina L, Vivancos-Ramón MV (2016) Electrochemical characterization of corrosion products in leaded bronze sculptures considering ohmic drop effects on Tafel analysis. Electroanalysis 28:833–845

34. Tétreault J, Robbiola E, Cano E, Van Bommel M, Scott DA, Dennis M, Barthés-Labrousse M-G, Minel L (2003) Corrosion of copper and lead by formaldehyde, formic and acetic acid vapours. Stud Conservat 48:237–250
35. Cano E, Lafuente D, Bastidas DM (2010) Use of EIS for the evaluation of the protective properties of coatings for metallic cultural heritage: a review. J Solid State Electrochem 14:381–391
36. Marušić K, Otmačić-Ćurković H, Horvat-Kurbegović Š, Takenouti H, Stupnišek-Lisac E (2009) Comparative studies of chemical and electrochemical preparation of artificial bronze patinas and their protection by corrosion inhibitor. Electrochim Acta 54:7106–7113
37. Kosec T, Kurkovic HO, Legat A (2010) Investigation of the corrosion protection of chemically and electrochemically formed patinas on recent bronze. Electrochim Acta 56:722–731
38. Doménech-Carbó A, Ramírez-Barat B, Petiti C, Goidanich S, Doménech-Carbó MT, Cano E (2020) Characterization of traditional artificial patinas on copper using the voltammetry of immobilized particles. J Electroanal Chem 877:114494
39. Doménech-Carbó A, Doménech-Carbó MT, Álvarez-Romero C, Pasíes-Oviedo T, Buendía M (2019) Screening of Iberian coinage in the 2th-1th BCE period using the voltammetry of immobilized particles. Electroanalysis 31:1164–1173
40. Doménech-Carbó A, Doménech-Carbó MT, Martínez-Lázaro I (2010) Layer-by-layer identification of copper alteration products in metallic works of art using the voltammetry of microparticles approach. Anal Chim Acta 680:1–9
41. Bongiorno V, Campodonico S, Caffara R, Piccardo P, Carnasciali MM (2012) Micro-Raman spectroscopy for the characterization of artistic patinas produced on copper-based alloys. J Raman Spectrosc 43:1617–1622
42. Basso E, Invernizzi C, Malagodi M, La Russa MF, Versan D, Lottici PP (2014) Characterization of colorants and opacifiers in roman glass mosaic *tesserae* through spectroscopic and spectrometric techniques. J Raman Spectrosc 45:238–245
43. Montoya N, Montagna E, Lee Y, Doménech-Carbó MT, Doménech-Carbó A (2017) Raman spectroscopy characterization of ten-cash productions from the late Chinese emperors to the Republic. J Raman Spectrosc 48:1337–1345
44. Debbichi L, Marco de Lucas MC, Pierson JF, Krüger P (2012) Vibrational properties of CuO and Cu_4O_3 from first-principles calculations, and Raman and infrared spectroscopy. J Phys Chem C 116:10232–10237
45. Doménech-Carbó A, Doménech-Carbó MT, Montagna E, Álvarez-Romero C, Lee Y (2017) Electrochemical discrimination of mints: the last Chinese emperors Kuang Hsü and Hsüan T'ung monetary unification. Talanta 169:50–56
46. Doménech-Carbó A, Doménech-Carbó MT, Álvarez-Romero C, Montoya N, Pasíes-Oviedo T, Buendía M (2017) Electrochemical characterization of coinage techniques the 17th century: the *maravedís* case. Electroanalysis 29:2008–2018
47. Budd P, Gale D, Pollard AM, Thomas RG, Williams PA (1993) Evaluating lead isotope data: further observations. Archaeometry 35:241–247
48. Attanasio D, Bultrini G, Ingo GM (2001) The possibility of provenancing a series of bronze punic coins found at tharros (Western Sardinia) using the literature lead isotope database. Archaeometry 43:529–547
49. Doménech-Carbó A, Doménech-Carbó MT, Peiró-Ronda MA, Osete-Cortina L (2011) Authentication of archaeological lead artifacts using voltammetry of microparticles: the case of the *Tossal de Sant Miquel* Iberian plate. Archaeometry 53:1193–1211
50. Doménech-Carbó A, Doménech-Carbó MT, Lstras M, Herrero M (2015) Detection of archaeological forgeries of Iberian lead plates using nanoelectrochemical techniques. The lot of fake plates from Bugarra (Spain). Forensic Sci Int 247:79–88
51. Domergue C (1990) Les mines de la Péninsule Ibérique dans l'Antiquité romaine. Roma
52. Doménech-Carbó A, Del Hoyo-Meléndez J, Doménech-Carbó MT, Piquero-Cilla J (2017) Electrochemical analysis of the first Polish coins using the voltammetry of immobilized particles. Microchem J 130:47–55

53. Doménech-Carbó MT, Di Turo F, Montoya N, Catalli F, Doménech-Carbó A, De Vito C (2018) FIB-FESEM and EMPA results on *Antoninianus* silver coins for manufacturing and corrosion processes. Sci Rep 8:10676
54. Di Fazio M, Felici AC, Catalli F, Doménech-Carbó MT, De Vito C, Doménech-Carbó A (2020) Solid-state electrochemical characterization of emissions and authorities producing Roman brass coins. Microchem J 152:104306
55. Doménech-Carbó A, Del Hoyo-Meléndez J, Labut M, Doménech-Carbó MT (2021) Electrochemical analysis of coffin portraits from the National Museum in Krakow. J Solid State Electrochem 25:2767–2776
56. Doménech-Carbó A, Bernabeu J (2019) Correlation between lead isotope analysis and solid-state electrochemistry for determining the provenance of archaeological bronze. J Solid State Electrochem 23:2803–2812
57. Doménech-Carbó A, Doménech-Carbó MT, Peiró-Ronda MA, Martínez-Lázaro I, Barrio J (2012) Application of the voltammetry of microparticles for dating archaeological lead using polarization curves and electrochemical impedance spectroscopy. J Solid State Electrochem 16:2349–2356
58. Di Turo F, Montoya N, Piquero-Cilla J, De Vito C, Coletti F, Favero G, Doménech-Carbó A (2017) Archaeometric analysis of Roman bronze coins from the *Magna Mater* temple using solid-state voltammetry and electrochemical impedance spectroscopy. Anal Chim Acta 955:36–47
59. Redondo-Marugán J, Piquero-Cilla J, Doménech-Carbó MT, Ramírez-Barat B, Al-Sekhaneh W, Capelo S, Doménech-Carbó A (2017) Characterizing archaeological bronze corrosion products intersecting electrochemical impedance measurements with voltammetry of immobilized particles. Electrochim Acta 246:269−279
60. Retter U, Widmann A, Siegler K, Kahlert H (2003) On the impedance of potassium nickel(II) hexacyanoferrate(II) composite electrodes—the generalization of the Randles model referring to inhomogeneous electrode materials. J Electroanal Chem 546:87–96
61. Chen G, Waraksa CC, Cho H, Macdonald DD, Mallouk TE (2003) EIS studies of Porous oxygen electrodes with discrete particles. I. Impedance of oxide catalyst supports. J Electrochem Soc 150:E423–E428
62. Lee SJ, Pyun SL (2007) Assessment of corrosion resistance of surface-coated galvanized steel by analysis of the AC impedance spectra measured on the salt-spray-tested specimen. J Solid State Electrochem 11:829–839
63. Doyle RL, Lyons MEG (2014) The mechanism of oxygen evolution at superactivated gold electrodes in aqueous alkaline solution. J Solid State Electrochem 18:3271–3286
64. Diaz-Morales O, Calle-Vallejo F, de Munck C, Koper MTM (2013) Electrochemical water splitting by gold: evidence for an oxide decomposition mechanism. Chem Sci 4:2334–2343
65. Hasse U, Wulff H, Helm CA, Scholz F (2013) Formation of gold surfaces with a strongly preferred {100}-orientation. J Solid State Electrochem 17:3047–3053
66. Jeyabharathi C, Ahrens P, Hasse U, Scholz F (2016) Identification of low-index crystal planes of polycrystalline gold on the basis of electrochemical oxide layer formation. J Solid State Electrochem 20:3025–3031
67. Burke LD, O'Mullane AP (2000) Generation of active surface states of gold and the role of such states in electrocatalysis. J Solid State Electrochem 4:285–297
68. Nowicka A, Hasse U, Sievers G, Donten M, Stojek Z, Fletcher S, Scholz F (2010) Selective knockout of gold active sites. Angew Chem Int Ed 49:3006–3009
69. Herrera-Gallego J, Castellano CE, Calandra AJ, Arvia AJ (1975) The electrochemistry of gold in acid aqueous solutions containing chloride ions. J Electroanal Chem 66:207–230
70. Ferragud-Adam X, Piquero-Cilla J, Doménech-Carbó MT, Guerola-Blay V, Company X, Doménech-Carbó A (2017) Electrochemical analysis of gildings in Valencia Altarpieces: a cross-age study since 15th until 20th century. J Solid State Electrochem 21:1477–1487
71. Martínez B, Piquero-Cilla J, Montoya N, Doménech-Carbó MT, Doménech-Carbó A (2018) Electrochemical analysis of gold embroidery threads from archaeological textiles. J Solid State Electrochem 22:2205–2215

72. Doménech-Carbó A, Scholz F, Schmitt RT, Usera J, García-Fuster AM, De la Fuente E, Cinnaya J, Piquero-Cilla J, Montoya N (2017) Electrochemical characterization of natural gold samples using the voltammetry of immobilized particles. Electrochem Commun 85:23–26
73. Doménech-Carbó A, Scholz F, Doménech-Carbó MT, Piquero-Cilla J, Montoya N, Pasíes-Oviedo T, Gozalbes M, Melchor-Montserrat JM, Oliver A (2018) Dating of archaeological gold by means of solid state electrochemistry. ChemElectroChem 5:2113–2117
74. Doménech-Carbó A, Scholz F, Brauns M, Tiley-Nel S, Oliver A, Aguilella G, Mpntoya N, Doménech-Carbó MT (2020) Electrochemical dating of archaeological gold based on refined peak current determinations and Tafel analysis. Electrochim Acta 337:135759

Additional Literature

75. Aldaz A, España T, Montiel V, Lopez-Segura M (1986) A simple tool for the electrolytic restoration of archaeological metallic objects with localized corrosion. Stud Conserv 31:175–176
76. Bastidas JM, López-Delgado A, Cano E, Polo JL, López FA (2000) Copper corrosion mechanism in the presence of formic acid vapor for short exposure times. J Electrochem Soc 147:999–1005
77. Beldjoudi T, Bardet F, Lacoudre N, Andrieu S, Adriaens A, Constantinides I, Brunella P (2001) Surface modification processes on European union bronze reference materials for analytical studies of cultural artifacts. Surf Eng 17:231–235
78. Bernard MC, Joiret S (2009) Understanding corrosion of ancient metals for the conservation of cultural heritage. Electrochim Acta 54:5199–5205
79. Cepriá G, Abadías O, Pérez-Arantegui J, Castillo JR (2001) Electrochemical behavior of silver-copper alloys in voltammetry of microparticles: a simple method for screening purposes. Electroanalysis 13:477–483
80. Cepriá G, Cepriá JJ, Ramajo J (2004) Fast and simple electroanalytical identification of iron oxides in geological samples without sample pretreatment. Microchim Acta 144:139–145
81. Constantinides I, Adriaens A, Adams F (2002) Surface characterization of artificial corrosion layers on copper alloy reference materials. Appl Surf Sci 189:90–10
82. Costa V (2001) The deterioration of silver alloys and some aspects of their conservation. Rev Conserv, International Institute of Conservation 2:19–35
84. Di Carlo G, Giuliani C, Riccucci C, Pascucci M, Messina E, Fierro G, Lavorgna M, Ingo GM (2017) Artificial patina formation onto copper-based alloys: chloride and sulphate induced corrosion processes. Appl Surf Sci 421:120–127
85. Di Turo F (2019) Limits and perspectives of archaeometric analysis of archaeological metals: a focus on the electrochemistry for studying ancient bronze coins. J Cult Herit 43:271–281
86. Degrigny C, Le Gall R (1999) Conservation of ancient lead artifacts corroded in organic acid environments: electrolytic stabilization/consolidation. Stud Conserv 44:157–169
87. Doménech-Carbó A (2013) Electrochemical analysis of metallic heritage artefacts: voltammetry of microparticles (VMP). In: Edwards HGM, Vandenabeele P (eds) Analytical archaeometry. RSC, London
88. Doménech-Carbó MT, Álvarez-Romero C, Doménech-Carbó A, Osete-Cortina L, Martínez ML (2019) Microchemical surface analysis of historic copper-based coins by the combined use of FIB-FESEM-EDX, OM, FTIR spectroscopy and solid state electrochemical techniques. Microchem J 148:573–581
89. Encinas-Bachiller P, Tascón-García ML, Vázquez-Barbado MD, Sánchez-Batanero P (1994) Electrochemical study of iron(II) and iron(III) compound mixtures in the solid state. Application to magnetite characterization. J Electroanal Chem 371:161–166
90. Fetisov VB, Ermakov AN, Belysheva GM, Fetisov AV, Kamyshov VM, Brainina KhZ (2004) Electrochemical dissolution of magnetite in acid solutions. J Solid State Electrochem 8:565–571

91. Gholenji FA, Adriaens A (2012) Electrochemical quantification of copper-based alloys using voltammetry of microparticles: optimization of the experimental conditions. J Solid State Electrochem 16:535–543
92. Grygar T (1997) Dissolution of pure and substituted goethites controlled by the surface reaction under conditions of abrasive stripping voltammetry. J Solid State Electrochem 1:77–82
93. Hasse U, Fricke K, Dias D, Sievers G, Wulff H, Scholz F (2012) Grain boundary corrosion of the surface of annealed thin layers of gold by OH·radicals. J Solid State Electrochem 16:2383–2389
94. Hughes R, Rowe M (1991) The colouring bronzing and patination of metals. Thames & Hudson, London
95. Izumi T, Watanabe I, Yokoyama Y (1991) Activation of a gold electrode by electrochemical oxidation-reduction pretreatment in hydrochloric acid. J Electroanal Chem Interf Electrochem 303:151–160
96. Jeyabharathi C, Hasse U, Ahrens P, Scholz F (2014) Oxygen electroreduction on polycrystalline gold electrodes and on gold nanoparticle-modified glassy carbon electrodes. J Solid State Electrochem 18:3299–3306
97. Mancey DS, Shoesmith DW, Lipkowski J, McBride AC, Noel J (1993) An electrochemical investigation of the dissolution of magnetite in acidic electrolytes. J Electrochem Soc 140:637–642
98. MacLeod ID, Richtie IM (1981) Detection of debasement in (forged) silver coins by means of corrosion potential measurements. Archaeometry 23:65–70
99. MacLeod ID (1991) Identification of corrosion products on non-ferrous metal artifacts recovered from shipwrecks. Stud Conserv 36:222–234
100. Mouhandess MT, Chassagneux F, Vittori O, Accary A, Reeves RM (1984) Some theoretical aspects of the electrodissolution of iron oxide (α-Fe_2O_3) in carbon paste electrodes with acidic binder. J Electroanal Chem 181:93–105
101. Neff D, Dillmann P, Bellot-Gurlet L, Beranger G (2005) Corrosion of iron archaeological artefacts in soil: characterisation of the corrosion system. Corros Sci 47:515–535
102. Plenderleith HJ, Werner AE (1971) The conservation of antiquities and works of art: treatment, repair and restoration, 2nd edn. Oxford University Press, London
103. Robbiola L, Portier R (2006) A global approach to the authentication of ancient bronzes based on the characterization of the alloy–patina–environment system. J Cult Herit 7:1–12
104. Sandu I, Marutoiu C, Sandu IG, Alexandru A, Sandu AV (2006) Authentication of old bronze coins I. Study on archaeological Patina. Acta Universit Cibinensis Sect F Chem 9:39–53
105. Scholz F, de Lara L, González G, de Carvalho LM, Hilgemann M, Brainina KhZ, Kahlert H, Jack RS, Minh DT (2007) Indirect electrochemical sensing of radicals and radical scavengers in biological matrices. Angew Chem Int Ed 46:8079–8081
106. Scott DA (2002) Copper and bronze in art: corrosion, colorants and conservation. II. Getty Publ, Los Angeles
107. Zakharchuk N, Meyer S, Lange B, Scholz F (2000) A comparative study of lead oxide modified graphite paste electrodes and solid graphite electrodes with mechanically immobilized lead oxides. Croat Chem Acta 73:667–704
108. Žežula I, Gálová M (1999) Application of abrasive stripping voltammetry in corrosion science. I. determination of the corrosion potentials of metals. J Solid State Electrochem 3:231–233

Chapter 11
Metallic Heritage: Electrochemistry of Metal Objects

11.1 Direct Electrochemistry of Metal Artifacts

In the preceding chapter, the application of electrochemical techniques has been described to study samples, extracted from the corrosion layers of metallic artifacts, i.e., separated from the original object. In this chapter, we will describe the application of electrochemical techniques directly in situ onto the object (sculptures, structural, and decorative architectural elements) and/or using the object as an element (working electrode) of an electrochemical circuit.

Two main fields will be considered, electrochemical techniques applied with analytical purposes (identification of materials, techniques, etc.), including diagnosis of conservation state, and methodologies of protection and techniques of stabilization. The second group of techniques is aimed at the preservation of metallic objects but differs in their specific 'philosophies'. The former is focused on the formation of protective coatings over the metal surfaces, while the second is addressed to eliminating corrosion products and corrosion-promoting species—essentially, dechlorination—via the application of electrochemistry, i.e., of either potential or current inputs. Some of these aspects were briefly introduced in Chap. 2 and will be treated here in more detail.

It is important to recognize that the objects recovered from a given environment (burial, shipwrecks) tend to equilibrate with the same reaching in most cases a stationary situation. This implies that the separation from the medium may produce an enhancement of the degradation of the artifact. For instance, buried iron objects can contain soluble salts that, once extracted and exposed to a humid atmosphere, can dissolve thus promoting the corrosion of the remaining core [1].

A. Doménech-Carbó and M. T. Doménech-Carbó, *Electrochemistry for Cultural Heritage*, Monographs in Electrochemistry,
https://doi.org/10.1007/978-3-031-31945-7_11

11.2 Diagnosis

11.2.1 Polarization Curves

Polarization curves constitute one of the most classical methods to study metal corrosion, typically applied to metal coupons. Ideally, the experiment consists of the measurement of the current flowing through the electrochemical cell when the oxidation of a metal surface in contact with a given electrolyte is forced by the application of different potentials. In common experiments, the potential is scanned from highly cathodic to highly anodic potentials at a low scan rate (typically 0.1–10 mV s^{-1}) to achieve steady-state conditions. The resulting data are displayed as log (current) versus (potential), i.e., Tafel-type plots of ln$I$ versus E. Figure 11.1 shows a typical potentiodynamic polarization curve obtained for a eurocent coin in contact with two differently aggressive media, mineral water and 0.10 M $NaClO_4$ aqueous solution, illustrating the differences between them.

In a typical polarization curve for a metal coupon in contact with neutral or alkaline electrolyte solutions, one can distinguish different regions. Starting at sufficiently cathodic potentials, the HER process is developed, being followed by a region of slow variation of the current with the applied potential (A). This is characterized by the mass transport control associated with hydrogen ions and is influenced by the ohmic solution resistance. In this region, often an inflection appears that reflects the change to mass transport control associated with other species (B). This is the beginning of

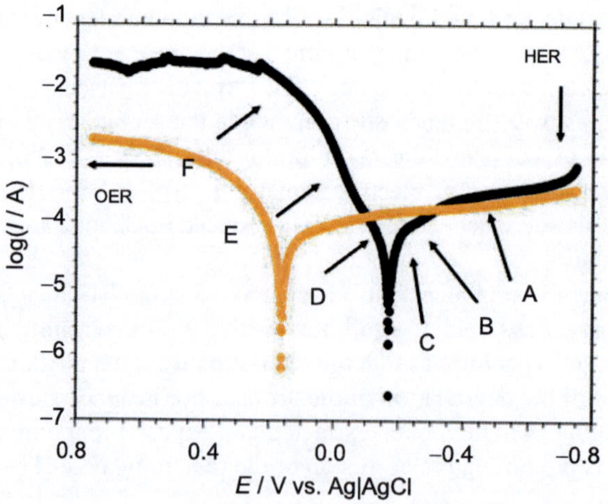

Fig. 11.1 Potentiodynamic polarization curves recorded for an aged eurocent coin in contact with mineral water (orange circles) and 0.10 M $NaClO_4$ aqueous solution (black circles). The potential was initiated at −0.75 V versus Ag/AgCl in the positive direction at a scan rate of 5 mV s^{-1}. The arrows marking the different (A–F, see text) regions correspond to the black curve. Note that the axes are opposite to the IUPAC convention

the active region where the metal is oxidized to metal ions under potential control. This can be described in terms of the Butler-Volmer kinetics as,

$$J = J_{corr}\left\{\exp\left(\frac{E - E_{corr}}{b_{an}}\right) - \exp\left(-\frac{E - E_{corr}}{b_{cat}}\right)\right\} \tag{11.1}$$

where E is the applied potential (difference) and J is the recorded current density. E_{corr} and Jcorr are the corrosion potential and corrosion current density, respectively, while b_{cat} ($= RT/\alpha_{cat}nF$) and b_{an} ($= RT/\alpha_{an}nF$) are the cathodic and anodic Tafel slope, respectively. Then, the current reaches a sharp minimum corresponding to the rest potential where the current is zero (but which cannot be displayed in a logarithmic scale). Around this minimum there are two regions (C, D) where there is a linear relationship between the logarithm of the current and the potential. These are the Tafel regions where

$$E = a \pm b \ln|J| \tag{11.2}$$

Here, a is a constant, and b equals b_{cat} or b_{an}. In practice, the corrosion current is accompanied by the charging process of the double layer (which can be seen as an interfacial capacitance associated to the charge density stored at the electrode/electrolyte solution interface). The smaller the ratio of the Faraday current to the capacitive one in the rest potential region, the more the zero-current potential during the sweep will deviate from the true rest potential. This means that, by applying a Tafel representation, the minimum will not be at the potential to which the anodic and cathodic sections are to be extrapolated, and even these sections will be distorted by the capacitive currents. This effect increases with the potential scan rate so that, to avoid this perturbation on the determination of the corrosion parameters, polarization curves are recorded at low potential scan rates (often below 1 mV s^{-1}). For the same reasons, instead of large perturbations, small-perturbation methods (typically EIS) are customary in corrosion studies.

The polarization curve may display an inflection (E) associated with mass transport control further yielding an almost constant value (F) associated to the formation of a (passive) oxide layer. This region eventually produces a moderate current decrease at more positive potentials reflecting the occurrence of a transpassive processes (oxidative dissolution of the oxide layer). Finally, if sufficiently anodic potentials are reached, the current varies abruptly as a result of the appearance of the OER process. Passivity is related to the oxide-covered state of the metal. For metals of very high oxide stability, the active region is often missing; for many active metals, there is very difficult to find a passive region.

The potential at which the sharp minimum of current appears is called the corrosion potential, E_{corr}. This parameter, as well as the corrosion current, I_{corr}, is strictly defined by the intersection of the two Tafel branches and can be calculated from the respective slopes and intercepts, from which the polarization resistance (or corrosion resistance) by unit of surface area is calculated as,

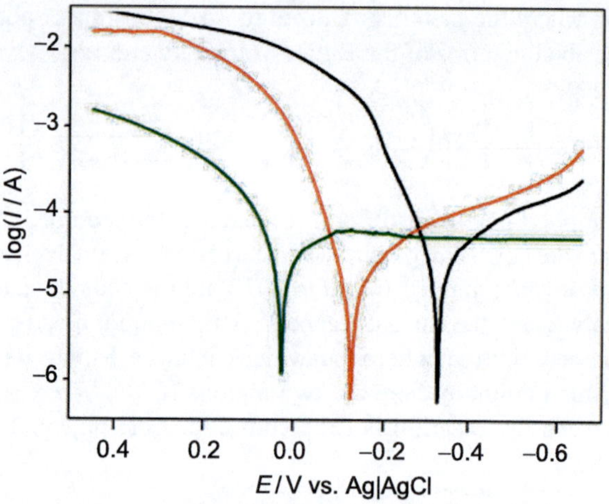

Fig. 11.2 Potentiodynamic polarization curves recorded for a eurocent coin in mineral water (green line), 0.10 M $NaClO_4$ (red line) and 0.05 M NaCl (black line) aqueous solutions. Potential scan initiated at −0.65 V versus Ag/AgCl in the positive direction; potential scan rate 5 mV s^{-1}. The coin was submerged until its half. Note that the axes are opposite to the IUPAC convention

$$R_p = \left(\frac{b_{cat} b_{an}}{b_{cat} + b_{an}}\right)\left(\frac{1}{J_{corr}}\right) \tag{11.3}$$

These parameters are of interest in studies on protective coatings, commented on in a further section.

In Fig. 11.2, the polarization curves are compared which were recorded for a eurocent (equivalent to a copper coupon) in contact with three different electrolytes: mineral water, 0.10 M $NaClO_4$, and 0.05 M NaCl. One can see that the corrosion potential is shifted to more negative potentials by increasing the corrosivity of the medium.

Since polarization curves involve the oxidation of the metal surface, their direct application to works of art and archaeological objects cannot be recommended. Notice that recording a polarization curve starting at the far cathodic end, the metal corrosion products can be reduced and, if hydrogen bubbling is produced, the corrosion layer can be removed. Then, when the region of the 'rest potential' is reached, the material has changed considerably relative to the original state of the artifact.

Accordingly, such measurements are usually applied to metal probes [2]. Eventually, they can be used for metal powders [3] and corrosion products [4] applying VIMP sampling. This technique can also be used for direct testing of archaeological objects in exceptional cases of abundant metal remains using 'smooth' electrolytes [5]. As previously noted, this technique is generally applied to metal probes submitted to artificial, accelerated aging tests, and/or testing protective coatings.

11.2.2 Electrochemical Impedance Spectroscopy

EIS has been extensively used in the field of metallic heritage. Traditionally, this technique has been used to test protective coatings [6] and passivation [7, 8] using redox probes (usually metal coupons of a few cm^2 of surface area). Accordingly, impedance parameters are normalized to surface area units; resistances are expressed as Ω cm^2 and capacitances as F cm^{-2}.

The performance of direct impedance measurements on metal artifacts is receiving increasing attention. In the case of artifacts of small size, there is the possibility of immersion into adapted electrochemical cells although, obviously, this is an exceptional practice in the archaeological domain [5, 9, 10]. An example of the observed impedance responses is shown in Fig. 11.3 for a eurocent coin immersed until its half into 0.10 M $NaClO_4$. Here, the experimental spectrum is compared with theoretical ones using two different equivalent circuits (compare with Figs. 2.6, 2.7, 2.8, 2.9 and 2.10 in Chap. 2). It is pertinent to note that a variety of equivalent circuits have been used to describe the impedance response of metal probes [6].

To minimize the deterioration of the objects, air-saturated mineral water was used as the medium for the ORR process as the redox probe applying the corresponding bias potential between −0.4 and −0.7 V versus Ag/AgCl [9, 10]. Again, the main difficulty is to model the impedance response. In principle, there is a metallic nucleus covered by a fine passive layer and a gross layer of corrosion more or less continuously covering the passive layer and having more or less porosity. Of course, as indicated in Chap. 2, the Kramers–Kronig (K–K) transforms have to be applied to confirm the 'true spectral' nature of the impedance measurements.

A representative example is provided by the impedance response of iron nails from the ancient Alcantarins convent in Ontinyent, (Spain), dated back to the 19th–twentieth century. A quantity of ca. 200 of these nails was found and stored at the Institut de Restauració del Patrimoni, Universitat Politècnica de València, Spain, where a minority fraction (8) was submitted to desalination protocols with 0.10 M NaOH and Na_2SO_3 NaOH baths, two of the recommended desalination procedures [11]. A representative impedance spectrum of the specimen in contact with air-saturated 0.10 M $NaClO_4$ is depicted in Fig. 11.4.

A scheme with the equivalent circuit used to model the recorded EIS spectra is shown in Fig. 11.5 [5]. This was inspired by studies on pore models [12] assuming that the metal nucleus is entirely covered by an impermeable passive layer of magnetite, and the corrosion layer contains pores filled by the electrolyte. The reported circuit includes a solution resistance (R_s), a charge transfer resistance at the passive layer/electrolyte solution interface (R_{pas}), and interfacial capacitance of the passive layer/electrolyte solution contact, Q_{pas}. This branch is in parallel with a second branch caused by the charge transfer resistance at the corrosion layer/electrolyte interface, R_{corr}, the capacitance associated with the corrosion layer/electrolyte interface, Q_{corr}, and an impedance associated with ion diffusion through the corrosion layer, D (Warburg-type element). The circuit is completed by a resistance due to the metal nucleus/conducting passive layer interface, R_g, and a second resistance, R_{in}, due to

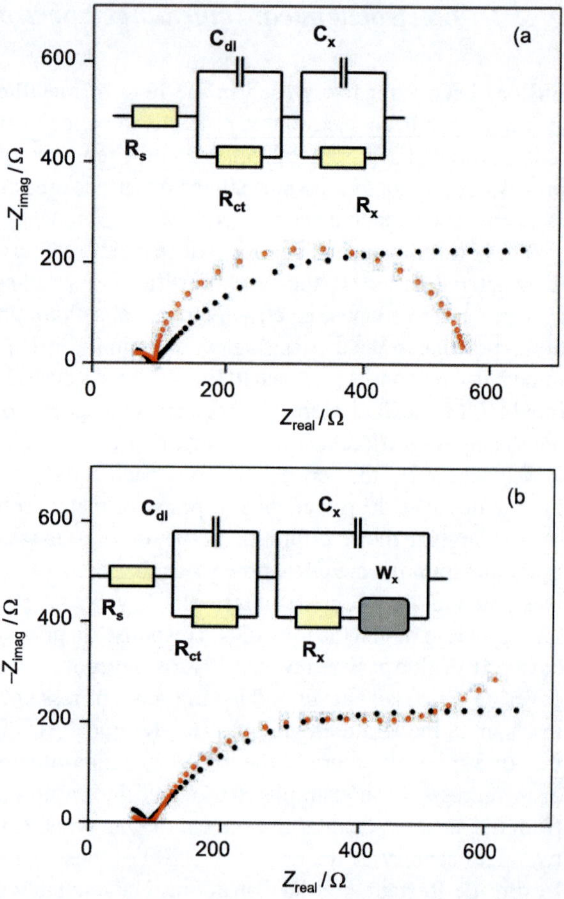

Fig. 11.3 Nyquist plots of $-Z_{imag}$ versus Z_{real} of the impedance spectra recorded for a eurocent coin immersed into 0.10 M $NaClO_4$ aqueous solution (black circles). Theoretical spectra are depicted as red circles corresponding to the best fit of experimental data to the equivalent circuits in **a** and **b**. Frequency range: 0.10–10^5 Hz; peak-to-peak amplitude 10 mV; and bias potential −0.60 V versus Ag/AgCl

the internal resistance of the voltmeter. These resistances incorporate the transport resistance in the metal nucleus and the conducting passive layer plus the artifact/connector contact and the transport resistance between the corrosion layer and the passive layer. These can be viewed as related to the contact potential between different metals and/or semiconductors [13].

As usual, the capacitive contributions are represented by constant phase elements. These account for the non-ideal capacitive behavior of the interfacial charge transfer mechanism associated with the non-uniform distribution of current. Non-ideality is attributed to surface roughness, impurities, dislocations, or grain boundaries [14–16].

There is another difficulty to be accounted for: the objects have in general an irregular shape, and in replicate experiments, they are differently immersed. This means that the absolute values of the impedance parameters can significantly vary from one experiment to another and that there is no possibility of an accurate measurement of the immersed area.

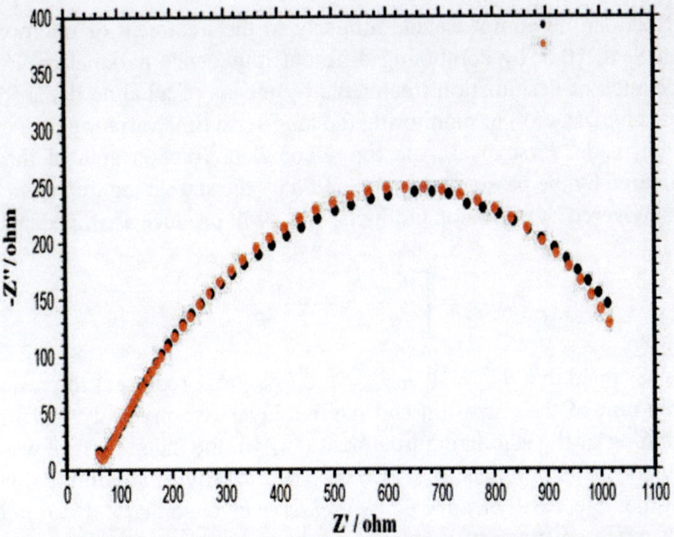

Fig. 11.4 Nyquist plots of $-Z_{imag}$ versus Z_{real} of the impedance spectra recorded for an iron nail from the ancient Alcantarins convent in Ontinyent, immersed into 0.10 M $NaClO_4$ aqueous solution (black circles). The theoretical spectrum is depicted as red circles corresponding to the best fit of experimental data to the equivalent circuit in Fig. 11.5. Frequency range: 0.10–10^5 Hz; peak-to-peak amplitude 10 mV; and bias potential −0.60 V versus Ag/AgCl

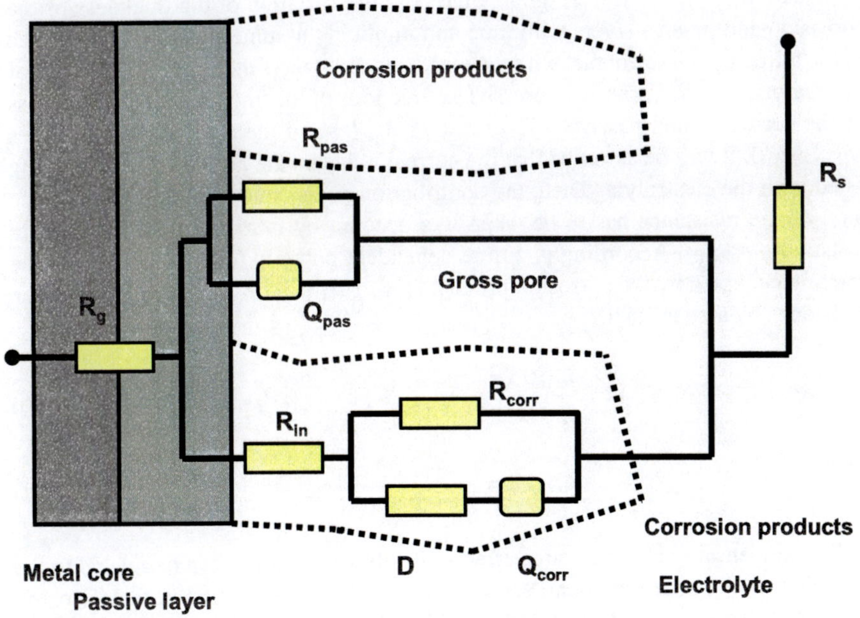

Fig. 11.5 Equivalent circuit used to describe the impedance measurements on iron nails in contact with mineral water using the ORR process as a redox probe reported in [5]

This effect can be compensated, similarly to the treatment of microparticulate deposits in Sect. 10.6, by combining different impedance parameters. For testing the performance of desalination treatments by means of alkaline baths ($NaHCO_3$, Na_2SO_3 plus NaOH, etc.) or monitoring the long-term time variation of such parameters can be made. First of all, the time-dependent fraction area of the metallic nucleus covered by the passive layer generated by the stabilizing treatment θ_{pas}, and the fraction covered by corrosion products, θ_{corr}, will produce an impedance of [17],

$$Z_{tot}(t) = \left[\frac{\theta_{corr}(t)}{R_{corr}} + \frac{1-\theta_{corr}(t)}{R_{pas}}\right]^{-1} \tag{11.4}$$

where it is assumed that $\theta_{pas} = 1 - \theta_{corr}$ and R_{corr}, R_{pas} represent the resistance by surface area unit of the corrosion and passive layers, respectively. Assuming that at the beginning of the stabilizing treatment ($t = 0$), the metal surface was entirely covered by corrosion products, $Z_{tot}(0) = R_{corr}(0)$. Ideally, at an infinite time, all the surfaces of the object will be covered by a passive layer so that $Z_{tot}(\infty) = R_{pas}(\infty)$. Then, the surface coverage of corrosion products at an intermediate time t can be calculated as:

$$\vartheta_{corr}(t) = \left(\frac{Z_{tot}(0)}{Z_{tot}(t)}\right)\left(\frac{Z_{tot}(t) - Z_{tot}(\infty)}{Z_{tot}(0) - Z_{tot}(\infty)}\right) \tag{11.5}$$

Equation (11.5), however, ignores the possible variation of the thickness of the corrosion and passive layers with time and implicitly assumes that the surface area of the artifact put into contact with the electrolyte is always the same. To account for this factor, R_{corr}, R_{pas}, can be expressed as functions of the time-dependent thickness of the corrosion and passive layers, $\delta_{corr}(t)$, $\delta_{pas}(t)$ and the respective resistivities (ρ_{corr}, ρ_{pas}). It will be assumed that the corrosion layers are porous, the pores being filled with the electrolyte. Then, the contribution of the solution-filled pores to the total ohmic resistance has to be taken into account by introducing the electrolyte resistivity, $\rho_{electrol}$. Accordingly, a time-dependent porosity (defined as the volume fraction occupied by the corrosion products [18, 19]) will be introduced, so that the resistance by area unit will be,

$$R_{corr}(t) = \frac{\delta_{corr}(t)}{\vartheta_{corr}(t)}\{\chi(t)\rho_{corr} + [1-\chi(t)]\rho_{electrol}\} \tag{11.6}$$

$$R_{pas}(t) = \frac{\delta_{pas}(t)}{[1-\vartheta_{corr}(t)]}\rho_{pas} \tag{11.7}$$

To compensate for surface area effects, the above quantities can be related to the capacitance by surface area unit associated with charge separation in the corrosion layer, $C_{corr}(t)$, and the electrolyte filling the pores, $C_{electrol}(t)$. The former can be expressed as,

$$C_{corr}(t) = \varepsilon_{corr}\frac{\vartheta_{corr}(t)}{\delta_{corr}(t)}\chi(t) \tag{11.8}$$

where ε_{corr} is the averaged dielectric permittivity of the corrosion products. The second capacitance can be written as,

$$C_{electrol}(t) = \varepsilon_{electrol}\frac{\vartheta_{corr}(t)}{\delta_{corr}(t)}[1 - \chi(t)] \tag{11.9}$$

$\varepsilon_{electrol}$ being the dielectric permittivity of the electrolyte. Assuming that the resistivities and permittivities are time-independents and considering that, in contact with aqueous electrolytes, $\varepsilon_{el} >> \varepsilon_{corr}$, the inverse of the surface-independent RC product will provide a time-characteristic frequency $f(t)$ given by:

$$f(t) = (R_{corr}C_{corr})^{-1} = \varepsilon_{electrol}[1 - \chi(t)]\{\rho_{corr}\chi(t) + \rho_{el}[1 - \chi(t)]\} \tag{11.10}$$

This frequency can be approximated by that at which there is a maximum in the (negative) phase angle in impedance spectra such as in Fig. 11.6. Accordingly, the variation in the porosity of the corrosion layer relative to the initial porosity during the stabilization/desalination process can be approximated, assuming that $\rho_{electrol} << \rho_{corr}$, by means of the relationship [5]:

$$\frac{\chi(t)}{\chi(0)} \approx \frac{f(0)}{f(t)} \tag{11.11}$$

As shown in Bode plots of $-\varphi$ versus log$f$ depicted in Fig. 11.6, the maximum phase angle is shifted from low to middle frequencies along the desalination period, reflecting the variation in parameters such as the porosity.

The in situ study of metal objects is receiving increasing attention, focused on the use of EIS by means of portable equipment. These involve two basic approaches [6]: the localized use of liquid electrolytes or the use of jellified electrolytes.

The first approach consisted of the use of contact probe cells in which the liquid electrolyte is supported by a cloth forming a film between the working and counter electrodes, for which different designs have been devised [6]. The second approach involves the use of gel and hydrogel electrolytes. The use of gel electrolytes such as those employed in medicine for electrocardiograms offers problems due to the difficulty to entering into porous surfaces, high impedance, and inhomogeneous distribution of the gel on the metal surface [20]. Accordingly, recent research is focused on hydrogels based on agar and agarose.

The first aspect to be treated is the selection of the electrolyte and cell. Different jellified electrolytes, cell designs, and electrode configurations have been described [21–27]. For a given hydrogel, equilibrating solution determines the degree of swelling and the conductivity. Reasonable conductivity has to be attained in principle without introducing aggressive ions such as chlorides.

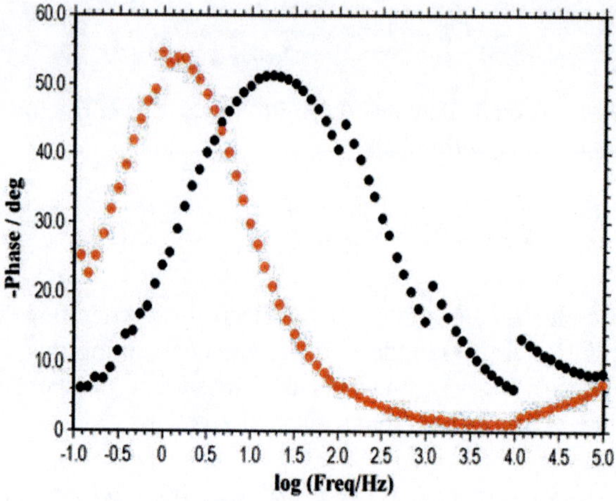

Fig. 11.6 Bode plots of $-\varphi$ versus log$f$ of the impedance spectra recorded for an iron nail from the ancient Alcantarins convent in Ontinyent, immersed into 0.10 M $NaClO_4$ aqueous solution at the beginning (red circles) and after 60 days (black circles) of desalination process performed in contact with a 0.10 M NaOH bath. Frequency range: 0.10–10^5 Hz; peak-to-peak amplitude 10 mV; and bias potential −0.60 V versus Ag/AgCl

The interpretation of data is a second crucial aspect in this field. A variety of responses and equivalent circuits have been reported. Figure 11.7 depicts the Nyquist and Bode plots obtained for two Roman coins (*Valens AES3* and *Rostrum Tridens AES*) using an agar cell [27].

Experimental impedance spectra can reasonably be fitted to the equivalent circuit depicted in Fig. 11.8a. Here, the first resistance/constant phase element association (R_{ct}/CPE_{dl}) represents the charge transfer resistance and double-layer capacitance of the metal/electrolyte interface, whereas the second association ($R_{corr} - CPE_{corr}$) can be ascribed to the patina impedance. There is, however, a variety of equivalent circuits reported for describing EIS on metal patinas with jellified electrolytes. Several ones [6] are schematized in Fig. 11.8. These circuits include additional elements to account for charge transport through the jellified electrolyte and the different interfaces.

11.2.3 Open-Circuit Potential

Open-circuit potential (OCP) measurements have been extensively used to characterize metallic materials in the classical examination of metal coupons submitted to corrosion tests. In fact, OCPs are routinely determined prior to most EIS measurements. The application of OCP tests in the field of cultural heritage was aimed at qualitative analysis using mineral water and Na_2SO_4, KNO_3, and $NaHCO_3$ electrolytes [28] and was recently applied to corrosion evaluation in reinforcements embedded in concrete [29]. Here, the OCP is measured in a cell where the working

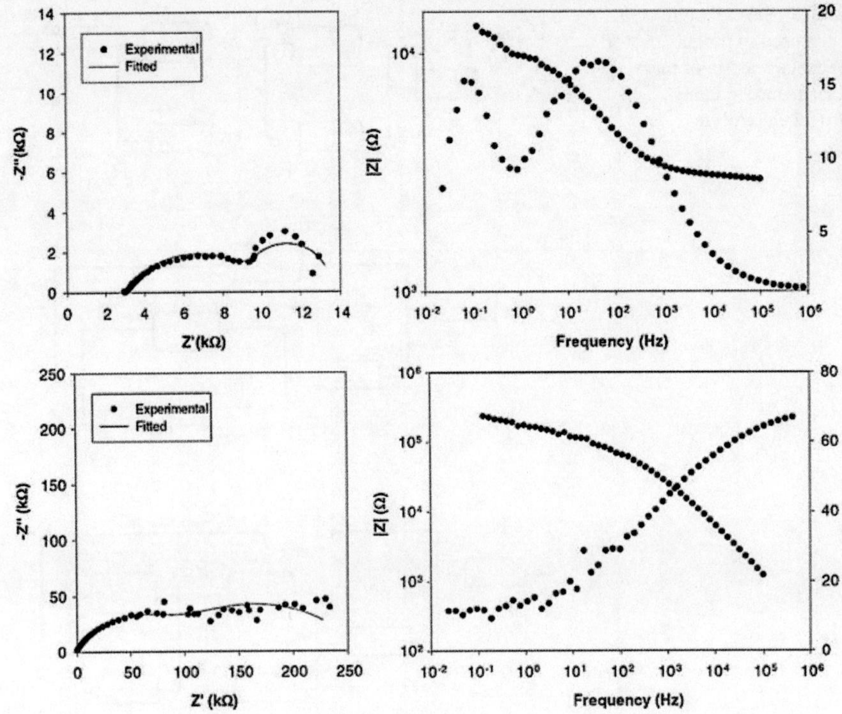

Fig. 11.7 Nyquist and Bode plots were recorded using an agar cell in contact with *Valens AES3* (upper) and *Rostrum Tridens AES* (lower) bronze Roman coins at the OCP. Circles: experimental data points; continuous lines: 'best' theoretical impedance spectra fitted to the equivalent circuit in Fig. 11.8a. Adapted from Ref. [27], with permission

electrode is connected to the iron/steel reinforcement and the reference electrode to the concrete surface. This method mimics the standardized one of the American Society for Testing and Materials (ASTM) [30]. The device is completed with a copper wire immersed into a $CuSO_4$ connected to the concrete surface by a wet sponge or paper.

In the heritage field, the main problem is the limitations imposed by the maintenance of the integrity of the objects and the difficulty of data interpretation. A recent approach extends OCP measurements to a limiting experimental situation: the measurement 'in dry', i.e., simply connecting the working and reference crocodile clamps to two points of the corroded metallic object [31–33]. This procedure can be viewed as a limiting case of the above methods where the electrolyte solution is constituted by the humidity layer impregnating the corrosion layer covering the metallic artifact.

The measurements on archaeological objects were performed by connecting the reference electrode clamp to a conducting region of the object. In the case of iron objects, adopting the description of the corrosion layers described in Sect. 10.2 [34]

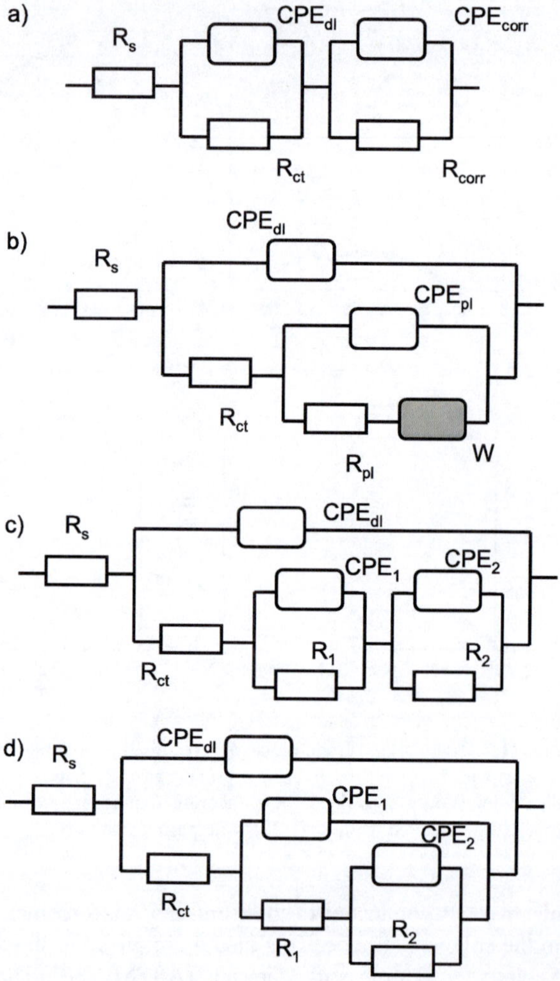

Fig. 11.8 Several equivalent circuits used to model impedance spectra of metal patinas recorded using hydrogel electrolytes

implies the connection to a region where the conductive oxide (usually magnetite) layer is exposed.

Figure 11.9 shows a schematic representation of the resulting system used for the theoretical interpretation of dry OCP measurements [32, 33]. One clamp is connected to a conducting region and the other to an area of the so-called dense product layer (DPL, see Chap. 10) of the corrosion layer. Both areas are in 'electrochemical' contact through the humidity layer impregnating the system so that it behaves like a 'battery' which driving force is the contact potential difference between the conducting phases and the semiconducting components of the DPL. When the two terminals of that 'battery' are connected by means of an external circuit, the current flows and the difference of potential can be measured. Using a voltmeter of large (ideally, infinite) impedance, the OCP variation with time can be recorded.

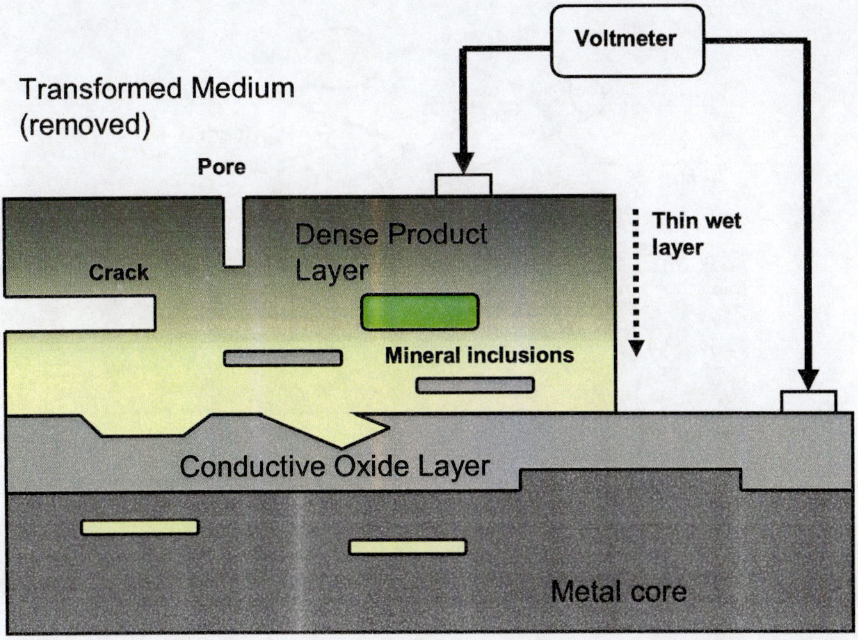

Fig. 11.9 Scheme for the interpretation of dry OCP measurements adopting the typical structure of the corrosion layers of archaeological artifacts iron buried in the soil. Adapted from Ref. [32], with permission

Figure 11.10 depicts a scheme of the simple experimental arrangement accompanied by the photographic image of a typical system where it can be applied. This is an iron nail recovered from the archaeological site of La Bastida de les Alcusses archaeological site (Moixent, València, Spain), which is dated back to the fourth century BCE. The studied specimens were a part of a set of ca. 100 nails under study in the *Museu de Prehistòria de València*, Spain.

As a first application, this method permits the mapping of the levels of corrosion in terms of a rate constant determined from the analysis of transients in the potential/time curves (vide infra). Optionally, two aluminum foils can be intercalated between the clamps and the object in order to avoid possible erosion in its surface. Although covered by a fine oxide layer, experimental data on iron and steel nails the aluminum foils do not modify the response recorded in their absence.

Using ordinary equipment, the method can reasonably be applied with minimal clamp separations of 2–5 mm, thus providing ‘images’ of areas around 0.5 cm^2. This mapping is illustrated in Fig. 11.10 where areas exhibiting different corrosion level are distinguished. This was based on the values of the decay constant k calculated from experimentally determined current/time curves. These can be fitted (see below) to an exponential decay which is slower as larger is the corrosion layer.

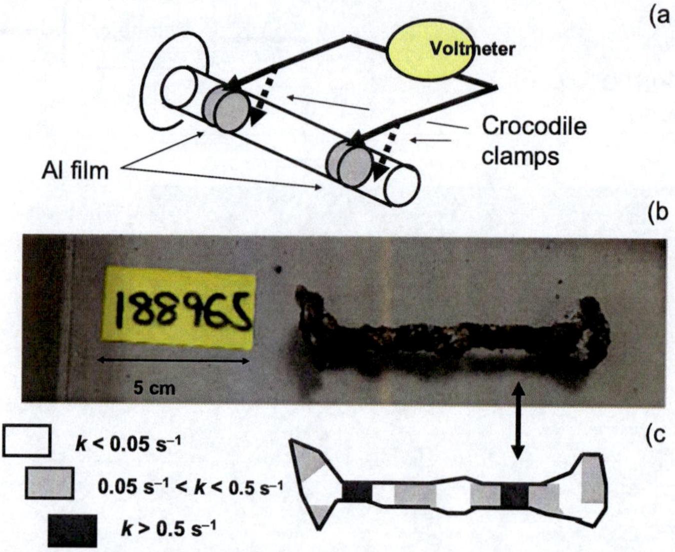

Fig. 11.10 **a** Schematic for 'dry' OCP measurements; **b** photographic image of an iron nail from La Bastida de les Alcusses (Spain) archaeological site (cat. n. 188,965); and **c** mapping of surface corrosion level from OCP data based on a rate constant determined from transient potential/time curves (see text). The double arrow marks the point of connection with the reference terminal of the potentiostat. Reproduced from Ref. [31], with permission

Figure 11.11 illustrates several of the potential/time curves recorded for a contemporary stainless steel nail and two different iron nails from the aforementioned archaeological site, one of which was previously submitted to a stabilizing protocol with NaOH (3% wt) plus tannic acid (5% wt) solution. The contemporary stainless steel nail presented an abrupt variation of the potential (V_D) with time corresponding to a variation of a few mVs.

The archaeological iron nails, however, displayed a more smooth variation of V_D with time tending to a constant, limiting OCP value after times ranging between a few seconds and 2–5 min. This long-time tendency curve was superimposed to potential/time transients covering different timescales at the subsecond level.

The V_D versus t curves recorded for homogeneous objects were scarcely sensitive to variations in the position of the clamps. No significant OCP curves were obtained between experiments performed with and without the Faraday cage. Similarly, no significant differences were obtained with and without interposed aluminum sheets.

If two equal clamps are connected to two points of the same 'dry' metal, no potential differences are established and no transient responses can be expected. However, under ordinary conditions the metal surfaces are covered by a fine layer of oxides of non-conducting or semiconducting nature impregnated/covered by a fine humidity layer. As a result, the metal surface can exhibit local differences in compactness, porosity, etc. This provides opportunity for the appearance of different local contact potential differences that eventually increased by the erosion produced

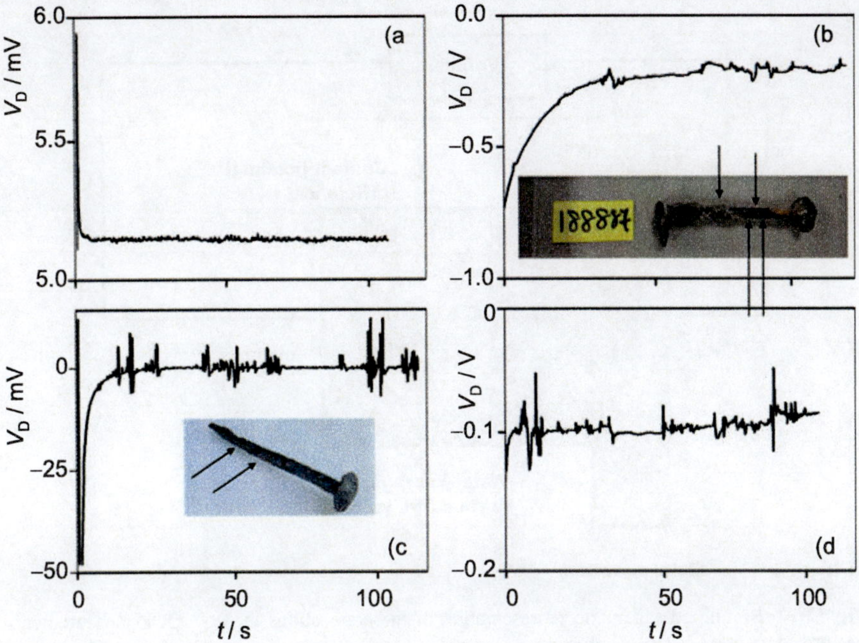

Fig. 11.11 Transient potential/time curves are recorded in 'dry' OCP measurements on: **a** a contemporary stainless steel nail and **b–d** iron nails from La Bastida de les Alcusses (Spain) archaeological site. Data correspond to a highly altered nail (cat. n. 188887), connected between light and high corrosion areas (**b**) and between two lightly corroded areas (**d**) and a nail (cat. n. 188960) submitted to stabilizing protocol with NaOH (3% wt) plus tannic acid (5% wt) solution. Notice the different potential scale in (**a, c**) and (**b, d**). Insets: photographs of the corresponding nails. The arrows mark the position of the respective connections. Reproduced from Ref. [31], with permission

by the clamp connection. In the case of 'clean' iron steel nails (Fig. 11.11a), the ideal contact potential difference is zero and the measured potential difference is initially less than 1 mV and rapidly tends to stabilize. In the case of more or less corroded objects (Fig. 11.11b–d), when the clamps connect a 'clean' conducting area with a corroded region, the measured potential differences reach larger values ultimately reflecting the metal/semiconductor contact potential difference.

Notice that the measurement process determines, to some extent, the modification of the composition of the surface region as a result of the current pass. This means that repeated experiments on the same area do not produce identical current/time curves.

Figure 11.12 depicts a simplified version of Fig. 11.9 showing the connections in 'dry' OCP measurements performed on corroded metal objects. The essential idea is that one clamp is put in contact with one conducting region (base metal plus fine layer of conducting oxides (CO) as previously noted). This is in contact with the DPL constituted by semiconducting compounds permeated by humidity

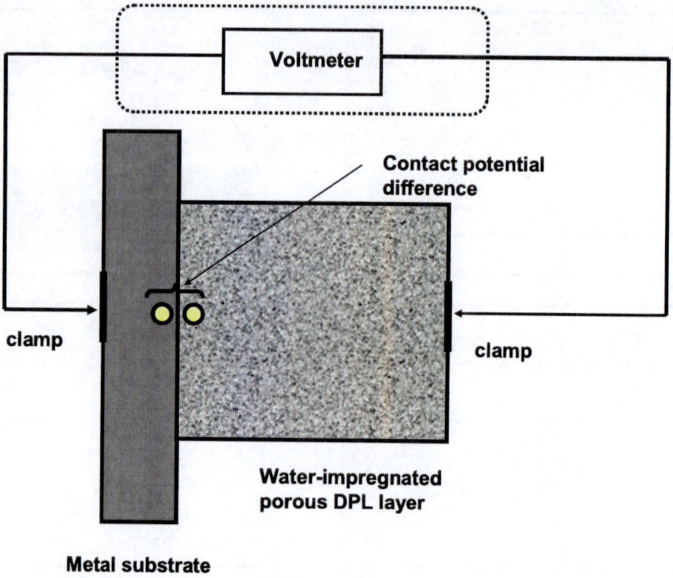

Fig. 11.12 Simplified schematic representation of the connections in 'dry' OCP measurements performed on corroded metal objects

which is put in contact with the other clamp. There is a contact potential difference established between the metal substrate and the semiconducting DPL which will act as the promoter of charge transport when the external circuit is closed. This layer can be modeled such as described for porous oxide coatings on metal surfaces [12] in terms of two charge transport routes: through the semiconducting pore walls and through the electrolyte solution filling the pores.

Accordingly, the equivalent circuit used to describe OCP measurements can be divided in two 'wet' and 'dry' branches. Three essential assumptions are made: (a) when the electrical circuit is closed through an external circuit containing the voltmeter, there a contact potential difference is established derived from the difference in the work functions (see [13]) between the CO and the DPL; (b) the humid layer impregnating the corrosion layer gives rise to double-layer charging effects; and (c) space charging phenomena [35, 36] can appear as a result of the semiconducting properties of most corrosion products.

The difference of potential established between the CO and the DPL layers, ϕ, promotes charging phenomena through the system. Here, it is possible to separate charge through the 'dry' and 'wet' branches. The former can be modeled as a series combination of the resistance of the CO phase (R_{CO}) plus a parallel association of the DPL resistance (R_{DPL}) and the capacitance of the CO-DPL interface ($C_{CO\text{-}DPL}$). This last circuit element can be associated with the aforementioned space charge charging effects. The 'wet branch' can be modeled as two parallel associations of resistance plus capacitor. The first one corresponds to double-layer charging effects (C_{dl}) in

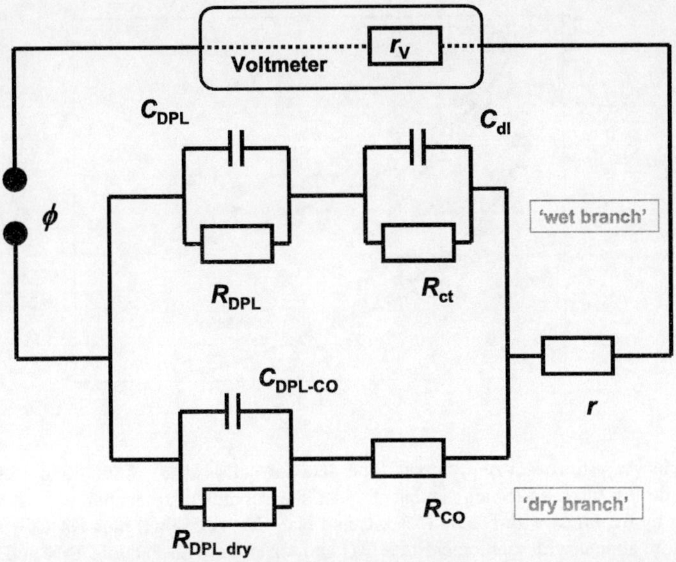

Fig. 11.13 Equivalent circuit is used to describe dry OCP measurements on archaeological artifacts of iron buried in soil. Adapted from Ref. [32], with permission

the humidity layer and the charge transfer resistance at the electrolyte-DPL interface (R_{ct}). The second includes the space charge charging (C_{DPL}) and the resistance of the humidity-impregnated DPL region (R_{DPL}).

To complete the equivalent circuit, the ohmic resistance of the electrolyte, clamps, external conductors, etc. (r), and the internal resistance of the voltameter (r_V) are incorporated into Fig. 11.13. Of course, this is a simplified scheme ignoring factors such as the heterogeneity of the solid phases, diffusion effects, thickness, porosity, and tortuosity, of the different layers. It is also assumed that the contact potential between conducting phases (clamps/aluminum foils/object) cancels.

In the simplest case, it is assumed that only charge transfer through the 'wet branch' is significant and that the DPL effects are negligible. Then, the system can be described on the basis of the following equations [32]:

$$\phi = I_V(r + r_V) + \frac{q}{C_{dl}} \tag{11.12}$$

$$I_V = I + \frac{dq}{dt} \tag{11.13}$$

$$I R_{ct} = \frac{q}{C_{dl}} \tag{11.14}$$

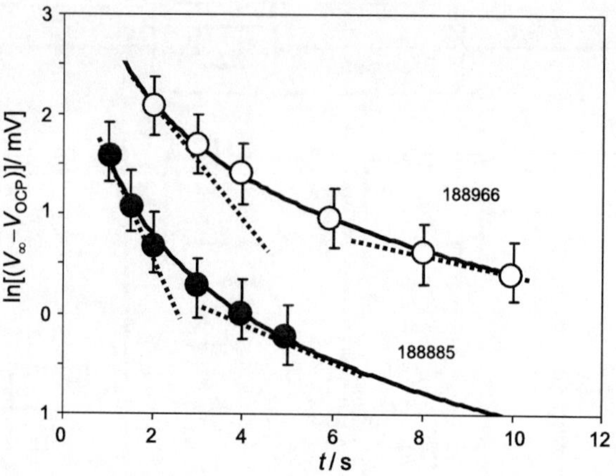

Fig. 11.14 ln(V_{OCP}(lim) − V_{OCP}) versus time data for nails cat. n. 188966 and 188885 from La Bastida de les Alcusses (Spain) archaeological site submitted, respectively, to stabilization/desalination by treatment with NaOH (3% wt) and $NaHCO_3$ (2.5% wt) plus Na_2CO_3 (2.5% wt) followed by treatment with tannic acid (5% wt) and coverage with Paraloid B44 and wax. The continuous lines correspond to the polynomial fit of experimental data, and the dotted lines represent the asymptotic variations consistent with Eqs. (11.23) and (11.25)

where I represents the intensity of the current flowing through the R_{ct} and I_V the current passing through the voltmeter. Then, the charging current through the double-layer capacitor will be dq/dt so that,

$$\phi = \frac{q}{C_{dl}} + (r + r_V)\left(\frac{q}{R_{ct}C_{dl}} + \frac{dq}{dt}\right) \tag{11.15}$$

The charge/time current will result from the integration of the equation,

$$\int_0^q \frac{(r + r_V)dq}{\phi - \frac{q}{C_{dl}}\left(1 + \frac{r + r_V}{R_{ct}}\right)} = \int_0^t dt \tag{11.16}$$

Then, the charge passed at a time t will be,

$$q = \frac{C_{dl}\phi}{1 + \frac{r + r_V}{R_{ct}}}\left\{1 - \exp\left[-\frac{t}{(r + r_V)C_{dl}}\left(1 + \frac{r + r_V}{R}\right)\right]\right\} \tag{11.17}$$

Then, the intensity passing through the resistance R_{ct} will be,

$$\begin{aligned} I &= \frac{q}{R_{ct}C_{dl}} \\ &= \frac{\phi}{R + r + r_V}\left\{1 - \exp\left[-\frac{t}{(r + r_V)C_{dl}}\left(1 + \frac{r + r_V}{R_{ct}}\right)\right]\right\} \end{aligned} \tag{11.18}$$

and the measured OCP, V_{OCP}, results as:

$$V_{\mathrm{OCP}} = \phi\left(\frac{r_{\mathrm{V}}}{R_{\mathrm{ct}} + r + r_{\mathrm{V}}}\right)\left\{1 + \frac{R_{ct}}{r + r_{\mathrm{V}}}\exp\left[-\frac{t(R_{\mathrm{ct}} + r + r_{\mathrm{V}})}{(r + r_{\mathrm{V}})R_{\mathrm{ct}}C_{\mathrm{dl}}}\right]\right\} \tag{11.19}$$

This corresponds to an exponential variation of the OCP with time. Interestingly, at time zero, the initial OCP, $V_{\mathrm{OCP}}(0)$, equals $\phi r_{\mathrm{V}}/(r + r_{\mathrm{V}})$, whereas, at infinite time, the OCP tends to a limiting value $V_{\mathrm{OCP}}(\mathrm{lim}) = \phi r_{\mathrm{V}}/(R_{\mathrm{ct}} + r + r_{\mathrm{V}})$.

Notice that, as far as the measured V_{OCP} depends on the internal resistance of the voltmeter, the potential measurements are instrument-dependent. This problem, strictly common to all or almost all electrochemical measurements, is in general negligible, but here may be significant. In the case in which r_{V} was much larger than all other resistances (this corresponds to 'ideal' voltmeters), the currents tend to zero and the measured V_{OCP} would be constant and equal to ϕ.

Since the initial potential possesses relatively large uncertainty, it is convenient to monitor the difference,

$$V_{\mathrm{OCP}} - V_{\mathrm{OCP}}(\mathrm{lim}) = \phi\left(\frac{R_{\mathrm{ct}}r_{\mathrm{V}}}{(R_{\mathrm{ct}} + r + r_{\mathrm{V}})(r + r_{\mathrm{V}})}\right)\exp\left[-\frac{t(R_{\mathrm{ct}} + r + r_{\mathrm{V}})}{(r + r_{\mathrm{V}})R_{\mathrm{ct}}C_{\mathrm{dl}}}\right] \tag{11.20}$$

again reflecting an exponential time dependence.

A second case assumes that space charge effects in the wet DPL branch are negligible, but the dry branch possesses sufficient conductivity to produce significant charge transport. Then, a similar treatment yields [32],

$$V_{\mathrm{OCP}} - V_{\mathrm{OCP}}(\mathrm{lim}) = \frac{\phi}{B}\left[Ar_{\mathrm{V}}(C_{\mathrm{dl}} + C_{\mathrm{DPL-CO}}) - \frac{r_{\mathrm{V}}(R_{\mathrm{ct}} + R_{\mathrm{DPL\ dry}})}{R_{\mathrm{ct}}R_{\mathrm{DPL\ dry}}}\right]\exp(-At) \tag{11.21}$$

where

$$A = \frac{R_{\mathrm{ct}}R_{\mathrm{DPL\ dry}} + (r + r_{\mathrm{V}})(R_{\mathrm{ct}} + R_{\mathrm{DPL\ dry}})}{R_{\mathrm{ct}}R_{\mathrm{DPL\ dry}}(r + r_{\mathrm{V}})(C_{\mathrm{dl}} + C_{\mathrm{DPL-CO}})};$$

$$B = \left(1 + \frac{(r + r_{\mathrm{V}})(R_{\mathrm{ct}} + R_{\mathrm{DPL\ dry}}}{R_{\mathrm{ct}}R_{\mathrm{DPL\ dry}}}\right) \tag{11.22}$$

The third case of interest is that in which, as in the first one, the dry region does not contribute significantly to charge transfer, whereas electrolyte double layer and semiconductor space charging effects are significant in the wet region. Assuming that the charge of the first capacitor is much faster than the charge of the second, one obtains [32],

$$V_{\mathrm{OCP}} - V_{\mathrm{OCP}}(\mathrm{lim}) = \left(\frac{\phi Y r_{\mathrm{V}}C_{\mathrm{dl}}}{X} - \frac{\phi r_{\mathrm{V}}}{XR_{\mathrm{ct}}}\right)\exp(-Yt) \tag{11.23}$$

where

$$X = \frac{R_{ct} + R_{DPL} + r + r_V}{R_{ct}};$$
$$Y = \frac{R_{ct} + R_{DPL} + r + r_V}{R_{ct}C_{dl}(R_{DPL} + r + r_V)} \tag{11.24}$$

assuming that the second capacitor starts charging when the first capacitor is almost completely charged, one can arrive at the relationship:

$$V_{OCP} - V_{OCP}(\lim) = \phi r_V \left(1 - \frac{1}{X}\right)\left(\frac{1}{r + r_V} - \frac{1}{ZC_{DPL}R_{DPL}}\right)\exp(-\frac{Zt}{r + r_V}) \tag{11.25}$$

where

$$Z = \frac{R_{DPL} + r + r_V}{C_{DPL}R_{DPL}} \tag{11.26}$$

Accordingly, one can expect that the time variation of $V_{OCP} - V_{OCP}(\lim)$ tends asymptotically to the exponential dependence expressed by Eq. (11.23) at short times, moving to that given by Eq. (11.25) at long times.

Experimental data for pristine iron nails such as in Fig. 11.11 agreed with the common prediction from Eqs. (11.20) to (11.22) of a linear variation of $\ln(V_{OCP} - V_{OCP}(\lim))$ with time measured just after the connection of the clamps. In the case of nails submitted to stabilization treatments, however, there was some deviation from linearity which can be interpreted in terms of two limiting linear tendencies, respectively, at short and long times. Figure 11.14 shows pertinent data for two stabilized nails [32].

For practical purposes, it is convenient to interpret the linear $\ln(V_{OCP} - V_{OCP}(\lim))$ versus t variations using the simplest expression (11.20), but taking the resistance and capacity values as apparent ones (R_{app}, C_{app}). Assuming $r << r_V$, the $R_{app}C_{app}$ term, the R_{app}/r_V ratio, and the contact potential (ϕ) can be approximated by the equations:

$$R_{app}C_{app} = k^{-1} \approx -\frac{1}{\text{slope}}\left(\frac{V_{OCP}(0)}{V_{OCP}(\lim)}\right) \tag{11.27}$$

$$\frac{R_{app}}{r_V} \approx 1 - \frac{V_{OCP}(0)}{V_{OCP}(\lim)} \tag{11.28}$$

$$\phi \approx V_{OCP}(\lim) \tag{11.29}$$

respectively. Notice that k represents an apparent rate constant as taken in Fig. 11.10.

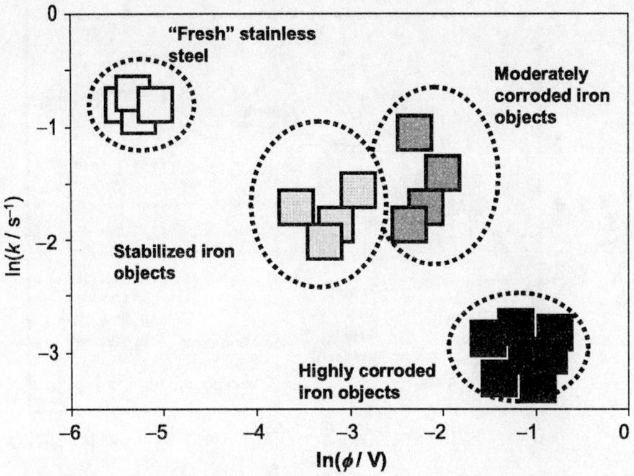

Fig. 11.15 Two-dimensional diagram based on the values of ϕ and k in double logarithmic scale grouping different objects including a series of iron nails coming from the La Bastida de les Alcusses archaeological site (Moixent, València, Spain), dated back to the fourth century BCE and iron sheet taken from *The Lighthouse at the End of the World* (Argentine) [31, 32]. Adapted from Ref. [31], with permission

The above parameters can be used for grouping the archaeological samples. This can be seen in Fig. 11.15, where the values of ϕ and k in a double logarithmic scale are used for grouping iron nails coming from the La Bastida de les Alcusses archaeological site (Moixent, València, Spain), an iron sheet taken from *The Lighthouse at the End of the World* (Argentine) and other objects [31, 32].

Remarkably, the values of R_{app}, C_{app}, obtained from long-time transients in OCP measurements appear to be consistent with capacitance values recorded in EIS experiments on porous anodic oxide films on steel, aluminum, etc. (typically, 10^{-7} F cm^{-2} [37], and charge transfer resistance of 'clean' metal surfaces ranges around 10^7 Ω cm^2 [38], taking into account the lower conductivity of the magnetite passive surface (resistivity between 10^{-2} and 10^{-4} Ω cm [39] and iron oxides (ca. 10^9 Ω cm for goethite [40])).

Short-time transients displayed, as shown in Fig. 11.11, a varied morphology. In principle, three types of transients can be distinguished, as depicted in Fig. 11.16.

These transients can be interpreted as resulting from the formation of metastable corrosion pits [41, 42]. This type of corrosion, which has received considerable attention, evolves in two steps, namely metal oxidation and recovery. In the first step, the metal is locally dissolved; in the second, the metal repassivates. This is schematically depicted in Fig. 11.17 for iron in contact with a corrosive electrolyte. The first step is driven by the relatively fast anodic Fe to Fe^{2+} oxidation with unconstrained diffusion of the generated cations into the electrolyte. This is the so-called activation-controlled pitting, ending when the pit radius reaches a critical value and the dissolution flux ceases. The stability of the pit is determined by its stability product. Below it, the pit is metastable and may repassivate; up to the pit stability product, the pit is stable

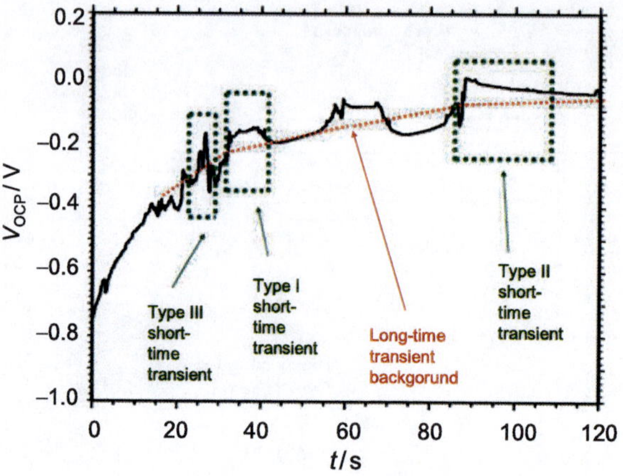

Fig. 11.16 Experimental V_{OCP} versus time data determined for corroded iron nail (cat. n. 188912) from La Bastida de les Alcusses (Spain) archaeological site. The long-time background transient (fine dotted line) and three types of short-time transients (gross dotted lines) are marked. Reproduced from Ref. [33], with permission

[43]. In all cases, the metal oxidation has to be coupled with a cathodic process, which must occur in the vicinity of the pit.

A second pit formation regime, depicted in Fig. 11.17b, occurs when the rate of oxidative metal dissolution exceeds the rate of diffusion of the generated ions into the electrolyte so that the ions can form insoluble compounds such as $FeCl_2$, $Fe(OH)_2$, and $Fe(OH)_3$. This defines the so-called diffusion-controlled corrosion regime [43].

The volume (v) of the pit can roughly be estimated from the charge passed, Q, as [38]:

$$v = \frac{QM}{z\rho F} \tag{11.30}$$

where M is the molar mass of the metal, z is the number of moles of electrons transferred in the anodic oxidative metal dissolution, ρ is the metal density, and F is the Faraday constant.

On the first examination, the short-time transients appearing in 'dry' OCP measurements can be interpreted as the result of pit formation on iron artifacts through the thin humidity layer impregnating their surfaces. In several short-time transients, divided into types I and II in Fig. 11.16, the transient is reflected in potentials always more anodic than the long-time background transient. This suggests that pit growth/repassivation rates are controlled by the electrolyte double-layer charge/discharge [33]. In type III pits, however, the transient is constituted by both cathodic and anodic excursions over the background V_{OCP} versus t curve. These can be viewed as systems where there are successive oxidative dissolution and reductive passivation/redeposition processes governed by solid-state electrochemistry (Chap. 2).

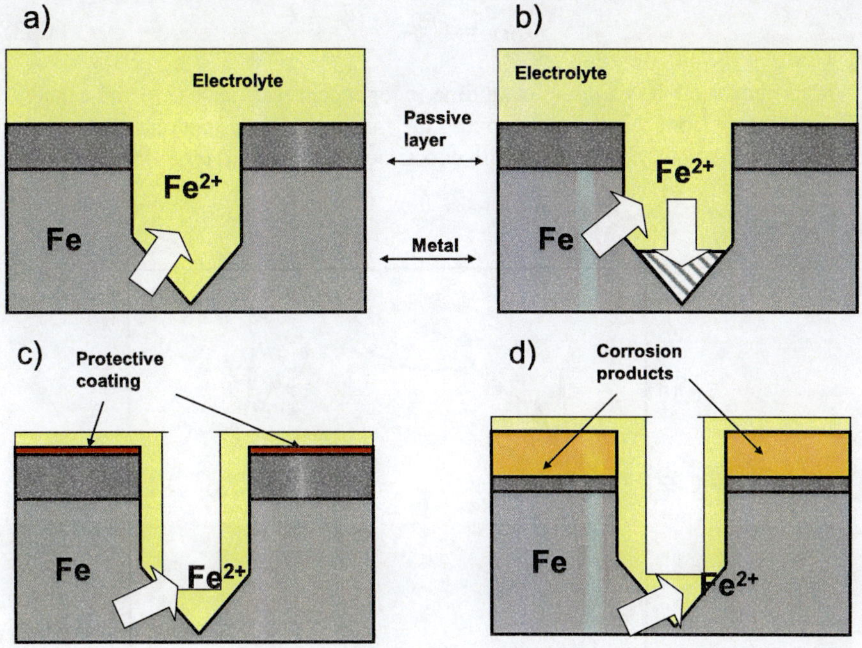

Fig. 11.17 Schematic representation of pit formation on iron surfaces in contact with electrolytes under conditions of 'ordinary' pitting experiments. **a** Activation-controlled pitting, **b** diffusion-controlled pitting, and 'dry' OCP response on **c** stabilized, and **d** non-stabilized, corroded iron objects. Reproduced from Ref. [33], with permission

In the growing pit step, the OCP varies until reaches the value corresponding to that existing in the passive state, E_o. This can be described as [38],

$$E_o - V_{OCP} = RC\frac{dV_{OCP}}{dt} \tag{11.31}$$

where C is the total interface capacitance (double layer, space charge, ...) and R the electron transfer resistance at a particular applied potential. Introducing the overpotential U_{OCP} superimposed to the long-time background, the pitting formation event, and assuming that the resistance (R_1) and the capacitance (C_1) of the system do not vary with time, one can obtain,

$$U_{OCP} = U_{OCP}^{lim}\left(1 - e^{-\frac{t}{R_1C_1}}\right) \tag{11.32}$$

This corresponds to the expression for a charging process with a time constant R_1C_1 which applies for transients of types I and II [33]. The second pitting step can be described assuming that, at a critical time t_c, the potential reaches a critical value, U_{OCP}^{crit}, and repassivation begins. The corresponding potential/time relationship will be,

$$U_{\mathrm{OCP}} = U_{\mathrm{OCP}}^{\mathrm{crit}} \mathrm{e}^{-\frac{t-t_c}{R_2C_2}} \tag{11.33}$$

In this equation, it is assumed that time-independent resistance (R_2) and capacity (C_2), differing from those characterizing the first pitting step, operate. This model provides a satisfactory agreement with experimental type I and type II transients, as illustrated in Fig. 11.18a.

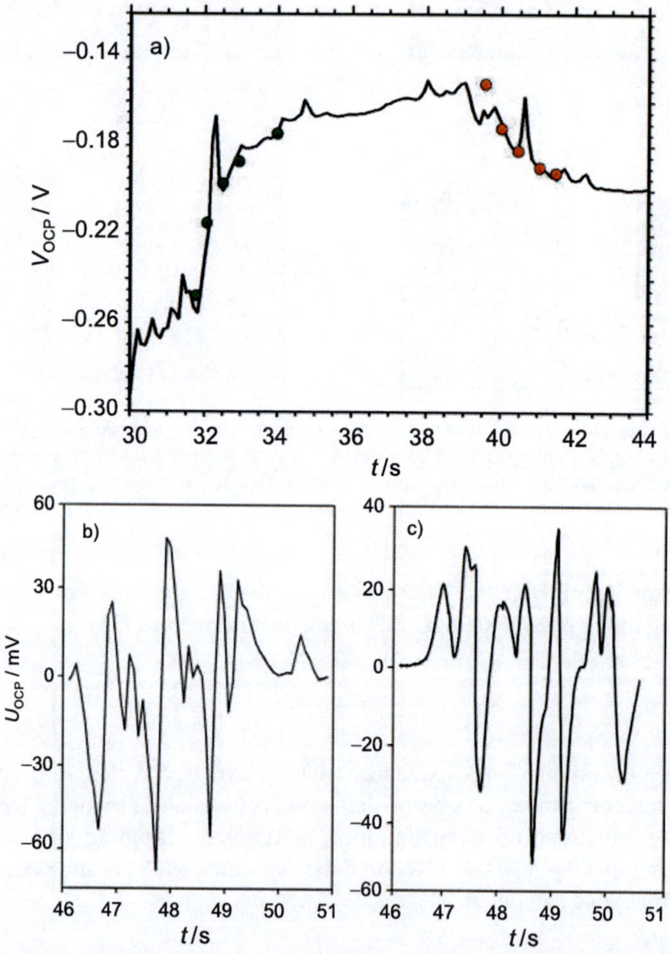

Fig. 11.18 **a** Experimental type I transient recorded for an iron nail cat. from La Bastida de les Alcusses (Spain) archaeological site (cat. n. 188,912) in 'dry' OCP measurements superimposed to theoretical points from Eqs. (11.31) to (11.32) taking $U_{\mathrm{OCP}}{}^{\mathrm{lim}} = 95$ mV, $RC = 0.75$ s (green cycles) and $U_{\mathrm{OCP}}{}^{\mathrm{lim}} = 95$ mV, $RC = 1.10$ s, $t_c = 38.0$ s (red cycles). **b** Experimental U_{OCP} versus time curve recorded for the same iron nail in a time interval of 5 s showing typical type III short-time transients. **c** Theoretical curve from Eq. (11.37) obtained from the superposition of seven symmetric transients taking $k_1R_1 = k_2R_2 =$ values of 5.0, 10.0, and 20.0 s, $t_c =$ of 0.1 and 0.2 s and Q values between 0.1 and 0.3. Adapted from Ref. [33], with permission

As previously noted, the OCP transients of type III can be treated in terms of successive anodic/cathodic cycles of oxidative dissolution/redeposition processes [33]. A simple approach is obtained by expressing the (over)potential as,

$$U_{\mathrm{OCP}} = R\frac{\mathrm{d}q}{\mathrm{d}t} \tag{11.34}$$

where q is the charge passed during each one of the generation/repassivation pitting steps and R is the pit resistance. Adopting an approach similar to that described in Sect. 3.9, the charge passed can be expressed as a logistic function of the time of the form,

$$\frac{\mathrm{d}(q/Q)}{\mathrm{d}t} = k\left(\frac{q}{Q}\right)\left(1-\frac{q}{Q}\right) \tag{11.35}$$

where Q denotes the maximum charge stored in the pit introduced in Eq. (11.30) and k is the reaction rate constant. Then, the charge/time and (over)potential/time functions will be [33],

$$q = \frac{Q\mathrm{e}^{kt}}{1+\mathrm{e}^{kt}} \tag{11.36}$$

$$U_{\mathrm{OCP}} = kQR\frac{\mathrm{e}^{kt}}{(1+\mathrm{e}^{kt})^2} \tag{11.37}$$

These equations can be taken as corresponding to pitting initiation and define an anodic excursion over the OCP background. Now, it is assumed the subsequent repassivation step is driven by some cathodic reaction. The cathodic excursion observed for type III transient in Fig. 11.16 can be described assuming that this process follows kinetics similar to that of the previous initiation but starting at a critical time, t_c. Accordingly, the net U_{OCP} versus time profile can be described as [33],

$$U_{\mathrm{OCP}} = k_1QR_1\frac{\mathrm{e}^{k_1t}}{(1+\mathrm{e}^{k_1t})^2} - k_2QR_2\frac{\mathrm{e}^{k_2(t-t_c)}}{(1+\mathrm{e}^{k_2(t-t_c)})^2} \tag{11.38}$$

In this equation, k_1, k_2 represent the rate constants for the successive initiation and repassivation steps, and R_1, R_2, the corresponding ohmic resistances. As can be seen in Fig. 11.18b, c, this simple model permits to satisfactorily reproduce the experimental short-time transients of type III using reasonable values of the rate constants.

11.2.4 Other Techniques

There are several other techniques used in corrosion studies that are of potential application in the study of metallic heritage. Future research should explore their capabilities in this field. These include Kelvin probe [44], scanning electrochemical microscopy (SECM) [45], and electrochemical noise measurements [46], the last being used in a minute study for diagnosis of deterioration purposes [47]. In the voltammetric domain, cavity microelectrode devices [48] and stripping voltammetry microprobe [49] assays offer capabilities in metallic heritage. See also additional literature cite at the end of this chapter.

11.3 Preservation of Metallic Heritage

11.3.1 Selective Removal and Anodization

Some procedures of metal treatment involve chemical/electrochemical handling of corrosion products. The metal replacement was historically used for gold- and tin-plating bronze as well as for copper removal from its alloys with noble metals. Controlled corrosion of metals is a technique used since antiquity to produce pigments such as ceruse and verdigris [1]. Within this wide group of techniques, the preparation of artificial patinas as well as gilding, including amalgamation, can ultimately be seen as protective strategies for metal surfaces. Gilding can be achieved upon immersion of copper or bronze coupons in a solution of any Au(III) salt via metal replacement or via Galvanic replacement connecting the copper object with the sacrificial anode (zinc, pyrite, …) through a metallic wire [50]. This technique offers the possibility of a 'smooth' cleaning of archaeological objects although without the control provided by potentiostatic and galvanostatic techniques.

Generically, the protection of a metal surface from corrosion involves the formation of a protective barrier. This involves a variety of protective coatings (see next section) in principle acting as barriers blocking the access of oxygen or other species to the metal surface. A complementary strategy is the use of sacrificial anodes such as zinc or aluminum for steel [8]. Here, the anode polarizes the base metal cathodically, hence decreasing the stability of its oxide layer.

Electrochemical anodization is a widely used technique adopted for the protection of metals in the technological context. This is achieved by the application of a given potential to the metal, which is in contact with a suitable electrolyte. The aim is the growth of a coating of passive products (usually metal oxides) onto the metal substrate. The passivating conditions can be extremely variable; for instance, steel can be passivated in either acidic or alkaline aqueous solutions, while metals such as aluminum or chromium can do so in neutral media [8].

In the cultural heritage field, extensive metal anodization has no significance, as far as this technique involves a drastic change in the composition of the object. Local anodization, however, may eventually be applied.

11.3.2 Protective Coatings

One of the most widely used strategies aimed to the conservation of metallic heritage consists of the application of protective coatings. There is an enormous variety of products used for this purpose, including wax, oils, polymers, and vegetal extracts, … These products should satisfy several requirements [6]:

(a) No or only little change in the visual appearance of the object.
(b) Reversibility, i.e., the possibility of removal to return the object to its original state.
(c) Respect the original object with no modification of the material of the original artifact, including patinas and/or corrosion layers.
(d) Long-term efficiency and easy maintenance, including the possibility of renewing the coating.

In principle, the protective coatings exert a barrier effect by simply blocking the arrival of humidity and oxygen to the metal surface. In this view, the barrier effect involves the displacement of water adsorbed onto the metal surface by organic molecules [51].

The protective efficiency of the coating (η) can be evaluated from polarization curves using the corrosion current density and/or corrosion resistance measured for the unprotected (j_{uncoat}) and coated (j_{coat}) metal immersed in a test solution.

$$\eta(\%) = 100 \times \left(1 - \frac{j_{coat}}{j_{uncoat}}\right) \tag{11.39}$$

Alternatively, this parameter can be evaluated from EIS measurements from the charge transfer resistances (determined for a suitable equivalent circuit) of the uncoated ($R_{ct\text{-}uncoat}$) and coated ($R_{ct\text{-}coat}$) metal immersed in a test electrolyte,

$$\eta(\%) = 100 \times \left(1 - \frac{R_{ct\text{-}coat}}{R_{ct\text{-}uncoat}}\right) \tag{11.40}$$

However, considerable research is devoted to design 'active' protective coatings causing chemical protection and self-repair [52]. These usually consist of a primer barrier, directly deposited onto the metal surface (more precisely, the metal primary patina of conductive oxides) and a topcoat, as schematized in Fig. 11.19. Here, the separation of cathodic and anodic reaction sites on iron covered by polymeric coatings permeable to oxygen and water is schematized [53]. In these systems, mechanisms of inhibition of diffusion rates and shifts of electrochemical interfaces can operate.

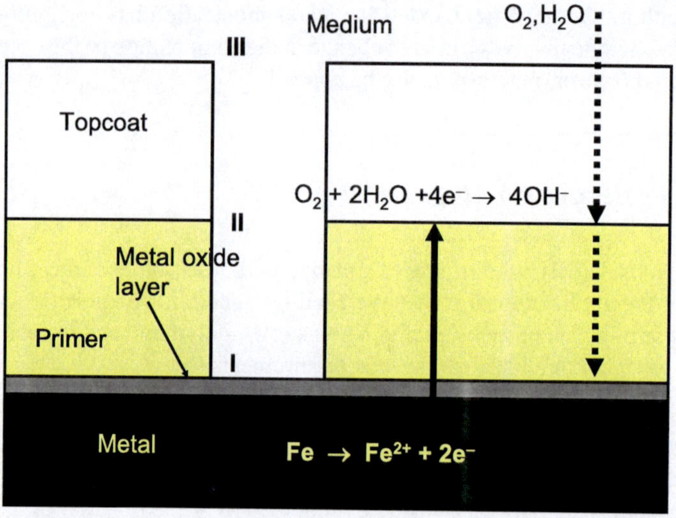

Fig. 11.19 Basic structure of a two-layer protective coating on iron showing the separation of cathodic and anodic reaction sites. I, II, and III represent, respectively, the conducting metal oxide/polymer primer, polymer primer/topcoat, and topcoat/medium interfaces

It is believed that the corrosion of steel occurs at the Fe/Fe_3O_4 interface by water diffusion through the porous conducting oxide, while Fe^{2+} ions diffuse through the oxide layer via vacancy hopping or through the grain boundaries then reacting with water or O^{2-} ions of the generated Fe_2O_3 oxide layer. It appears that electronic effects associated to polymers such as polyaniline slow corrosion by inhibiting the diffusion of Fe^{2+} and O^{2-} ions but also displacing the electroactive interface from the metal/electrolyte solution interface (denoted as I in Fig. 11.19) to the polymer/electrolyte solution interface (II).

Typical coatings consist of a polymeric coating resin (primer or binder) deposited onto the metal surface and a top coat. Really, the metal surface is the primary patina of metal oxides eventually formed by mechanical abrasion or blasting and/or chemical or electrochemical pretreatments. The primer serves as the basic barrier against the water and ionic ingress and is adhered to the substrate via chemical bonding and mechanical interlocking. The topcoat provides resistance to environmental aggression (by light, humidity, mechanical, thermal, and microbial stress). Intermediate coatings may often be used.

The use of self-healing coatings is an increasingly growing research field. These can be divided into autonomous, in which the protective properties are regenerated by the coating with no need for external intervention, and non-autonomous, where there is a need for an external input to stimulate the regenerative process. Autonomous self-healing coatings involve the use of corrosion inhibitors and/or repairing agents restoring (defect filling) the polymer film. Non-autonomous self-healing coatings can use dynamic bonds and shape-memory polymers, but also multiaction self-healing coatings driven by mechanic, pH, etc. stimuli have been devised [52].

The mechanisms of self-healing are a matter of debate. An example of the complexity of this type of system is the coatings of conducting polymers, typically polyaniline. These are redox polymers existing in different interconvertible redox forms. Here, the protective effect has been associated with the separation of cathodic and anodic reaction sites. The idea is that if oxygen permeates the coatings until the metal/primer interface, the cathodic reaction occurs at that interface, giving rise to blistering and delamination. In contrast, if oxygen does not permeate the primer, the cathodic reactions take place at the primer/topcoat interface with lower metal degradation [51]. The self-healing effect can be attributed to the coupling of the metal oxidation with the reduction of the polymer at the metal coating interface and the coupling of oxygen reduction with polymer oxidation at the polymer electrolyte interface. The first process is accompanied by the release of polymer counteranions giving rise to insoluble metal salts that passivate it [54].

11.3.3 Electrochemical Treatments, Dechlorination

Metal objects can be submitted to different electrochemical treatments for cleaning, stabilization, and consolidation purposes [1]. The term cleaning is applied to procedures aimed to remove lightly bound products covering the metal surface. These are 'smooth' procedures that can be simply performed by contacting the object with a piece of a less noble metal (zinc, aluminum), both being immersed into a suitable electrolyte solution.

Electrochemical cleaning involves the application to the object of a reductive potential input relative to a second electrode (platinum, graphite) in contact with a suitable electrolyte solution. The aim is to reduce the metal compounds in the corrosion layer to the metal under controlled conditions. Alternatively, one can apply a relatively large cathodic potential producing hydrogen bubbles that act mechanically on the corrosion products.

These procedures involve the immersion of the object into an electrolytic bath, a process that may involve several technical problems depending on the shape and size of the object, the nature of the metal, and the corrosion products, etc. Of course, cleaning fragile objects is critical [55]

The terms stabilization and dechlorination are employed in a flexible form to describe the electrochemical processes devoted to the elimination of harmful ions (mainly chloride and sulfate) of the metallic objects in order to stop the corrosion catalyzed by such species. The stabilization can also be seen as a process of increasing the long-term resistance to corrosion by generating a stable passive layer. The frontier between cleaning and stabilization/dechlorination is, accordingly, diffuse. Given the recognized importance of chloride ions as corrosion promoters, the study of dechlorination methods has received considerable attention [56, 57], the studies on the diffusion of chloride ions being of particular interest [58, 59].

The use of electrolytic techniques for cleaning and stabilization of metal objects varied significantly throughout the twentieth century because of the aggressive character of some treatments and the reproduction of the corrosion after short time [60]. Current research is devoted to optimizing the conditions of operation under potentiostatic control, use of pulsed potential inputs as well as to combining with other techniques such as laser ablation, etc. The variety of corrosion patterns makes necessary, however, a detailed case-by-case study.

An interesting aspect of these techniques is the performance of local stabilization. Since the earlier approaches [61], different pencil electrodes have been reported. Ideally, we can use a drop of electrolyte solution deposited onto the region to be treated and press an inert pencil electrode. Then a potential difference was established between the dry metal surface and the working electrode by means of a potentiostatic device. More elaborated devices consist of a membrane tip mounted into a glass cylinder containing the electrolyte and the working, reference, and auxiliary electrodes [62]. Application of a cathodic potential to the tip produces the reduction of metal corrosion products to the metal. Typical applications include the cleaning of silver objects covered by silver tarnish (Ag_2S), where the reduction process can be described as,

$$Ag_2S_{solid} + 2H^+_{aq} + 2e^- \rightarrow 2Ag_{solid} + H_2S_{aq} \tag{11.41}$$

Commercial types of equipment are currently available. Of course, the use of these devices requires a previous analytical study of the object and an accurate selection of the operational conditions.

References

1. Doménech-Carbó A, Doménech-Carbó MT, Costa V (2009) Doménech-Carbó A, Doménech-Carbó MT, Costa V (2009) Electrochemical Methods in Archaeometry, Conservation and Restoration (Monographs in Electrochemistry series, Scholz F, Ed). Springer, Berlin-Heidelberg, chapter 5.
2. Hernández-Escampa M, Gonzalez J, Uruchurtu-Savarin J (2010) Electrochemical assessment of the restoration and conservation of a heavily corroded archaeological iron artifact. J Solid State Electrochem 40:345–356
3. Žežula I, Gálová M (1999) Application of abrasive stripping voltammetry in corrosion science. I. determination of the corrosion potentials of metals. J Solid State Electrochem 3:231–233
4. Doménech-Carbó A, Doménech-Carbó MT, Peiró-Ronda MA, Martínez-Lázaro I, Barrio J (2012) Application of the voltammetry of microparticles for dating archaeological lead using polarization curves and electrochemical impedance spectroscopy. J Solid State Electrochem 16:2349–2356
5. Doménech-Carbó A, Lastras M, Rodríguez F, Cano E, Piquero-Cilla J, Osete-Cortina L (2014) Monitoring stabilizing procedures of archaeological iron using electrochemical impedance spectroscopy. J Solid State Electrochem 18:399–409
6. Cano E, Lafuente D, Bastidas DM (2010) Use of EIS for the evaluation of the protective properties of coatings for metallic cultural heritage: a review. J Solid State Electrochem 14:381–391

7. Mcdonald DD, Sun A (2006) An electrochemical impedance spectroscopic study of the passive state on Alloy-22. Electrochim Acta 51:1767–1779
8. Burleigh TD, Dotson TC, Dotson KT, Gabay SJ, Sloan TB, Ferrel SG (2007) Anodizing steel in KOH and NaOH solutions. J Electrochem Soc 154:C579–C586
9. Doménech-Carbó A, Capelo S, Piquero-Cilla J, Doménech-Carbó MT, Barrio J, Fuentes A, Al-Sekhaneh W (2016) Dating archaeological copper using electrochemical impedance spectroscopy. Comparison with voltammetry of microparticles dating. Mater Corr 67:120–129
10. Di Turo F, Parra R, Piquero-Cilla J, Favero G, Doménech-Carbó A (2019) Crossing VIMP and EIS for studying heterogeneous sets of copper/bronze coins. J Solid State Electrochem 23:771–781
11. Schmutzler B, Eggert G (2010) The chloride left behind (dis)solving an analytical problem. In Eggert G, Schmutzler B (eds) Achaeological Iron conservation colloquium 2010. State Academy of Art and Design Stuttgart
12. Poljacek SM, Risovic D, Cigula T, Gojo M (2012) Application of electrochemical impedance spectroscopy in characterization of structural changes of printing plates. J Solid State Electrochem 16:1077–1089
13. Venkatraman MS, Cole IS, Emmanuel B (2011) Model for corrosion of metals covered with thin electrolyte layers: pseudo-steady state diffusion of oxygen. Electrochim Acta 56:7171–7179
14. Li WS, Cai SQ, Luo JL (2004) Chronopotentiometric responses and capacitance behaviors of passive film formed on iron in borate buffer solution. J Electrochem Soc 151:B220–B226
15. Liu W, Zhang H, Qu Z, Zhang Y, Li J (2010) Corrosion behavior of the steel used as a huge storage tank in seawater. J Solid State Electrochem 14:965–973
16. Toledo-Matos LA, Pech-Canul MA (2011) Evolution of an iron passive film in a borate buffer solution (pH 8.4). J Solid State Electrochem 15:1927–1934
17. Lee SJ, Pyun SL (2007) Assessment of corrosion resistance of surface-coated galvanized steel by analysis of the AC impedance spectra measured on the salt-spray-tested specimen. J Solid State Electrochem 11:829–839
18. Turgoose S (1993) Structure, composition and deterioration of unearthed iron objects, in current problems in the conservation of metal antiquities. Tokyo National Research Institute of Cultural Properties, Tokyo, pp 35–52
19. Venkatraman MS, Cole IS, Emmanuel B (2011) Corrosion under a porous layer: a porous electrode model and its implications for self-repair. Electrochim Acta 56:8192–8203
20. Angelini E, Grassini S, Parvis M, Zucchi F (2012) An in situ investigation of the corrosion behaviour of a weathering steel work of art. Surf Interface Anal 44:942–946
21. Ramírez Barat B, Cano E (2019) Agar versus agarose gelled electrolyte for in situ corrosion studies on metallic cultural heritage. Cehm ElectroChem 6:2553–2559
22. Cano E, Crespo A, Lafuente D, Ramírez Barat B (2014) A novel gel polymer electrolyte cell for in-situ application of corrosion electrochemical techniques. Electrochem Commun 41:16–19
23. England AH, Clare TL (2014) Synthesis and characterization of flexible hydrogel electrodes for electrochemical impedance measurements of protective coatings on metal sculptures. Electroanalysis 26:1059–1067
24. Grassini S, Corbellini S, Parvis M, Angelini E, Zucchi F (2016) A simple Arduino-based EIS system for in situ corrosion monitoring of metallic works of art. Measurement 114:508–514
25. Elsener B, Alter M, Lombardo T, Ledergerber M, Wörle M, Cocco F, Fantauzzi M, Palomba S, Rossi A (2016) A non-destructive in-situ approach to monitor corrosion inside historical brass wind instruments. Microchem J 124:757–764
26. Letardi P, Salvadori B, Galeotti M, Cagnini A, Porcinai S, Santagostino Barbone A, Sansonetti A (2016) An in situ multi-analytical approach in the restoration of bronze artifacts. Microchem J 125:151–158
27. Di Turo F, De Vito C, Coletti F, Mazzei F, Antiocchia R, Favero G (2017) A multi-analytical approach for the validation of a jellified electrolyte: application to the study of ancient bronze patina. Microchem J 134:154–163

28. Degrigny C, Guibert G, Ramseyer S, Rapp G, Tarchini A (2009) Use of E_{corr} vs time plots for the qualitative analysis of metallic elements from scientific and technical objects: the SPAMT Test Project. J Solid State Electrochem 14:425–435
29. Sassolini A, Colozza N, Papa E, Hermansson K, Cacciotti I, Arduini F (2019) Screen-printed electrode as a cost-effective and miniaturized analytical tool for corrosion monitoring of reinforced concrete. Electrochem Commun 98:69–72
30. ASTM Standard C876 (2009) Standard test method for corrosion potentials of uncoated reinforcing steel in concrete. ASTM International, West Conshohocken
31. Doménech-Carbó A, Peiró-Ronda MA, Vives-Ferrándiz J, Duffó GS, Farina S (2021) 'Dry' electrochemistry: a non-invasive approach to the characterization of archaeological iron objects. Electrochem Commun 125:106992
32. Doménech-Carbó A, Peiró-Ronda MA, Vives-Ferrándiz J, Duffó GS, Farina S, Doménech-Carbó MT (2022) Modeling 'dry' OCP measurements to characterize archaeological iron corrosion I: long-time transients. J Electroanal Chem 911:116210
33. Doménech-Carbó A, Peiró-Ronda MA, Vives-Ferrándiz J, Duffó GS, Farina S, Doménech-Carbó MT (2022) Modeling 'dry' OCP measurements to characterize archaeological iron corrosion II: short-time transients. J Electroanal Chem 911:116–211
34. Neff D, Dillmann P, Bellot-Gurlet L, Béranger G (2005) Corrosion of iron archaeological artefacts in soil: characterisation of the corrosion system. Corros Sci 47:515–535
35. Bisquert J (2002) Theory of the impedance of electron diffusion and recombination in a thin layer. J Phys Chem B 106:325–333
36. Fabregat-Santiago F, Garcia-Belmonte G, Bisquert J, Zaban A, Salvador P (2002) Decoupling of transport, charge storage, and interfacial charge transfer in the nanocrystalline TiO_2/electrolyte system by impedance methods. J Phys Chem B 106:334–339
37. Zheng ZJ, Gao Y, Zhu M (2014) Studying the fine microstructure of the passive film on nanocrystalline 304 stainless steel by EIS, XPS, and AFM. J Solid State Electrochem 18:2201–2210
38. Guan L, Cai J, Li Y (2020) Characteristics of the random transients caused by clustered pitting corrosion for Al–Mg microelectrode. Mater Corros 71:86–92
39. Vella L, Emerson D (2012) Electrical properties of magnetite- and Hematite-Rich Rocks and Ores. In: ASEG extended abstracts 22nd international geophysical conference and exhibition, Brisbane, Australia, 2012
40. Cornell R, Schwertmann U (2003) The iron oxides—structure, properties, occurrences and uses, 2nd ed. Wiley-VCH Verlag, Weinheim, p 664
41. González-García I, Burstein GT, González S, Souto RM (2004) Imaging metastable pits on austenitic stainless steel in situ at the open-circuit corrosion potential. Electrochem Commun 6:637–642
42. Berthomé G, Malki B, Baroux B (2006) Pitting transients analysis of stainless steels at the open circuit potential. Corros Sci 48:2432–2441
43. Jafarzadeh S, Chen Z, Bobaru F (2019) Computational modeling of pitting corrosion. Corros Rev 37:419–439
44. Williams G, McMurray HN, Worsley DA (2002) Cerium(III) inhibition of corrosion-driven organic coating delamination studied using a scanning Kelvin probe technique. J Electrochem Soc 149:B154–B162
45. Fernández-Pérez BM, Izquierdo J, González S, Souto RM (2014) Scanning electrochemical microscopy studies for the characterization of localized corrosion reactions at cut edges of coil-coated steel. J Solid State Electrochem 18:2983–2992
46. Jez M, Mitoraj M, Godlewska E, Jakubowska M, Bas B (2014) Evolution of corrosion behaviour of selected metallic samples by electrochemical noise measurements. J Solid State Electrochem 18:1635–1646
47. Martínez-Lázaro I, Doménech-Carbó A, Doménech-Carbó MT, Pastor MT, Amigó V (2010) Electrochemical criteria for evaluating conservative treatments applied to contemporary metallic sculpture. A case study. J Solid State Electrochem 14:437–447

48. Cachet-Vivier C, Keddam M, Vivier V, Yu LT (2013) Development of cavity microelectrode devices and their uses in various research fields. J Electroanal Chem 688:12–19
49. Gulppi M, Pavez J, Zagal JH, Sancy M, Azocar M, Scholz F, Paez MA (2015) Stripping voltammetry microprobe (SPV): substantial improvements of the protocol. J Electroanal Chem 745:61–65
50. Lechtman H (1984) Pre-columbian surface metallurgy. Sci Am 250:56–63
51. Santos JR, Mattoso LHC, Motheo AJ (1998) Investigation of corrosion protection of steel by polyaniline films. Electrochim Acta 43:309–313
52. Zhang F, Ju P, Pan M, Zhang D, Huang Y, Li G, Li X (2018) Self-healing mechanisms in smart protective coatings: a review. Corros Sci 144:74–88
53. Schauer T, Joos A, Dulog L, Eisenbach CD (1998) Progr Org Coat 33:20–27
54. Spinks GM, Dominis AJ, Wallace GG, Tallman DE (2002) Electroactive conducting polymers for corrosion control. J Solid State Electrochem 6:85–100
55. Scott DA, Seeley NJ (1987) The washing of fragile iron artifacts. Stud Conservat 32:73–76
56. Watkinson D (1996) Chloride extraction from archaeological iron: comparative treatment efficiencies. In: Roy A, Smith P (eds) Archaeological conservation and its consequences. International Institute for Conservation, London, pp 208–212
57. Selwyn LS (2004) Overwiew of archaeological iron: the corrosion problem, key factors affecting treatment, and gaps in current knowledge. In: Proceedings of the metal 2004, National Museum of Australia, Canberra, pp 294–306
58. North NA, Pearson C (1978) Washing methods for chloride removal from marine iron artifacts. Stud Conservat 23:174–186
59. Selwyn LS, McKinnon WR, Argyropoulos V (2001) Models for chloride ion diffusion in archaeological iron. Stud Conservat 46:109–120
60. Barrio J, Chamón J, Pardo AI, Arroyo M (2009) Electrochemical techniques applied to the conservation of archaeological metals from Spain: a historical review. J Solid State Electrochem 13:1767–1776
61. Aldaz A, España T, Montiel V, Lopez-Segura M (1986) A simple tool for the electrolytic restoration of archaeological metallic objects with localized corrosion. Stud Conservat 31:175–176
62. Degrigny C, Jeanneret R, Witschard D, Baudin C, Bussy G, Carrel H (2016) A new electrolytic pencil for the local cleaning of Silver Tarnish. Stud Conservat 61:162–173

Additional Literature

63. Adriaens A, Dowsett M, Leyssens K, Van Gasse B (2007) Insights into electrolytic stabilization with weak polarization as treatment for archaeological copper objects. Anal Bioanal Chem 387:861–868
64. Ahmad S (2003) Reinforcement corrosion in concrete structures, its monitoring and service life prediction—a review. Cement Concrete Com. 25:459–471
65. Alviz-Meza A, Chipoco Haro DA, Velarde MI (2021) Scientific insights of electrochemical replacement approach to describe the origin of Pre-Columbian Peruvian gilded copper-based objects. J Phys Conf Ser 2118:012015
66. Angelini E, Grassini S, Parvis M, Zucchi F (2012) An in situ investigation of the corrosion behaviour of aweathering steel work of art. Surf Interface Anal 44:942–946
67. Appleby AJ, Zagal JH (2011) Free energy relationships in electrochemistry: a history that started in 1935. J Solid State Electrochem 15:1811–1832
68. Cano E, Bastidas DM, Argyropoulos V, Fajardo S, Siatou A, Bastidas JM, Degrigny C (2010) Electrochemical characterization of organic coatings for protection of historic steel artefacts. J Solid State Electrochem 14:453–466

69. Chiavari C, Rahmouni K, Takenouti H, Joiret S, Vermaut P, Robbiola L (2007) Composition and electrochemical properties of natural patinas of outdoor bronze monuments. Electrochim Acta 52:7760–7769
70. Dalard F, Gourbeyre Y, Degrigny C (2002) Chloride removal from archaeological cast iron by pulsating current. Stud Conservat 47:117–121
71. de Vivies P, Cook D, Drews MJ, Gonzalez NG, Mardikian P, Memet JB (2007) Transformation of akaganéite in archaeological iron artefacts using subcritical treatment. In: Degrigny C, Van Langh R, Joosten I, Ankersmit B (eds) Proceedings of the international conference on metals conservation, Amsterdam, Netherlands, Amsterdam, pp 17−21
72. Degrigny C, Le Gall R (1999) Conservation of ancient lead artifacts corroded in organic acid environments: electrolytic stabilization/consolidation. Stud Conserv 44:157–169
73. Guilminot E, Baron G, Memet JB, Huet N, Le Noc E (2007) Electrolytic treatment of archaeological marine chloride impregnated iron objects by remote control. In: Degrigny C, Van Lang R, Joosten I, Ankersmith B (eds) Metal 07. Proceedings of the Interim meeting of the ICOM-CC metal WG, vol 3. Rijksmuseum Amsterdam, Amsterdam, pp 38−43
74. Grassini S (2013) Plasma treatments for the cleaning and protection of metallic heritage artefacts. In: Dillmann P, Watkinson D, Angelini E, Adriaens A (eds) Corrosion and conservation of cultural heritage metallic artefacts. Woodhead Publishing, pp 552–569
75. Guo J, Hoyt N, Williamson M (2021) Multielectrode array sensors to enable long-duration corrosion monitoring and control of concentrating solar power systems. J Electroanal Chem 884:115064
76. He L, Liang J, Zhao X, Jiang B (2011) Corrosion behavior and morphological features of archaeological bronze coins from ancient China. Microchem J 99:203–212
77. Hernández RdPB, Hernández IV, Aoki B, de Melo THG, HG (2011) Electrochemical impedance spectroscopy investigation of the electrochemical behaviour of copper coated with artificial patina layers and submitted to wet and dry cycles. Electrochim Acta 56:2801–2814
78. Khan PF, Shanthi V, Babu RK, Muralidharan S, Barik RC (205) Effect of benzotriazole on corrosion inhibition of copper under flow conditions. J Environ Chem Eng 3:10–19
79. Liu J, Li Y, Wu M (2008) Electrochemical methods for chloride removal from simulated cast iron artefacts. Stud Conservat 53:41–48
80. Mansfeld F (2009) Fundamental aspects of the polarization resistance technique—the early days. J Solid State Electrochem 13:515–520
81. Mardikian P, Gonzalez N, Drews MJ, Nasanen L (2010) The use of subcritical solutions for the stabilization of archaeological iron artifacts. In: Eggert G, Schmutzler B (eds) Archaeological iron conservation colloquium 2010. State Academy of Art and Design Stuttgart
82. Novakovic J, Vassiliou P, Georgiza E (2013) Electrochemical cleaning of artificially Tarnished Silver. Int J Electrochem Sci 8:7223–7232
83. Organ RM (1968) Design for scientific conservation of antiquities. IIC/Butterworths, London
84. Otmacic Curkovic H, Kosec T, Marušić K, Legat A (2012) An electrochemical impedance study of the corrosion protection of artificially formed patinas on recent bronze. Electrochim Acta 83:28–39
85. Palomar T, Ramírez Barat B, García E, Cano E (2016) A comparative study of cleaning methods for Tarnished Silver. J Cult Herit 17:20–26
86. Plenderleith HJ (1956) The conservation of antiquities and works of art. Oxford University Press, London
87. Ramírez-Barat B, Cano E (2015) The use of agar gelled electrolyte for in situ electrochemical measurements on metallic cultural heritage. Electrochim Acta 182:751–762
88. Ramos-Negrón OJ, Escobar-Jiménez RF, Arellano-Pérez JH, Uruchurtu-Savarín J, Gómez-Aguilar JF, Lucio-García MA (2019) Electrochemical noise analysis to identify the corrosion type using the stockwell transform and the shannon energy: Part II. J Electroanal Chem 855:113597
89. Rocca E, Mirambet F (2010) The electrochemical techniques for the diagnosis and restoration treatments of technical and industrial heritage: three examples of metallic artefacts. J Solid State Electrochem 14:415–423

90. Rosalbino F, Scavino G, Mortarino G, Angelini E, Lunazzi G (2011) EIS study on the corrosion performance of a Cr(III)-based conversion coating on zinc galvanized steel for the automotive industry. J Solid State Electrochem 15:703–709
91. Santana JJ, González-Guzmán J, Fernández-Mérida L, González S, Souto RM (2010) Visualization of local degradation processes in coated metals by means of scanning electrochemical microscopy in the redox competition mode. Electrochim Acta 55:4488–4494
92. Schmutzler B, Eggert G (2010) The chloride left behind (dis)solving an analytical problem. In: Eggert G, Schmutzler B (eds) Achaeological iron conservation colloquium 2010. State Academy of Art and Design Stuttgart
93. Shchukin DG (2013) Container-based multifunctional self-healing polymer coatings. Polym Chem 4:4871–4877
94. Torgoose S (1982) Post-excavation changes in iron antiquities. Stud Conservat 27:97–101
95. Versan D, Berzioli M, Caglio S, Casoli A, Lottici PP, Medeghini L, Poldi G, Zannini P (2014) An integrated multi-analytical approach to the study of the dome wall paintings by Correggio in Parma catedral. Microchem J 114:80–88
96. Walker R (2001) Instability of iron sulfides on recently excavated artefacts. Stud Conservat 46:141–152
97. Xu Y, Liu L, Xu C, Wang X, Tan MY, Huang Y (2020) Electrochemical characteristics of the dynamic progression of erosion-corrosion under different flow conditions and their effects on corrosion rate calculation. J Solid State Electrochem 24:2511–2524
98. Zerbino J, Gassa L (2003) Electrochemical impedance spectroscopy study of cuprous oxide films formed on copper: effect of pH and sulfate and carbonate ions. J Solid State Electrochem 7:177–182
99. Zhang Z, Yuan X, Zhao Z, Li X, Liu B, Bai P (2021) Electrochemical noise comparative study of pitting corrosion of 316L stainless steel fabricated by selective laser melting and wrought. J Electroanal Chem 894:115351
100. Belkaid S, Ladjouzi MA, Madani S (2011) Effect of biofilm on naval steel corrosion in natural seawater. J Solid State Electrochem 15:525–537

Chapter 12
Electrochemical Metal Dating

12.1 Time and Electrochemistry, an Overview

History, archaeology, geology, paleontology, and other sciences have among their basic targets the elaboration of chronological schemes and, ultimately, the dating of objects and/or events. Neglecting the subtle problems associated to the definition of time (or times), absolute dating is conceived assuming that geological and historical phenomena have occurred within the frame of an 'external' time, flowing uniformly, i.e., within the concept of the time definition of Newtonian physics [1].

Seen as an analytical problem [2], dating involves the heuristic definition of a timescale and the adoption of quantitative methods to calculate the units of that scale [3], i.e., the adoption of a 'clock' measuring time intervals and fixing of an event to define the time zero of the scale. The geological, historical, etc., clocks need a zero-setting by a given event (the fabrication of an artifact, the death of a living organism, etc.) and the time passed since this instant is determined by the variation of any measurable physico-chemical property.

There is a large variety of dating methods [4–6], from palynology, dendrochronology, acanthochronology, etc., to radiocarbon dating, isotope abundance ratios, luminescence, etc. A part of these methods are associated with chemical processes such as obsidian hydration, lignin degradation, amino acid racemization, and corrosion advancement. Ideally, all dating methods should satisfy several conditions [1]:

(a) Universality: The involved physico-chemical phenomenon must occur in all regions at all times.
(b) Continuity and uniformity: The time dependence of the physico-chemical phenomenon under study should be continuous, with no breaks, and uniform, submitted to a unique, well-defined rate law.
(c) Accessibility to measurement: The physical quantity can experimentally be measured and allow for age determination.

A. Doménech-Carbó and M. T. Doménech-Carbó, *Electrochemistry for Cultural Heritage*, Monographs in Electrochemistry,
https://doi.org/10.1007/978-3-031-31945-7_12

Accordingly, the determination of the age of an object and/or event requires the knowledge of the universal rate law and/or the empirical construction of a calibration curve from experimental data for objects/events of known age. This process, however, is made difficult by the fact that the above conditions only partially hold for existing methodologies. As a paradigmatic example, radiocarbon dating is conditioned by reservoir effects, variations in the atmospheric $^{14}C/^{12}C$ ratio, isotope fractionation, and other constraints so that refined calibrations have to be introduced for dating [7, 8].

In principle, there are no restrictions about the type of technique (among them, electrochemical ones) usable for dating purposes. In all cases, however, specific constraints relating to the locus, time, historical conditions, etc., can affect the age determination. Metal corrosion phenomena, for instance, are quite sensitive to the physico-chemical properties of the object and the environment, and the conditions (temperature, humidity, contaminants, etc.) in which the process occurred. This implies that all dating methods are applied under the acceptance of a series of simplifying assumptions that define a basic methodology subsequently refined.

Another important constraint in fields such as cultural heritage and paleontology is the restriction of the amount of samples disposable for physico-chemical analyzes. It is exactly this aspect, as underlined in previous chapters, that reinforces the interest in electrochemical techniques for dating purposes.

It is pertinent to note that *absolute* dating is not necessarily the ultimate analytical goal in using electrochemical methods. In most cases, electrochemical data can be used to establish *relative* chronologies to characterize historical periods, etc. Several examples of this time–electrochemistry linkage have been described in preceding chapters. Here, the focus will be on electrochemical dating methodologies. The earlier proposal was made by Scholz, Brainina, Zakharchuk, et al. for dating ceramic materials based on their electrocatalytic capabilities on selected electrochemical processes (see Sect. 8.6) [9]. The current chapter describes current achievements and includes future perspectives in this 'young' research field. The reader should be aware that several of the described approaches have been included here as prospective studies and, consequently, have some speculative character.

12.2 Dating of Corroded Metals

12.2.1 Antecedents

In the archaeological domain, metal dating is of obvious interest. However, this target is usually met by radiocarbon dating of the organic remains and/or luminiscence, hydroxylation rate, etc. dating of ceramic objects, which accompany the metal artifacts. In fact, direct methods for dating metals are limited to lead, based on the Meissner superconducting effect [10], and gold based on He, U, and Th analysis [11, 12], in principle requiring relatively large amounts of sample.

Metal aging involves corrosion phenomena (with the peculiarities of the aging of noble metals) and other coupled features, in particular, decuprification, destannification, and dezincation appearing in bronze and brass corrosion [13–17]. These phenomena appear to be generalized so that the Zn/Cu, Sn/Cu, and Pb/Cu ratios in the patina and the base metal of bronze and brass alloy were proposed as chronological indicators [18–21].

Solid-state electrochemical techniques permit a series of proposals for dating based on the monitoring of the advancement of metal aging. These are based on the VIMP methodology by its inherently high sensitivity that permits limiting of the sampling to a few nanograms from the metal patina. Two basic strategies are applied, the measurement of the response of metals and/or corrosion products, and the analysis of their electrocatalytic effects on redox probes. Although voltammetry is the reference technique, also impedance measurements can be devised [22–25].

12.2.2 Lead

Doménech-Carbó et al. [26] proposed different electrochemical criteria for dating lead artifacts based on the record of the voltammetric responses of nano- or submicrosamples attached to graphite electrodes using the 'one-touch' sampling protocol (see Sect. 10.3). The age markers are (i) the voltammetric signatures due to the reduction of Pb(IV) species; (ii) the ratio between the voltammetric signals associated with the reduction of porous and compact litharge components of the corrosion layers; and (iii) the variation of the electrocatalytic features associated to HER, OER, and salicylic acid oxidation. The methodology has been developed for a set of archaeological samples recovered from a restricted area (Valencian Community, Spain). All samples were corroded under burial conditions in calcareous soils and exhibited apparently similar conservation states.

Figure 12.1 shows the voltammetric responses of a sample of contemporary lead and a sample extracted from a *Lampadarium* fragment recovered in the archaeological site of Punta de l'Illa (Cullera, Spain), of Visigothic age (sixth century CE) immersed in aqueous acetate buffer at pH 4.85. The voltammogram of the recent lead is dominated by a prominent cathodic wave at −0.55 V (I) followed by a weak shoulder at −0.95 V (II). The samples of archaeological lead displayed a more complex voltammetric profile where the signal II becomes clearly enhanced and preceded by cathodic shoulder at ca. −0.70 V (III). In the region of potentials positive versus Ag/AgCl, a series of overlapping peaks at 1.10 (IVa), 0.90 (IVb), and 0.75 V (V) appear. These signals were almost completely absent in recent lead samples.

These signals can be interpreted considering that lead corrosion generally produces a primary compact, semipermeable patina of litharge (PbO) which is subsequently covered by a secondary, porous patina of the same compound with different compactness and crystallinity. Lead passivation studies suggested that the primary patina acts as a semipermeable membrane favoring the growth of a porous, less

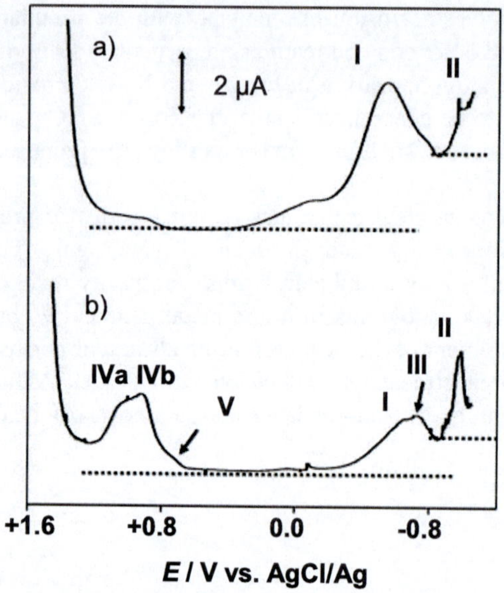

Fig. 12.1 Square wave voltammograms of graphite electrodes modified with: **a** contemporary lead and **b** a sample from the Punta de l'Illa (Cullera, Spain) archaeological site, in contact with 0.50 M HAc/NaAc buffer at pH 4.85. Potential scan initiated at 1.45 V in the negative direction; potential step increment 4 mV; square wave amplitude 25 mV; frequency 5 Hz. The dotted lines mark the baseline adopted for peak current measurement. Adapted from Ref. [26], with permission. Note that the axes are opposite to the IUPAC convention

compact secondary patina resulting in more negative electrode potentials [27, 28]. Then, peaks I–III can be attributed to the reduction of different PbO layers to Pb, a process widely studied in Chap. 3. In turn, the signals IV–V can be assigned to the reduction of PbO_2 (or possibly PbO_n $(1 < n < 2)$) species to PbO [29, 30]. This process can be represented as [29]:

$$PbO_{2solid} + 2xH^+_{aq} + 2xe^- \rightarrow PbO_{2-xsolid} + xH_2O \quad (12.1)$$

where $(0.6 < x < 1)$.

The first dating method is based on the assumption that, under fixed corrosion conditions, both the compact and porous litharge patinas grow with time. According to the treatment made in Sect. 3.9, the peak currents of processes I and II (I_I, I_{II}) are proportional to the net amounts (mass, mole) of compact and porous patina (m_I, m_{II}). There is, however, a practical problem, the overlapping of the corresponding signals. Then, the most reasonable option is to express the ratio between the peak currents measured using the baselines depicted in Fig. 12.1, as:

$$\frac{I_{II}}{I_I} = \frac{g_{II}m_{II} + h_I m_I}{g_I m_I} \quad (12.2)$$

where g_I, g_{II}, and h_I are electrochemical constants.

In principle, it is possible to assume that the mass by a surface unit of corrosion layer, y, varies following a power law of the type [10],

$$\frac{dy}{dt} = ky^{-\alpha} \tag{12.3}$$

Characterized by a rate constant k and an exponent α. Integration between $t = 0$ and t, and between $y = 0$ and y, of the above equation yields,

$$y = [(1+\alpha)kt]^{\frac{1}{1+\alpha}} \tag{12.4}$$

In the case in which the corrosion layer contains two components, I and II, we can consider two extreme—extremely simplified—situations. If component I forms a compact cover layer during the manufacturing of the object which does not further expand, y_I can be taken as a constant, essentially identical for all objects similarly handled, and then, only the mass per surface unit of component II increases with time. This is case A, for which the peak current ratio will be,

$$\left(\frac{I_{II}}{I_I}\right)_A = \frac{h_I}{g_I} + \left(\frac{g_{II}}{g_I}\right)[(1+\alpha)kt]^{\frac{1}{1+\alpha}} = H + Gt^{\frac{1}{1+\alpha}} \tag{12.5}$$

The second case (B) assumes that the composition of the patina evolves continuously from the metal phase (or from a compact, electrochemically inaccessible primary patina). Two possibilities can now be considered. If component II is formed at the expense of component I (case B1), then,

$$\frac{dy_I}{dt} = k_I y_I^{-\alpha} - k_2 y_I^{-\beta}; \quad \frac{dy_{II}}{dt} = k_{II} y_I^{-\beta} \tag{12.6}$$

This case can be simplified by assuming that the formation of the primary, compact patina is fast. Then, the secondary patina begins to be formed and this event marks the zero time of the clock. Then,

$$\frac{dy_{II}}{dt} = k_{II}(y_o - y_{II})^{-\beta} \tag{12.7}$$

and the peak current ratio follows as:

$$\left(\frac{I_{II}}{I_I}\right)_{B2} = \frac{h_I}{g_I} + \left(\frac{g_{II}}{g_I}\right)\left\{\frac{y_o - [y_o^{1+\beta} - (1+\beta)k_{II}t]^{\frac{1}{1+\beta}}}{y_o}\right\} \tag{12.8}$$

If the components I and II are independently formed, in case B2, we can write:

$$\frac{dy_I}{dt} = k_I y_I^{-\alpha}; \quad \frac{dy_{II}}{dt} = k_{II} y_{II}^{-\beta} \tag{12.9}$$

Then follows:

$$\left(\frac{I_{\mathrm{II}}}{I_{\mathrm{I}}}\right)_{\mathrm{B2}} = \frac{h_{\mathrm{I}}}{g_{\mathrm{I}}} + \left(\frac{g_{\mathrm{II}}}{g_{\mathrm{I}}}\right)\left\{\frac{[(1+\beta)k_{\mathrm{II}}t]^{\frac{1}{1+\beta}}}{[(1+\alpha)k_{\mathrm{I}}t]^{\frac{1}{1+\alpha}}}\right\} \tag{12.10}$$

Figure 12.2 shows the calibration graph corresponding to the variation of the $I_{\mathrm{II}}/I_{\mathrm{I}}$ ratio versus the object's age in square wave voltammograms of lead samples in conditions such as in Fig. 12.1 for a set of genuine unrestored archaeological lead samples [26]. The experimental $I_{\mathrm{II}}/I_{\mathrm{I}}$ versus t plot can satisfactorily be fitted to Eq. (12.5) taking $G = 3.4 \times 10^{-4}$ year$^{-0.934}$, $H = 0.22$, and $\beta = 0.07$. This last value is coincident with that calculated from Meissner effect on lead artifacts [10]. Figure 12.2 includes data points for archaeological objects of known age previously submitted to restoration and recognized 'modern' forgeries of Iberian objects. The corresponding data points for all these lead samples fall clearly far from the band in the diagram where the genuine, unrestored samples are located.

A similar treatment can be applied to the ratio between the signals corresponding to the reduction of PbO_n and the reduction of PbO forms. Since the PbO–PbO_2 aerobic oxidation is thermodynamically favored, the ratio between the corresponding peak currents (or peak areas) will vary with time following a pattern such as represented by Eqs. (12.2)–(12.6), as is supported by available data [26].

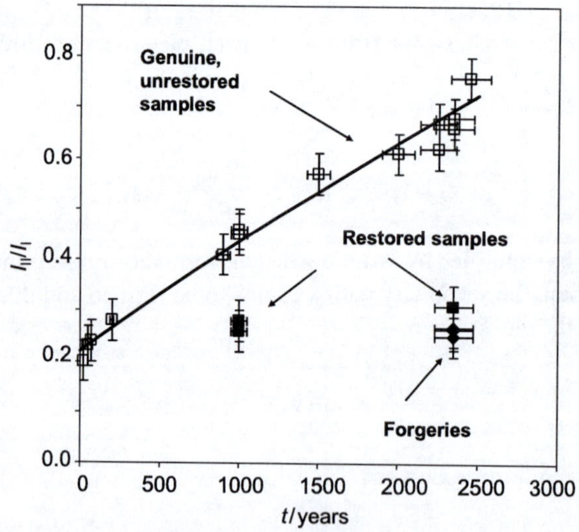

Fig. 12.2 Variation of the $I_{\mathrm{II}}/I_{\mathrm{I}}$ ratio with the corrosion time t in square wave voltammograms of lead samples in conditions such as in Fig. 12.1 for a set of genuine unrestored archaeological artifacts (squares), restored archaeological objects (solid squares), and recognized forgeries (solid rhombs). The continuous line represents the theoretical curve inserting $\beta = 0.07$; $K = 3.0 \times 10^{-5}$ year^{-1}, $h_{\mathrm{I}}/g_{\mathrm{I}} = 0.22$ and $g_{\mathrm{II}}/g_{\mathrm{I}} = 5.1$ into Eq. (12.6) Adapted from Ref. [26], with permission

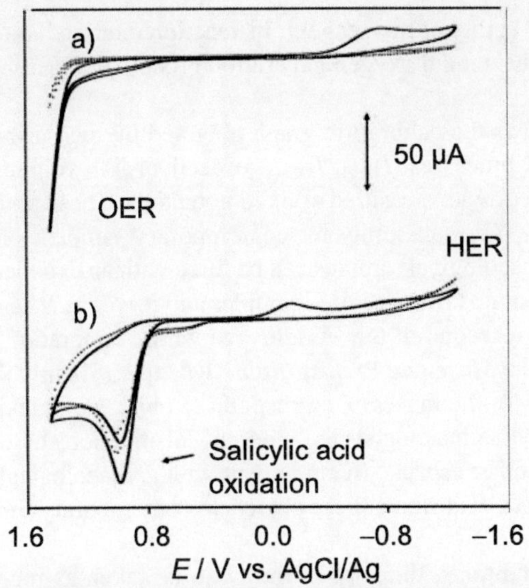

Fig. 12.3 Cyclic voltammograms recorded at graphite electrodes modified by a sample of contemporary lead (dotted lines) and an archaeological sample (continuous lines) from *the Tossal de Sant Miquel* (Llíria, Spain) of Iberian age immersed into **a** 0.50 M potassium phosphate buffer, pH 7.00 and **b** 5.0 mM salicylic acid solution in 0.50 M potassium phosphate buffer, pH 7.00. Potential scan rate 100 mV s^{-1}. Adapted from Ref. [26], with permission. Note that the axes are opposite to the IUPAC convention

Another way for age estimation was tested using electrocatalytic effects on the oxidation of salicylic acid in the HER and OER processes in experiments such as in Fig. 12.3, corresponding to sample-modified graphite electrodes in contact with aqueous phosphate buffer at pH 7.00, as well as ORR process in acetonitrile (MeCN) solution [26].

Figure 12.3 shows the cyclic voltammograms recorded for a sample of contemporary lead (dotted lines) and a sample from an Iberian plate dated back to the 4th to second century BCE (continuous lines) recovered in the Tossal de Sant Miquel (Llíria, Spain) archaeological site. Here, the electrolyte was 0.50 M potassium phosphate buffer at pH 7.00 optionally adding salicylic acid. One can see that archaeological samples produce the reinforcement of the extreme OER current and the decrease of the corresponding HER current. The addition of salicylic acid determines, however, a decrease in the OER current.

The promotion of the OER process can be attributed to the presence of PbO_2 forms, in line with the extended use of PbO_2 as electrode material with high oxidizing ability, in particular on OER [31]. In turn, lead metal catalyzes the HER process [32]. Accordingly, the quotient between the catalytic currents for HER and OER reactions could be representative of the Pb(/PbO_2 ratio in the sample-modified electrode. The presence of salicylic acid, apart from the effects of its electrochemical oxidation process, increased data repeatability. Possibly, this feature can be associated with the radical scavenging ability of this reagent. Its reaction intermediate hydroxyl radicals generated during the anodic oxygen generation process will cause a decrease in the OER current [33].

Figure 12.4 shows the calibration graph obtained by measuring the currents for the HER and OER processes (I_{HER}, I_{OER}, respectively) in voltammograms such as in Fig. 12.3 using currents measured at fixed potentials. These potentials were fixed to approximate I_{HER}/I_{OER} ca. unity for contemporary samples. Now, experimental data points for archaeological samples can be fitted with an exponential function and become clearly separated from restored samples and forgeries. These were one object of Iberian age and a second of the Middle Age whose restoration was documented in the archives of the *Museu de Prehistòria de València* (Spain). The forgeries, also from the repository of the museum, correspond to objects mimicking Iberian plates but identified by the archaeologists as 'modern' falsifications by obvious pitfalls in the epigraphy and other aspects. In agreement with archaeological criteria, voltammetric data permitted to attribute these objects to contemporary times (dotted line in Fig. 12.4).

Under these conditions, the I_{OER} current can be taken as representative of the Pb(IV) species existing in the corrosion layers while the I_{HER} current would be representative of the eventual lead extracted during the sampling process and, mainly, of the total lead electrochemically generated by reduction of all lead oxides.

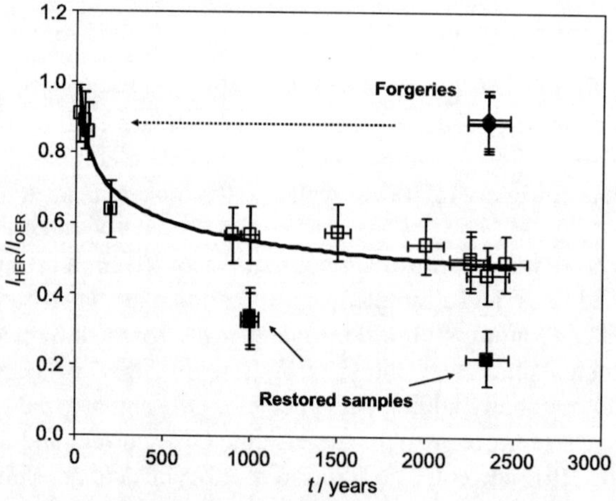

Fig. 12.4 Variation of the I_{HER}/I_{OER} ratio with the age in cyclic voltammograms of lead-modified electrodes in contact with aqueous phosphate buffer at pH 7.00 for a set of genuine archaeological samples (squares), restored archaeological samples (solid squares), and recognized forgeries (solid rhombs). The continuous line represents the fit of experimental data to Eq. (12.11) taking $K = 1.5$, $\xi = -0.15$. Adapted from Ref. [26], with permission

These currents can be approached by combining two equations similar to Eq. (12.4) so that,

$$\left(\frac{I_{HER}}{I_{OER}}\right)_{B2} = \left(\frac{g_{HER}}{g_{OER}}\right)\left\{\frac{[(1+\gamma)k_{HER}t]^{\frac{1}{1+\gamma}}}{[(1+\mu)k_{OER}t]^{\frac{1}{1+\mu}}}\right\} = Kt^{\xi} \qquad (12.11)$$

which is in turn equivalent to Eq. (12.10). As can be seen in Fig. 12.4, experimental data can satisfactorily be fitted to an exponential function consistent with Eq. (12.11).

A dating procedure based on EIS measurements was also described [34]. Microparticulate deposits of sample-modified graphite electrodes in contact with 0.10 M $NaClO_4$ aqueous solution applying the OCP as the bias potential for each sample. The impedance response consisted of an apparently unique capacitive loop and the maximum (minus) phase angle (see Sect. 10.6 for the detailed analysis of similar—but not equivalent—systems) appears to vary monotonically with the age of the specimens.

Current studies are focused on lead artifacts submitted to moderate corrosion in contact with buffered calcareous soils. Under these conditions, litharge is the main component of lead patinas. Obviously, considering cases of patinas containing significant amounts of other corrosion products (cerussite, hydrocerussite, etc.) is of interest. A particular field of interest is dating lead artifacts recovered from marine environments, where data in progress indicate that lead chloride, carbonate, and sulfate forms are produced.

Future research should include a refined description of voltammetric and impedance data, consideration of in-depth analysis (see next section), and handling a major number of archaeological samples coming from different provenances. Of course, combining electrochemistry with other techniques (isotopic data in particular) is of interest.

12.2.3 Copper and Bronze

The first approach to the electrochemical dating of copper and copper-based alloys was based on the estimate of the averaged tenorite/cuprite ratio in the metal patina using the VIMP methodology [35]. The essential idea is that cuprite forms in general the primary corrosion patina of copper-based alloys (see Chap. 10) and that its oxidation to tenorite,

$$2Cu_2O_{solid} + O_{2g} \rightarrow 4CuO_{solid} \qquad (12.12)$$

is a thermodynamically favored process. The tenorite/cuprite ratio can be estimated from the ratio of the intensity (peak currents, peak areas) of the respective voltammetric signals at ca. −0.10 (I) and −0.40 V versus Ag/AgCl (II) recorded in contact

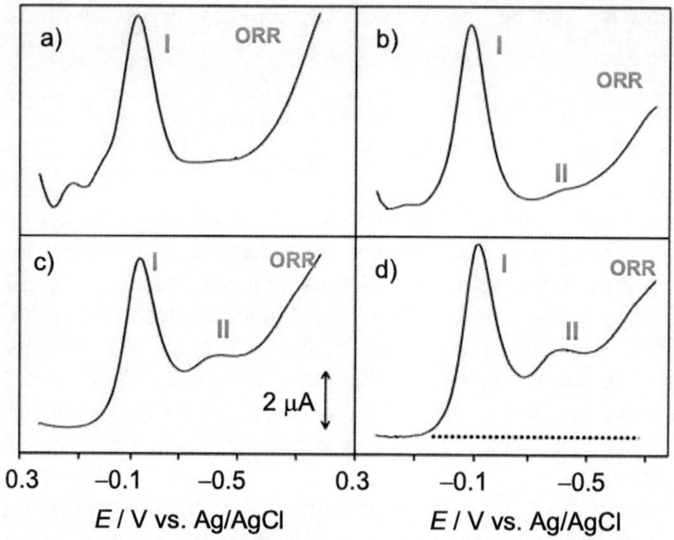

Fig. 12.5 Square wave voltammograms for samples extracted by means of the one-touch protocol of: **a** 'fresh' and **b** 'aged' Eurocent coins; **c** halfpenny British coin (1890); **d** Philippus versus Spanish *maravedí* (1709) in contact with air-saturated 0.25 M aqueous HAc/NaAc buffer at pH 4.75. Potential scan initiated at 0.25 V in the negative direction; potential step increment 4 mV; square wave amplitude 25 mV; frequency 5 Hz. The dotted line in **d** marks the baseline used for peak current measurements. Adapted from Ref. [35], with permission. Note that the axes are opposite to the IUPAC convention

with aqueous acetate buffer as illustrated in Fig. 12.5. Here, the square wave voltammograms recorded for microparticulate deposits of samples from Eurocent coins ('fresh' and 'aged') and a halfpenny British coin produced in 1890, and a Spanish *maravedí* minted in 1709 are depicted. One can see in this figure that the height of the tenorite-related cathodic process II relative to the cuprite-associated reduction I increases on increasing the age of the coin.

The application of this methodology requires, obviously, that the objects were moderately corroded under reasonably uniform conditions so corrosion products were limited to cuprite and tenorite. There are obvious constraints for dating: the differences in the composition of the base alloy and the metallographic properties must produce differences in the corrosion pattern (see Chap. 10), and the tenorite/cuprite ratio varies with the depth, making it sensitive to the more or less abrasive sampling. Additionally, the presence of organic matter as a result of handling and the heterogeneities in the corrosion process can influence the voltammetric response. This last aspect can be seen in Fig. 12.6, in which a microphotograph of the surface of a bronze coin minted in 1770 is shown.

Figure 12.7 shows the calibration curve built from a set of moderately corroded copper and bronze coins taking the averaged peak current ratio (I_{II}/I_{I}) from at least

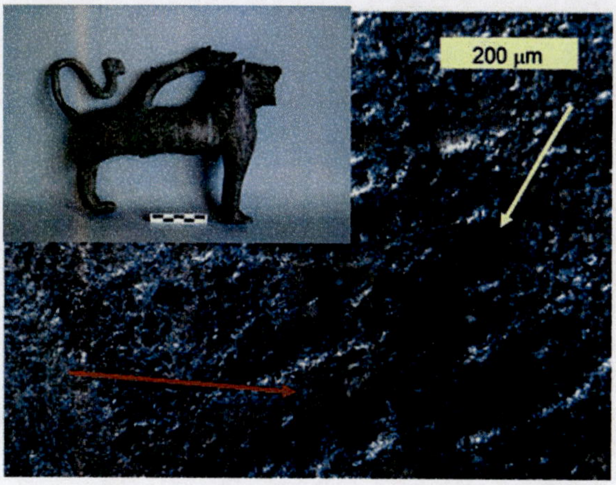

Fig. 12.6 Optical microscopy image (40×) of a Spanish bronze coin minted in 1770. The arrows mark regions where cuprite (red arrow) and tenorite (yellow arrow) appear. Inset: water pitcher of the Caliphal period from Valencia (Spain), dated back ca. 950 CE, *Museu Municipal de Xàtiva*, Spain. Reproduced from Ref. [35], with permission

three independent measurements for each one. Here, experimental data suggest that Eq. (12.8) or similar applies, being reasonably fitted to equations of the type [35, 36]:

$$\left(\frac{I_{\mathrm{II}}}{I_{\mathrm{I}}}\right) = H + \frac{At^a}{1 \pm Bt^b} \tag{12.13}$$

suggesting an intermediate behavior between the extreme models B1 and B2 described in Sect. 12.2.2. Since these equations involve a large set of adjustable parameters, they can be considered semiempirical approximations. Future research

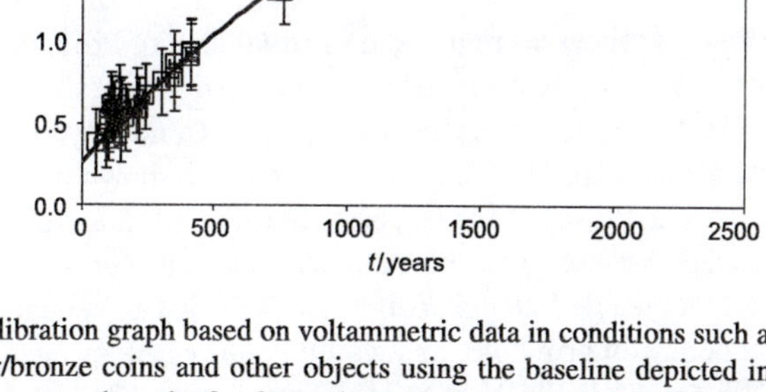

Fig. 12.7 Age calibration graph based on voltammetric data in conditions such as in Fig. 12.5 for a series of copper/bronze coins and other objects using the baseline depicted in this figure. The continuous line corresponds to the fit of experimental data to Eq. (12.13) inserting $H = 0.25$, $A =$ 2.3 year, $B = 1000$ year, $a = 0.97$, $b = 0.93$. Adapted from Ref. [35], with permission

must provide a much more accurate description of metal corrosion phenomena for properly modeling experimental voltammetric data. In particular, modeling must take into account the in-depth variation of the composition of the corrosion layers; a first approximation to this matter will be presented in Sect. 12.2.6 (see below).

The uncertainty of age time estimates, Δt using voltammetric measurements approximated by Eqs. (12.4)–(12.13) can be related to both the uncertainty in the peak current ratio measurements, and the uncertainty in the adjustable kinetic parameters, in turn, derived from the analysis of the calibration data. Application of the usual theory of error propagation leads to different age-dependent Δt values depending on the mathematical form of the calibration equation [26, 35, 36].

As in the case of lead, EIS data were also tested as the source of age calibration using sample-modified electrodes in contact with air-saturated mineral water and the reduction of dissolved oxygen as the redox probe [36].

12.2.4 Leaded Bronze

The practice of adding lead to bronze alloys, known since antiquity, lowers its cost and improves its fluidity and mold-filling capability. The low solubility of lead in copper, however, produces peculiar structural features and corrosion patterns (see Sect. 10.2). Additionally, lead can also be used as decorative/protective coating on bronze statuary.

The VIMP of leaded bronze results in well-defined voltammetric responses characterized by the appearance of intense signals for the reduction of copper and lead corrosion products. This can be seen in Fig. 12.8a, b, where the square wave voltammograms of two samples of leaded bronze from the Roman archaeological sites of Gadara (Jordan, 4th CE) and Valeria (Spain, 1st CE) in contact with aqueous acetate buffer at pH 4.75 are shown [37]. The voltammograms show well-defined peaks for copper corrosion products and lead ones. The reduction of cuprite plus minerals of the malachite, atacamite, and/or brochantite groups appear superimposed as signal C_1, while tenorite reduction often appears as signal C_3. Lead(II) corrosion products display cathodic signals labeled as C_2 while plattnerite and substoichiometric Pb(IV) forms yield the peak C_4. The relevant point to emphasize is that the total area of Cu-related reduction processes relative to the total area of Pb-localized processes decreases by decreasing the age of the object submitted to sampling.

As was seen in Sect. 12.2.2, signal C_4 shows a general tendency to increase relative to signal C_2 on increasing age. An illustrative example is shown in Fig. 12.8c where it is accompanied by a FESEM image corresponding to one typical Pb globular formation in the leaded bronze of a modern sculpture currently on the campus of the Universitat Politècnica de València (Fig. 12.8d) [37]. This tendency suggested a possible aging tendency. In principle, this would be the result of the major tendency of lead ($E^{o}(Pb^{2+}/Pb) = -0.126$ V vs. NHE) than copper $E^{o}(Cu^{2+}/Cu) = 0.340$ V vs. NHE) to oxidize. Then, lead exerts as a protective component against copper

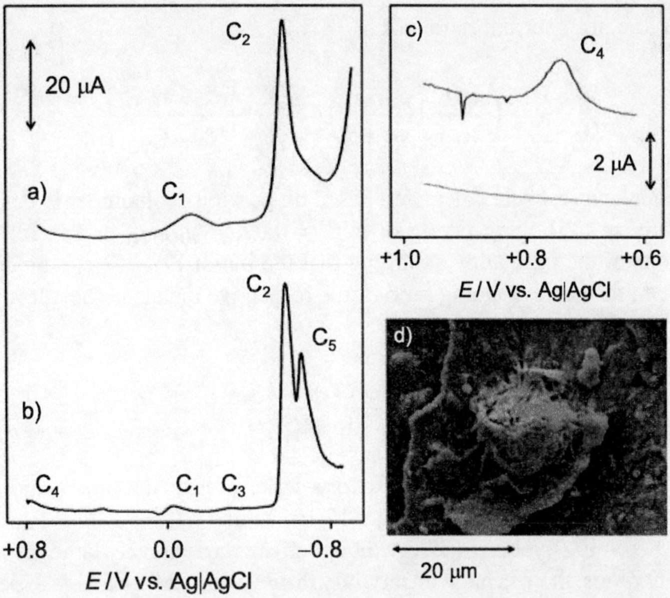

Fig. 12.8 Square wave voltammograms of samples of leaded bronze from the archaeological sites of **a** Gadara (Jordan, 4th CE) and **b** Valeria (Spain, 1st CE). Electrolyte: 0.25 M HAc/NaAc aqueous solution at pH 4.75; potential scan initiated at 0.85 V in the negative direction; potential step increment 4 mV; square wave amplitude 25 mV; frequency 5 Hz. Inset **c** shows the region around 0.8 V versus Ag/AgCl of the voltammograms recorded for an Iberian votive figurine (Despeñaperros, Spain 4th–2th BCE) and a figurine from the *Sant Cristòfol* church (Picassent, Spain, 18th–19th CE), respectively, upper and lower curves. Inset **d** shows the FESEM image of a sample extracted from the sculpture *Crónica del viento* by Martín Chirino, 1991 (Museu de l'Universitat Politècnica de València). Reproduced from Ref. [37], with permission. Note that the axes in the voltammograms are opposite to the IUPAC convention

oxidation. However, local corrosion cells in the material are formed resulting in the formation, evidenced by experimental data, of copper and lead corrosion products.

Assuming that a fine layer of the primary metal corrosion products (mainly compact cuprite and litharge) is initially formed, the zero of time will be marked by the beginning of the growth of a porous secondary patina of corrosion composed by porous cuprite (plus tenorite) and litharge. Assuming, as before, that potential rate laws apply for each one of these processes, one can express the age variation of the net amount of MO corrosion product by the surface unit in the secondary patina as,

$$y_{MO} = [Y_{MO}{}^{1+\delta} + k_{MO}t]^{\frac{1}{1+\delta_{MO}}} \tag{12.14}$$

where Y_{MO} denotes the amount of MO corrosion product when the formation of the secondary corrosion layers begins and k_{MO}, δ_{MO}, the rate constant and kinetic exponent, respectively. Introducing the usual coefficients of electrochemical response,

g_{CuO}, g_{PbO}, the peak current ratio will be [37],

$$\frac{I\,(\mathrm{C}_1)}{I\,(\mathrm{C}_2)} = \left(\frac{g_{\mathrm{CuO}}}{g_{\mathrm{PbO}}}\right)\left(\frac{y_{\mathrm{CuO}}}{y_{\mathrm{PbO}}}\right) = \frac{\left[Y_{\mathrm{CuO}}^{1+\alpha} + k_{\mathrm{CuO}}t\right]^{\frac{1}{1+\alpha}}}{\left[Y_{\mathrm{PbO}}^{1+\beta} + k_{\mathrm{PbO}}t\right]^{\frac{1}{1+\beta}}} \tag{12.15}$$

Theoretical curves were calculated based on previous voltammetric data on lead [26] and bronze [35] aging taking $\alpha \approx \beta \approx 0.07$, $g_{\mathrm{CuO}}/g_{\mathrm{PbO}} = 1.0$, for different compositions of the base alloy assuming that the initial $Y_{\mathrm{CuO}}/Y_{\mathrm{PbO}}$ ratio equals the ratio between the corresponding ratio of the respective metals in the alloy, $Y_{\mathrm{Cu}}/Y_{\mathrm{Pb}}$. Then, Eq. (12.15) can be rewritten as,

$$\frac{I\,(\mathrm{C}_1)}{I\,(\mathrm{C}_2)} = \frac{\left[Y_{\mathrm{Cu}}/\,Y_{\mathrm{Pb}} + k^{*}_{\mathrm{CuO}}t\right]^{0.93}}{\left[1 + k^{*}_{\mathrm{PbO}}t\right]^{0.93}} \tag{12.16}$$

Figure 12.9 shows experimental data for a series of leaded bronze samples superimposed to theoretical lines from Eq. (12.19) taking $k^{*}_{\mathrm{Cu}} = 1.0 \times 10^{-3}$ years^{-1} and $k^{*}_{\mathrm{Pb}} = 1.0 \times 10^{-1}$ years^{-1} [37]. Although given the relatively large dispersion in peak current ratios, there is large uncertainty in age estimates, available experimental data [37, 38] are consistent with the previous treatment, thus providing a reasonable approximation to age estimates.

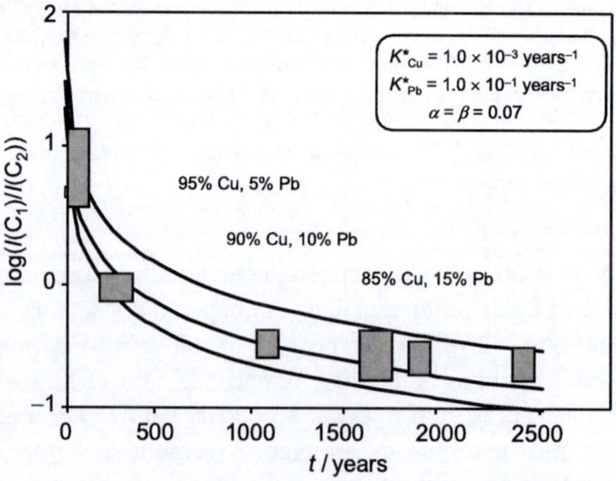

Fig. 12.9 Theoretical lines for different selected compositions of leaded bronze inserting $k^{*}_{\mathrm{Cu}} = 1.0 \times 10^{-3}$ years^{-1} and $k^{*}_{\mathrm{Pb}} = 1.0 \times 10^{-1}$ years^{-1} into Eq. (12.16) and experimental data of the $I(\mathrm{C}_1)/I(\mathrm{C}_2)$ peak current ratio recorded in negative-going voltammograms for different archaeological and modern samples in conditions such as in Fig. 12.8. Adapted from Ref. [37], with permission

12.3 Gold Dating

Dating gold is of particular archaeological interest because of its wide use in jewelry, cult figures, coinage, etc., with high material and symbolic value. Due to its noble character, the degradation upon aging is in principle much less important than for other less noble metals and much less sensitive to the local conditions of degradation. The aging of gold results in a loss of brightness without apparent generation of corrosion products. Based on Volta potential measurements, gold aging has been attributed to the formation of oxides of impurity metals [39] or physical adsorption of oxygen [40], hydrophobic recovery being an important factor to be accounted for [41].

As described in Chap. 10 (Sect. 10.7), in contact with HCl electrolytes, gold displays well-defined electrochemistry based on its oxidation to gold oxides/oxo-hydroxides [Eqs. (10.25)–(10.29)] and the oxidative dissolution to Au(III)-chloride complexes [Eq. (10.31)]. This process is accompanied by the oxidation of surface active sites [Eq. (10.30)] associated with coordinatively unsaturated gold atoms and lattice defects, impurities, etc.

The first criterion for dating purposes is the abundance of surface defects [42]. The proportion of defect sites relative to 'ordinary' gold will increase with the age as a result of different phenomena, in particular, oxygen adsorption and concomitant exchange with subsurface atoms [43, 44]. This phenomenon has been studied in the context of spillover phenomena in solid-state catalysts [45]. Figure 12.10 shows a schematic representation of these processes [42].

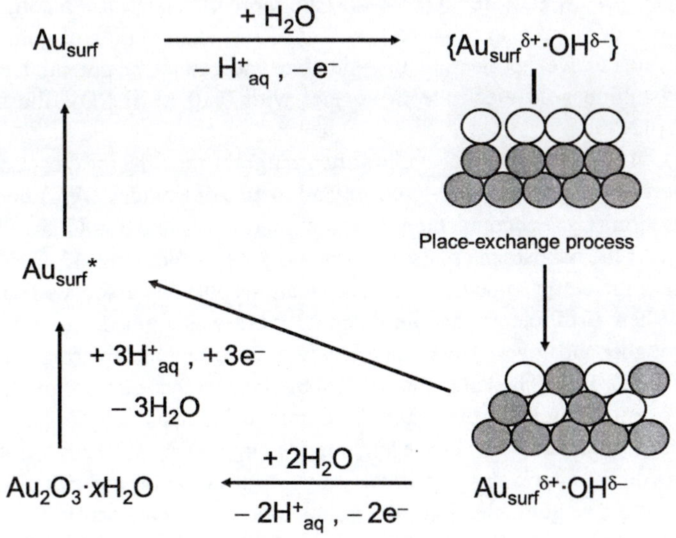

Fig. 12.10 Schematic representation of the generation and propagation of surface defects in gold surfaces. Reproduced from Ref. [42], with permission

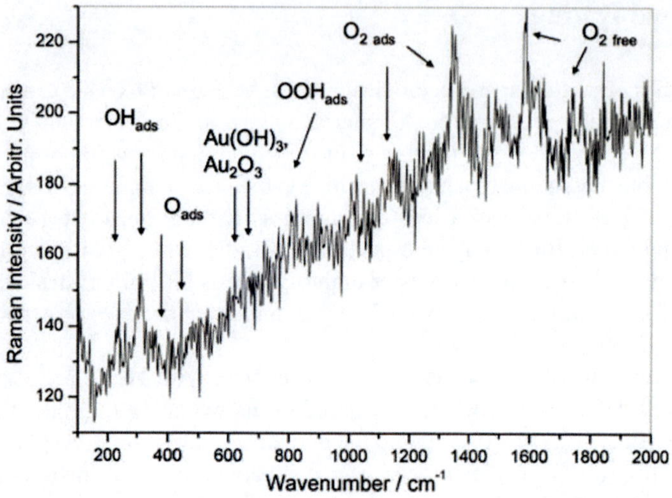

Fig. 12.11 Raman spectrum was recorded for a gold medal fabricated in 1903. Raman features attributable to different oxygen species according to the literature [46–48] are marked by arrows. Adapted from Ref. [42], with permission

The importance of oxygen adsorption is underlined by Raman spectroscopy studies [46–48]. Figure 12.11 depicts the Raman spectrum recorded for a gold medal fabricated in 1903 where spectral features associated with O_2 molecules, oxygen atoms, ·OH and ·OOH radicals adsorbed onto gold, as well as those associated with $Au(OH)_3$ and Au_2O_3 and 'free' O_2 [46–48] are more or less clearly recorded.

The above ideas are also consistent with the AFM examination of gold surfaces during the successive application of oxidative and reductive potential inputs to a microcrystalline gold electrode in contact with 0.10 M H_2SO_4, illustrated in Fig. 12.12 [42].

Figure 12.12a shows a typical cyclic voltammogram showing the electrochemical features described in Sect. 10.7; gold oxidation to gold oxides (A_{Au}) and subsequent reduction (C_{Au}) accompanied by the oxidation of active gold sites (marked by arrows). At the nanoscopic scale, the original gold surface (Fig. 12.12b) shows a rough pattern including some isolated major features possibly associated with oxide deposits. When a sufficiently anodic potential input was applied, the gold surface becomes rougher and covered by rounded features accompanied by irregular, peaked features (Fig. 12.12c). The latter can be attributed to the formation of oxide humps. The subsequent application of a cathodic potential input partially regenerates the original rough gold surface where several humps remain (Fig. 12.12d). Consistently with studies on the attack of ·OH radicals on gold surfaces [44] and electrochemical generation of active gold sites [43], the gold surfaces are quite sensitive to oxygen species.

The electroactivity of active gold sites is usually reinforced in successive potential scans as denoted by the increase of the intensity of the corresponding voltammetric

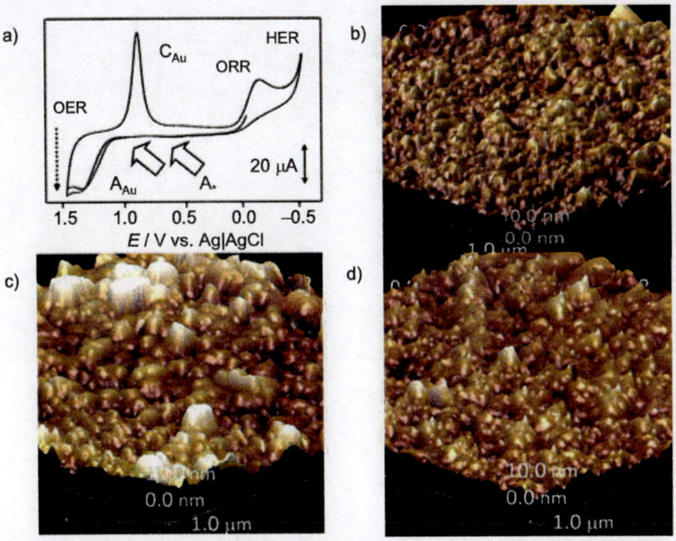

Fig. 12.12 **a** Cyclic voltammogram of a polycrystalline gold electrode immersed in 0.10 M H_2SO_4 air-saturated aqueous solution at a potential scan rate of 50 mV s^{-1}. **b–d** AFM topographic images (1 × 1 × 0.01 μm) of a gold plate in contact with the same electrolyte solution **b** before and **c**, **d** after application of successive potential inputs of **c** 1.25 V for 2 min and **d** 0.10 V during 2 min. The arrows mark weak signals for the oxidation of active gold sites. Reproduced from Ref. [42], with permission. Note that the axes in the voltammogram are opposite to the IUPAC convention

signal (A* in Fig. 12.12a) in the second scan relative to the first one, as illustrated in Fig. 12.13. Here, the 1st and 2nd anodic linear potential scans displayed for graphite electrodes modified with samples from a *dinar* from Santa Elena hoard dated back to 1018 and a *Carolus IV escudo* (Spain, 1794) are depicted. One can see that the relative height of the signal for the oxidation of active sites (A_2), practically absent in the initial scan, becomes clearly defined in the second one. In turn, the main gold oxidation peak (A_1) clearly diminishes from the 1st to the 2nd potential scan.

Interestingly, voltammograms in Fig. 12.13 reveal that the ratio between the intensity of the signal A_2 in the second scan, $I(A_2(2))$, and the intensity of the signal A_1 in the first scan, $I(A_1(1))$ (measured using the baselines depicted in Fig. 12.13a), increases with the age of the object. Voltammetric data in conditions such as in this figure for a set of gold coins of known age permitted the construction of the calibration curve depicted in Fig. 12.14 corresponding to the $I(A_2(2))/I(A_1(1))$ versus time t representation [42]. The insets in this figure show the images of two coins from the repository of the *Museu de Prehistòria de València* (Spain) which kindly facilitated the study.

Experimental data from replicate voltammetric experiments revealed, however, that the $I(A_2(2))/I(A_1(1))$ ratio varied with the absolute value of $I(A_1(1))$. This is in agreement with the general problem of the net amount of sample transferred onto the

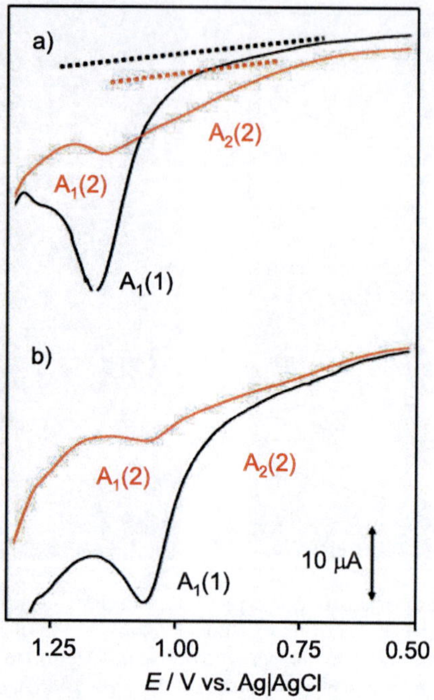

Fig. 12.13 Linear potential scan voltammograms of gold samples from **a** a *dinar* from Santa Elena hoard dated back to 1018 and **b** a *Carolus IV escudo* minted in Spain in 1794 attached to a graphite electrode in contact with 0.10 M HCl. The first (black lines) and second (red lines) positive-going potential scans are superimposed. Potential scan rate 50 mV s^{-1}. The dotted lines in **a** depict the baselines used for current measurements. Reproduced from Ref. [42], with permission. Note that the axes are opposite to the IUPAC convention

electrode surface in VIMP sampling [1, 23]. Accordingly, $I(A_2(2))/I(A_1(1))$ values normalized to $= I(A_1(1)) = 60$ μA were taken in Fig. 12.14 [42].

The calibration graph in Fig. 12.14 can in principle be interpreted as reflecting adsorption kinetics, satisfactorily fitting to Lagergren's equation [49]. This equation corresponds to a pseudo-first-order rate law of the form,

$$\frac{I(A_2(2))}{I(A_1(1))} = H - Ge^{-kt} \tag{12.17}$$

Experimental data in Fig. 12.14 can satisfactorily be fitted to Eq. (12.17) taking $H = 0.15 \pm 0.05$, $G = 0.58 \pm 0.08$, and $k = (5.8 \pm 0.4) \times 10^{-4}$ years^{-1}.

The apparent adsorptive-driven character of the gold aging prompted the proposal of a second electrochemical dating methodology [50]. The essential idea was that, in contact with HCl electrolytes, the equilibrium potential for the oxidative dissolution of gold described in Chap. 10 (Eq. (10.3)) was,

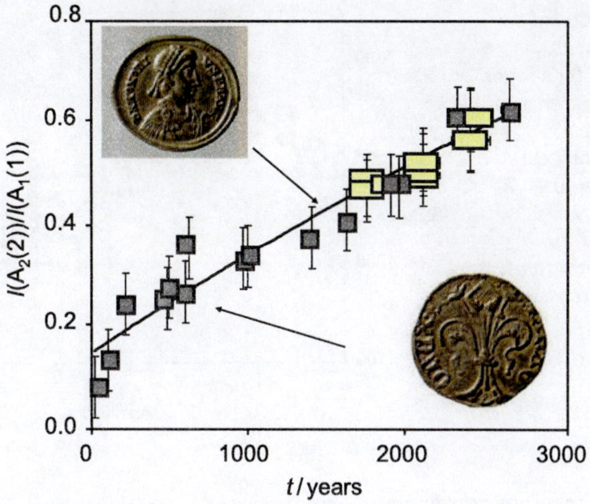

Fig. 12.14 Calibration graph corresponding to the age variation of the $I(A_2(2))/I(A_1(1))$ normalized (see text) ratio in linear potential scan voltammograms in conditions such as in Fig. 12.13 for a set of gold samples of known age. Continuous line corresponds to the theoretical curve obtained from Eq. (12.20) taking $H = 0.15$, $G = 0.50$, and $k = 5.8 \times 10^{-4}$ years^{-1}. Insets: photographic images of samples of a medieval *Florin* (upper image, Alfons V, Valencia, Spain, 1416–1418 CE) and a Roman *Solidus* (lower image, Honorio, Mediolanum, 394–395) whose positions in the diagram are marked by arrows. Adapted from Ref. [42], with permission

$$E^{\circ\prime}_{\mathrm{Au}} = E^{\circ}(\mathrm{Au}^{3+}/\mathrm{Au}) + \frac{RT}{3F}\ln\frac{a_{\mathrm{AuCl}x}}{a_{\mathrm{Au}}\beta_x a^{x}_{\mathrm{Cl}-}} \tag{12.18}$$

where a_J represents, as usual, the thermochemical activity of the J-species. In this equation, it was assumed, for simplicity, that a unique Au(III) complex of stoichiometry $\mathrm{AuCl}_x{}^{(3-x)+}$ is formed whose formation constant is β_x (corresponding to the complexation equilibrium $\mathrm{Au}^{3+}{}_{\mathrm{aq}} + x\mathrm{Cl}^{-}{}_{\mathrm{aq}} \rightarrow \mathrm{AuCl}_x{}^{(3-x)+}{}_{\mathrm{aq}}$).

For a 'clean' gold surface, $a_{\mathrm{Au}} = 1$, but for an aged surface it is reasonable to expect that $a_{\mathrm{Au}} < 1$ due to the formation of oxygen adsorbates. This will produce, under equilibrium conditions, a shift of the formal electrode potential. This can be approximated by taking $a_{\mathrm{Au}} = 1 - a_{\mathrm{ox}}$, a_{ox} being the activity of the O-adsorbates. Accordingly, the new formal potential, $E^{\circ\prime}{}_{\mathrm{Au\{O\}}}$, will be,

$$E^{\circ\prime}_{\mathrm{Au\{O\}}} = E^{\circ\prime}_{\mathrm{Au}} - \frac{RT}{3F}\ln(1 - a_{\mathrm{ox}}) \tag{12.19}$$

Since a_{ox} will vary with the age of the gold object, the $E^{\circ\prime}{}_{\mathrm{Au\{O\}}} - E^{\circ\prime}{}_{\mathrm{Au}}$ potential difference will systematically vary with the age under fixed electrochemical conditions. Since a_{ox} should increase with aging time, the formal electrode shift should increase with the age.

Of course, this approach is conditioned, as the precedent one, by the fact that the composition of the base metal (archaeological gold contains minority silver,

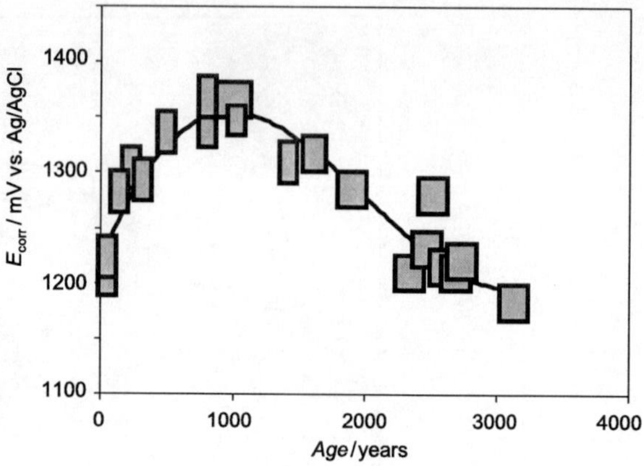

Fig. 12.15 Variation of the corrected potential E_{corr} versus aging time for a set of gold samples of known age. The continuous line represents the theoretical curve inserting $B = 0.41$, $k_a = 0.0012$ years^{-1} and $k_b = 0.0021$ years^{-1} into Eq. (12.22). Adapted from Ref. [50], with permission

copper, and other minor components) and its metallographic structure will influence the voltammetric response (and the O adsorption characteristics). Here, there is an additional problem: Eqs. (10.21) and (10.22) refer to equilibrium potentials and these are not accessible to voltammetric measurements when electrochemical reversibility is not attained. Based on Tafel-type analysis of voltammetric curves applied to the peak $A_1(1)$ (see Fig. 12.13), an apparent, corrected $E^{o\prime}{}_{Au\{O\}}$, E_{corr}, was estimated [50]. The representation of such E_{corr} values versus age for a set of gold samples of known age is shown in Fig. 12.15.

Experimental data in this figure do not yield a monotonically increasing or decreasing tendency. This can be rationalized by assuming that (at least) two processes associated with the adsorption of oxygen species on gold operate: the adsorption on the external metal surface and the subsequent diffusion/interchange of oxygen functionalities into subsurface regions (see Fig. 12.10). On the first examination, this situation can be described in terms of two consecutive processes,

$$Au_{surf} + \{O\} \xrightarrow{k_a} Au\{O\}_{surf} \xrightarrow{k_b} Au\{O\}_{sub} \quad (12.20)$$

where k_a and k_b are the respective rate constants. Expressing the thermochemical activity of the O-adsorbates as the ratio between the actual surface density of such species (Γ) and the maximum attainable surface density (Γ_o), and assuming that conditions of first-order kinetics apply, the net activity of the oxygen functionalities adsorbed on gold will be,

$$a_{ox} = \frac{\Gamma}{\Gamma_o}\left(e^{-k_a t} - e^{-k_b t}\right) \quad (12.21)$$

so that the corrected potential will vary with time as,

$$E_{corr} = E^{o\prime}_{Au} - B\left[1 - \frac{\Gamma}{\Gamma_o}\left(e^{-k_a t} - e^{-k_b t}\right)\right] \quad (12.22)$$

where B is an adjustable constant. As can be seen in Fig. 12.15, experimental data can be fitted to Eq. (12.25) taking $B = 0.41$ mV, $k_a = 0.0012$ years^{-1}, and $k_b = 0.0021$ years^{-1}.

12.4 Prospective on Metal Dating

In principle, electrochemical dating of tin and zinc artifacts could be made following a scheme similar to that described for lead in Sect. 12.2.2 because corrosion layers of SnO_2 and ZnO, both having well-defined voltammetric responses [51, 52], monopolize the respective corrosion patinas.

The case of silver, however, is more complicated because the corrosion leads to AgCl but, depending on the environmental conditions, to silver oxides and/or silver tarnish (Ag_2S), in all cases with well-defined electrochemistry [53]. Possibly, in cases where silver tarnish was the main corrosion product, the above methodology was applicable.

There are cases of alloyed objects in which the methodology described for leaded bronze was potentially useful. In fact, chronological schemes were described from voltammetric data on copper-containing [54], and lead-containing [55] silver objects, silver-containing [56] and lead-containing [57] bronzes. Note that such schemes can be applied to achieve information on the chronology of archaeological strata [58].

The voltammetric methods described in the precedent sections can be refined and expanded considering the peculiarities of the VIMP sampling and its concomitant possibility of a layer-by-layer analysis of metal corrosion layers as described in Sect. 10.4. As described here, the application of successive cathodic scans to the set of microplates adhered to a graphite electrode can in be formally seen as composed of k laminas of average area A and thickness δ. If a unique corrosion product is formed whose compaction, crystallinity, porosity, etc., vary with depth, the net charge passed in the Nth reductive scan, $q(N)$, during an n-electron cathodic process where the corrosion product is reduced to the metal, can be expressed as [59]:

$$q(N) = nkFA\delta(N)\rho(N) \tag{12.23}$$

where $\delta(N)$ is the thickness of the layer removed in the Nth voltammetric run and $\rho(1)$ is the average molar density of the corrosion product in this lamina. Assuming that the deposit of metal does not interfere with the response in successive scans and that the number of laminas anchored to the graphite electrode remains essentially constant, the variation of $q(N)$ with N will be representative of the gradient of compactness, crystallinity, etc., of the corrosion product in the patina. The total charge passed after N successive scans scan will be:

$$Q(N) = \sum q(N) = nkFA \sum \delta(N)\rho(N) \tag{12.24}$$

In principle, the density of the electroactive corrosion product can be expressed as a continuous function of the depth z so that the depth $z(N)$ reached in the Nth potential scan will equal to the sum of the $\delta(N)$ values. There are a variety of density/depth functions potentially applicable [59]. In the case of a potential variation of the patina properties with depth, a density function such as,

$$\rho(z) = \rho_{\text{inn}} - \Delta\rho\left(1 - \frac{z}{z_{\text{inn}}}\right)^{\alpha} \tag{12.25}$$

can be taken. Here, $\Delta\rho$ represents the difference between ρ_{inn}, the density of the core of the corrosion layers, and ρ_{surf}, the corresponding density at the external surface of the patina, and α, $\Delta\rho$ and α an exponent whose value has to be decided from experimental data. In the first scan, the peak current/peak area is representative of the properties of the more external corrosion layer (see Sect. 10.4). The charge passed in this scan will be:

$$\begin{aligned} q(1) &= nkFA\int_{z=0}^{z_1}[\rho_{\text{inn}} - \Delta\rho(1 - \frac{z}{z_{\text{inn}}})^{\alpha}]\text{d}z \\ &= nkFA\left[(\rho_{\text{inn}} - \Delta\rho)z_1 + \frac{\Delta\rho}{1+\alpha}\left(\frac{z_1}{z_{\text{inn}}}\right)^{1+\alpha}\right] \end{aligned} \tag{12.26}$$

where z_1 is the depth reached in the initial scan. Following the same line of reasoning, the charge passed during the Nth scan will be:

$$\begin{aligned} q(N) = nkFA\Big[&(\rho_{\text{inn}} - \Delta\rho)(z_N - z_{N-1}) \\ &+\frac{z_{\text{inn}}\Delta\rho}{1+\alpha}\left[\left(\frac{z_N}{z_{\text{inn}}}\right)^{1+\alpha} - \left(\frac{z_{N-1}}{z_{\text{inn}}}\right)^{1+\alpha}\right]\Big] \end{aligned} \tag{12.27}$$

It can easily be demonstrated that, on computing the cumulated charge, several z-terms cancel, and the total charge passed at the Nth scan will be:

$$Q(N) = nkFA\left[(\rho_{\text{inn}} - \Delta\rho)z_N + \frac{z_{\text{inn}}\Delta\rho}{1+\alpha}\left(\frac{z_N}{z_{\text{inn}}}\right)^{1+\alpha}\right] \tag{12.28}$$

This equation predicts a potential variation of $Q(N)$ with z. Assuming that the depth reached in the Nth scan, z_N, is as proportional to the scan number (i.e., $z_N = \varepsilon N$), Eq. (12.22) can be rewritten as:

$$Q(N) = nkFA(\Delta\rho\varepsilon N + \frac{\varepsilon^{1+\alpha}}{1+\alpha}N^{1+\alpha}) \tag{12.29}$$

This equation predicts a polynomial variation of $Q(N)$ on N in repetitive voltammetry. For dating purposes, the relevant points to emphasize are that the values of ε and α are those determining the gradient of properties of the corrosion patina with the depth and that this gradient will vary with the age of the object.

Figure 12.16 depicts the $Q(N)$ versus N plot obtained from square wave voltammetric data (8 successive cathodic scans) recorded, using aqueous acetate buffer at pH 4.75 as the electrolyte solution, for a sample from a *blanca* medieval Spanish coin (*Enrique IV* king) produced between 1454 and 1474, [48]. Here, the sum of the successive peak currents measured in voltammograms similar to those depicted in Fig. 10.12 is represented versus N. Since, as modeled in Sect. 3.9, the peak currents in linear potential scan voltammetry can be taken as proportional to the charge passed, assuming that the same condition applies for the difference current in square wave voltammetry, the cumulated peak currents will be also proportional to the cumulated charge. As can be regarded in Fig. 12.10, the $Q(N)$ versus N plot can be approximated by a curve tending to a limiting value Q(lim).

For dating purposes, the slope of the central, almost linear region in the $Q(N)$ versus N curve and the $Q(2)/Q(\text{lim})$ ratio appear to be suitable for building a calibration graph, as illustrated in Fig. 12.17 [60]. This figure summarizes the data recorded for a set of bronze coins from the repository of the *Museu de Prehistòria de València* (Spain), but incorporating a wide set of data is currently in progress. $Q(2)$ was preferred to $Q(1)$ because the first scan, as representative of the more external corrosion layer, is more sensitive to organic matter, local inhomogeneities, etc., than the second one. Interestingly, two different calibration curves were obtained for bronze and silver-containing bronze coins.

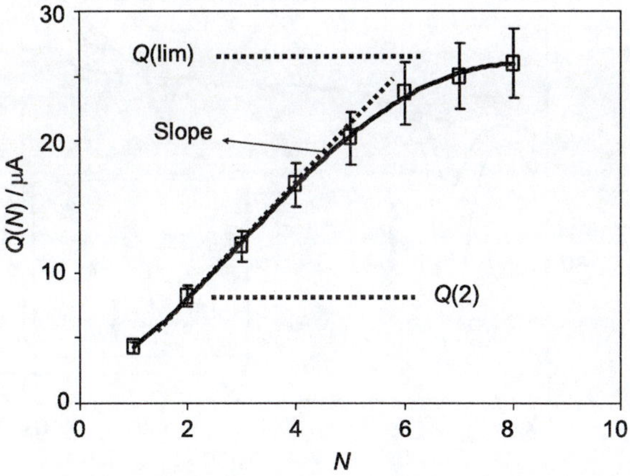

Fig. 12.16 $Q(N)$ versus N plot from square wave voltammetric data in conditions such as in Fig. 10.12 for 8 successive cathodic scans applied to a sample from a *blanca* medieval Spanish coin (Enrique IV, 1454–1474) in contact with aqueous acetate buffer at pH 4.75 [60]. The continuous line marks the fit of a polynomial function to the data taking N and $Q(N)$ as continuous variables. The error bars correspond to relative uncertainties of 10%. Dotted lines define several parameters of interest for dating purposes (see text)

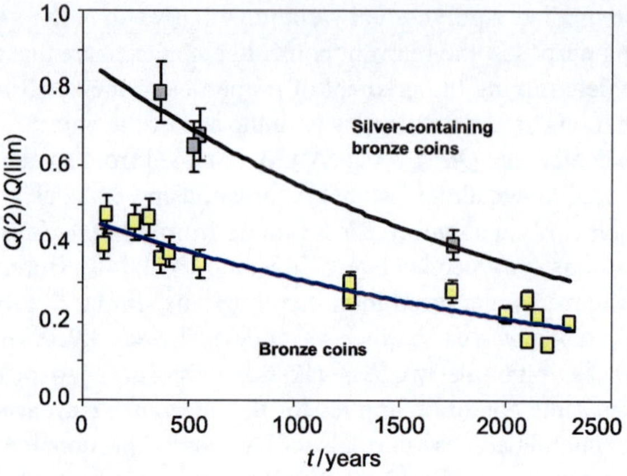

Fig. 12.17 Age calibration graph constructed from voltammetric data for a set of coins from the repository of the *Museu de Prehistòria de València* (Spain) based on data analysis summarized in Fig. 12.10 [60]. Here, the $Q(2)/Q(\text{lim})$ ratio is plotted versus the age of the coins. The continuous lines represent the fit of data to polynomial functions. The error bars correspond to relative uncertainties of 10%

Repetitive voltammetry can also be exploited for gold dating combining the gold-related signals and those associated with copper and silver (see Sect. 10.7). Figure 12.18 presents a representative example of this type of voltammetric pattern.

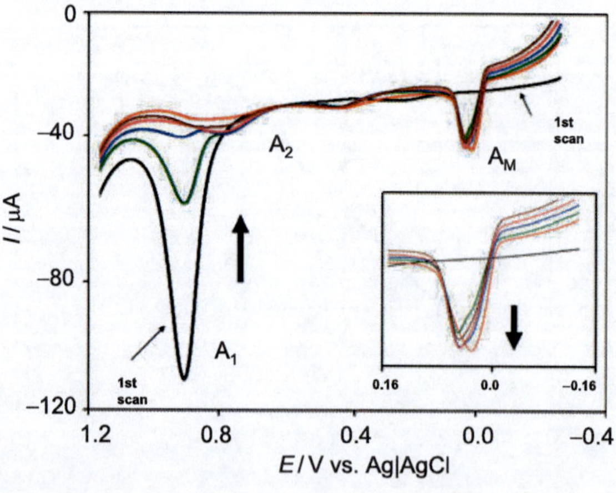

Fig. 12.18 Successive linear potential scan voltammograms were recorded on a graphite electrode modified with a sample from a gold ring fabricated in 1988 in contact with air-saturated 0.10 M HCl. Inset: detail of the region of potentials where the copper plus silver signals appear. The arrows mark the tendency in successive scans. Potential scan rate 50 mV s^{-1}. Adapted from Ref. [50], with permission. Note that the axes are opposite to the IUPAC convention

One can see that the main signal for gold oxidation (A_1) decreases in successive anodic scans while the signals for active gold sites (A_2) and the copper plus silver (see Sect. 10.7) signal A_M increase apparently tending to a limiting value. This response can be interpreted by considering that, as occurring for other metals (see Sect. 10.4), VIMP sampling on gold surfaces extracts a set of plates that become anchored to the graphite electrode. Figure 12.19 shows the secondary electron images of gold plates from a medal fabricated in 1903 transferred onto a graphite bar in contact with 0.10 M HCl aqueous solution before and after application of a potential step of 1.25 V versus Ag/AgCl for 5 min.

The application of the anodic input determines the delamination of the gold plate, in agreement with the oxidative dissolution nature of process A_1. The delamination will expose (see Sect. 10.4) progressively deeper regions of the gold plate so that repetitive voltammetry provides in-depth information on the external region of the gold object. The different evolution of the peaks A_1 and A_M in successive scans can then be interpreted as reflecting the increase of the copper plus silver concentration

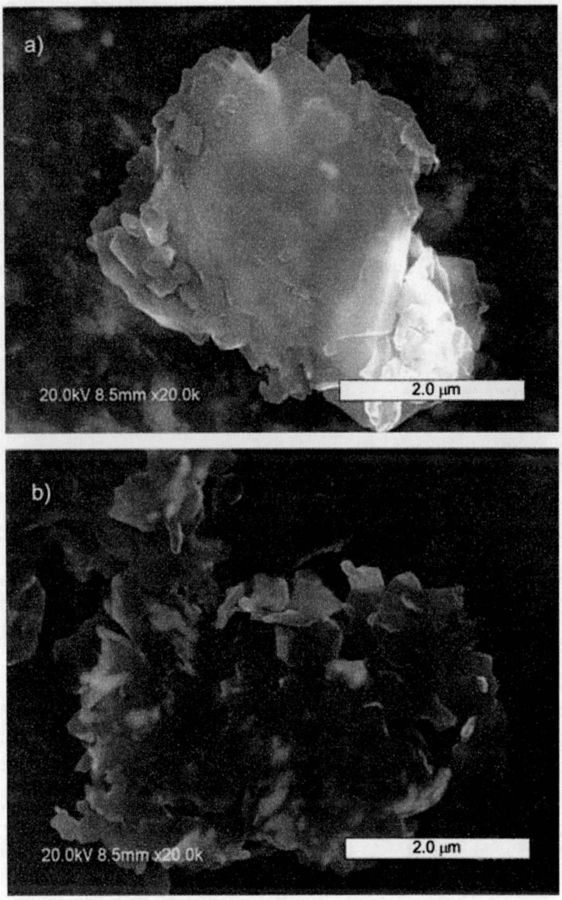

Fig. 12.19 Secondary electron images of two gold plates from a medal fabricated in 1903 abrasively transferred onto a graphite bar in contact with 0.10 M HCl aqueous solution **a** before and **b** after application of a potential step of 1.25 V for 5 min. Although the images correspond to different gold plates, they show representative views of gold laminas before and after application of oxidative potential inputs in HCl. Reproduced from Ref. [42], with permission

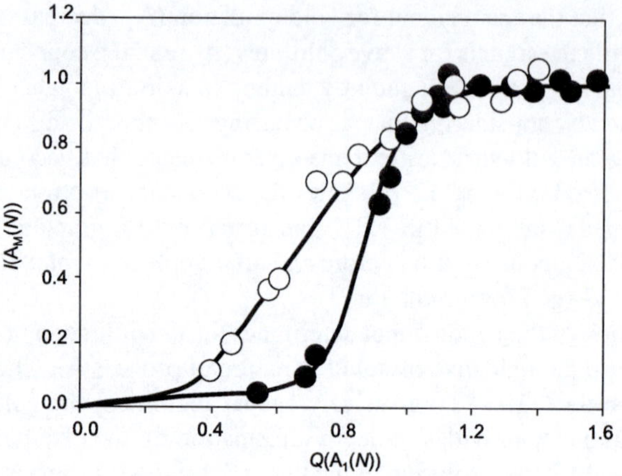

Fig. 12.20 Generalized $I(\mathrm{A_M}(N))$ versus $Q(\mathrm{A_1}(N))$ plots, after normalization, after repetitive voltammetry (experimental conditions such as in Fig. 12.18) on graphite electrodes modified with samples from a gold ring fabricated in 1988 (solid circles) and a Roman bracelet from the Coca (Segovia, Spain) archaeological site dated back ca. 450 CE (circles) [61]

from the external region to the gold bulk. This feature would reflect the phenomenon of desilvering (and decuprification), in principle common for gold objects under 'ordinary' conditions of aging. Currently available data [61] suggest that plots of the peak current of the process $\mathrm{A_M}$ in successive scans, $I(\mathrm{A_M}(N))$, versus the cumulated peak current of the process $\mathrm{A_1}$ in the Nth scan, $Q(\mathrm{A_1}(N))$, provide s-shaped curves in the initial subsurface region (Fig. 12.20): The slope in the central part of the curve, normalized to the concentration of copper plus silver in the metal bulk, decreases on increasing the age of the object, apparently following an exponential decay (Fig. 12.21). At the expense of refinements and an accurate evaluation of the influence of metal composition and manufacturing techniques, this method provides promising results for gold dating.

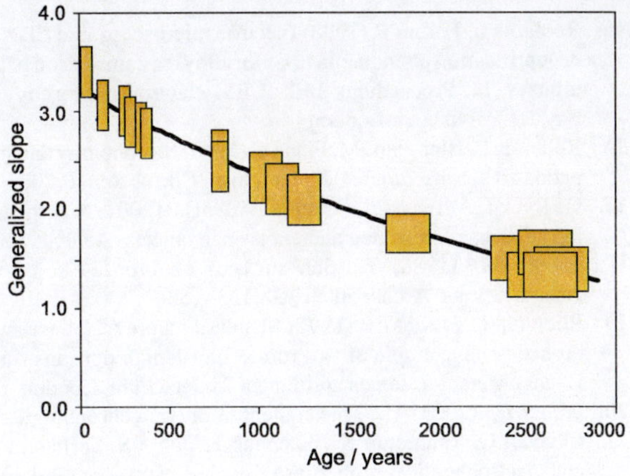

Fig. 12.21 Calibration graph based on the slope of the central, linear-like region in the generalized $I(A_M(N))$ versus $Q(A_1(N))$ representations such as in Fig. 12.20. The continuous line represents the fit of the data points to exponential decay

References

1. Doménech-Carbó A (2017) Electrochemical dating: a review. J Solid State Electrochem 21:1987–1998
2. Doménech-Carbó A (2015) Dating—an analytical task—ChemTexts 1:article 5
3. Burchfield JD (1998) The age of the earth and the invention of geological time. Geological Society, London, Special Publications 143:137–143
4. Aitken MJ (1990) Science-based dating in archaeology. Longman, New York chapter 1
5. Geyh MA, Schleicher H (1990) Absolute age determination—physical and chemical dating methods and their application. Springer, Berlin-Heidelberg
6. Bowen R (2011) Radioactive Dating Methods. In: Vertés A, Nagy S, Klencsár Z, Lovas RG, Rösch F (eds) Handbook of nuclear physics, 2nd edn. Springer, Heidelberg, Chapter 17
7. Stuiver M, Reimer PJ, Braziunas TF (1998) High-precision radiocarbon age calibration for terrestrial and marine simples. Radiocarbon 40:1127–1151
8. Reimer PJ, Bard E, Bayliss A, Beck JW, Blackwell PG, Ramsey CB, Buck CE, Cheng H, Edwards RL, Friedrich M, Grootes PM, Guilderson TP, Aflidason H, Hajdas I, Hatté C, Heaton TJ, Hoffmann DL, Hogg AG, Hughen KA, Kaiser KF, Kromer B, Manning SW, Niu M, Reimer RW, Richards DA, Scott EM, Southon JR, Staff RA, Turney CSM, van der Plicht J (2013) IntCal 113 and Marine 13 radiocarbon age calibration curves 0–50,000 years cal BP. Radiocarbon 55:1869–1887
9. Scholz F, Schröder U, Meyer S, Brainina KhZ, Zakharchuk NF, Sobolev NV, Kozmenko OA (1995) The electrochemical response of radiation defects of non-conducting materials an electrochemical access to age determinations. J Electroanal Chem 385:139–142
10. Reich S, Leitus G, Shalev S (2003) Measurement of corrosion content of archaeological lead artifacts by their Meissner response in the superconducting state; a new dating method. New J Phys 5:99.1—99.9
11. Eugster O (1996) Dating gold artifacts applications for noble gas analysis of gold. Gold Bull 29:101–104
12. Eugster O, Kramers J, Krähenbühl U (2009) Detecting forgeries among ancient gold objects using the U, Th-^{4}He dating method. Archaeometry 51:672–681
13. Weisser TS (1975) The de-alloying of copper alloys, in Conservation in archaeology and the applied arts. Preprints of the contributions to the Stockholm Congress, International institute for conservation, pp 207–214
14. Leoni M, Diana M, Guidi G, Pierdominici F (1991) Sul fenomeno della destannazione dei manufatti bronzei di provenienza archeologica. La Metallurgia Italiana 83:1033–1037

15. Robbiola L, Portier R (1998) Electron microscopy and EDX analysis in the investigation of the decuprification phenomena in CuSn alloys: a comparison between archaeological and synthetic bronzes. In: Proceedings 14th ICEM electron microscopy, vol III, pp 289–290. Institute of Physics Publishing, London
16. Robbiola L, Blengino JM, Fiaud C (1998) Morphology and mechanisms of formation of natural patinas on archaeological CuSn alloys. Corros Sci 40:2083–2111
17. De Ryck I, Adriaens A, Pantos E, Adams F (2003) A comparison of microbeam techniques for the analysis of corroded ancient bronze objects. Analyst 128:1104–1109
18. Meeks ND (1986) Tin-rich surfaces on bronze—some experimental and archaeological considerations. Archaeometry 28:133–162
19. Robbiola L, Hurtel LP (1997) Standard nature of the passive layers of buried archaeological bronze—the example of two roman half-length portraits. In: MacLeod I, Pennec S, Robbiola L (eds) Metal 95. James and James Science Publ, London, pp 109–117
20. Welter JM (2003) The zinc content of brass: a chronological indicator. Techné 18:27–36
21. Chiavari C, Rahmouni K, Takenouti H, Joiret S, Vermaut P, Robbiola L (2007) Composition and electrochemical properties of natural patinas of outdoor bronze monuments. Electrochim Acta 52:7760–7769
22. Doménech-Carbó A (2011) Tracing, authentifying and dating archaeological metal using the voltammetry of microparticles. Anal Methods 3:2181–2188
23. Doménech-Carbó A, Scholz F (2019) Electrochemical age determinations of metallic specimens—the utilization of the corrosion clock. Acc Chem Res 52:400–406
24. Doménech-Carbó A (2020) Voltammetry of immobilized particles for the future. J Solid State Electrochem 24:2063–2065
25. Doménech-Carbó A (2021) Electrochemistry in archaeology and art conservation. Israel J Chem 61:113–119
26. Doménech-Carbó A, Doménech-Carbó MT, Peiró-Ronda MA (2011) Dating archaeological lead artifacts from measurement of the corrosion content using the voltammetry of microparticles. Anal Chem 83:5639–5644
27. Pavlov D, Monakhov B (1987) Effect of Sb on the electrochemical properties of $Pb/PbSO_4/H_2SO_4$ and $Pb/PbO/PbSO_4/H_2SO_4$ electrodes. J Electroanal Chem 218:135–153
28. Laitinen T, Monahov B, Salmi K, Sundholm G (1991) Ring-disk electrode studies of soluble intermediates formed during the polarization of Pb in H2SO4. Electrochim Acta 36:953–963
29. Pavlov D (1981) Semiconductor mechanism of the processes during electrochemical oxidation of PbO to PbO_2. J Electroanal Chem 118:167–185
30. Cai W-B, Wan Y-Q Liu, H-T Zhou, W-F (1995) A study of the reduction process of anodic PbO_2 film on Pb in sulfuric acid solution. J Electroanal Chem 387:95–100
31. Zhao G, Zang Y, Lei Y, Lv B, Gao J, Zhang Y, Li D (2010) Fabrication and electrochemical treatment application of a novel lead dioxide anode with superhydrophobic surfaces, high oxygen evolution potential, and oxidation capability. Environ Sci Technol 44:1754–1759
32. Panizza M, Cerisola G (2003) Electrochim. Acta Influence of anode material on the electrochemical oxidation of 2-naphthol: Part 1. Cycl Voltam Potent Step Experim 48:3491–3497
33. Ai S, Wang Q, Li H, Jin L (2005) Study on production of free hydroxyl radical and its reaction with salicylic acid at lead dioxide electrode. J Electroanal Chem 578:223–229
34. Doménech-Carbó A, Doménech-Carbó MT, Peiró-Ronda MA, Martínez-Lázaro I, Barrio J (2012) Application of the voltammetry of microparticles for dating archaeological lead using polarization curves and electrochemical impedance spectroscopy. J Solid State Electrochem 16:2349–2356
35. Doménech-Carbó A, Doménech-Carbó MT, Capelo S, Pasíes T, Martínez-Lázaro I (2014) Dating archaeological copper/bronze artifacts using the voltammetry of microparticles. Angew Chem Int Ed 53:9262–9266
36. Doménech-Carbó A, Capelo S, Piquero-Cilla J, Doménech-Carbó MT, Barrio J, Fuentes A, Al-Sekhaneh W (2016) Dating archaeological copper using electrochemical impedance spectroscopy. Comparison with voltammetry of microparticles dating. Mater Corros 67:120–129

37. Doménech-Carbó A, Doménech-Carbó MT, Redondo-Marugán J, Osete-Cortina L, Barrio J, Fuentes A, Vivancos-Ramón MV, Al-Sekhaneh W, Martínez B, Martínez-Lázaro I, Pasíes-Oviedo T (2017) Electrochemical characterization and dating of archaeological leaded bronze objects using the voltammetry of immobilized particles. Archaeometry 60:308–324
38. Doménech-Carbó A, Del Hoyo-Meléndez J, Labut M, Doménech-Carbó MT (2021) Electrochemical analysis of coffin portratits from the National Museum in Krakow. J Solid State Electrochem 25:2767–2776
39. Plumb RC, Thakkar N (1965) Volta potential studies of the aging of gold surfaces. J Phys Chem 69:439–441
40. Antes LL, Ackerman N (1951) Contact potential variations on freshly condensed metal films at low pressures. J Appl Phys 22:1395–1398
41. Rysiazhnyi V, Slavicek P, Cernak M (2014) Aging of plasma-activated copper and gold surfaces and its hydrophilic recovery after water immersion. Thin Solid Films 550:373–380
42. Doménech-Carbó A, Scholz F, Doménech-Carbó MT, Piquero-Cilla J, Mpntoya N, Pasíes-Oviedo T, Gozalbes M, Melchor-Montserrat JM, Oliver A (2018) Dating of archaeological gold by means of solid state electrochemistry. ChemElectroChem 5:2113–2117
43. Burke LD, O'Mullane AP (2000) Generation of active surface states of gold and the role of such states in electrocatalysis. J Solid State Electrochem 4:285–297
44. Nowicka A, Hasse U, Sievers G, Donten M, Stojek Z, Fletcher S, Scholz F (2010) Selective knockout of gold active sites. Angew Chem Int Ed 49:3006–3009
45. Doménech-Carbó A, Sabaté F, Sabater J (2018) Electrochemical analysis of catalytic and oxygen interfacial transfer effects on MnO_2 deposited on gold electrodes. J Phys Chem C 122:10939–10947
46. Diaz-Morales O, Calle-Vallejo F, de Munck C, Koper MTM (2013) Electrochemical water splitting by gold: evidence for an oxide decomposition mechanism. Chem Sci 4:2334–2343
47. Desilvestro J, Weaver MJ (1986) Surface structural changes during oxidation of gold electrodes in aqueous media as detected using surface-enhanced Raman spectroscopy. J Electroanal Chem 209:377–386
48. Yeo BS, Klaus SL, Ross PN, Mathies RA, Bell AT (2010) Identification of hydroperoxy species as reaction intermediates in the electrochemical evolution of oxygen on gold. ChemPhysChem 11:1854–1857
49. Plazinski W, Rudzinski W (2009) Kinetics of adsorption at solid/solution interfaces controlled by intraparticle diffusion: a theoretical analysis. J Phys Chem C 113:12495–12501
50. Doménech-Carbó A, Scholz F, Brauns M, Tiley-Nel S, Oliver A, Aguilella G, Montoya N, Doménech-Carbó MT (2020) Electrochemical dating of archaeological gold based on refined peak current determinations and Tafel analysis. Electrochim Acta 337:135759
51. Doménech-Carbó A, Doménech-Carbó MT (2005) Electrochemical characterization of archaeological tin-opacified lead-alkali glazes and their corrosion processes. Electroanalysis 17:1959–1969
52. Doménech-Carbó A, Sánchez-Ramos S, Yusá-Marco DJ, Moya-Moreno M, Gimeno-Adelantado JV, Bosch-Reig F (2004) Electrochemical determination of boron in minerals and ceramic materials. Anal Chim Acta 501:103–111
53. Capelo S, Homem PM, Cavalheiro J, Fonseca ITE (2013) Linear sweep voltammetry: a cheap and powerful technique for the identification of the silver tarnish layer constituents. J Solid State Electrochem 17:223–234
54. Doménech-Carbó A, Doménech-Carbó MT, Pasíes-Oviedo T, Bouzas-Bello MC (2012) Modeling corrosion of archaeological silver-copper coins using the voltammetry of immobilized particles. Electroanalysis 24:1945–1955
55. Doménech-Carbó A, Del Hoyo-Meléndez J, Doménech-Carbó MT, Piquero-Cilla J (2017) Electrochemical analysis of the first Polish coins using the voltammetry of immobilized particles. Microchem J 130:47–55
56. Doménech-Carbó A, Doménech-Carbó MT, Álvarez-Romero C, Pasíes-Oviedo T, Buendía M (2019) Screening of Iberian coinage in the 2th–1th BCE period using the voltammetry of immobilized particles. Electroanalysis 31:1164–1173

57. Doménech-Carbó A, Doménech-Carbó MT, Álvarez-Romero C, Montoya N, Pasíes-Oviedo T, Buendía M (2017) Electrochemical characterization of coinage techniques the 17th century: the *maravedís* case. Electroanalysis 29:2008–2018
58. Di Turo F, Montoya N, Piquero-Cilla J, De Vito C, Coletti F, Favero G (2017) A. Doménech-Carbó, Archaeometric analysis of Roman bronze coins from the *Magna Mater* temple using solid-state voltammetry and electrochemical impedance spectroscopy. Anal Chim Acta 955:36–47
59. Doménech-Carbó A, Donnici M, Álvarez-Romero C, Daniele S, Doménech-Carbó MT (2021) Multiple-scan voltammetry of immobilized particles of ancient copper/bronze coins. J Solid State Electrochem 25:195–206
60. Doménech-Carbó A, Donnici M. Unpublished results
61. Doménech-Carbó A, Scholz F, Brauns M, Tiley-Nel S, van Bennekom J, van Bork E, Barrio J, Martínez-Caballero S, Oliver A, Aguilella G, Martínez B, Doménech-Carbo MT (2023) Electrochemical dating of archaeological gold ased on repetitive voltammetry monitoring of ilver/copper in depth concentration gradients. Microchem J 90:103661

Additional Literature

62. Bartlett PN, Dunford T, Ghanem MA (2002) Templated electrochemical deposition of nanostructured macroporous PbO_2. J Mater Chem 12:3130–3135
63. Blakelock ES (2016) Never judge a gold object by its surface analysis: a study of surface phenomena in a selection of gold objects from the Staffordshire hoard. Archaeometry 58:919–929
64. Burke LD, Nugent PF (1997) The electrochemistry of gold: I the redox behaviour of the metal in aqueous media. Gold Bull 30:43–53
65. Cherevko S, Kulyk N, Chung C-H (2012) Utilization of surface active sites on gold in preparation of highly reactive interfaces for alcohols electrooxidation in alkaline media. Electrochim Acta 69:190–196
66. Doménech-Carbó A, Doménech-Carbó MT, Gimeno-Adelantado JV, Bosch-Reig F (2006) H-point standard addition method applied to solid state stripping voltammetry. Application to the determination of lead and tin in archaeological glazes. Anal Bioanal Chem 385:1552–1561
67. Doyle RL, Lyons MEG (2014) The mechanism of oxygen evolution at superactivated gold electrodes in aqueous alkaline solution. J Solid State Electrochem 18:3271–3286
68. Doménech-Carbó A (2012) Electrochemical techniques, in analytical archaeometry, selected topics. In Edwards HGM, Vandenabeele P (eds) The royal society of chemistry, chapter 7, pp 210–241
69. Manas A (2020) Gold's red shift: colorimetry of multiple reflections in grooves. Gold Bull 53:147–158
70. Nowicka AM, Hasse U, Hermes M, Scholz F (2010) Hydroxyl radicals attack metallic gold. Angew Chem Int Ed 49:1061–1063
71. Scholz F (2011) The electrochemistry of particles, droplets, and vesicles – the present situation and future tasks. J Solid State Electrochem 15:1699–1702
72. Woodbourne S, Pienaar M, Tiley-Nel S (2009) The dating of the gold graves from Mapungubwe Hill. J African Archaeol 7:95–101

Index

A. Doménech-Carbó and M. T. Doménech-Carbó, *Electrochemistry for Cultural Heritage*, Monographs in Electrochemistry,
https://doi.org/10.1007/978-3-031-31945-7